TIME, CREATION AND THE CONTINUUM

TIME, CREATION
AND THE CONTINUUM

theories in antiquity and the early middle ages

Richard Sorabji

Cornell University Press

Ithaca, New York

First published 1983 by Cornell University Press
Second printing, 1986.
First printing, Cornell Paperbacks, 1986.

International Standard Book Number 0-8014-1593-4 (cloth)
International Standard Book Number 0-8014-9358-7 (paper)
Library of Congress Catalog Card Number 82-48714

Printed in the United States of America

*The paper in this book is acid-free and meets the guidelines
for permanence and durability of the Committee on Production
Guidelines for Book Longevity of the Council on Library Resources.*

TO JON STALLWORTHY

It is harder to get agreement between two
philosophers than between two waterclocks

Seneca *Apoc.* 2

Contents

II TIME AND ETERNITY

III TIME AND CREATION

Acknowledgments

I should like to thank my colleague in mathematics, Clive Kilmister, who gave a seminar with me at King's College, London, in the autumn of 1978, on ancient and modern theories of time, and David Bohm and Basil Hiley, from the Birkbeck physics department, who were kind enough to participate and to explain their own theories of time, as well as to attend to some of the Neoplatonist theories.

I am grateful to the Society for the Humanities at Cornell University for inviting me as a research fellow in the first half of 1979. I had the benefit of presenting some of my early ideas to a seminar attended by two mediaeval specialists, Norman Kretzmann and Edith Sylla, and also of taking part in a conference, arranged at the Society by Norman Kretzmann, on atomism and the continuum in ancient and mediaeval thought. A further chapter was criticised by the Cornell Philosophy Discussion Club on a return visit in 1981.

Since 1979, I have profited from discussion in three years of seminars held at the Institute of Classical Studies in the University of London. Among those involved, I am especially indebted to Henry Blumenthal, Myles Burnyeat, Nicholas Denyer, Stephen Everson, Pamela Huby, Lindsay Judson, A.C. Lloyd, A.A. Long, John Procope, David Sedley, Robert Sharples and Fritz Zimmermann, the last of whom has been kind enough to supply me with copious translation of, and instruction on, Islamic texts.

Even with this help, I felt the danger of writing on so large a period of thought, and I therefore sought the expert criticism of John Rist and Gerard O'Daly on selected portions of the manuscript. I thank them both for their extremely helpful comments, which were supplemented by those of the Cornell University Press's anonymous referee.

It may appear below that I am disagreeing most frequently with the people to whom I am indebted most. This could be true of Norman Kretzmann, from whom I have learnt so much about mediaeval philosophy and indeed so much good philosophy. It could also be true of A.C. Lloyd, who has made available to me his immense knowledge of Neoplatonism, and who is the outstanding exception to the remark I shall make below, that analytic philosophers have neglected this school of thought. Finally, I have ventured to disagree a number of times with Myles Burnyeat, who has for twelve years been a constant philosophical companion, a never-failing source of stimulus and a formidable critic. Going beyond the call of friendship, he offered me comments of a characteristically trenchant sort on the entire manuscript, and I wish to acknowledge my enormous debt to him.

The typing has once again been undertaken by Mrs D. Woods. It was financed by the British Academy's Small Grants Research Fund in the

Humanities, as was some of the expert criticism I received.

Some of the material has appeared, or will appear, in a different version in other places, but, in almost every case, this book will contain a more up to date version. Thus some of the material from Part V on atomism has appeared in the proceedings of the conference held at Cornell: *Infinity and Continuity in Ancient and Mediaeval Thought*, edited by Norman Kretzmann (Cornell University Press, Ithaca, New York, 1982). An earlier version of Chapter 10, 'Myths about non-propositional thought', has appeared in the Festschrift for G.E.L. Owen, *Language and Logos*, edited by Martha Nussbaum and Malcolm Schofield (Cambridge University Press, 1982). Chapter 26, 'Stopping and Starting', is a revision of what has now been twice revised, but in the new version I have come to somewhat different conclusions. It appeared first in *Proceedings of the Aristotelian Society* (supp. vol. 50, 1976), and then, revised, in J. Barnes, M. Schofield, R. Sorabji (eds), *Articles on Aristotle*, vol 3, London 1979. Finally, some of the material in Part I on the reality of time was used in a lecture to the British Academy and appeared in its *Proceedings* for 1982, and some of Chapters 13 and 14 on infinity and the creation was used in an inaugural lecture, published in 1982 by the Publications Office, King's College, London. I thank the editors and publishers who have agreed to my use of the material in this book.

I am grateful to Colin Haycraft for the epigraph, and to the following for the illustrations: Monreale mosaic: Scala; photograph of the author: Simon Sharma.

Finally, I have pleasure in dedicating the book to my friend since childhood, Jon Stallworthy.

R.R.K.S.

Abbreviations

AA	J. Barnes, M. Schofield, R. Sorabji (eds) *Articles on Aristotle*, 4 vols, London 1975-9
ABG	*Archiv für Begriffsgeschichte*
AGP	*Archiv für Geschichte der Philosophie*
APQ	*American Philosophical Quarterly*
ATR	*Anglican Theological Review*
BJPS	*British Journal for the Philosophy of Science*
CAG	*Commentaria in Aristotelem Graeca*, Berlin
CE	*Cronache Ercolanesi*
CMG	*Corpus Medicorum Graecorum*
CQ	*Classical Quarterly*
Diels	H. Diels, *Doxographici Graeci*, 1879
DK	H. Diels and W. Kranz (eds) *Die Fragmente der Vorsokratiker*, 5th ed. 1934
EP	*Les Études Philosophiques*
GRBS	*Greek, Roman and Byzantine Studies*
HSCP	*Harvard Studies in Classical Philology*
HTR	*Harvard Theological Review*
Int.PQ	*International Philosophical Quarterly*
IOS	*Israel Oriental Studies*
JAOS	*Journal of the American Oriental Society*
JHP	*Journal of the History of Philosophy*
JHS	*Journal of Hellenic Studies*
JP	*Journal of Philosophy*
JRS	*Journal of Roman Studies*
JTS	*Journal of Theological Studies*
PAAJR	*Proceedings of the American Academy for Jewish Research*
PAPA	*Proceedings of the American Philosophical Association*
PAPS	*Proceedings of the American Philosophical Society*
PAS	*Proceedings of the Aristotelian Society*
PBA	*Proceedings of the British Academy*
PCPS	*Proceedings of the Cambridge Philological Society*
PG	*Patrologia Graeca*, ed. Migne
PIASH	*Proceedings of the Israel Academy of Sciences & Humanities*
P.Lit.	*Philosophy & Literature*
PL	*Patrologia Latina*, ed. Migne
PQ	*Philosophical Quarterly*
PR	*Philosophical Review*
PS	*Philosophical Studies*

PW	A. Pauly, G. Wissowa and W. Kroll *Realencyclopädie der klassischen Altertumswissenschaft*, 1893-
REG	*Revue des Études Grecques*
RIP	*Revue Internationale de Philosophie*
Sci.Am.	*Scientific American*
SP	S. Sambursky and S. Pines, *The Concept of Time in Late Neoplatonism*, Jerusalem 1971
SVF	*Stoicorum Veterum Fragmenta*, ed. von Arnim
TAPS	*Transactions of the American Philosophical Society*

Introduction

The central theme of this book is time. Is time real? Does it depend on consciousness? Could it exist without a universe full of change? Did the universe, or did time itself, have a beginning? Does anything escape from time? Could human beings? Or should they rather expect after death an endless recurrence of past history? Does God escape from time? If so, how can he create the world, or intervene in it, or have knowledge of it? Does creation imply a temporal beginning? Does time flow or stay still? In either case, does it come smoothly or in atomic fragments? These are only some of the questions that were discussed in antiquity and the early middle ages. I shall be considering both the discussions of these questions and in many cases the questions themselves.

Time is a subject which connects with many others. The above questions have implications for how to live, how to face death, how, if at all, to conceive of God. Often too the subject of time brings one to the heart of a philosopher's ideas, because his metaphysical beliefs are so bound up with it. I shall be interested in these interconnexions more than in, say, the definitions of time offered by one thinker after another. The question whether time is continuous or atomic was originally discussed as only one facet of the debate between atomists and their opponents, and consequently the last part of the book will be concerned not just with the continuity of time, but with the continuum in general. The idea of creation has temporal implications which will be the subject of Part III; but it also has non-temporal implications of causal agency, which will be considered in Part IV. The subject-matter will, then, be wider than time, and this fact is marked by the title: *Time, Creation and the Continuum*.

Ancient Greek philosophy lasted over a thousand years, from 500 B.C. to at least A.D. 550, and it had a strong impact on Islamic thought between A.D. 750 and 1200. I shall write about the whole period of Greek philosophy and touch upon some aspects of Islamic philosophy. One result of this choice of period is that I shall talk about some philosophical traditions which have, with a few exceptions,[1] been almost wholly neglected by present-day philosophers in the Anglo-Saxon countries, but which I believe to be full of philosophical interest. The period down to Aristotle is well known, and recently there has been a renaissance in studies of the period which runs from the death of Aristotle in 322 B.C. to A.D. 205, when Plotinus, the founder of Neoplatonism, was born. But the great period of interplay between Neoplatonists and Christians up to the perishing of Athenian Neoplatonism in the mid-sixth century A.D. has largely been left to philologists, theologians,

---

[1] The most notable exception as regards Neoplatonism is A.C. Lloyd.

historians and others. Even they have often joined philosophers in
condemning the Neoplatonism of the end of that period as 'arid', 'scholastic',
'moribund', or 'dead'. Venom is reserved particularly for the Athenian, as
opposed to the Alexandrian, Neoplatonists, of whom Gibbon says, in *The
Decline and Fall of the Roman Empire*, that Plato would have blushed to
acknowledge them. The following is a trenchant statement of an almost
universal view:

> It is clear that in all this, Neoplatonism had become scholastic. After this time
> there are no major figures, and when Justinian closed the philosophical schools
> in 529 A.D., there was no fruitful philosophy to which to put a stop. The spirit
> of Greek philosophical thought, which, it is clear, had steadily become weaker
> during this period, had finally died.[2]

The normal verdict has been challenged, on historical grounds, by one
outstanding historian: Alan Cameron.[3] I hope my discussion will support
what he says by providing concrete philosophical illustration. During the
period of Neoplatonism, Philoponus made brilliant use of the concept of
infinity, in order to prove that the universe must have had a beginning
(Chapter 14). Boethius offered a remarkable solution to the question how
God's knowledge of our future acts could leave them free (Chapter 16).
Particularly ingenious defences of the reality of time were produced by
Iamblichus, Augustine, Damascius and Simplicius (Chapters 2 to 5).
Various aspects of McTaggart's flowing time series were brought out for the
first time by Iamblichus, Augustine, Damascius and Ammonius (Chapter 3).
Augustine applied all his ingenuity to defending the possibility of a
changeless God creating time and the universe (Chapter 15), and Gregory of
Nyssa, before Berkeley, analysed the creation in terms of idealism (Chapter
18). Origen tried to prove the theory of recurrent worlds, but Augustine
replied that that would lead to. despair, and instead went half way to
Plotinus' ideal of escape from the world of time (Chapter 12).

Time is a subject on which there was cross-fertilisation between Platonists
and Christians. On some aspects of it they were violently opposed, while on
others they came surprisingly close together. The neglect of the period by
present-day philosophers is quite understandable, since so many of the
Neoplatonists write in an obscure and off-putting way. But it actually
becomes easier to read them when one does so in connexion with Christians
like Augustine or Boethius, who are often handling the same themes, but in a
much more readable way.

I would be distorting, however, if I were to find more philosophical interest
in this late period than in the earlier ones. In fact, I shall discuss all the
periods, and it is only that the interest of the late period is in more need of
emphasis. This is particularly true of the topic of causality; I shall argue
(Chapter 20) that the Neoplatonist contributions to this topic are remarkable

[2] D.W. Hamlyn, 'Philosophy after Aristotle', in D.J. O'Connor, ed., *A Critical History of
Western Philosophy*, London 1964, 78. For other examples of the normal view, see: G. Downey,
'Justinian's view of Christianity and the Greek Classics', *ATR* 40, 1958, 3-12; John Glucker,
Antiochus and the Late Academy, Hypomnemata 56, Göttingen 1978, 329.

[3] Alan Cameron, 'The last days of the Academy of Athens', *PCPS* n.s.15, 1969, 7-29, esp. 25-9.

– not more remarkable than those of earlier periods, but much less fully appreciated. In some cases, the earlier periods make a unique contribution. Thus I shall present Zeno's paradoxes (Part V) as a major catalyst, and I shall describe Aristotle (Part I) as sharper than any of his successors in the discussion of certain questions, even if his very sharpness keeps him clear of some of the more exotic and interesting theories of time produced later. Again, the Hellenistic period, which officially stretches from 322 to 30 B.C., has the most immediate interest for the *physics* of time, with its discussion of time-atoms.

I think readers may agree that, on certain aspects of time, the ancient discussions are fuller than modern ones: the sheer range of arguments and of possibilities considered is greater. It is only with Einstein and the Theory of Relativity that a comparably massive advance has been made. Relativity Theory itself is something on which I shall touch only occasionally; but certain more recent speculations in the physics of time will be found to have analogues in antiquity.

As regards Islamic thought, it will repeatedly be found that it developed Ancient Greek philosophy in very interesting ways. I shall also try to argue (Chapter 25) that occasionally Greek thought can throw light on the meaning of Islamic philosophy.

I shall be going outside the sphere in which I have any specialised background, not only in discussing Neoplatonism, but still more in discussing some of the Church Fathers who were particularly influenced by Greek thought, and more again in discussing early Islamic thought. Oversights are inevitable in so large an undertaking, but I think there is value in seeing these historically connected periods together, for the light they throw on each other. Other people could profitably bring other kinds of background to such a study, but my own background is in analytic philosophy and the earlier history of Greek thought. The philosophical interest will often explain my selection of material, although historical interest will do so too.

I have made no attempt to offer any original suggestions about thirteenth- and fourteenth-century philosophy in the Latin-speaking West. But I shall occasionally refer to these later developments, and I hope that scholars of this period may find some interest in my discussion of earlier figures. The influence on this period of Philoponus, for example, is not always noticed, and his infinity arguments have been credited to St Bonaventure eight hundred years later (Chapter 13), although Islamic scholars are very well aware of his importance.

Even though much of the earlier Greek material is available in translation, the end of the Greek period has been less well served. I am glad to say that some of the omissions are now due to be remedied with translations, to be published by Duckworth, of Philoponus' *de Aeternitate Mundi contra Proclum*, of his fragmentary *contra Aristotelem*, of his *Corollary on Place* and *Corollary on Void* and perhaps of some of his other works, along with the *Corollary on Time* and *Corollary on Place* of his rival Simplicius.[4] I should also refer to the contribution

---

[4] Simplicius has been better served at this point than Philoponus, since his *Corollary on Time* has been translated into German by E. Sonderegger, *Simplikios zur Zeit, Hypomnemata* vol. 70,

of Shmuel Sambursky, who has done more than anyone to bring out the interest of Neoplatonist physics.[5] In Islamic studies, of course, there is still more opportunity for progress, since so many texts are not yet edited, let alone translated.

Göttingen 1982, while portions of his *Corollary on Time* and *Corollary on Place* are available in English in S. Sambursky and S. Pines, *The Concept of Time in Late Neoplatonism*, Jerusalem 1971 (SP), and S. Sambursky, *The Concept of Place in Late Neoplatonism*, Jerusalem 1982. Various extracts from Philoponus in translation are cited in Chapter 13, but I plan to give a fuller account of what is available for Philoponus in a collection of papers on him to be published by Duckworth, in an article on him in the *Theologische Realenzyklopädie*, and in a subsequent book.

[5] For his main translations, see note 4, but translations for this period are also included in his *The Physical World of Late Antiquity*, London 1982.

PART I
THE REALITY OF TIME

CHAPTER ONE

Is Time Real?

The paradoxes

Is time real? In *Physics* 4.10, Aristotle cites a set of paradoxes designed to show that it is not. These paradoxes and their variants stimulated the invention or application of some of the boldest theories about the nature of time for the next nine hundred years of Greek antiquity, and into the mediaeval period: time comes along in indivisible atoms, or in divisible leaps, time is unreal, time is in the mind, or at least its divisions are, there is a kind of time which does not flow, the whole of time is present, or the present corresponds to the minimum perceptible period. Some of these theories are of particular interest, because counterparts of them have been re-invented by psychologists and physicists in the last hundred years, and even in the last decade.

In at least one instance, I think we can see that Aristotle was too brilliant a thinker to need any elaborate theory of the nature of time in order to dismantle the paradoxes. As a result, his solutions, though more effective, tend to be less interesting than those of many successors. But this claim will need to be argued, because, with one exception,[1] Aristotle does not follow his statement of the paradoxes with a statement of solutions, and so we have to gather from elsewhere in his text how he would have been likely to solve them. He certainly does not intend to accept them.

Of course, we know that time exists. For one thing, there is something self-defeating about denying its existence; for that very denial requires time in which it may take place. The denial of time is self-defeating in the same way as the denial that one exists or thinks, a denial whose self-defeating character was exploited by Descartes. For another thing, any purposive agent must have a rudimentary idea of the difference between the future desired state of affairs and the present actual state; in other words, he must have some crude awareness of time, and any being capable of considering the existence of time is likely to be a purposive agent.

Therefore in so far as Aristotle's paradoxes suggest that time does not exist (and at least one of them is most naturally put that way), we know that they are wrong. But that does not put an end to the matter. For one thing, it has not yet been ruled out that time might exist, but with a lower degree of

---

[1] The exception is that he does discuss in the next chapter whether the now is for ever the same or for ever different. But he does not discuss further the paradox of the parts of time, or of the ceasing instant.

reality than we would have expected.[2] For another thing, however wrong the suggested conclusion may be, it does not follow that the paradoxes should be ignored. Few philosophers will be able to rest happily until they can see how to answer them; and if they were to rest happily, they would forfeit much that can be learnt about the nature of time. Nor is the subject only one for specialists: it concerns everyone. The fear of death, for example, is very much bound up with considerations about time, and particularly with considerations about what I shall be calling McTaggart's A-series (the term will be explained in Chapter 3). We are not concerned that our death will take place (say) in the twenty-first century, because that century is the twenty-first, but rather because it is *soon*. Again, we feel more horror at the idea of our *future* non-existence than at the idea of our *past* non-existence. Evidently, we value not only life, but the fact of having more of it *to come*. The ideas of 'soon', 'future' and 'to come', unlike the idea of 'twenty-first century', belong with McTaggart's A-series, and if it could be shown at least that the *A-series* did not exist, that alone should lend us tranquillity. I do not believe in fact that either A- or B-series can be shown not to exist; at most we may try in some contexts to attach less importance to the A-series.

Let me start with a translation of *Phys.* 4.10, 217b29-218a30:

> The next subject to come after those discussed is time. First, it is a good plan to raise some puzzles about it by means of some commonplace ideas, and to ask if it is one of the things that exist (*ontōn*) or not, and then what its nature is. From the following one might suspect that either it does not exist (*estin*) at all, or hardly and obscurely.

> 217b33 Some of it has occurred and is not, while some is going to be and is not yet, and time is composed of these two, whether infinite time or the time one is at any moment taking up. Now what is composed of the non-existent would be thought to be unable to partake of existence (*ousia*).

> 218a3 Further, when a divisible thing exists, if it does, either all or some of its parts must exist. But time is a divisible thing of which some has occurred, while some is going to be, and none is. For *now* is not a part of time, since a part can serve as a measure of the whole, and the whole must be composed of parts, whereas time is not thought to be composed of nows.

So far the argument is that time does not exist, for none of its *parts* exists, since neither the past nor the future exists; while if someone pleads the existence of *now*, we reply that, being sizeless, now is not a *part* of time, even if it does exist. Certain points call for clarification.

In the word 'now', Aristotle often combines two ideas, though sometimes one idea occurs without the other. The first idea is that now is *present*, the second idea is that it is an *instant*. An instant is not a very short period, but rather the beginning or end (the boundary) of a period. It therefore has no size, for it is not a very short line, but rather the boundary of a line. The idea that there are sizeless instants should be no more controversial than the idea

<hr />

[2] A recent version of this view in physics is that of David Bohm, who suggests that time depends on a more fundamental and nontemporal reality: *Wholeness and the Implicate Order*, London 1980, 210-12.

that periods have beginnings and ends. An example of an instant would be two o'clock, which is the end of one hour and the beginning of another. I shall, however, return later to some of the reasons which may make the idea of an instant seem problematic, and I shall acknowledge that the idea involves a certain (harmless) degree of idealisation.

But why should the present be treated as a mere instant? Aristotle has an argument for this in *Phys.* 6.3, 234a9-19. The central idea is that if the present were an extended stretch of time, it would overlap with the past and future. No doubt the common intuition that the present is distinct from the past and future would have to be abandoned if it could be shown to be unsatisfiable. But Aristotle shows that it *is* satisfiable, precisely by construing the present as an instant. At best, he recognises that ordinary language has a secondary use of the word 'now' to refer to what is *near* the present instant.[3] But in view of his 'overlap' argument, he cannot think this use altogether legitimate. Let me now quote Aristotle's next argument:

> 218a8 Again, it is not easy to see whether the now which appears to divide past and future always remains one and the same or is for ever different. Suppose it is for ever different: then if none of the ever differing parts of time are simultaneous with each other, except where one is contained and the other contains as the shorter time is contained in the longer; and if a now which does not exist but existed previously must have ceased to exist at some time, in that case nows will not exist simultaneously with each other, and the earlier now must always have ceased to exist. Now it cannot have ceased to exist during itself, because it exists then, but it cannot have ceased to exist in another now either. For let it be impossible for nows to be next to each other, as it is for points. If in that case it cannot have ceased to exist in the next now, but has ceased to exist in some other now, then it will have existed simultaneously with the infinitely many nows intervening between itself and the later one; but that is impossible.

The new puzzle starts from the idea that the present instant is ever different. In that case, we may ask when the present instant (say, two o'clock) ceases to exist. It cannot cease to exist while it is existing, for that would involve a contradiction. It cannot cease at the very *next* instant, for instants never are next to each other, any more than are geometrical points on a line. The only remaining alternative seems to be that two o'clock ceases to exist at some later instant, say, at one second past two; but then (absurdly) it would have remained in existence at all the infinitely many intervening instants. Two points call for clarification.

First, the sense of 'exist' here is 'be present'. Secondly, Aristotle is right that instants are never next to each other. To see why, we can imagine trying to name the instant next to two o'clock. Will it occur a millionth of the way through the ensuing second? But there is an instant closer to two than that: the instant a two millionth of the way through the ensuing second. Nor is that latter instant immediately next to two, for we can take ever smaller fractions *ad infinitum.* Nor is it any good saying that the next instant, like the house next door in a terraced row, is *no* distance away. For then it will not be

---

[3] *Phys.* 4.13, 22a20-4.

distinct from two, since unlike the house next door, it will not even have any *parts* which are separated by a distance.

Aristotle now blocks off the remaining escape route: the suggestion that the present instant does not cease to exist, but is always the same. There would then, he objects, be only one instant, whereas we need at least two to specify the boundaries of any period. Worse, we should be stuck at the same instant as the people of ten thousand years ago. I shall quote the remaining text:

> 218a21 Yet it is not possible either for it to remain always the same. For no finite divisible thing has only a *single* boundary, whether it be continuous in one dimension or in more; and the now is a boundary, and one can take a finite time.

> 218a25 Again, if being simultaneous rather than temporally earlier or later is being in one and the same now, and if earlier and later things alike are in this now, then things that happened ten thousand years ago would be simultaneous with things that have happened today, and nothing would be earlier or later than anything else.

Aristotle's solution of the 'ceasing instant'

There have been many conjectures about how Aristotle would solve his paradoxes. But, so far as I know, the crucial passages for the paradox of the ceasing instant have not been cited. Aristotle's solution, I believe, would come in two parts. The first point is that we must distinguish between the present and the perfect tense: we can never say, using the present tense, that the present instant *is* ceasing to exist. But we can say of what we once called the present instant that it *has* ceased to exist. When? The second part of Aristotle's answer would be: at any subsequent instant you like, however close – a millionth of a second later, or a two millionth. There will be no first subsequent instant.

It so happens that Aristotle phrases his paradox in terms of the *perfect* tense, asking when the present instant *has* ceased to exist. To answer the paradox so phrased, one needs only the second part of the solution. However, the paradox becomes harder if phrased in terms of the *present*, so that it asks when the present instant *ceases*. And to answer that, both parts of the solution are required.

It should not be very controversial that Aristotle would favour the second part of the solution, which uses the idea of any subsequent instant, however close. He is very much alive to the fact that a given instant has no immediate successor, and that you can for ever choose closer ones. It is the first point which is less familiar, that we can use the perfect tense but not the present. But this point is suggested by the following passage, which appears in a different context and in a different work.[4]

> For besides what has been said, there are also paradoxes about coming into existence and ceasing to exist. It is thought that in the case of a substance, if it

[4] *Metaph.* 3.5, 1002a28-b11.

now exists without having existed previously, or later fails to exist after previously existing, it must be in process of coming into existence or ceasing to exist [Aristotle expresses this by the *present* tense: *gignesthai, phtheiresthai*]. But with regard to points, lines and surfaces, when they exist at one time without existing at another, they cannot be in process of coming into existence or ceasing to exist. For as soon as bodies come into contact or are divided, the boundaries simultaneously become one if they touch and two if they are divided. Hence when the bodies have been put together, one boundary does not exist but *has* ceased to exist [*ephthartai*, perfect tense, b3, my italics], and when they have been divided, the boundaries exist which did not exist before (for the point, being indivisible, was not divided into two). And if the boundaries are in process of coming into existence or ceasing to exist, from what are they coming into existence?

It is similar with the now in time; for this too cannot be in process of coming into existence or ceasing to exist, and yet is thought to be ever different, which shows that it is not a substance. Clearly, it is the same with points, lines and planes, for the same account holds, since all alike are boundaries or divisions.

Aristotle has in mind that when, for example, two lines are joined, we get only one point of junction, so that at least one of the previous end points must have ceased to exist. I have chosen the passage because, as well as points, it also mentions the present instant and the question of its cessation, and mentions this explicitly. But for the general idea I could have cited a number of passages. Thus the idea that a thing can exist at one time and fail to exist at another, without ever being in process (present tense) of coming into existence or ceasing to exist, is applied to indivisible entities in general, and to certain other entities too.[5] Moreover, the use of the perfect tense, as opposed to the present, is authorised in more than one passage. In the passage quoted, it is authorised at 1002b3, when Aristotle says that one of the previously exposed end points *has* ceased to exist. The perfect is authorised again, when Aristotle says, talking of essence:[6]

This then either must be everlasting, or must be capable of ceasing to exist without ever being in process [*phtheiresthai*, present tense] of ceasing, and must *have* come into existence [*gegonenai*, perfect tense, b15, my italics] without being in process [*gignesthai*, present tense] of coming.

[5] Aristotle counts as exempt from a *process* of coming into existence: processes themselves (*Phys.* 5.2, repeated 7.3, 247b13), the occurrence of contact or of sundering, and the resulting points, lines and planes; the existence of units, of instants, of pleasure, seeing, hearing, perceiving and in general of indivisible wholes (*Metaph.* 1002a28-b11; 1044b21; 1060b18; *Cael.* 280b26; *NE* 1174b10-13; *Sens.* 446b4), the existence of relations between two or more things (*Phys.* 225b11-13; 246b11-12; 247b4), and the occurrence of coincidences in which two or more states of affairs are accidentally conjoined (*Metaph.* 1026b22-4 and 1027a29). He further says that white and certain other forms or essences do not undergo a process of coming into existence (*Metaph.* 1039b26; 1043b14; 1044b21). But this is qualified in two ways: first, he considers that certain forms escape such a process merely by being everlasting (*Metaph.* 1034b16-18; 1043b14). This may be true of the forms of natural substances, but not (*Metaph.* 1034b18-19; 1039b20-7; 1043b15-23; 1044b21-4; 1060b23-8; 1070a13-26) of forms such as white or of forms of artifacts. The second qualification is that a stick does undergo a process of coming to be white, even though white does not undergo a process of coming to exist in the stick (*Metaph.* 1044b23).

[6] *Metaph.* 8.3, 1043b14-16.

It may seem that at *Phys.* 6.6, 237b10, Aristotle forgets himself and maintains that if a thing *has* come into existence, it must previously have been in process of coming into existence. On the contrary, the passage only confirms what has been said, because it explicitly exempts indivisible entities from the general statement:

> Hence it is apparent that what has come into existence [*gegonos*] must previously have been in process of coming into existence [*gignesthai*] ... *in the case of things which are divisible and continuous* [my italics].

I believe that this is how Aristotle would solve the paradox of the ceasing instant. Moreover, I think that the solution is not only brilliantly ingenious, but also entirely effective. It renders unnecessary the alternative solutions of the following nine hundred years.

It is not only from the present tense but also from the aorist that the perfect is dissociated in these examples.[7] That is to say, if something *has* ceased (perfect tense), it does not follow straight off that there is a particular time at which it ceased (aorist). These facts about tense should be familiar from other kinds of example. Thus a man of fifty *has* ceased being a child, but it does not follow that there is a particular time at which he *ceased*, or at which it would be true to say, 'he is ceasing'. Admittedly, the inapplicability of 'ceased' and 'ceasing' in this example is based on quite different considerations – the gradualness of coming of age. But the example is enough to make the logical point that the perfect does not imply the present or aorist.

Recommended solutions for the 'parts of time'

Let us now return to the paradox of the non-existent parts of time. I want to suggest two ways of solving this paradox, although only the first, I believe, gets to the heart of the matter. It does so by asking what is meant when we say that there *is* a past and a future, or that they *exist*. The author of the paradox may not have thought out what is meant, but the truth behind his paradox depends on the fact that 'exist' sometimes means 'be present', and that neither past nor future exists in *this* sense. Of course, when the point is spelt out in this way, the non-existence of past and future appears quite unremarkable, but, until it is spelt out, the author of the paradox may well feel that his claim of non-existence is more significant. He is in any case still entitled to ask whether there is some *other* sense in which the past and future do exist, and this question puts an onus on us. For it is not enough simply to affirm that there is such a sense, if we cannot make our claim plausible by spelling that sense out.

It is more natural in English to put the point by saying that there *is* a past and a future (there is no one English equivalent of Aristotle's verb 'to be'), and the sense of this claim can be brought out by considering a *contrast*: what could be meant by someone who said the opposite? He might be a pessimist, who was expecting at any moment that final great implosion of matter which, according to some astronomers, will bring the universe to an end. Or

[7] I am grateful to Michael Leahy and students at Kent University for discussion of this point.

he might be a sceptic, who took seriously a possibility close to that suggested by Russell, that the universe has been created a moment ago, with all the fossil layers in place, and our brains stocked with illusory 'memories'. If he further accepted the view, which will be discussed in Chapter 6, that there can be no time if there are not changes going on in the universe, then he could draw a conclusion about time. Instead of merely saying 'there was no universe in the past', or 'there will be no universe in the future', he could say 'there is no past', or 'there is no future'. It is in contrast to this denial that we can see a sense for the claim that there is a future and a past; and not only does this claim turn out to have a sense, but it is, furthermore, a claim that any rational person ought to endorse. I believe, therefore, that this constitutes a solution to the paradox: it is only in the irrelevant sense of being present that the past and future do not exist. In the sense that matters, there is a past and there is a future, and so there is time.

It will now be clear why it is less natural to put the point in English by saying that the past *does* exist, for that invites the retort that it merely *did* so *once*, that is, it *was* present. No such retort is plausible when it is said that *there is* a past, that is, that the present has antecedents.

Aristotle does not get as far as discovering this solution to the paradox; but he takes a big step in the right direction, for at least he recognises, in a quite different context, that a different sense of the verb 'exist' applies to temporal entities. Thus at *Phys.* 3. 6, 206a21-3, he says:

> But existence is said of things in many ways, and an infinity can be said to exist in the same way as a day or a competition (*agōn*) exists, namely through one thing coming into being after another.

Aristotle's point is still a long way from the one needed. It is that the Olympic games can be said to be going on, even if only some *part* of them, say the discus, is going on. It is different with something like a book: I should provoke justified derision if I said my next book existed, on the grounds that some part existed, such as the first three pages. Aristotle would need to develop his point about a competition or a day, if he were to use it for the purpose of solving his paradox. He would need to say that time can exist through one *instant* becoming present after another. In that case, there need never be a *part* of time which is present in order for time to exist. Even then, his argument would not necessarily satisfy the opposition unless he could show *why* the existence of the whole needs no more than the existence of one *instant* at a time.

In fact, Aristotle does not, and would not, develop his point in this way. For one thing, if time existed through one *instant* coming into being after another, this would strongly suggest the view, which he rejects, that extended time can be composed of unextended instants. For another,[8] we have seen that coming into being (*gignesthai*, 206a21) is not something he thinks appropriate to unextended entities. As it is, his mind is not on the whole of time, but on something like a day. The kind of existence which is appropriate to a day and which he is trying to analyse, is still existence in the sense of

---

[8] I owe this second point to Marc Cohen.

being present. He is therefore a long way from the sense of 'exist' which I have suggested may be relevant to solving the paradox. And in any case the paradox is not what he is discussing. So the most I would say is that he provides more of the materials that are needed for solving the paradox than his successors provide, when he recognises the general point that temporal entities may exist in a different sense of the word. We shall see in Chapter 5 that some of his successors were to exploit these materials nearly nine hundred years later.

There is a second solution of the paradox about the parts of time, which reveals a further defect in the paradox, although it does not afford an entirely adequate answer. It insists that the present can have a length and hence can be a 'part' of time. But showing this is more complicated than is sometimes recognised.[9] For one cannot afford merely to point to ordinary language, and to remark that we often do treat the present as having a length. For the 'overlap' argument suggests that ordinary language incorporates a confusion, since it also tries to insist that the present does not overlap with future and past. So the 'overlap' argument must be met, and I suggest the following answer to it. In differing contexts, we think of the present as having different lengths, and in some contexts we think of it as having no length at all. But where we do think of it as having a length, we avoid overlap, by thinking of the future as if it did not begin immediately, or by thinking of the past as not having just ended. For example, a person whose leg is in plaster may say, 'I cannot get about so well just now.' His use of 'now' picks out a period which begins with his leg being damaged and ends perhaps with the removal of the plaster. For that context, and that purpose, the future is not thought of as beginning until the plaster is removed, although with another turn in the conversation, it might be treated as beginning sooner, with a correspondingly shorter present, or with an instantaneous one.

Although I think this answer all right as far as it goes, it does not go far enough, because we are entitled, when we choose, to treat the present as instantaneous. And we do not want to have to admit that, at least when we do that, Aristotle's paradox has force, and time has to be treated as non-existent. For example, a person can say, in a flash of insight, 'now I see', or, idealising his time signals as well as his instants, he can tell us that nine o'clock will be present at the beginning of the third stroke. Yet again, a man on his death bed can divide his life without remainder into a (wasted or enjoyable) past and an all-too-short future. For his purposes, he need not think of himself as having any present except as the boundary between the past and future. We do not want to concede that, in taking up this attitude, he is treating time as non-existent.

[9] Cf. R. Suter, 'Augustine on time with some criticism from Wittgenstein', *RIP* 16, 1962, 387-94, and J.R. Lucas, *A Treatise on Time and Space*, London 1973, 20-5, who comments that we need not be ashamed of our ordinary practice. What I am looking for is an explanation of *how* we can avoid being ashamed. J.N. Findlay intends the somewhat different line, I think, that we might just *legislate*, without explanation, what we will treat as present ('Time: a treatment of some puzzles', *Australasian J. Psych. & Phil.* 19, 1941, 225-7; repr. in A.G.N. Flew, *Logic and Language* series 1, Oxford 1951, 45-7; and in J.C.C. Smart, *Problems of Time and Space*, New York, 1964, 346-8). Wittgenstein's own discussions are superior to these: the best known are in *Philosophical Investigations*, Oxford 1953, I §89; *The Blue Book*, Oxford 1958, pp.6 and 26-7; but Peter Winch has drawn my attention to a much richer one in *Philosophische Bemerkungen*, sections 5 and 7.

Solutions rejected

Many solutions to Aristotle's paradoxes have been attempted other than the ones I have advocated. Before I turn to their history, there are just three attempts I should mention, in order to get them out of the way. Some people maintain that there are no such things as instants. I suggested that it should be no more controversial that there are instants than that there are beginnings and ends of periods, and that an example of an instant would be two o'clock. It may be protested, however, that two o'clock, like other boundaries of periods, has to be defined in terms of events. It is the instant when the sun or the designated clock hand stands in a certain alignment. But an actual clock hand has a certain thickness. To define an exact instant of alignment, we should have to idealise, taking three sizeless points, say, three centres of mass, within the parts of the clock. We idealise again when we talk of an alignment between these points, since a straight line is one along which an *ideal* measuring rod has to be placed the smallest number of times, or along which a light ray would travel in *ideal* conditions. So the idea of an instant such as two o'clock is an idealisation, and it should be admitted that for ordinary purposes people may not need to concern themselves with idealised instants. None the less, I do not think that that would make it right to say that they do not exist. The degree of idealisation involved in talking of instants is not very great; no greater than is familiar from the talk of point masses in Newtonian mechanics. Nor do I think it possible to dismiss Aristotle's paradoxes on the grounds that they are inapplicable to our world, so long as our world is not idealised. For we should surely be disturbed by the paradoxes, just so long as they apply to our world *as idealised* so as to contain instants.

I should add in parenthesis that talk of sizeless points in modern science does not always involve an idealisation, as it does for Newtonian mechanics. For there is evidence that some sub-atomic particles differ from others in being genuinely pointlike. That is, they have no size, but do have position, along with such other properties as mass, electric charge and intrinsic angular momentum. If electrons and positrons, for example, did have a positive size, this should reveal itself because, when they collide, they annihilate each other and give rise to other particles, sometimes of one kind, sometimes of another. The frequency of different kinds of outcome would be affected if the electron and positron had a positive size. Study of actual frequencies has not disconfirmed the hypothesis that they are pointlike, and has established that if they have a positive size, their diameter cannot be greater than 10^{-16} centimetres.[10]

A second attempt which I should like to get out of the way questions whether instants can be *present*. A difficulty may be raised on the basis of the suggestion, which I shall discuss in Chapter 3, that to judge something present is to judge something along these lines: 'it is simultaneous with this … ', where the word 'this' can be followed by a variety of nouns, for example, 'judgment', 'thought', 'utterance', 'sense-experience'. Thus if a person suddenly recognises some fact and thinks, 'now I see', he will, on this view,

[10] *Sci. Am.* 242, March 1980, 'Science and the citizen', pp. 68-9.

be thinking something like: 'I see (recognise) simultaneously with this thought.' The difficulty can be brought out more clearly when we turn from thoughts to utterances, for utterances have to take up time. So even if a person has taken care to anticipate the arrival of an idealised instant, he cannot get it, being sizeless, to coincide exactly with his utterance, but at best only to fall within it. Since the instant is not exactly coincident with his utterance, can it be genuinely present, given the view that presentness is to be understood in terms of simultaneity with such things as utterances?

The main answer to this ought to be that the proposed analysis of presentness is not right as it stands. In Chapter 3 I shall advocate a revision, which will eliminate the reference to utterances and thoughts. But it can be added that even if the reference to utterances were mandatory, the account could still appeal to idealised utterances which would match the idealised instants. An idealised utterance would present a point, say a beginning point, so sharp and abrupt that it could coincide perfectly with an idealised instant.

A third attempt at solution would argue that the present instant is the boundary between past and future, and that once it is conceded that a boundary exists, existence must also be granted to the things bounded. I hesitate to take such a short way with the problem. First, it fails to diagnose why it seems so plausible to say that the past and future do not exist. Secondly, even if a boundary presupposes a thing bounded, it remains to be shown that boundary and thing bounded must be equally real. The opposition offered an argument, one interpretation of which would be that the things bounded are indeed less real. Thirdly, the suggested solution can be turned on its head, and the Neoplatonist Damascius reveals that that is what actually happened. Evidently the paradox-mongers argued that, since they deny the reality of what is bounded, they equally deny the reality of the present instant which does the bounding.[11]

---

[11] In reply, Damascius only makes the hypothetical point that, *if* they once accepted the reality of the boundary, they would have to accept the reality of the bounded. But instead of pressing them to do that, he says that the present which he wants to introduce is not a mere boundary in any case (Simplicius *in Phys.* 796, 28-30; 799, 29-30; in SP, 78, ll.23-4; 86, 1-2, with English translation).

CHAPTER TWO

Solutions from Diodorus to Augustine

Diodorus Cronus

Aristotle had a much younger contemporary, Diodorus Cronus, whose date of death has recently been brought down to around 284 B.C.[1] Since Aristotle died in 322 B.C., it is not certain how far they overlapped. Diodorus is not known to have written down his ideas. He was instead famed for his oral arguments as a dialectician, and his five daughters were called dialecticians as well. There is a story that he died in despair after proving unable immediately to answer some logical conundrums at a banquet given by Ptolemy Soter. He was also known for his wit, and is said to have called his slaves 'Nevertheless', 'His', 'On the one hand' and 'On the other', in order to make a point about language. The nickname Cronus, meaning old codger, was certainly not earned by the sparkling Diodorus, but was merely inherited from his teacher. Although he finished up in Alexandria, Diodorus first taught in Athens, where he influenced the founders of the three main Hellenistic schools, the Stoics, the Sceptics and the Epicureans.

Diodorus delighted in paradoxes, many of which he took from Aristotle. Several will be discussed in later chapters, and it will be seen that he sometimes used atomist theory in order to deal with them. It is not certain whether he tried to solve Aristotle's paradoxes of time. But there is a certain likelihood that he did, since many of the paradoxes he is known to have tackled are related to Aristotle's. For example, there is a connexion between Aristotle's paradox of the ceasing instant and Diodorus' question when a wall ceases to exist – while it is intact, or after it has disintegrated.[2] I shall only claim, however, that Diodorus' atomism gave him the *materials* for solving the paradoxes of time. And in this chapter I shall discuss his atomism only so far as is necessary for showing that it supplied these materials.

Diodorus' ideas on atomism are recorded by Sextus Empiricus. An atom, in Greek thought, differs from a geometrical point in that, although it is indivisible, it is supposed to have a positive size. (We shall see eventually that some Islamic and fourteenth-century Western thought differed in this regard). Diodorus supported atomism, but he had clearly taken to heart the objections raised by Aristotle in *Phys.* 6. One objection, which Aristotle had borrowed in turn from a passage in Plato, was that if you believe in atomic

---

[1] David Sedley, 'Diodorus Cronus and Hellenistic philosophy', *PCPS* n.s. 23, 1977, 74-120. I am following Sedley for biographical details.

[2] ap. Sextum *M* 10.347-9.

spaces or bodies, you will be committed to atomic movements.[3] For an atomic body could not slide part by part from one atomic space into the next, vacating part of the first space, while occupying part of the next. This is impossible on the atomist hypothesis, because neither atomic bodies nor atomic spaces have parts. The atomic body could only *have* moved with an atom-length jerk, disappearing from the first atomic place and reappearing in the next, without ever being in transit. And Aristotle considered this impossible.

Diodorus responded with an argument which we shall encounter again in Chapters 22 and 24, and which is presumably intended to show that, even on Aristotle's view, motion will have to occur in this peculiar way, by a thing disappearing from one place and reappearing in another. For if the place of a thing is (as Aristotle held)[4] its *immediate* surroundings, then it cannot move in the place where it is (that place is too tight-fitting to allow movement). But neither, obviously, can it move in the place where it is not. It can at best *have* moved from one to the other, with a jerk.[5] The only difference will be that Aristotle is not committed to his jerks being of atomic length, or a multiple of atomic length. In view of this reply, I think Sextus may do Diodorus less than justice when he speaks as if his jerks follow only from his belief in indivisible bodies and spaces.[6] Aristotle had argued that jerks would follow from that. But Diodorus seems to be responding that jerks, even if not jerks of atomic length, follow also from some more widely held assumptions shared by Aristotle himself.

Diodorus' further response was to ask what was wrong with the idea of jerks. Admittedly they involve a thing's *having* moved (perfect tense), without there ever being a stage at which it *is* moving (present tense). But then there is nothing unusual, he said, about being able to use the perfect without the present: we can say of Helen that she *has* had three husbands without anyone ever being able to say, 'she now has three husbands'.[7]

Some people would object that discontinuous motion would in effect amount to the annihilation of a body and the creation of a new one; and admittedly there are difficulties about reidentifying subatomic particles because of certain aspects of relativistic quantum field theory.[8] But the difficulties do not arise from a mere discontinuity in the path traced. The predictability of discontinuous paths should supply criteria enough for reidentifying particles, especially if the discontinuity of path was hard to detect. And with more distinctive bodies such as human ones, or the crown jewels, there would be ample criteria for reidentification.

Exception may be taken to my saying that Diodorus supported atomism. For since he was a dialectician, this may encourage the belief that he had no doctrines of his own, but simply argued on either side of a question and sought to embarrass others. But I do not believe that this can be the case. For Sextus Empiricus presents Diodorus' ideas on atomic bodies, spaces and

[3] Plato *Prm.* 138D2-E7; Arist. *Phys.* 6.1, 231b18-232a17; 6.4, 234b10-20; 6.10, 240b8-241a6.
[4] *Phys.* 4.4. At least, this is one of two definitions.
[5] Sext. *M* 10.85-90; 143; cf.48 and *PH* 2.245; 3.71.
[6] *M* 10.85.
[7] *M* 10.91-2; 97-101.
[8] M.L.G. Redhead, 'Wave-particle duality', *BJPS* 28, 1977, 72.

movements as his 'personal doctrine' (*oikeion dogma*, *M* 10.86), and as something he 'taught' (*edidaske*, 10.97; 143). It is the sceptic Sextus who likes to collect arguments on *either* side of every case. But Sextus treats Diodorus as a man with a doctrine in 10.86, and contrasts him with the Pyrrhonian sceptics (*hoi apo tēs skepseōs*). Moreover, he reports no arguments by Diodorus on the other side, against atomism. Diodorus, we have seen, did not present the divorce of the perfect tense from the present as an embarrassment for atomism, but rather as something we ought to accept.

Besides atomic movements I want to argue that Diodorus accepted atomic times. A time-atom, in Greek thought, is like an instant in being indivisible, but differs because, surprisingly, it is supposed to have a positive size. Aristotle had argued that atomists are committed not only to atomic movements, but also to atomic times, for one kind of atom implies another.[9] But it is not generally recognised that Diodorus accepted time-atoms,[10] and indeed Wehrli once suggested that time-atoms were later introduced by Strato precisely as a *riposte* to Diodorus.[11]

Diodorus' reference to time-atoms comes in a passage of Sextus which relates arguments by Diodorus against the view that anything ever *is* moving. But as Diodorus has not been explicitly named for some paragraphs, it needs to be confirmed that the reference is still to him. *M* 10.85 introduces a 'weighty' argument by Diodorus; 10.112 some 'less weighty and more sophistical' arguments of his. Then the crucial passage is introduced, at the end of 10.118, by the statement that these less weighty arguments should be set aside, and use made instead of '*those* arguments'. Nicholas Denyer has suggested[12] that what follows is being viewed merely as a version of the 'weighty' argument of Diodorus. And this suggestion is made plausible by the use of the word '*those*'. Moreover, the ensuing argument does include an idea which forms part of the 'weighty' argument, and which is repeatedly ascribed to Diodorus, namely, the idea that a thing does not move either when it is in the first place or when it is in the second.[13] The following is a translation of 10.119-20.

If a thing is moving, it is moving now. If it is moving now, it is moving in the present (*enestōti*). If it is moving in the present, it must be moving in a partless (*ameres*) time. For if the present has parts, it will inevitably be divided into a past and a future part, and so will no longer be present.

If a thing is moving in a partless time, it traverses partless places. But if it traverses partless places, it is not moving. For when it is in the first partless

---

[9] *Phys.* 6.1, 231b18.

[10] Not even by those readiest to detect time-atomism in antiquity: S. Sambursky and H.-J. Krämer. Jürgen Mau remarks in passing that Diodorus had an atomism of time, space and matter, but this does not look like a considered conclusion, since the texts he cites do not support him as regards time ('Über die Zuweisung zweier Epikur-Fragmente', *Philologus* 99, 1955, 107). However, Nicholas Denyer has independently ascribed time-atomism to Diodorus on the basis of the same passage as myself, in 'The atomism of Diodorus Cronus', *Prudentia* (Auckland) 13, 1981, 33-45. I am grateful to him, moreover, for discussions in which he defended our common interpretation from attack at two points to be recorded below.

[11] F. Wehrli, *Die Schüle des Aristoteles*, vol.5. 1st ed. only, 1960, 63.

[12] op. cit.

[13] Sext. *M* 10.86-90; *PH* 2.245; 3.71. For further evidence see n.55 of Sedley, op. cit.

place, it is not moving, since it is still in the first partless place. And when it is in the second partless place, again it is not moving, but rather *has* moved. Therefore a thing never *is* moving.

There is more than one reason for construing the partless times here not as sizeless instants, but as time-atoms. To take the partless places first, they at least appear to be atoms, rather than sizeless points, as emerges from three considerations. First, if they were mere points, Diodorus would not allow that one could *have moved* across them; to have 'traversed' a sizeless point is not to have made any progress at all. Secondly, as I shall argue in Chapter 24, Aristotle had successfully proved for points, but not for atoms, that it is impossible to arrange them successively in an order of first and second in the way that the partless places are arranged here.[14] Thirdly, we have in any case been told a little earlier that Diodorus accepts space-atoms. The partless places, then, are atoms; and if so, the partless times which are putatively adequate for traversing them will surely be atomic too. That the times are atomic is confirmed by a further consideration: if Diodorus had been thinking of instants rather than of time-atoms, he would have had a quicker proof available that nothing can be moving (present tense). For it is much more obvious that an instant, being sizeless, leaves no room for moving.

One objection deserves consideration. It might be thought that the time-atoms mentioned here are needed only on the assumption that something 'is moving in the present', and that this assumption comes to be rejected with the denial that anything ever is moving. But this is not the structure of the argument: the need for partless times does not arise from the assumption that something is moving in the present. It is argued on the different ground that a divisible present would be divided into past and future.[15]

I have for now picked out only two aspects of Diodorus' view: his acceptance of atomic times and movements. My reason is that it is these two aspects that supply the materials for solving Aristotle's paradoxes of time. What are these materials?

The first half of Aristotle's answer to the paradox of the ceasing present is that you can say that it has ceased, but never that it is ceasing. There is reason to think that Diodorus would have been more than happy with this part of the solution. For, as we have seen, he argues strenuously that the perfect tense can be used in divorce from the present tense, even in the case of the verb 'to move', where Aristotle thinks such a usage absurd. As for the remaining part of the solution, Diodorus' atomism would give things a different twist. For if asked when the present has ceased he would be likely to say, 'at the very *next* time-atom'.[16]

As for the paradox of the non-existent parts of time, Diodorus' atomism would give him a ready answer. For if the present is a time-atom, it will have a positive size, and so can be counted as a *part* of time. In that case, one part of time will exist after all, in the sense of being present. Moreover, Diodorus

---

[14] *Phys.* 6.1, 231b6-10; 8.8, 264a4.

[15] I owe this reply to Denyer.

[16] An alternative solution for Diodorus would be to say that the present time-atom does cease (present tense), and that it does so at the later of the two instants which bound it.

will not be vulnerable to the objection that his present time-atom will overlap fatally with the past and future. For if it is genuinely atomic it will be indivisible, and hence incapable of overlap.

I am not saying that Diodorus' solutions would be satisfactory. That depends on whether there are time-atoms, and it is not yet clear whether we can even make sense of such an idea. For if a period has a positive size, how can it avoid having an earlier part and a later part, and so being divisible, i.e. non-atomic? I shall return to this question in Chapter 24.

The Stoics

I shall also postpone to Chapter 24 consideration of Epicurus' treatment of time. Epicurus lived from 341 to 270 B.C., and set up his school in Athens in 306 B.C. His career thus overlapped with Diodorus' in the same city. I shall be agreeing with the view that Epicurus postulated time-atoms, so that he would have had available the same solution as Diodorus. He would not, however, have minded ascribing a very low degree of reality to time, since, as will be seen in Chapter 7, he considered it only an attribute of an attribute and a mere appearance. For the present, however, I want to consider the Stoics.

Diodorus' pupil Zeno of Citium (not to be confused with the earlier Zeno of Elea, the propounder of paradoxes) set up a rival to Epicurus' school. He founded the Stoa in Athens around 300 B.C. In contrast to Diodorus and Epicurus, the Stoics denied time-atoms and insisted that time was infinitely divisible. They thus had to face all over again the question whether any time can be said to be present, or whether the supposed present does not consist wholly of past and future. Plutarch, writing from a Platonist viewpoint in the first to second century A.D., was to attack them on this issue, saying that they annihilated the present and pointing out that they did not avail themselves of the solution of a minimal time period (*elachistos chronos*), or partless now (*to nun ameres*), *de Communibus Notitiis* (*CN*) ch. 41, 1081c.

The Stoics considered not only the status of the present, but also that of the past and future, and thus in effect addressed themselves to the whole of Aristotle's puzzle about the parts of time. Plutarch's chapter is one source of information, but a fuller source is Arius Didymus, a Stoic grammarian of the first century B.C. He reports the views of three Stoics on the present question: Chrysippus (*c.* 280-*c.* 206 B.C,) was the third head of the Stoic school in Athens; Apollodorus of Seleucia on the Tigris flourished around 130 B.C.; and Poseidonius of Apamea, head of the Stoic school in Rhodes, lived from *c.* 135 to *c.* 55 B.C. The relevant sections, collected in Diels, p. 461, read as follows:

Apollodorus in his *Physics* defines time ... Some of it is past, some present (*enestēkos*), and some future. But all time is present (*enestanai*), just as we say the year is present (*enestēkenai*), circumscribing a wider band (*kata meizona perigraphēn*). And the whole of time is said to exist (*huparchein*), even though none of its parts exists exactly (*apartizontōs*).

Poseidonius: ... as regards *when* in time, some is past, some future and some present. The last consists of a part of the past and a part of the future surrounding the division between them. But the division is point-like (*sēmeiōdē*). *Now* and suchlike are thought of broadly (*en platei*) and not exactly (*kat' apartismon*). And *now* is spoken of also with reference to the least perceptible time surrounding the actual division.

Chrysippus ... says most clearly that no time is wholly (*holōs*) present (*enistatai*). For since continuous things are infinitely divisible, time as well is all infinitely divisible in this way. Hence no time is present (*enestanai*) exactly (*kat' apartismon*), but is described as present only broadly (*kata platos*). He says that only the present (*enestōta*) exists (*huparchei*), while the past and future subsist (*huphestanai*), but do not exist in any way, except that in which accidental predicates are said to exist. E.g. walking about exists in me when I walk about, and does not exist when I lie or sit down.

Plutarch describes Chrysippus' view similarly (*CN* ch. 41, 1081F):

Chrysippus ... in his treatise *On Vacuum* and elsewhere, says that the past and future do not exist (*huparchei*), but subsist (*huphestēkenai*); only the present (*enestēkos*) exists. But in the third, fourth and fifth books *On Parts* he holds that some of the present is future and some past. Hence it results that such time as exists for him is divided into what does not exist. Or rather no time at all is left existing, if the present has no part which is not either future or past.

One feature of these passages is that they make Chrysippus and Poseidonius both allude to our common conception of the present as having a span. We talk of the present day, for example, or the present year. Poseidonius adds a further point: we can mean by the present the least perceptible period surrounding the present instant. This idea of a least perceptible period is likely to have been introduced by Epicurus, as an analogue of his least perceptible space.

Chrysippus' view

But what is Chrysippus' position? Plutarch indicates that Chrysippus' statements were divided between different works. Perhaps he first declared that only the present existed, but then, when he came to write on *parts*, realised that some revision was called for. For since the broadly conceived present has a span, it will overlap with the past and future. But what moral did he draw? For the point about overlap might be used in two opposite ways. It might be said that therefore a certain portion of the past and future do after all have existence, because they have presentness. Alternatively, it might be maintained that the present cannot really overlap with the past and future. In that case, the present which was previously declared to be the only existent part of time, will have after all to be viewed as sizeless, and so not a *part* of time at all. This would be to concede the force of Aristotle's paradox, and to allow that in some sense time does not exist. Which did Chrysippus intend? I shall argue for the latter.

There is independent evidence that the Stoics thought of time as less than fully real. For they distinguished three grades of reality. Only bodies could be called existents (*onta*). Incorporeal entities could be called somethings (*tina*), but not existents (*onta*). After that there were mere conceptions (*ennoēmata*), which were nothings (*outina*).[17] Time was placed in the intermediate category,[18] along with place, vacuum, and statements (*lekta*),[19] that is, things stated.

Another way of distinguishing degrees of reality is by means of the verb 'to subsist', that is, to underlie (*huphistasthai*). This verb is used in the quoted passages on time to characterise the status of past and future. The idea of subsisting is used in parallel to that of being a mere 'something'. For Galen records that the Stoics distinguished the subsistent (*huphestos*) from the existent (*on*).[20] Further, Sextus and Diogenes Laertius quote the Stoic definition of statements (*lekta*) which describes them as subsisting (*huphistamena*).[21] There are also related compound verbs – *paruphistasthai*, to subsist alongside something else, and *prouphistasthai*, to subsist before something else does – and these are used of statements and of place.[22] Finally, Proclus reports that time was treated as subsistent (*huphistamenon*).[23]

So far everything suggests that time has an intermediate status as a something (*ti*), which subsists (*huphistatai*), without being an existent (*on*). Moreover, Plutarch gives us a reason why incorporeals are not treated as existents. The reason is that it is a mark of the existent to act and be acted on.[24] This fits with Proclus' comment that time and the other incorporeals are downgraded as being inactive (*adranē*).[25]

I think that this is the right interpretation, but there are some complications. One problem is that we should expect not only the word *on*, but also the other word for existent, *huparchon*, to be withheld from time and from the other three incorporeals. Yet we saw Chrysippus allowing *huparxis* both to the present[26] and to my walking, so long as I am walking, even though neither of these is corporeal. One thing that may have motivated him to upgrade these two things is a feeling that both of them are in some sense

---

[17] Some of the relevant texts are collected in *SVF* 2.329-35, and 521, viz., Alexander *in Top.* 301, 19; 359, 12; Sext. *M* 1.17; 10.218; Seneca *Letters* 58, 12; Anonymi *Proleg. in Cat.* p.34b (Brandis, Schol. in Aristotelem); Philo *Legum Allegoriae* 3.175; Plut. *CN* 30, 1074D; Procl. *in Tim.* (Diehl) 3.95, 10-14. See also Sext. *PH* 2.86; *M* 8.32-4; Plut. *adv. Col.* 1116B-C. There is a particularly helpful discussion by A.A. Long, 'Language and Thought in Stoicism', in his anthology *Problems in Stoicism*, London 1971. See also J.M. Rist, *Stoic Philosophy*, Cambridge 1969, ch.9; Pasquale Pasquino, 'Le statut ontologique des incorporels', in Jacques Brunschwig (ed.), *Les Stoïciens et leur logique*, Paris 1978.

[18] Sext. *M* 10.218 (time is incorporeal and a 'something'); Diog. Laert. *Lives* 7.141 (incorporeal); Plut. *CN* 1074D (not an existent, *on*, not *huparchei*); *adv. Col.* 1116B-C (a something, but not an existent, *on*); Procl. *in Tim.* (Diehl) 3.95, 10-14 (incorporeal, inactive, not an existent, *on*, subsisting, *huphistamenon*, in bare thought).

[19] All four are listed at Sext. *M* 10.218 as incorporeal somethings (*tina*), and at Plut. *adv. Col.* 1116B-C as somethings, but not existents (*onta*). In line with this, Plut. *CN* 1074D lists various *lekta* as not being existents (*onta*) for the Stoics, while Sext. *M* 10.3 and *PH* 3.124 records Stoic definitions of place and void which treat the occupying body as the only existent (*on*).

[20] *de Meth. Med.* (Kühn, 10, p.155).

[21] Sext. *M* 8.70; Diog. Laert. *Lives* 7.63.

[22] Sext. *M* 8.12 and 261 of statements; Simpl. *in Cat.* 361 of place.

[23] *in Tim.* (Diehl) 3.95, 10-14.

[24] *CN* 1073E.

[25] *in Tim.* (Diehl) 3.95, 10-14.

[26] I do not think that *huparxis* is allowed to the past or future, except in the sense that, *when they are present*, they have *huparxis*.

actual. If so, he need not be intending to go the whole way and to give them the same status as corporeals.

There is a further complication. Sextus records numerous sceptical arguments designed to show that there is no such thing as time, place, vacuum, or statement, or the various functions of statement, such as sign, proposition, or proof. In these controversies the Stoics are always represented as being on the other side, and as maintaining that there are such things.[27] What causes confusion is that these arguments are often conducted in terms of whether the things in question exist, where the words for existence are the standard verb 'to be' (*einai* from which come *on* and *onta*) and also *huparchein*. We thus get the surprising situation that the Stoics, who downgraded the four incorporeals as not being existents (*onta*), and sometimes as not enjoying *huparxis*, now appear as defending the *einai* and *huparxis* of the incorporeals after all. We should refuse to be mystified, however, for the controversy is not conducted in strict Stoic terminology. This is because it does not concern the Stoic school alone. If we transposed the discussion into their canonical terms, we should say that many of the sceptical arguments are designed to show that time and the other incorporeals are mere nothings. Thus the arguments often aim at establishing that the very conception of the various incorporeals is incoherent. The Stoic reply, correctly formulated, would be that on the contrary the things in question are somethings. And indeed this way of talking is sometimes used. But since the attack is directed against non-Stoic schools as well, a different terminology becomes standard, and the Stoics in reply argue for the existence (*einai*, *huparxis*) of the incorporeals.

Sometimes, Sextus makes it quite clear that the question at issue is simply whether we are talking about a something rather than a nothing. Thus he tells us that the Stoics made time and the other incorporeals into *somethings*,[28] and he replies by arguing that the category of something does not exist (*einai*), but is nothing (*ouden*).[29] Again, in connexion with place he protests that no adequate reason has been given to show that it is *something* (*huparchein ti ton topon*).[30]

One final complication is that Sextus ascribes the view that incorporeals lack *huparxis* only to *some* Stoics, such as Basileides, as if *other* Stoics allowed *huparxis* to incorporeals.[31] But we need to know in what sense these other Stoics conceded *huparxis*. Perhaps they were again merely rebutting the claim that there was no such thing as these incorporeals. Or perhaps they were merely thinking of such concessions as that the present and my walking, when I walk, can be elevated to the status of *huparxis*. In either of these cases, they would not be abandoning the intermediate status of the incorporeals as somethings.

From all this I conclude that the Stoics wanted to assign time an intermediate level of reality. In that case, I would suggest, they had every reason to welcome Aristotle's paradox. For on the one hand, the paradox

[27] On time see *M* 10.169-247; *PH* 3.136-50; on place *M* 10.3-36; *PH* 3.119-30; on vacuum *M* 9.332; 10.3; *PH* 3.124; on statement *M* 8.70-84; *PH* 2.107-9; on sign *M* 8.141-4; 159-61; 171; 275-98; *PH* 2.104-33; on proposition *M* 8.80; 83-4; *PH* 2.108-9; on proof *M* 8.337-481; *PH* 2.122; 144-203.

[28] *M* 10.218; 234.   [29] *M* 10.234-7.

[30] *M* 10.19.   [31] *M* 8.257-64.

would not assign to time too low a level of reality. For it only gives to time the same status as is assigned to the past and future, and Chrysippus granted to past and future a certain 'subsistence'. On the other hand, the paradox would not allow to time too high a level of reality. For it implies that the only 'part' of time, namely the present, to which Chrysippus had conceded a limited kind of *huparxis*, is not a *part* after all. It thus makes clear what for the Stoics was never in doubt anyway: that no part of time is existent (*on*). Their satisfaction with Aristotle's paradox should be all the greater, because Aristotle uses the very same word as themselves (*on*, or other parts of the verb *einai*), to express the non-existence of time. But now it becomes clear that, despite the sameness of the Greek word, there is a difference which can be brought out more easily in English. For the paradox in its original version was prepared to treat the past and future, and hence to treat time, as simply non-existent. But this could be a misleading way to express the Stoics' view, which might be better put by saying that they wanted to represent time as not *real*, or as less than fully *real*. This would do more justice to their *reason* for downgrading time – that it neither acts nor is acted on – and also to the fact that they do not want time to be a mere nothing.

Two rival interpretations of Chrysippus

That concludes the case for my interpretation, but there are rival ones, which make Chrysippus *resist*, rather than endorse, Aristotle's paradox, and resist it by exploiting the idea of the 'broad' present. It has been suggested, for example, that Chrysippus' solution is the one which I described in Chapter 1 as going part of the way, but as not going far enough. It shows how the present can have a positive length without overlapping with the past and future, provided that we *push back* the boundaries of the past and future. This idea, however, has been ascribed to Chrysippus only tentatively, because, as must be admitted, there is no mention of pushing back in the relevant text.[32]

Another way of exploiting the 'broad' present, which I have referred to already, would be to admit that it *does* overlap with the past and future, but to argue boldly that this merely bestows on some of the past and future the reality which belongs to the present. The disadvantage of this interpretation is that we should expect to hear of more argument in Chrysippus on two issues: first, on how it can be legitimate to violate our ordinary assumption that the present may not overlap with the past and future, and secondly, on why such an overlap would add reality to some of the past and future, rather than subtracting reality from the present.

A final defence of these rival interpretations might point to the fact that the Stoics recognise a broad conception of a thing's *place*,[33] of which they use very

---

[32] G.E.L. Owen cites Plut. *CN* 1081E-1082A, in which Chrysippus is reported as saying of the present that some is future and some past. Owen suggests the possibility of reinterpreting this to mean almost the opposite of what it seems to say: any part of the present *can 'equally' be taken as* past or future, but he carefully says that he cannot be sure ('Aristotle on time', in Peter Machamer and Robert Turnbull, eds., *Motion and Time, Space and Matter*, Ohio State University 1976, 18-19, repr. in *AA* 3, 153).

[33] Sext. *M* 10.15; 95; 108; *PH* 3.75; 119.

much the same language,[34] and they consider this broad conception perfectly legitimate. Can I be right, then, that Chrysippus would come to treat as illegitimate the corresponding broad conception of *time*? A man's broadly conceived place might be, for example, Alexandria, while his exact place fits him perfectly, and is just the same size as himself. In describing the broadly conceived place as not exact,[35] the Stoics do not mean that it is an improper conception, but only that it is not the tight-fitting one. There is more, for the Stoics do not confirm the low reality of place by an argument based on the narrow conception of it. Is it then reasonable of me to suppose that they confirm the low reality of time by an argument based on the narrow conception of the present?

I would reply that we cannot infer from the Stoic treatment of place to their treatment of time, for place and time are not alike in the relevant respects. For one thing, it is genuinely embarrassing to allow the present to overlap with the past and future, whereas there is nothing obviously objectionable about broadly conceived places overlapping, so there is not the same reason to expect the notion of broadly conceived place to be rejected. Sextus himself probably senses that the broad conception of place is not easily to be discredited by paradox. For although he attacks the Stoics whenever he can, he says that he will not raise arguments against the broad conception of place,[36] and contents himself with smearing it as an improper usage.[37] Yet again, even if the Stoics had rejected the broad conception of place, they could not have gone on from there to construct an argument about place parallel to the argument which downgrades the reality of time. For they would not have been able to cite any ordinary presumption that more distant places are less real, parallel to the common presumption that the past and future are less real than the present. I conclude that Chrysippus' treatment of place provides no evidence for his attitude to time.

Poseidonius and Apollodorus

I have been considering the Stoics generally, but, of course, they did not all agree with their most famous representative, Chrysippus. Poseidonius, for example, in the passages quoted above, refers to a shortest perceptible present. One possibility is that he solves the paradox by putting this forward as an existent part of time, but the texts do not say enough for us to be sure. Certainly, Apollodorus takes a line of his own in the quoted passages, and tries to defend the reality of time. He is, moreover, quite right in his idea that time can be said to exist (*huparchein*) in some sense, even if none of its parts exists exactly (he has in mind, presumably, that none of its parts is *present* unless presentness is taken broadly). But his explanation of *how* time can be

---

[34] *kata platos, en platei* (broadly); *kata perigraphēn* (circumscribing). S. Sambursky interprets the latter, wrongly I think, as referring to the year passing on its circuit (*Physics of the Stoics*, London 1959, 142).

[35] They talk of *akribeia* and of *apartismos*. The latter word is connected with the idea of fitting, and the corresponding verb is already used of a body and its place by Aristotle (*Phys.* 3.5, 205a32).

[36] *M* 10.15; *PH* 3.119.   [37] *PH* 3.119.

said to exist in these circumstances is not satisfactory. We speak in ordinary language of quite a broad span, a year, or even a millennium, as being present. But Apollodorus' suggestion that we could extend this to the *whole* of time would deprive the word 'present' of the necessary contrast with past or future.

Transmission of the paradox

After Apollodorus, the problem was kept vividly before people's minds by several sources. One such source was a Neopythagorean writing under the name of Archytas. The original Archytas of Tarentum, a Pythagorean and friend of Plato, lived around 400 B.C. But the genuine writings of Archytas and of other Pythagoreans were augmented, perhaps in the third to first centuries B.C., by a large group of spurious works.[38] One of these, *Peri tō katholou logō* (Doric dialect), which was probably written in the second half of the first century B.C., presents a theory of categories, and includes a discussion of time.[39] In 1914 a late MS of a non-Doric version was found, and its comments on time are also preserved by Simplicius,[40] who, however, mistakenly takes the author to be the original Archytas. The discussion of time includes a version of Aristotle's paradoxes.

Next, in the first to second centuries A.D., Plutarch narrated Aristotle's puzzle about the parts of time and ascribed it to his own teacher.[41] Later again, towards the end of the second century, Sextus Empiricus made his massive compilation of sceptical arguments, and these include several versions of the same puzzle.[42] The aim was to show that there are equally strong arguments on either side of any issue, so that one should suspend judgment, which will in turn bring tranquillity of mind. Sextus claimed to trace the origins of his school back to Pyrrho in the fourth century B.C. Without taking this claim too literally, we can assume that the sceptical arguments in his collection had been used in the interim by his predecessors.

The paradox also entered into Christian thought: modified versions of it are found in Eusebius and Basil, both of Caesarea in the fourth century A.D.,[43] and I imagine that it may have been known still earlier to Tatian in the second. At any rate, there is a passage in Tatian's indiscriminate attack on the Greek philosophers which looks as if it corresponds to Plutarch's. Plutarch had used the paradox in order to contrast God's eternity with the paradoxical character of time; but Tatian, while possibly substituting a different paradox, seems to take the contrast further: only eternity is real; the

---

[38] For a survey of the historical questions involved see Holger Thesleff, *An Introduction to the Pythagorean Writings of the Hellenistic Period*, Åbo, 1961.

[39] See Thomas Alexander Szlezak, *Pseudo-Archytas über die Kategorien*, Peripatoi vol.4, Berlin and New York 1972; Philippe Hoffmann, 'Jamblique Exégète du Pythagoricien Archytas: trois originalités d'une doctrine du temps', *EP* 1980, 307-23.

[40] in Cat. 352, 24-356, 7; and *in Phys.* 785, 12-787, 28. The texts are conveniently assembled in SP, 24-38, with an English translation, which should, however, be used with caution.

[41] Plut. *On the E at Delphi* 392E-F.

[42] *M* 6.63 and 67; 10.182-4; 192; 197-200; 242-3; *PH* 3.144-6. Cf. also *M* 8.81-4; *PH* 2.109; 144, where Sextus divides not only present time, but also present sayings and things said into past and future parts.

[43] Eusebius *de Laud. Const.* 6, p.615 (PG 20, 1340C-1344A). Basil *in Hex.* Homily 1, §5

passage of time, involving past, present and future, is a subjective illusion.

> Why do you divide time (*chronon*) for me, and say that some of it is past, some
> present and some future? For how can the future become past, if the present
> has being? It is like people on board ship who think through their ignorance
> that the hills are on the run when the ship is moving. That is how you fail to
> recognise that it is you who are running past, whereas eternity (*aiōn*) stands
> still as long as its maker wishes.[44]

Curiously enough, Tatian's complaint that the division into past, present
and future is a subjective illusion is not so very far from the view I shall
associate with the next figure to be discussed: the Aristotelian Alexander of
Aphrodisias (fl.*c.* A.D. 205).

Alexander of Aphrodisias

Simplicius says of Alexander that not even the most industrious of Aristotle's
interpreters handed down any solutions to the paraxodes of time.[45] None the
less, Alexander provides some interesting *materials* for a solution. For in a
short treatise *On Time*, which is preserved in Latin and Arabic, and which is
translated in a paper by Robert Sharples,[46] Alexander says of the instant,

> its generation is in the *mind* (*eius generacio est in mente*).[47]

This fits in with a generally idealist treatment which he gives to instants and
to the division of time. Time in itself is unitary; it is only in our *minds* or
thoughts that we divide it up by means of instants. I shall return to
Alexander's reasons for this view in Chapter 7. For the present I shall quote
two longer extracts from the Latin version:

> And I say that time is a number [i.e. a plurality], although it is one connected
> continuum, only because it is many in our *thought* (*estimacio*). For there are not
> times [plural], except in potentiality and in thought, not in actual being ... The
> instant is indeed in time, as the point is in the line, but there is a difference
> between them, because the point is in the line in actuality, but the instant is there
> in thought, not in actuality ... When time exists in actuality it is only divided into
> years and months or days in thought ... But if it were divisible in actuality, there
> would be between its parts an interval which was not time, just as a motion is
> one, and it cannot be that we pass from one motion to another <without> there
> being between the two motions something which is not a motion. And it is not
> possible that between two times there should be an interval which is not time.
> Time therefore is in actuality one, although it is divided in potentiality.

---

[44] Tatian *ad. Gr.* 26 (PG 6, 861).

[45] *in Phys.* 795, 34.

[46] Robert Sharples, in collaboration with F.W. Zimmermann, 'Alexander of Aphrodisias, *On Time*', *Phronesis* 1982.

[47] *de Temp.*, translated into Latin, probably from some Arabic version in the twelfth century by Gerard of Cremona, ed. G. Théry, in 'Autour du décret de 1210: II, Alexandre d'Aphrodise, aperçu sur l'influence de sa noétique', *Bibliothèque Thomiste* 7, 1926, 92-7 (see 95, l.36) The ninth-century Arabic translation is edited by A. Badawi, in *Commentaires sur Aristote perdus en Grec et autres epîtres*, Beirut 1971, 19-24.

If time is not used up, then indeed in itself it is one, and is multiplied for us only in the mind, according as the mind measures and estimates it. And similarly we shall say that 'exists' and 'will exist' have no reality except in our estimations. In itself, however, [time] is one, continuous and sempiternal, in accordance with a single ordering.[48]

The relevance of Alexander's view to the paradox of the ceasing instant is this: there is not really any ceasing or beginning of the instant. It begins or ceases only in our *minds*. Such a solution, had he offered it, would admittedly not be decisive, because a question could still be raised about the cessation in our minds and existence in our minds of the instant: how does the mind see these as related to each other? But what Alexander might say is that although this new question is as hard as the original, at least it only affects our *thoughts* about time, and not time as it really is.

Alexander's view of time is equally relevant to Aristotle's other paradox[49] For although the past no longer exists and the future does not yet exist, Alexander would presumably say that the division into past and future does not affect time itself. We shall see in Chapter 5 that some relevance, although perhaps not this one, of Alexander's view to the paradox of the parts of time was later picked up and exploited by Damascius and Simplicius – and this in spite of Simplicius' report that Alexander handed down no solutions.

Although I have called Alexander's view idealist, he is not talking about time itself, but about the *division* of time, for example its division into past, present and future, when he says that such divisions exist only in thought. None the less, it will be seen in Chapter 7 that there is a hint of idealism applied to time itself, when Alexander repeats Aristotle's claim that there would not be time without soul, and buttresses it with an additional argument.[50]

For a more thoroughgoing idealism in regard to time, we must wait till Augustine (A.D. 354-430). I turn to him next, in a brief departure from chronological order, postponing consideration of the Neoplatonist Iamblichus to the next chapter.

Augustine

Augustine explicitly considered the paradox of the parts of time and gave it its most eloquent and arresting exposition. The eleventh book of his *Confessions* can be recommended as delightful reading. Moreover, Augustine offered a striking new solution. The only way to save the existence of time would be to treat past, present and future as three mental states, for then they could all exist at one and the same instant within the mind. The past would be memory, the present attention, the future expectation. Tentative as he is about this hypothesis, he does go so far as to say that time is an extension (*distentio*) of something, and that it would be surprising if it were not (*mirum si non*) an extension of the *mind* (*animus*). The following are the relevant passages of the *Confessions*.

---

[48] op. cit. 94, ll.22-42 and 97, ll.4-9.
[49] I am grateful to Robert Sharples for first drawing my attention to this further relevance.
[50] *de Temp.* (Théry), 95, ll. 11-15; cf. Alexander ap. Simpl. *in Phys.* 759, 20-760, 3.

As is now evident and clear, the future and past do not exist; nor is it proper to say, 'there are three times, past, present and future'. But perhaps it would be proper to say, 'there are three times, a present of the past, a present of the present and a present of the future'. For these three things do exist in the mind (*anima*), and I do not see them elsewhere. The present of the past is memory, the present of the present is attending (*contuitus*), the present of the future expectation. If we may speak this way, I see three times, and I acknowledge that there are three ...

Thus it seems to me that time is nothing other than an extension (*distentio*), but of what it is an extension I do not know. It would be surprising if it were not an extension of the mind (*animus*) itself ...

In you [my mind], I say, I measure times. What I measure is the effect (*affectio*), itself present, which passing things create in you, and which remains when they have passed away. In measuring times, I do not measure the things which passed in the act of creating that effect, but rather the effect itself. Therefore either times *are* that effect, or I do not measure times at all.[51]

Augustine weaves some extra elements into the original version of Aristotle's paradox, for he argues not only that the present is sizeless (XI.15), but also as pseudo-Archytas and Basil of Caesarea had said, that it is fleeting (XI.14). Moreover, he adds the question how we can *measure* the time of uttered syllables as long or short, if neither the time nor the sound is in existence (XI.15-16; 21-2; 26-8). His solution, that past, present and future can all be available at once as a *distentio* in the mind, has the paradoxical effect of making time more like eternity. Time is frozen for inspection, and available all together. In XI.18, Augustine insists that, wherever the past and future exist, they do not exist *as* past and future, but as present, or else the paradox of their non-existence would recur. In XI.28, he claims that a long future is really a long expectation, and a long past is really a long memory, and the length referred to must presumably be the length of the *distentio* in the mind, which is not to be thought of as long-lasting (or else the paradoxes would recur), but as long in a sense appropriate to extensions whether spatial, temporal, mathematical, or (as in this case) mental. The effect of making time sound more like eternity cannot, however, be intended by Augustine: he wishes to *contrast* God's eternity with our time. Nor does he want to present time as a static B-series (to use McTaggart's terminology): we shall see in Chapter 16 that it is unique to God that he sees the past and future not *as* past and future. Nor again does Augustine want to say that time is unreal: he finds it, with its distractions, all too real. The one implication that would perhaps be welcome is that time, being a mental entity, might with sufficient mental progress be shed, and I shall consider this implication in Chapter 11.

Augustine's psychological account of time is not, so far as I know, taken up by his immediate successors, nor even by himself in later writings. The closest I have found to an echo comes in Leibniz, who makes use of the paradox of the parts of time in his fifth letter to Clarke (§ 49) in 1716. Leibniz's conclusion is that time must be ideal rather than real, and he explains what

[51] *Conf.* XI.20; 26; 27.

he means by a comparison with genealogical lines (§ 47), which he regards as ideal, although expressing real truth.

Although the psychological analysis of time is the best known feature of Augustine's treatment, there are other dimensions to his discussion. First, he has unusual views on the relation of time to *change* and *movement*. In *Confessions* XI, while preparing the way for the idea that time is an extension found in the mind, he attacks the rival view that time (or times, plural – he uses both ways of talking) are celestial motions (XI.23). His counter-claim is the strong one, already found in Basil of Caesarea,[52] that time would not be affected if the heavenly bodies stood still, as the sun is said to have done for Joshua, or if they accelerated. But does not time require *some* change or other to continue? The answer 'yes' is presupposed in Book XI,[53] and it is further stated in XII.8, 11 and 12 that times, and in *City of God* XI.6 and XII.16 that time does require change.

City XII.16 contains a further and very unexpected idea, namely the theory of angelic time. Hours, days, months and years, which are ordinarily called times, began with celestial motion. But we can also distinguish a kind of time which depends on the *mental* movements of the angels. This theory had already been offered in two works straddling the *Confessions*: *de Genesi ad Litteram Liber Imperfectus* III.8 and *de Genesi ad Litteram* V.5.12. The reference, or the main reference, here is to the six days of creation, on the first three of which, according to Genesis, celestial motions had not yet been created. The mental movements of the angels are described in *de Gen. ad Lit.* IV.22.39 and *City* XI.7, and they are not connected with memory and expectation. Rather, the angels appear to alternate between seeing the six types of created thing as they exist in God's Wisdom, and seeing them as they are in themselves, returning each time to praise of the Creator, and finishing with contemplation of his rest. But should we understand these as movements and as time? The *de Gen. ad. Lit.* is particularly insistent that the six days should be thought of as one, and that there are no delays or temporal intervals separating the things created. In them the words 'before' and 'after' refer to a causal, not a temporal order.[54] If we stress this point of view, we may wonder whether it is still appropriate to describe all this as *time*, but all three passages are insistent that it is. What we can assume is that, although it began three 'days' before celestial movement, this does not mean that it is older in terms of the more familiar kind of time.

[52] It has been shown by J.F. Callahan that this discussion mirrors Basil of Caesarea *adv. Eunom.* 1.21: 'Basil of Caesarea: a new source for St. Augustine's theory of time', *HSCP* 63, 1958, 437-54.

[53] That time requires change is presupposed by Augustine's view that there was no time before the creation (XI.13 and 30). We have something close to an explicit statement in XI.14, where Augustine says that if nothing passed there would be no past time, if nothing arrived no future time, and if nothing existed no present time. We can further guess that if time involves memory and expectation, it is almost bound to involve change. Finally, although the potter's revolving wheel is presented in XI.23 not as giving *existence* to time, when celestial motion ceases, but as something which could still be *measured* by time, it may none the less be a sign of the importance Augustine attaches to change that he does not raise the more radical question, which I shall consider below in Chapter 6, whether we could measure a period in which all change ceased, even that of the potter's wheel.

[54] See especially *de Gen. ad. Lit.* IV.26.43; 32.49; 33.51; 33.52; 34.53; 35.56; V.5.15.

There was no hint of this theory in *Confessions* XII. There Augustine describes the 'heaven of heaven', which includes the angels.[55] But what he stresses is that the heaven of heaven does *not* change (XII.19). Nor does it have a past to remember or a future to expect (XII.11). He tries to give it a status intermediate between time and eternity. Thus, on the one hand, it is rapt in changeless contemplation, but, on the other, it is still *capable* of change (XII.11 and 19), and so is not coeternal with God (XII.11 and 12). None the less, since it does not change in fact, it does not suffer times (XII.19), is not stretched out into times (XII.11), but is beyond times (XII.9), and partakes in (XII.9), or enjoys (XII.12), God's eternity, without being coeternal with him. Can this account be made compatible with the description elsewhere of the angels as engaging in mental movements? Presumably it can, in so far as those mental movements are taken as being not so much movements as distinguishable coexistent states of mind.

To sum up, Augustine presents us with the following points of view. Time is a dimension of the mind. It does not strictly require the persistence of celestial motion, although it does require change. Besides the times which began with celestial motion, there is also a quasi-time which depends on the quasi-change in angelic minds. But since this change is only a quasi-change, the angels can also be viewed as poised between time and eternity.

---

[55] For uncertainty in Origen and Augustine as to whether the heaven of heaven contains or consists of intelligent beings like angels, or intelligible patterns used as models for the physical creation, see Chapter 15, and the references to Pépin and Solignac given there.

CHAPTER THREE

Iamblichus' Solution:
Static and Flowing Time

The Neoplatonist Iamblichus from Syria, who died around A.D. 325, tried to solve the paradox about the parts of time by distinguishing between a static time and a flowing time. Some such distinction has played a major role in twentieth-century discussions of time, and Iamblichus' distinction has been compared with that introduced by McTaggart in 1908.[1] One point of interest will be how close a comparison there is. I shall start with the modern distinction.

The modern distinction

McTaggart divided our time-words into two groups. In the first group he placed 'past', 'present' and 'future', to which we can add 'now', 'today', 'tomorrow', 'yesterday', 'ago', 'soon', 'hence', 'over', 'still to come', and (a point which is sometimes neglected) the tenses 'was', 'is' and 'shall be'. In the second group, he placed 'earlier', 'simultaneous', 'later', to which we can add 'before', 'after', '1 December 1978 A.D.'. Philosophers since McTaggart have sought to distinguish these groups by reference to three features. First, as McTaggart himself emphasised, words in the first group imply a certain flow: what is at one time future ceases to be future and becomes present and then past. No comparable flow is associated with the second group of words. If one battle is later than another, or later by 1978 years than the birth of Christ, it is *changelessly* so.

A second feature of the first group of words is that they are indispensable for guiding our emotions and actions.[2] Suppose the following announcement

---

[1] J.M.E. McTaggart, 'The unreality of time', *Mind* n.s. 17, 1908, 457-74, revised in *The Nature of Existence*, London 1927, vol. 2, ch. 3. The comparison is made by S. Sambursky, 'The Concept of Time in Late Neoplatonism', *PIASH* 2, Jerusalem 1968, 153-67, reproduced in SP, 12-21. It has been suggested to me that McTaggart was anticipated by Henri Bergson and William James, but I do not think that this is so. In his *Essai sur les données immédiates de la conscience*, Bergson thinks that if we try to conceive time as a static geometrical line, we are really thinking of space. He does not agree that there really is a static temporal series at all (pp. 90-1; 98-110 of Pogson's English translation, *Time and Free Will*, London 1910). He also interchanges McTaggart's static and flowing expressions (p.227) without noting the difference.

[2] This has been most fully argued by Richard Gale, in *The Language of Time*, London 1968, ch. 4; and 'Tensed statements', *PQ* 12, 1962, 53-9. Arthur Prior had already pointed out that if one is *pleased* on such and such a date that something is *over*, what one is pleased at is not the fact that its conclusion precedes such and such a date, but the fact that it is over. These are, therefore, not the same fact. ('Thank goodness that's over', *Philosophy* 34, 1959, 17).

were made, couched in the language of the second group: 'A hand grenade is [tenseless] thrown into King's College on 1 December 1978'. This would not serve to guide the actions and emotions of people working in King's College, unless they could judge, 'But 1 December 1978 is *today*'. If on the contrary they possessed none of the concepts associated with the first group of words, they could not judge whether the explosion was a hundred years into the past or a hundred years into the future. Their actions and emotions would not have been adequately guided.

The third feature which has been said to belong to the first group of expressions has proved the most controversial: it has been claimed that they are all definable in terms of the word 'this'. According to one account, 'now' might be paraphrased as 'simultaneously with this utterance'; 'past' as 'prior to this utterance'; 'future' as 'later than this utterance'. 'Today' means 'the day of this utterance', and to use a tense and say that something *will* be so amounts to saying 'its being so is later than this utterance'. 'Utterance' is only one of the proposed completions for the word 'this'. Bertrand Russell, who introduced this kind of definition, at first suggested a formula in terms of 'this sense-datum';[3] H. Reichenbach, who offered a very lucid elaboration of Russell's idea, preferred the phrase 'this token', where a token is a particular instance of some sentence.[4] Accordingly, he called such words 'token-reflexive'. But one might instead simply say 'this thought', or 'this event', leaving the context to determine the event intended.

There are, of course, *non*-temporal words among the token-reflexives, such as 'here', 'there', 'I', 'you', and the words 'this' and 'that' are included in the group. A token-reflexive word need not necessarily be defined in terms of the word 'this', but, the suggestion is, it can only be defined with the aid of other token-reflexives, and 'this', which is itself classed as a token-reflexive, is probably the most convenient member of the group.

The proposed connexion with the word 'this' was not accepted by McTaggart, and it can give rise to at least three kinds of objection: one concentrates on *secondary* uses of the word 'present'. Arthur Prior, for example, has pointed out that, in addition to saying that a battle *is* present, we can say that it *was* present.[5] Prior further adduces McTaggart's objection that among the events which were *once* present we can include events before there were any utterances or any conscious life in the universe. Indeed, the absence of conscious life, it could be added, was itself once a fact about the present. The objection is that in saying that any of these things *were* present we do not mean, 'they were simultaneous with *this* utterance'. But the answer to this kind of objection must surely be that uses of '*was* present' are to be understood as more or less remote extensions from uses of '*is* present'. If we wish to defend the kind of analysis under discussion (and below I shall suggest modifying it), then we can paraphrase the claim that some earlier event in the history of the universe was once present along the following lines:

---

[3] Bertrand Russell, 'The experience of time', *Monist* 25, 1915, 212-33.
[4] Hans Reichenbach, *Elements of Symbolic Logic*, New York 1948, §§50-1.
[5] Arthur Prior, 'On spurious egocentricity', ch. 2 of *Papers on Time and Tense*, Oxford 1968, repr. from *Philosophy* 1967.

if someone at the time of the event had said, 'it is simultaneous with this utterance', he would have expressed a true proposition. As for the absence of utterances, that was once a fact about the present, in that it was simultaneous with events which were present in the sense just described.

There is a second objection which ought simply to be admitted, and that is that the possibility of paraphrase in terms of 'this' is not confined to the *flowing* uses of time-words. This is partly because not all uses of a word like 'present' are flowing uses. Thus we can say that an event of some kind is at *some* time present, without revealing whether it is so in the past, present, or future. Or we can say that it is *either* past, *or* present, *or* future. In either case, we can still paraphrase the use of the word 'present' in terms of the word 'this', but the use is not a *flowing* one, because what is said is *changelessly* true, if true at all, nor is the use to the same extent action-guiding. What it tells us even entails and is entailed by what could be said in purely *static* terminology: an event of that kind *occurs* (tenseless). It may be thought to follow that this piece of static terminology can, with equal legitimacy, be paraphrased in terms of the word 'this'. But of this last step I am not so sure, for although there is a mutual entailment between what is said in the two terminologies, it is less natural to say that the *meaning* is the same. After all, the use of the static terminology does not draw attention to the fact that what occurs must be present at some time, and must now be either past, present, or future; and this makes a paraphrase in terms of 'this' less appropriate.

My attitude to the second objection is none the less to accept it in some form or other and to regard it as harmless. A third kind of objection, however, will call for a certain modification. It focuses on the *noun* ('thought', 'utterance') which follows the word 'this' in the paraphrases. Arthur Prior, for example, has protested that when we say, 'thank goodness that's over', we are not expressing relief at the unpleasant thing being earlier than some *utterance*.[6] Again, in Chapter 1 some difficulty arose over how the present instant, being sizeless, can be perfectly simultaneous with a protracted *utterance*. But if that is all the third kind of objection has to say, then the problems can easily be met. I have already spoken of the possibility of thinking in terms of an *idealised* utterance which will match the idealised instant by presenting some absolutely sharp and sizeless boundary. As for Prior's difficulty, it is not the word '*utterance*' which is crucial for the paraphrase, but the word '*this*'. We are expressing relief at the thing being earlier than some *time*, and we can only designate this time with the aid of the word '*this*', or some other token-reflexive. But we can perfectly well designate it without using the concept of an *utterance*. We are pleased that the thing is earlier than the time of 'this – ', where the blank can be filled by naming *any* current event. Since there will always be an utterance going on, whenever we say 'thank goodness that's over', the utterance is merely a convenient event to select.

There is, however, a more radical fault with paraphrases that bring in reference to utterances and other such events, and that is that they import extra ideas over and above those involved in thoughts of the past, present or

[6] Prior, op. cit. in n.2.

future. This can be illustrated by imagining a person waking up in confusion after a long sleep and thinking, 'what is the time now?' He need not be in a position to think, 'what is the time of this waking?' – he may be too confused to realise that he is waking – or 'what is the time of this thought?' – he may lack the concept of a thought. If he is sufficiently short of concepts and information, he may not be in a position to complete any thought of the form, 'what is the time of this – ?' Another way of bringing out that there is an unwanted surplus in the proposed paraphrases is to consider counterfactual situations. 'I am now tired' differs from 'I am tired at the time of this utterance/event', in that the first proposition expressed could still be true, even if there were no utterance or event of the kind indicated in the second. The remedy I would suggest is to paraphrase our waking man's thought by the words, 'what time is this?', thereby eliminating the reference to wakings and other events. I do not mean that this form of words could not be used to express some *different* question – it could[7] – but only that it *can* be used to express the question asked, 'what time is it now?', and to express it without remainder. Moreover, the question cannot be paraphrased without the use of 'this', or of some other token-reflexive.

I am not proposing the paraphrase of 'now' in terms of 'this time' as a pattern of analysis that would do for *all* token-reflexives. I doubt, for example, whether 'I' can be analysed without remainder in terms of 'this person'.[8] And I suspect that further light could be thrown, not only on 'I', but also on 'now' and on 'this', by David Kaplan's work, which explains token-reflexives not in terms of *each other*, but in terms of his notion of direct reference.[9] But that would go beyond present needs.

To return to the distinction between McTaggart's two kinds of time-word, it has figured prominently in twentieth-century debates. McTaggart spoke of his distinction as one between two temporal series, the A-series, past, present, future and the B-series, earlier, simultaneous, later. He considered the flowing series the more fundamental, and yet he believed that it involved a contradiction. Hugh Mellor has upheld McTaggart's idea that the A-series involves a contradiction, but has sought to rehabilitate the B-series, and this has been the commoner preference.[10] The attacks on the A-series have been diverse. Some have supposed that the flowing words for time are dispensable without loss, but that claim is refuted, I believe, by the considerations above about their role in guiding emotion and action. Mellor agrees about that, and so allows that we need A-*words* and A-*judgments*, but denies that there are any A-*facts*, or that there is any past, present and future, corresponding to them; there is only the B-series. Others have argued, less startlingly, that at any rate the flowing A-words are inessential to the nature of time; they merely express some relationship between time and ourselves. Russell called these words 'egocentric'. It has also been maintained that Relativity Theory shows

---

[7] It could, for example, be used to mean ,'What is the time just referred to?'

[8] Might not a madman, thinking he was a pumpkin, assent to 'I am here', but not to, 'this *person* is here'?

[9] In preparation: the typescript which I have seen is entitled, 'Demonstratives', and dated March 1977.

[10] D.H. Mellor, *Real Time*, Cambridge 1981.

the A-series to be impossible, or that at least it gives a complete account of time without making use of any of the flowing A-conceptions.[11] In Chapter 6, I shall defend the A-series from some of these attacks. For although the connexion with the word 'this' makes the A-series in a way egocentric, I shall argue that we cannot conceive of time existing without its flowing aspect.

Iamblichus' distinction

We can now consider how Iamblichus conceives his two kinds of time, and the two nows corresponding to them, although it will be seen that he is not easy reading. Many of the Neoplatonist texts on time have been made available in English by Sambursky and Pines in SP. My occasional disagreements should not obscure my great debt to them both. For many of the Iamblichus texts, there is an excellent translation, with commentary, by John Dillon.[12] Simplicius reports Iamblichus' view as follows:

> It looks then as if in these words Iamblichus is postulating one ungenerated now before the things which participate in it, and then [the nows] which are transmitted from this one to the participants. Just as with now, so also with time. There is one time before temporal things, and there are several times which come into being in what participates, so that in them one time is past, one present, one future.[13]

In the surrounding pages, Simplicius refers to Iamblichus' distinction four times.[14] And in another work, he ascribes to Iamblichus a generative time and a generated and flowing time.[15]

There is a small obscurity, because in the direct quotations from Iamblichus we often find reference not to two times, but to time and what participates in it, that is, events and things. But in fact there are two direct quotations, to be translated below, in which Iamblichus distinguishes between a higher now and a lower, or a plurality of lower, nows.[16] In any case, it fits with Iamblichus' general metaphysical views that there should be a higher and a lower time. For Plato had distinguished largeness itself, the largeness in Simmias, and Simmias who is large.[17] And correspondingly Iamblichus distinguishes what he calls an unparticipated universal, the participated universal which it generates, and the particular which participates.[18] This should give us, when applied to time, just what Simplicius describes: a generative time, a generated time, and the events and things which participate in the latter. If Iamblichus sometimes mentions the

---

[11] For a history of these arguments, see M. Čapek, *The Philosophical Impact of Contemporary Physics*, Princeton, 1961, 158-60.

[12] *Iamblichi Chalcidensis in Platonis Dialogos Commentariorum Fragmenta*, Leiden 1973, frs 61-8.

[13] *in Phys.* 793,3-7 (SP 38,32-40,4).

[14] *in Phys.* 787,10-11 (translated below); 793,12-22 (translated above); 793,28-9; 794,21 (SP 34, 21-3; 40, 10-22; 40, 28-9; 42,27).

[15] Generative (*genesiourgos*) Simpl. *in Cat.* 352,14 (SP 26,19); generated (*genētos*) and flowing (*rheōn*) *in Cat.* 354,14 (SP 28,17).

[16] ap. Simpl. *in Cat.* 355,8-14 and 355,27-356,1 (SP 30, 11-17 and 30,33-32,10).

[17] *Phaedo* 102A-D.

[18] Procl. *Elements of Theology* 23-4; 100; *in Tim.* (Diehl) 2.105,15-28; 240,4ff.; 313,19-24.

participants rather than the lower time, Simplicius reassures us by saying that the lower time *resides in* the participants.[19]

Iamblichus repeatedly describes the lower time both in reports and in direct quotation as flowing.[20] By contrast with what flows, the superior now is repeatedly said to be static and always the same in form.[21] And what is true of the superior now is true also of the superior time: it too is static and always the same in form.[22]

Flow is pictured in two equivalent ways in the following passage. Time is said to flow, but then we are told that it is located in events and things, and that these too are flowing. Tomorrow's concert, for example, can be thought of as flowing in relation to the higher now and time, and as touching them first with its beginning, then with its middle, and finally with its end:

> And where should we think that the flow (*rhoē*) and shifting (*ekstasis*) of time occurs? We shall say in the things which participate in time. For these are always coming into being and cannot take on the stable nature (*ousia*) of time [elsewhere: cannot take on the indivisible nature] without changing, but touch that nature with ever different parts of themselves.[23]

Comparison with McTaggart

We can now see that Iamblichus is motivated in much the same way as McTaggart when he talks of time flowing. This is suggested in the last two quotations by the talk of events coming part by part into contact with time and the now, and by the fact that the time which flows is the only kind which is divisible into past, present and future. It is also suggested by a passage which finishes up in almost the same words as the one just translated, but which begins by connecting the movement of time with *becoming past*.

> So if [pseudo-Archytas] should say that the now is past simultaneously with being thought and spoken of, we must postulate this kind of shifting (*ekstasis*) in the things which participate in time. For the things which come into being cannot take on the indivisible nature without changing, but touch that nature with ever different parts of themselves.[24]

It is further confirmation of Iamblichus' thinking like moderns that he

---

[19] *in Phys.* 792,33; 793,6 (SP 38,26; 40,2). Iamblichus himself says that the flow and shifting of time resides in the participating things, ap. Simpl. *in Phys.* 787,18; *in Cat.* 354,20 (SP 34,29; 28,25).

[20] Time flows: *in Cat.* 354,14 (SP 28,17); *in Phys.* 792,32 (SP 38,26); *in Phys.* 787,18 (direct quotation SP 34,30). There is a flow of coming to be: *in Cat.* 354,27 (direct quotation SP 28,32). And 'to be always flowing is most characteristic of participation in the now at the level of coming to be': *in Cat.* 355,7-8 (direct quotation SP 30,10-11).

[21] Direct quotation: *in Cat.* 354,18 and 25; 355,3, 9, 13, 21, 26 and 32 (SP 28,22 and 30; 30,6; 30,12; 30,16; 30,26; 30,32; 32,5) Report: *in Cat.* 353,21; *in Phys.* 787,12 and 23; 793, 2 (SP 26, 31; 34,23 and 35; 38,31).

[22] Direct quotation: *in Cat.* 354,32; *in Phys.* 787,19; 792, 34; Procl. *in Tim.* (Diehl) 3.33,3 (SP 28,38; 34,32; 38,28; 46,15).

[23] *in Phys.* 787,17-20 (SP 34,29-33); similar passages at 792,32-5 and *in Cat.* 354,20-3 and 30-3 (SP 38, 26-9; 28, 25-8; 28, 34-40).

[34] *in Cat.* 355,29-356,1 (SP 32,1-10).

connects the idea of flow with the idea that the now becomes ever different and ceases to be. Thus after reporting this way of talking about the now, Iamblichus says:

> But to come to be ever different and to cease and to be ever flowing (*rhein*), is most characteristic of participation in the now at the level of coming to be.[25]

But if one half of Iamblichus' distinction is like the modern one, this is not true of the other half, the static time. Here Iamblichus is a long way from McTaggart's point that the relations of earlier, simultaneous and later apply changelessly, when they apply at all. His idea is a much more Platonist one, that the higher time is elevated above the sensible world, and is not the sort of thing in connexion with which talk of division and flow makes sense. I hesitate to say that he treats it as a Platonic *Form*, because, since time is, according to Plato, a mere *image* of eternity, he is unwilling to class it among the 'primary realities' and 'really real things', and John Dillon has produced evidence that that means it is not a Platonic Form.[26] On the other hand, if Simplicius can be trusted, Iamblichus does treat the higher time as an essence (*ousia*), and the motion of which time is the 'number' as a Form (*eidos*).[27] Because Iamblichus has no inkling of one half of McTaggart's distinction, I think it would be wrong to say that he anticipated the distinction itself.

Application to the paradox concerning the parts of time

Now that we have some idea what Iamblichus' distinction is, the question remains how he applies it to the paradox concerning the parts of time. Iamblichus' solution is to concede that the inferior kind of time is unreal, although not for the reason alleged in the paradox, since the inferior present is not an indivisible instant, but a whole divisible stretch of time. Indivisibility is found rather in the superior kind of time and of now, but that does not make them unreal. On the contrary, precisely because they are not subject to division or flow, they are immune to the paradox. For it cannot be maintained of the superior time that some is no longer and some is yet to come. Simplicius summarises the strategy as follows:

> Iamblichus thinks we should understand partlessness (*ameres*) and unreality (*anupostaton*) to apply to different times.[28]

The wording in terms of partlessness and unreality is derived from the formulation of pseudo-Archytas, through whom the paradox had been transmitted to Iamblichus:

> Which is why time either does not exist (*esti*) at all, or is obscure and hardly

[25] *in Cat.* 355,7-8 (SP 30,10-11).
[26] Dillon fr. 64 = Procl. *in Tim.* (Diehl) 3.33,1-34,7, also, in curtailed form, in SP 46,14-37. Dillon's evidence in his note ad loc. is drawn from Syrianus.
[27] *in Phys.* 786,21; 787,30 (SP 32,29; 36,5).
[28] *in Phys.* 787,10-11 (SP 34,21-3).

exists. For how could it exist (*huparchoi*) in reality (*kat' alētheian*), if its past is no longer, its future is not yet, and its now is partless (*ameres*) and indivisible (*adiaireton*)?[29]

That the present of inferior time is not a partless instant for Iamblichus emerges from a further report by Simplicius, although it does not come out so clearly in the translation of Sambursky and Pines:

> He wants not only a now [i.e. a mere instant] to be present (*enestēkenai*), but also a time between two boundaries [i.e. a stretch of time].[30]

This report suggests that Iamblichus regards the present associated with the inferior time not as a partless instant, but as a whole period, which will mean that it can be treated as a *part* of time.

My description of Iamblichus' solution is confirmed elsewhere. In the following passage, Iamblichus denies the inference which it would be most natural to draw from pseudo-Archytas' words, that there is a *single* now, flowing yet static, divisible yet indivisible. Instead, he apportions flow and divisibility to *one* now, stability and indivisibility to *another*:

> How then does the same thing come to be ever different, yet remain the same in form (*kat' eidos*)? How is it divided, yet indivisible? How does it alter, yet combine its beginning and end in one? It is because there is a difference between the now in which things participate[31] in nature, and which is inseparable from things which come to be, and the [now] which is separate and by itself. One of these stands still (*hestēken*) in the same form (*eidei*) consistently (*hōsautōs*); the other is seen in continuous motion.[32]

Earlier in the same quotation, Iamblichus proposes to treat time in the same way as the now,[33] while later in the quotation, he returns to his distinction of two nows and repeats that one is unextended, the other moving:

> Hence if someone takes *now* as being a *part* of time, he will take it as being naturally bound up with change. But if he interprets it in the way that certain people have recognised as not being a [period of] time at all, then it will rather be an *origin* of time existing separately and remaining (*diamenei*) the same in form (*eidei*). And when, therefore, it is said that past time exists no longer, and that the future is not yet, we must recognise that these things are said of the nows which go forth, and travel along with motion, and alter together with movement. But when things are embraced in the now and separated off in it and never shift away (*existasthai*) from their proper origin, then they remain (*diamenei*) always in the now.[34]

---

[29] Ps.-Archytas ap. Simpl. *in Cat.* 353,13-15 (SP 24, 26-9).

[30] *in Phys.* 793,22-3. SP 40, 22-3, translate: 'He wants not only the Now to be in the present, but also the time in the interval between two limits.'

[31] This statement in l.15 resolves an ambiguity in l.8, where there was talk of things participating in the now itself. Here, with the distinction of two nows, it emerges that they participate only in the lower one.

[32] *in Cat.* 355,8-14 (SP 30,11-17).

[33] *in Cat.* 354,28-33 (SP 28, 33-40).

[34] *in Phys.* 355,29-356,1 (SP 32,1-10).

One last confirming passage deserves to be translated. Here Simplicius repeats his interpretation, and then quotes Iamblichus, who does not explicitly contrast his two times or two nows, but who does say that indivisibility and movement are incompatible properties, which therefore require different subjects.

So since time past is no longer, the future is not yet, and only the present (*enestēkos*) is thought to exist (*einai*), and since this is partless if taken at the now, and departs into non-existence simultaneously with existing, for these reasons he [pseudo-Archytas] says that the peculiarity of time is to be partless (*ameres*) and unreal (*anupostaton*). But Iamblichus does not agree with those who see partlessness in connexion with this flowing and generated time, and who call it unreal because it never is, but is always coming into being. 'For whatever is coming into being', he says, 'and whatever is changing in any way cannot be partless. For every change is always divisible because of its continuity. The partless, however, is *eo ipso* stationary (*histatai*) and in a state of *being*; if it were always *coming* into being, it would no longer preserve its form, while if it is expressly said to preserve its form, it cannot be always coming into being. So if [pseudo-Archytas] should say that the now is past, simultaneously with being thought and spoken of, we must postulate this kind of shifting (*ekstasis*) in the things which participate in time. For the things which come into being cannot take on the indivisible nature without changing, but touch that nature with ever different parts of themselves, and so falsely get their own character attributed to that nature. Hence ever changing numerical difference is a mark of the difference of participating things, but the form remains the same and displays the sameness of the partless now'.[35]

Application to the paradox of the ceasing instant

The latter part of this quotation makes clear that Iamblichus would have had a response not only to Aristotle's paradox of the parts of time, but also to the paradox of when the present ceases, if he had discussed it. Pseudo-Archytas had spoken of the ceasing of the present in the following words:

But the present (*enestōs*) time has passed simultaneously (*hama*) with being thought of and with having been present (*enestakōs*).[36]

It is this to which Iamblichus is responding in the most recent quotation, when he says that becoming past occurs in the things which participate in time, but does not characterise the 'indivisible nature' itself. This at best *implies* an answer to the question when the higher now ceases, namely that it simply does not cease.

But what about the lower now? We have seen that Iamblichus regards this as being a *stretch* of time, rather than an indivisible instant. In that case, it will no longer be so difficult for him to say when the present ceases. He can say, if he wishes, 'at its final instant'.

---

[35] *in Cat.* 354,9-26 (SP 28,12-31), partly translated above.
[36] *in Cat.* 352,29-30 (SP 24,7).

Proclus, Damascius and Simplicius

Iamblichus' distinction of two times was influential. Proclus (*c.* A.D. 411-485), as the leading Neoplatonist in Athens in the next century, upheld the distinction, and supported it by reference to a fresh suggestion about measurement.[37] If moving time is to be measured, there will have to be something stationary to measure it. A measuring apparatus which simply kept pace with moving time could not record its passage. Damascius and Simplicius, the last two leading Neoplatonists in Athens, will be discussed in Chapter 5. But although Damascius' main solutions apply to the inferior kind of time, he did accept a superior 'intellectual' time (*noeros chronos*),[38] and his pupil Simplicius did the same.[39]

Was Iamblichus' distinction anticipated?

Was Iamblichus the first to distinguish a static and a flowing time and a static and flowing temporal 'now'? Many anticipations have been suggested, but I doubt if any is genuine. I do not want to deny that in earlier authors we can find some passages that suggest flow and others that suggest stability. But what I am looking for is an awareness that there is a distinction to be made. Flow metaphors without awareness of a distinction can be found much earlier, and we do not have to wait (as has been suggested) for Simplicius, a Neoplatonist still later than Iamblichus.[40] Iamblichus himself thinks that his distinction is already to be found in Archytas;[41] but Simplicius rightly disagrees, and says that the reference in Archytas is to ordinary time in the physical world.[42] Indeed, the strain in Iamblichus' interpretation becomes apparent when, in a passage already translated,[43] he

---

[37] Procl. *in Tim.* 3.32,7-16 (SP 60,5-13). The reference to measurement is in 1.14 (SP 1.11).

[38] Damascius *in Prm.* (printed by Ruelle as if it were part of a single work *Dubitationes et Solutiones de Primis Principiis in Platonis Parmenidem*), vol. 2, 31,10-20

[39] *in Phys.* 632,33-633,6 and 638,29-34.

[40] *Pace* W. Wieland, 'Die Ewigkeit der Welt', commenting on Simplicius *in Phys.* 1163, 2, in *Die Gegenwart der Griechen im neueren Denken* (H.-G. Gadamer Festschrift), Tübingen 1960, 314, n.23. Heraclitus, admittedly, is not talking of time, so much as of things in time, when he says that nothing stays put and that you cannot step into the same river twice (ap. Platonem *Crat* 402A). But Plato, taking up the image, makes Socrates say that the names Rhea and Kronos given to the parents of the gods are names for flowing streams (*rheumatōn onomata*, 402B). And his idea is that 'Kronos' suggests *chronos* (time), so that evidently he thinks of time as a flowing stream. Seneca comments on Heraclitus, 'we too are carried past by no less rapid a current' (*cursus*, *Letters* 58.23). Plutarch describes time as 'like a flowing stream (*rheuma*) carrying everything past' (*de Def. Or.* 432A). The Aristotelian Boethus of Sidon in the first century B.C. is reported as saying, 'time looks as if its nature is always to flow (*rhein*)' (ap. Simpl. *in Cat.* 434,2-3). The Stoic Marcus Aurelius in the second century A.D. says, 'there is a kind of river (*potamos*) of things coming into being, and time is a violently flowing stream (*rheuma biaion*, 4.43; cf. 2.17; 5.23; 6.15; 7.19). There is a flow metaphor in Gregory of Nyssa, although it is applied only to the time of the present world, not to the new world expected shortly (*in Psalmos*, PG 44, col. 609C). There are also, as Wieland remarks, references to time moving which do not use the image of flow.

[41] Simpl. *in Phys.* 787,3-10; 792,22-3 (SP 34,13-21; 38,15).

[42] *in Phys.* 787,30; 788,9-10; 788,29-32 (SP 36,5; 36,22; 38,9-11).

[43] *in Cat.* 355,8-14 (SP 30,11-17).

discusses the characteristics ascribed by pseudo-Archytas to 'the now', and asks how such conflicting characteristics can be ascribed to (what is presented as if it were) one and the same thing:

> How then does the same thing come to be ever different, yet remain the same in form?

It is *Iamblichus'* suggestion that we must be dealing with two 'nows' rather than one.

Another person for whom priority has been claimed is Plato.[44] For in the *Parmenides* Plato describes time as travelling,[45] and talks of something (the One) travelling with it from the past via the now to the future,[46] which implies that the now stands still and is overtaken. On the other hand, Plato also says that the now is always present to the One,[47] which implies that the now travels along with it. It looks as if Plato needs a static and a travelling now. However, it must be insisted that, if he does, he does not himself recognise and state this implication.

Other dichotomies have also been read into Plato, but they are still further from the one we are looking for. In Chapter 17, I shall explain how Plutarch and Atticus, with some encouragement from earlier Epicureans, thought that in the *Timaeus* Plato distinguished between a disorderly time before the creation of the *kosmos* and an orderly time afterwards. Other Platonists detected a quite different distinction at *Timaeus* 38C. Plato moves there from a general account of time to a discussion of the planets viewed as instruments of time. The fullest analysis comes in Proclus, who is later than Iamblichus.[48] Proclus takes it that Plato intends a primary time and a secondary time. The primary time, he says, is the unique number and measure of the great revolution in which all heavenly bodies eventually become realigned in their original positions. But there is in addition a plurality of numbers, measures and times corresponding to individual planetary revolutions. The rudiments of Proclus' idea may be found earlier in Porphyry, who taught Iamblichus; for Porphyry distinguishes the time which corresponds to the movement of the World Soul from the diverse times of the sun, moon and Venus.[49] But none of these distinctions is between a static and a flowing time.

There is a contrast which is harder to distinguish from that of Iamblichus: the contrast between time and eternity. What makes this especially hard to distinguish is the fact that it is sometimes expressed in terms of two different kinds of 'now', and that makes it sound just like Iamblichus' dichotomy. But I shall explain in Chapter 8 why the 'now' of eternity must not, in my view, be thought of as being in *any* sense temporal, whereas the higher 'now' of

[44] Colin Strang, in 'Plato and the Instant', *PAS* supp. vol. 48, 1974, 69, referring to Plato *Prm.* 151E3-152E3, and to Arist. *Phys.* 4.11.

[45] *Prm.* 152A3.

[46] *Prm.* 152B4-D4.

[47] *Prm.* 152D8.

[48] *in Tim.* (Diehl) 3.19,2-32; 25,27-27,8; 52,16-58,4 (some portions are translated in SP).

[49] *Sententiae* 44. Plotinus, however, who was Porphyry's teacher, did not express himself in terms of two different times, but only in terms of accidental and essential properties of time, measuring being an accidental property (3.7.12 (25-38); 3.7.13 (18-30)).

Iamblichus is still associated with *some* kind of time. We need not, in any case, have any doubt about Iamblichus' view here, for he himself treats eternity as distinct from his two kinds of time, since he draws a *threefold* distinction among them. We have already seen evidence of this threefold distinction in Iamblichus' view that his higher time is not one of the primary realities because it is only an *image* of eternity. The threefold scheme is also recorded by Simplicius, with a quotation in which the moving (i.e. lower) time and the partless (i.e. higher) time are mentioned by Iamblichus alongside eternity.[50]

A large question remains, whether Aristotle can claim priority in distinguishing between static and flowing time. But I shall reserve it for the next chapter.

Other aspects of the modern distinction in antiquity

I have presented Iamblichus as the first to say that there is a static and a flowing time, and as anticipating certain aspects of McTaggart's distinction, although not the distinction itself. There are yet other aspects of the modern distinction which were appreciated in antiquity, and the period of Neoplatonism and Christianity was particularly fruitful in making these observations. A.C. Lloyd has found some hints still earlier, in the Stoic period, of Russell's idea that the tenses are egocentric, but he rightly presents them as no more than the merest hints.[51] A more definite reference to egocentricity is made by Damascius, and subsequently in Islamic thought. In Chapter 5, we shall see Damascius arguing that the division of time into past, present and future is merely *relative to us* (*pros hēmas*), and giving the reason that different people have a different present.[52] Rather similar terms are later used by certain Islamic theologians, in order to bring out the egocentricity of such terms as 'was' and 'past':

> The primitive meaning of the two words ['was' and 'will be'] is the existence of one thing and the non-existence of another. The third element which is the connexion between the two words is a necessary relation to us. The proof is that if we should suppose a destruction of the world in the future and afterwards a second existence for us, we should then say 'God was without the world', and this would be true, whether we meant its original non-existence or the second non-existence, its destruction after its existence. And a sign that this is a subjective relation is that the future can become past and can be indicated

[50] *in Phys.* 793,12-22 (SP 40,10-22).

[51] A.C. Lloyd, 'Activity and description in Aristotle and the Stoa', *PBA* 56, 1970-1, 227-40. Thus one source says (*Etymologicum Magnum* 820) that the present has its being simultaneously with one's utterance (*hama tōi legesthai*), although the point is only to emphasise its brevity. Again, the Stoic name for the main past tense in Greek is 'aorist', literally '*indefinite*'. And Lloyd reminds us that the Stoics would count a simple proposition as definite (*hōrismenon*), only if it was expressed by a demonstrative such as 'this', as in 'This man is walking', as opposed to 'Socrates is walking' (Sext. *M* 8.96). Perhaps, then, the aorist tense is thought of as indefinite in comparison with the present, because some analogy is sensed between the use of the present tense, and the use of the pronoun 'this'.

[52] ap. Simpl. *in Phys.* 798,6-9 (SP 82,3-6).

by the word 'past'.[53]

Another aspect of the flowing time concepts was also anticipated, I shall argue in Chapter 16. For it was recognised already by Augustine and Ammonius that judgments that something is future are liable to change when the thing becomes past. Augustine and Ammonius, therefore, while allowing a changeless God to have full knowledge of a future thing, denied that he would know it *as* future, for that would involve a change in his thought. Once again, the discussion was continued in Islam, but this time by the so-called 'philosophers', opponents of the 'theologians' just referred to. They preferred the conclusion that, since God cannot change, there is something he does not know, namely, a thing's presentness, pastness, or futurity.

I should finish with a warning about a passage from Tatian that was quoted in Chapter 2.[54] Since it treats the passage of time as a subjective illusion, someone might be tempted to think that it anticipates the idea that only McTaggart's static B-series is real, not the A-series. But what Tatian contrasts with the passage of time is not McTaggart's B-series, but *eternity*. And in the analogous passage in Plutarch, it is clearly *God*'s eternity which is meant. So I do not think we can safely find any McTaggartian anticipations here.

---

[53] So Ghazālī, ap. Averroem *Tahāfut al-Tahāfut* (Bouyges) p.72, translated by S. van den Bergh, London 1969, vol. 1, p.41.
[54] Tat. *ad. Gr.* 26 (PG 6, 861); cf. Plut. *On the E at Delphi* 392E-393B.

CHAPTER FOUR

Aristotle on Static and Flowing Time

Chapter 3 would be wrong, if we were to accept a suggestion made by Norman Kretzmann, in a paper to which I am very much indebted for first drawing my attention to the interest of the paradoxes.[1] On this suggestion, Aristotle had already, six and a half centuries before Iamblichus, recognised the difference between McTaggart's flowing A-series and static B-series and applied it to the paradoxes of time. Kretzmann's hypothesis is that the paradoxes are designed to attack the view that time is essentially *'passing'*,[2] or as he puts it, that time consists entirely of the past, the now and the future.[3] Aristotle's view of what is essential to time can, on the contrary, be stated entirely in terms of the static B-series.[4] The puzzles can be solved, on this interpretation of Aristotle, by treating the dynamic or flowing A-series as inessential. Kretzmann is here extending a suggestion of Fred Miller, according to whom Aristotle treats time in dynamic or flowing terms in *Phys*.4.11, but switches to static terminology in the next two chapters, 4.12 and 13.[5] According to Miller, Aristotle makes the dynamic terminology posterior, but because he cannot shake free of it sufficiently, he fails to see how the static terminology would have enabled him to solve the paradoxes. Instead, Aristotle offers an unsatisfactory solution.[6]

These suggestions are of wider interest, because they are contributions to quite an extensive debate. On one side, W.D. Ross, G.E.L. Owen and Stephen Clarke assign Aristotle a flowing or dynamic view of time, heavily dependent on the A-series. On the other side, W. Wieland accords him a static view.[7] But we must be cautious in interpreting discussions, because talk of dynamic and static conceptions of time can mean something entirely different.[8]

[1] Norman Kretzmann, 'Aristotle on the instant of change', *PAS*, supp. vol. 50, 1976, 91-114.

[2] ibid., *passim*, esp. 91-2; 107 | [3] ibid., 98; 107.| [4] ibid., 110. |

[5] Fred D. Miller, 'Aristotle on the reality of time', *AGP* 56, 1974, 132-55, esp. 145-7.

[6] ibid., 153-5.

[7] W.D. Ross, *Aristotle's Physics*, a revised text with introduction and commentary, Oxford 1936, 67. G.E.L. Owen, 'Aristotle on Time' in Peter K. Machamer and Robert G. Turnbull (eds) *Motion and Time, Space and Matter*, Columbus, Ohio 1976, 15-16 (repr. in *AA* 3, 150-1). Stephen Clarke, *Aristotle's Man*, Oxford 1975, 115. W. Wieland, *Die aristotelische Physik*, Göttingen 1962, 327.

[8] We have already seen that Iamblichus' conception of static time is not the same as McTaggart's. S. Samburky and V. Goldschmidt have both called the Stoics' conception of time 'dynamic', meaning only that in various ways time is continuous and infinitely divisible for them. (S. Samburky, *Physics of the Stoics*, London 1959, 2nd ed. 1971, 104; Victor Goldschmidt, *Le Système stoicien et l'idée de temps*, Paris 1953¹, 32).

An attraction of Kretzmann's suggestion is that it seems to offer a way of solving the first of Aristotle's paradoxes, the paradox according to which none of the parts of time exists. It would be admitted that the past and future do not exist, but besides talking of the past and future, we can talk of the static relations of earlier, simultaneous and later that hold between events. •The past may have gone, but the relation of later-ness which the battle of Hastings bears to the birth of Christ is unchanging, and has not passed away. Hence such static relations do exist, and they give existence to time.

It is less obvious how an appeal to the static relations would solve the problem of when the present instant ceases. It may seem that it would do so, if the flowing terminology could be abolished altogether; for then no questions about presentness and ceasing could be raised. But no one has suggested, so far as I know, that Aristotle wanted to abolish the flowing terminology. Kretzmann himself maintains something less: not that Aristotle thought the flowing terminology could be abolished, but that he thought it inessential. In talking of inessentiality, he is attributing to Aristotle the idea that passage is merely the way in which time presents itself *to us*. On this view, when we say that the present instant passes, we are describing not time, but merely its *appearance*. In that case, the paradox of ceasing can still be raised, for we must face the question when the appearance of ceasing occurs. But at least the puzzle will not cast doubt on the existence of time itself, but only on the coherence of our experience of it.

There are several difficulties with this suggestion. One that has been put to me is that an analogue of the puzzle of the ceasing instant can be raised without reliance on the flowing terminology. For we can ask of a static point which is present, say, half way along some static geometrical line just where it comes to be absent. And similarly we can ask of midnight on 31 December 1979 just where in the static B-series it comes to be absent. The notion of coming to be absent somewhere is to be taken as a static one, in that it will be *changelessly* true, if true at all, that a given instant comes to be absent at a given point.

Another difficulty is that it is hard to believe Aristotle would allow the universal experience of mankind to be illusory in relation to time. He holds that what seems so to *everybody* must be so.[9] He would surely, then, wish to solve the puzzle as it applies to our experience of time, and would not rest content with showing that it fails to apply to time itself.

I have in any case suggested that Aristotle supplies a quite different solution to the puzzle of the ceasing instant, a solution which in no way downgrades the flowing terminology, or reduces it to a report of appearance.

Aristotle's insensitivity to the static/flowing distinction

I should like to raise a difficulty which is of more general interest, because it bears upon Aristotle's whole analysis of time. I am doubtful that he ever sensed the difference between static and flowing terminologies sufficiently clearly to envisage a solution that turned on the distinction, and I shall offer

[9] *NE* 10.2, 1172b35-1173a2; cf. *Phys.* 2.5, 196b14.

my reasons for saying so.

First, it is significant that he runs together in his conception of the now the static idea of an instant and the flowing idea of presentness. Moreover, he does not succeed in disentangling them when it would help him to do so.

For example, he does not make it clear how we are to escape from the threat raised in *Phys.*4.10, that we are in danger of reducing time to a single, instant, the same for ourselves as for the people who lived ten thousand years ago. For he fails to make the necessary point that what is single is not the instant, but the character of presentness which attaches to the infinitely many instants.

Again, Aristotle does not give a satisfactory answer in *Phys.*4.11 to his question whether the now is for ever the same or for ever different.[10] What needs to be said is that the instants are ever different, but the character of presentness is ever the same. Pseudo-Archytas comes closer to saying this. His formulation is that the now is *numerically* diverse, but remains the same *in form*:

> But the now always was and will be, and will never cease in its coming to be ever different and numerically diverse, while the same *in form* (*eidei*).[11]

Aristotle's formulation is almost the opposite of this, for it is the now's *form* (*logos*,[12] *einai*[13]) which he makes ever different,[14] and its *substratum*[15] which he makes ever the same. The main passage is worth quoting, because it reveals what an elaborate answer Aristotle is driven to by his failure to appreciate the simpler answer in terms of presentness and instants:

> Here too [with the now as with moving bodies], the substratum [literally: that being which the now is the now] is the same; for it [the substratum] is the before and after in change. But the being (*einai*) is different; for it is in so far as the before and after can be counted that the now exists.[16]

I think that the meaning is this: a full definition of the now would specify a *substratum* or *subject*: earlier and later stages in a change. But these stages are countable as first, countable as second, and so on, in a manner described earlier in Aristotle's chapter.[17] The *formal* description of the now will say that the earlier and later stages are countable as first, countable as second, and so on. And it is this which makes the formal description ever different. This elaborate answer to the question whether the now is ever the same or ever different might have been dropped, if Aristotle had distinguished more clearly between the static notion of an instant and the flowing notion of presentness.

---

[10] *Phys.* 4.11, 219b9-33.

[11] Simpl. *in Cat.* 353,9-10 (SP 24,21-3) quotes Archytas. Iamblichus retains the idea of '*same in form*' for the *higher* now (ap. Simpl. *in Cat.* 355,8-14 (SP 30,11-17), quoted in Chapter 3); Simplicius applies it to time (*in Phys.* 777,10-13 (SP 68,25-9)).

[12] See *Phys.* 4.11, 219b20; 220a8.   [13] See *Phys.* 4.11, 219a21; b11; b14; b27.

[14] *Phys.* 4.11, 219b27-8; cf. 219b14.   [15] *Phys.* 4.11, 219b14-15; b26.

[16] *Phys.* 4.11, 219b26-8.

[17] *Phys.* 4.11, 219a22-b2.

The next piece of evidence for Aristotle's insensitivity to the distinction between static and flowing is that he makes no less than four comparisons between the now and a moving body,[18] but surprisingly never manages to articulate the sense in which the now is moving. The first analogy is that a moving object remains the same in respect of its *substratum* (it is still a stone, for example) even though it changes in its description or *logos* (it is now here, now there). The point of analogy turns out to be not the *motion* of the stone, but its *sameness*.[19] The second analogy between the now and a moving body is that we recognise before and after by means of the now, as we recognise before and after in motion by means of the moving body. Next, time and the now depend on each other for existence, as do motion and the moving body. Finally, the now both joins together and divides past and future, which is what the moving body does in relation to motion. None of these four comparisons does what we should expect, that is, to bring out a sense in which the now is moving.

The passage immediately following may seem to call more clearly for a moving now. There Aristotle contrasts the now with a geometrical point.[20] You can think of the single point which divides a line as in a way double, if you *pause* in your thoughts, and think of it as the end of one part and the beginning of another. But the now is unlike this, being ever different (*aiei heteron*) because of the motion of the moving body with which it has been compared. Again, we expect to hear that the now itself is moving, but again we are disappointed. The moral turns out to be that we should compare the now not with a single geometrical point, but with two geometrical points, that is, with perfectly static ones at either end of a line.[21]

I do not mean to deny that there are remarks which treat the now as being in motion. One such remark is that the now is at another and another (*en allōi kai allōi*).[22] The word 'at' encourages us more than do the commoner formulas[23] to think of the now as something which changes position. But this only emphasises Aristotle's insensitivity, because he never explains the relation between this treatment of the now as moving and the treatment of it elsewhere[24] as static. Much less does he ever say that there are two nows, one flowing, one static. In the passage just quoted, for example, Aristotle does not say that the now is at another and another *now*. If he had done so, he would have been forced to recognise two uses of 'now', one associated with motion and one with rest, but he does not.

These are perhaps the most important passages, and they already illustrate Aristotle's tendency to jumble static and flowing terminology; but

---

[18] *Phys.* 4.11, 219b16-220a9.

[19] *Phys.* 4.11, 219b16-25. Only so is the analogy applicable. For it could not be maintained that the substratum of the now (the earlier and later stages of change) are a *moving* thing. The point is rather that this element in the definition of the now is always the *same*, even though there is variation according as a stage is counted as first, second, or third.

[20] *Phys.* 4.11, 220a9-18.   [21] *Phys.* 4.11, 220a14-18.

[22] *Phys.* 4.11, 219b13-14.

[23] E.g. 'ever different in formal description' or 'countable as two'.

[24] See *Phys.* 4.11, 218b27; 219a27; a30, which treat nows as countable, and 220a14-16, which compares nows with the *two* points at either end of a line, rather than with a *single* point within a line.

there are yet other instances, which can conveniently be assembled in a footnote.[25]

Aristotle does not make flow inessential

My claim has been that Aristotle does not have as clear a sense of the difference between static and flowing terminology as Kretzmann's interpretation would attribute to him. He does sometimes speak of the now as if it flowed, but the further claim I want to make is that, when he does so, he does not consider the flowing terminology inessential for time in the manner suggested by Kretzmann. Thus in *Phys*.4.11, 220a1, he says that time and the now depend on each other for their existence.[26] When he repeats the point in *Phys*.8.1, 251b19-23, he defines the now by means of the flowing terms 'past' and 'future'. We may further recall the passage in *Phys*.3.6, 206a22, where Aristotle suggests that a day owes its existence to the dynamic process of one thing coming into being after another. Finally, when he defines time as the countable aspect of change in respect of before and after,[27] there is at least an indirect reference to flow. For he assumes that the counting involves seeing that a past now is different from a present now,[28] a perception which presupposes the flow of time. All this suggests to me that, without sharply distinguishing the flowing from the static, Aristotle none the less treats flowing conceptions as essential to the existence of time.

Is there any evidence on the other side, in favour of Aristotle's treating flow as inessential to time? He treats it as inessential to the notions of *before* and *after*. For he claims that the temporal before and after are posterior to the *spatial* before and after, which he treats as containing no idea of flow.[29] But what he says about *before* and *after* will not carry over to the idea of *time*, which includes much more than the notions of before and after.

If Aristotle had really made flow inessential to time, a gap would have opened up between his physics and his logic. For tense, which is a flowing conception, is made a defining characteristic of verbs at *Int*.3, 16b6-9. This must be a conscious decision. For Parmenides and Plato had on the contrary

---

[25] (i) First, the terms 'before' and 'after' belong with McTaggart's static B-series, and Aristotle encourages us to think of them statically, by saying that the temporal use of 'before' and 'after' derives from the use of the terms for spatial positions (*Phys*. 4.11, 219a14-19). So far the treatment is static; for spatial positions do not flow, even if we have in mind a possible movement between them in calling one before and one after. But then Aristotle switches and defines before and after by reference to the flowing terms 'past', 'future' and 'now'. Moreover, he does this in a context where the switch causes him inconvenience; for he has to explain that in the case of past time what is before is further from the now, while in the case of future time it is nearer (*Phys*. 4.14, 223a4-15). (ii) A further example of oscillation occurs over the expression 'now'. Sometimes Aristotle gives an account of it in flowing terms such as 'past' and 'future' (*Phys*. 4.13, 222a10-11; b1-2; cf. 8.1, 251b20-3). But sometimes he switches and describes it in static terms, as when he refers to the *before* and *after* in time (*Phys*. 4.11, 219b12), or in motion (*Phys*. 4.11, 219b26-8). (iii) Finally, *Phys*. 4.13 is relevant, for here Aristotle defines a list of temporal predicates taken indiscriminately from the static and flowing sides of the dichotomy.

[26] This is not withdrawn by the word *sumbebēken* at 220a22. By saying that the now is an accident (*sumbebēken*) of time, he here means only that it is not *identical* with time.

[27] *Phys*. 4.11, 219a22-8.

[28] cf. also 4.11, 218b23-219a1.

[29] *Phys*. 4.11, 219a14-19.

tried to articulate a use of the verb 'is', in which it is not to be contrasted with 'was' and 'shall be', in other words, a tenseless use.[30] It is notorious that Aristotle treats many statements and beliefs as containing tenses or equivalent token-reflexive devices, for he claims at *Cat.*5, 4a23-8 that statements and beliefs can shift their truth value, and change from being true to being false. It is, however, only entities containing tenses, or equivalent devices, which can thus shift their truth value: 'Theaetetus is now dying' can do so; 'Theaetetus' death belongs (tenseless) to 369 B.C.' cannot. This point about Aristotle has been strongly emphasised by Jaakko Hintikka, who considers Aristotle oblivious to any other kind of statement.[31] I do indeed doubt if Aristotle ever clearly conceived the idea of a tenseless statement. If he had done so, he ought to have mentioned it at *Int.*9, 19a39, at least on the interpretation of that passage for which I have argued elsewhere.[32] For he says, as I read it, that the prediction of a sea battle is not *yet* true or false, before the battle occurs or becomes inevitable, the implication being that it will become true or false at latest when the battle happens or fails to happen. But what will become true, after being neuter, when the sea battle happens, cannot be the *future* tensed statement, 'there *will* be a sea battle on 1 January 1990'. That statement, with its future tense, must then be false. At best, what becomes true will be the *tenseless* statement, 'there is (tenseless) a sea battle on 1 January 1990'. Aristotle's silence about this confirms that he had not thought out the idea of a tenseless statement. My point is that if Aristotle treats tense, a flowing conception, as essential to verbs and to statements, this fits with his treating flowing conceptions as essential also to time.

It should be clear how fruitful I think Kretzmann's interpretation is, despite any disagreement. It opens up questions about Aristotle's treatment of time and tense in general which go well beyond his solution to the paradoxes. I think it is also interesting that the application to the paradoxes of a static/flowing distinction which Kretzmann suggested for Aristotle is to be found, in a different form, in the Neoplatonist Iamblichus. If I am right, it is he who should receive the credit for inventing this kind of solution.

---

[30] Parmenides *The Way of Truth* fr. 8, l.5, in DK; and Plato *Tim.* 37E3-38B5.

[31] Jaakko Hintikka, *Time and Necessity*, Oxford 1973, chs 4 and 5. In ch. 8, however, Hintikka takes the point further than I would be willing to take it: see Richard Sorabji, *Necessity, Cause and Blame*, London 1979, ch.8.

[32] Sorabji, op. cit., ch.5.

CHAPTER FIVE

Solutions by the Last Athenian Neoplatonists: Divisible Leaps

In A.D. 529, the Christian Emperor Justinian struck at paganism by putting an end to teaching in the Neoplatonist school at Athens. The Neoplatonists were still working on Aristotle's paradoxes at that time and the head of the school, Damascius, offered solutions of his own. He took the paraxodes to affect only the flowing kind of time, not the 'intellectual' time, which he also recognised.[1] He none the less thought it important to solve the paradoxes, as they applied to flowing time. Simplicius, having enumerated the paradoxes, reports that Damascius tries to solve them by treating the present not as an instant, but as a stretch.[2] More startlingly, Damascius holds that time progresses in 'leaps', in such a way that the half does not come before the whole, and claims that the present is such a leap. We shall see that the idea of leaps has a long history, but first we must notice Damascius' description of the temporal leap, which is recorded by Simplicius verbatim:

'I am impressed', Damascius says, and in these words, 'at how they solve Zeno's problem by saying that the movement is not accomplished in indivisible units, but rather progresses in a whole stride (*bēma*) at one go. The half does not always precede the whole, but sometimes the movement as it were leaps (*huperallesthai*) over both whole and part. But those who said that only an *indivisible* now existed did not recognise the same thing happening in the case of time. For time always accompanies movement and as it were runs along with it, so that it strides along together with it in a whole continuous jump (*pēdēma*) and does not progress one now at a time *ad infinitum*. This must be the case because motion obviously occurs in things, and because Aristotle shows clearly that nothing moves or changes in a now but only *has* moved or changed, whereas things do change and move in time. At any rate, the leap (*halma*) in movement is a *part* of the movement which occurs in the course of moving and will not be taken at the now; nor, being present, will it occur in the non-present (*enestōti*). So that in which the present movement occurs is the present time, and it is infinitely divisible, just as the movement is, for each is continuous, and every continuum is infinitely divisible.'[3]

---

[1] For intellectual time, see Damascius *in Prm.* (= *Dub. et Sol.* (Ruelle), vol. 2, 31, 10-20, as above, French translation by A.-Ed. Chaignet, Paris 1898); for flowing time, Dam. ap. Simpl. *in Phys.* 776, 10-12; 798, 23-4 (SP 66,22-5; 82,23); and for his use of the distinction, Marie-Claire Galperine, 'Le temps intégral selon Damascius', *EP* 1980, 325-41.

[2] *in Phys.* 795, 33-5.

[3] *in Phys.* 796,32-797,13 (SP 78,27-80,5).

The origin of the idea of divisible leaps

So far as I know, the history of the idea of infinitely divisible leaps has not been written. I shall try here and in Chapter 25 to trace it from its origins in Aristotle up to the period of Islamic thought. Aristotle denies Empedocles' theory that light takes time to travel from the sun. Light can fill a whole space simultaneously (*hama*). Similarly, a pond which is not too large can freeze all over its surface at one go (*athroon*), and not 'the half before the whole'; and you get the same effect with heat. It cannot happen with motion, but it can with qualitative change and with growth or diminution.[4] Theophrastus, Aristotle's successor, reaffirms that this kind of discontinuity is found, and not only in qualitative change. He uses the point in order to show that there can be a first instant of having changed – something which is impossible in the case of *continuous* changes.[5] The example of light leaping is offered as an illustration of Theophrastus' point; it recurs in Alexander of Aphrodisias and it influenced Philoponus.[6]

We hear of a fresh development in Sextus Empiricus.[7] He starts with the view, which he attributes to the Stoics, that space, time and motion are all infinitely divisible. But he reports that in spite of calling motion infinitely divisible, some had imagined it as occurring at one go, and not the half before the whole. Presumably, the moving body disappears from one position and reappears a little further on, without an intervening time-lapse. The leap can be thought of as infinitely divisible, because the distance traversed would be infinitely divisible, and another jump could be across a shorter distance, indeed across a distance as short as you like. It is clear from Sextus and Damascius[8] that these leaps had been used in order to solve a version of Zeno's paradox of the half-distances. This paradox (in one version) is that nothing can reach any destination because, in order to do so, it must first go half way, then half the remaining distance, and so on *ad infinitum*. The proposed answer is that things do not have an infinity of half-distances to traverse, because they progress by leaps, disappearing from one spot and reappearing further on, without ever having been half way.

We can now return with a leap to Damascius, whose new move is to postulate that leaps are found also in *time*. One possible source of confusion is that he gives the name 'leap' not, as we might expect, to the instantaneous transition from one time to another, but to the intervening period between two instantaneous transitions,[9] a period in which, as we shall see, time is

---

[4] *Sens.* 6, 446a20-447a11; *Phys.* 8.3, 253b13-31.

[5] ap. Simpl. *in Phys.* 107, 12-16; and ap. Themistium *in Phys.* 197, 4-8.

[6] Philop. *in DA* 327,3; 328,34; 330,14-15; 26; 344,33-345,11; Alex. *in Sens.*, ad loc.; and in a treatise preserved in Arabic, entitled 'That to act is a wider expression than to move, according to Aristotle', translated into French by A. Badawi, *La Transmission de la philosophie Grecque au monde Arabe*, Paris 1968, 153ff.

[7] *M* 10.123-42; cf. *PH* 3.76-8.

[8] In Damascius the connexion with Zeno is explicit. In Sextus, after infinitely divisible leaps have been attacked, it is argued that the only alternative to them for the Stoics is to succumb to some version of Zeno's half-distance paradox (*M* 10.139-41; cf. *PH* 3.76).

[9] E.g. Dam. *in Prm.* (= *Dub. et Sol.* 2.242,24-6, SP 92,27-9), where he describes each leap as bounded by two instants, the terminus of one leap being the beginning of the next.

supposed to stand still.

The distinction of divisible leaps from atomic ones

Divisible leaps have caused much confusion among commentators. As we shall see, they are supposed to combine divisibility in some way with indivisibility, and one recent commentator has understandably found that baffling.[10] Many have not distinguished divisible leaps from those of the atomists which do not admit divisibility at all.[11] Sambursky, for example, takes the leap to be an indivisible time-atom.[12] But this interpretation does not do justice to the closing words of the quoted passage, which tell us that it is infinitely divisible. Sambursky's interpretation is based partly on what I believe to be a misinterpretation of the phrase, 'the same thing', in the sentence which I have translated:

> But those who said that only an *indivisible* now existed did not recognise the same thing happening in the case of time.

I take 'the same thing' to be divisible leaps, the point being that as well as indivisible nows there are *divisible* ones, namely, leaps. Sambursky takes 'the same thing' in an opposite way to be indivisibility, which is then said to belong not only to nows or instants but also to periods of time. But I do not think this is a possible reading when the context is so concerned with *divisibility*. The passage begins by saying that a leap of motion was originally postulated precisely as an *alternative* to 'indivisible' units, and it finishes by insisting that an infinitely divisible leap is needed to accommodate infinitely divisible motion. There are two other passages, both of them in Damascius' commentary on Plato's *Parmenides*, which discuss leaps and both describe leaps as *divisible*.[13]

Divisibility and indivisibility combined

On the other hand, as I have said, divisibility is combined in these leaps with indivisibility.[14] This is an important fact, because without the element of indivisibility, we shall see, Damascius would be open to the charge which has been levelled against him,[15] that the paradoxes of time arise all over again for his leap-like present. I think we can see clearly how it was possible for divisibility to be combined with indivisibility, at least if we stick to the original idea that leaps occur, not in time, but in *motion*. We must distinguish

---

[10] Erwin Sonderegger, *Simplikios zur Zeit, Hypomnemata* vol. 70, Göttingen 1982, p.134.

[11] This is true not only of S. Sambursky (see n.12), but also of the Islamic commentators whom I mention in Chapter 25.

[12] SP 18, extracted from *PIASH* 2, 1968, 153-67.

[13] *in Prm. = Dub. et Sol.* 2.236,9-11; 2.237,21; 2.242,13-16 (SP 88,5-7; 90,13; 92,13-16).

[14] In the passage quoted at the beginning, divisibility and indivisibility seem to be combined, when Damascius describes the movement as being over both part (so there is a part) and whole at one go. In one of the other passages just mentioned, the leap is described as being in some sense indivisible: *in Prm. = Dub. et Sol.* 2.241 (SP 90,35; 92,10; 92,13; 92,16-17; 92,25).

[15] Sonderegger, loc. cit.

between the transition itself, the distance covered and the time between transitions. The transition is indivisible in time, because the moving body takes no time to disappear from one spot and reappear further on. But the distance traversed is infinitely divisible and so is the time between transitions, so that any transition could in principle have been shorter or sooner.

With leaps of *time*, however, a different explanation is needed, and Damascius reveals something of what he thinks in a passage from his commentary on Plato's *Parmenides*:

> Or, as we were saying just now, time always flows and progresses by leaps, and each leap is a whole together and indivisible into parts at the level of progress by intervals. There is a different measure for different leaps, a shorter one for the slower star and a longer for the faster, which is why it traverses the same circle more quickly. And the times must differ, since the motions do. So also do the rests, for ... . To resume, these leaps are temporal measures marked off by demiurgic sections and in this way at least indivisible into parts. Each is a whole together, and must be said to display the halting (*epischesis*) of time in its advance, and to be called a 'now' not in the sense of a boundary of time, but in the sense of a time which is demiurgically indivisible into parts, even if it is divisible in our thought, and that infinitely. For bodies too are all infinitely divisible; yet there are in bodies demiurgic sections which are indivisible into parts.[16]

How are we to construe this? Damascius seems to think that all bodies, and hence the stars in particular, move by leaps,[17] and stellar leaps are so arranged that they serve as a clock, marking off corresponding leaps in time. The combination of divisibility and indivisibility which we have seen to belong to the leaps of *motion* is supposed to be transferred to the *temporal* leaps. But it is harder to see with temporal leaps how they can be indivisible. Presumably, time is thought of as standing still during a stellar rest, and as suddenly moving on at its end, without ever advancing part way. That is the theory, but why should we not give sense to the idea of a period of celestial rest having progressed part way, by making use of *additional* clocks which are out of phase with the celestial clock by amounts as small as we please? The tiny period of celestial rest might be divided into sections as small as we liked by the 'ticks', 'tocks' and 'tings', or at least by the tiny transitions, of terrestrial clocks, even though we might have to use indirect means for discriminating such closely spaced interruptions. If this is to be avoided, Damascius might need to synchronise all the leaps throughout the universe: that would yield a sense in which temporal leaps (that is, rests) were indivisible.

But what, then, about the opposite requirement, that there should also be a sense in they are divisible? Damascius calls them divisible *in thought* in the passage just quoted, and this sounds close to what is sometimes said about the chronons postulated by certain twentieth-century physicists. For

---

[16] *in Prm.* = *Dub. et Sol.* 2.241,29-242,15 (SP 90,33-92,16).

[17] ap. Simpl. *in Phys.* 797,3; 797,9; *in Prm* = *Dub. et Sol.* 2.236,17-18 and 23-4; 242,1-2 (SP 78,33; 80,1; 88,15-16; 88,23; 90,37).

although chronons are called atomic, rather than divisible, this turns out to be in the sense that no smaller period could ever be measured, and, according to some expositions, that leaves us free to apply the idea of infinite divisibility to a chronon; it is merely that the idea will not correspond to anything physical.[18] Alternatively, Damascius might ask us to think of divisibility in more concrete terms: the celestial rests *could* have been punctuated by terrestrial transitions, although they have not been so punctuated *in fact*.

This suggestion about how Damascius might combine divisibility with indivisibility in the case of *temporal* leaps does not give us anything so clear cut as we had with leaps of *motion*. For there it was possible to exploit the distinction between the divisibility of the *distance* traversed and the indivisibility of the *time* of transition. It must also be said that Damascius himself is insufficiently clear on the subject. For at least twice he appears to allow divisibility in a sense so strong that it would prevent him from solving Aristotle's paradox. A leap (i.e. a rest), on this conception, can after all be part way through.[19] On other occasions, however, he clearly rules this out through his description of the leap.[20]

Application to the paradox of the parts of time

What is clear is that Damascius' leaps differ from atomic ones by combining divisibility with indivisibility. Let us see how he uses them to solve the paradox of the parts of time. It is worth translating the passages which claim that the extended, leap-like present provides a solution, especially as the translation given by Sambursky and Pines for the second passage gives the *opposite* impression, that Damascius ends by plumping for an *unextended* present. Simplicius begins his discussion by saying:

> The philosopher Damascius tries to solve these puzzles, then, by deciding to take the present not as a now without parts. For that kind of now is not time, but a boundary of time.[21]

He is referring forward in particular to the following words of Damascius:

---

[18] G.J. Whitrow, *The Natural Philosophy of Time*, 2nd ed., Oxford 1980, 201.

[19] One such passage is the one quoted at the beginning of this chapter. Discussing leaps of motion, Damascius claims that an infinitely divisible movement will be going on in the *present* and that it needs an extended and infinitely divisible present to accommodate it, since a mere instant offers no room for moving, only for *having* moved. The other passage is one in which Damascius argues for the divisibility (*meriston*) of the temporal leap by saying that Plato allows a thing, as it progresses through time, to be in touch with one part of the extended now, while letting go of another part: *in Prm.* = *Dub. et Sol.* 2.237 (SP 90,12-14), echoing Plato *Prm.* 152C3-4.

[20] The leap (i.e. rest) is stationary (*hestōs*, *hestēken*, *stasis*, *enmenein* and, in the last passage quoted, *epischesis*). It does not come into being through a progressive process bit by bit. On the contrary, we should describe it not as coming to be in that way, but rather as being (*on, estin, einai, huphestanai, estin energeiāi*): *in Prm.* = *Dub. et Sol.* 2.236 (SP 88, 25-37); 2.242 (SP 92,11); 2.242 (SP 92, 16-23), the last phrase echoing Plato *Prm.*152B2-E2; and Dam. ap. Simpl. *in Phys.* 799, 18-20; 799,28-30 (SP 84,24-6; 84,35-86,2), the final lines corresponding to Simplicius' report, *in Phys.* 796,29-30 (SP 78,23-4).

[21] *in Phys.* 796,27-9 (SP 78,21-3).

But since the time which stands still in the now is thought to have subsistence and being, and since it is thought to be possible to solve the puzzle in this way, because that time is a *part* of divisible time, [our opponents, the paradox-mongers] show that the now is *not* a part of time by two dialectical arguments: ... Let that be taken as well said, for [it makes no difference, since] it is not in *that* now that time stands still. For the now defined [by them] does not have being in actuality, since if anyone grants being to the now which is a mere boundary of time, he will have to grant being to that which is bounded by it [which would defeat the paradox].[22]

Damascius here explains that the now which he has in mind is not a sizeless instant. Presumably, his leap-like now could have provided a solution on its own. For the present leap, being extended, is a *part* of time as Damascius repeatedly points out. It can therefore be maintained, contrary to the paradox, that at least one *part* of time exists, in the sense of being present. Moreover, this extended present will not be vulnerable to the objection that it overlaps partly with the past and partly with the future, so long as Damascius can find a suitable sense in which it is indivisible and has no parts. The leap-like present could even be used to solve the paradox of the ceasing present, for it is described as having a terminal instant,[23] and this should surely be a convenient point to refer to, if we are asked *when* the present ceases.

In fact, however, this is not the way in which Damascius uses his leap-like present. Instead, he combines it with at least one further idea. Moreover, Simplicius' personal view is that it is the *next* part of the solution which is the satisfactory part, and that it could settle matters on its own, while the leaps ought to be dropped altogether.[24] So let us look at Damascius' next suggestion. It concerns the sense in which time *exists*, namely that it 'has its being (*einai*) in coming to be (*gignesthai*)'. The following are Damascius' own words:

Moreover, a thing which is never at any time collected into one, but has its being (*einai*) in coming to be: that is what time is, in the form of day, night, month, or year. For none of these comes all in one go; nor does a competition (*agōn*), although a competition can be present (*pareinai*) so long as it is accomplished (*epiteleisthai*) a part at a time. Nor does a dance come all in one go, for this too happens a part at a time, and yet a person is said to be dancing the present (*enestōsan*) dance. In this way too, then, the whole of time exists (*ephestanai*) by coming to be, but not by being (*on*).[25]

Damascius has borrowed the example of a competition or games (*agōn*) from Aristotle, and he is making a new use of Aristotle's point, mentioned in Chapter 1, that such things have their being in a process of coming to be. He goes on to apply his conclusion to the paradox of the parts of time:

If, then, in saying this, my thought is not simply treading on thin air

[22] ap. Simpl. *in Phys.* 799,18-30 (SP 84,24-86,2).
[23] E.g. Dam. *in Prm.* = *Dub. et Sol.* 2.236,18-19; 2.242,24-6; cf. Dam. ap. Simpl. *in Phys.* 798,24-5 (SP 88,16-17; 92,27-9; 82,24-5).
[24] *in Phys.* 797,27-36 (SP 80,22-31).
[25] ap. Simpl. *in Phys.* 797,36-798,4 (SP 80,22-82,1).

(*kenembatein*), I think I can on the basis of this thought solve the problems relating to time. For the first problem says: since neither the past nor the future exists (*esti*), and since the time which is always being taken up, even infinite time, is composed of these, what is composed of non-existents (*mē onta*) would be thought incapable of sharing in existence (*ousia*). Clearly, the person who raises this difficulty is not taking to heart the flow (*rhoē*) of becoming, and is not separating things which have their being (*einai*) in coming to be from things which have their reality (*huphestōta*) all at once. Of the former one could say the reverse, that if part of them had not already happened, while some was yet to be, then some or all of their parts would have being (*einai*), [which is absurd] because being is quite inappropriate to these things; it is coming to be which belongs. And this is the way in which they have reality (*ho tēs hupostaseōs tropos*), since the entire form (*eidos*) is essentially flowing (*rheon*).[26]

Damascius is saying that for the whole of time to exist it is not necessary that the whole should be accomplished at one go; on the contrary, time exists through a process of coming into being. But what is it that Damascius thinks of as coming into being? – presumably the leap-like parts of time. Here, then, the two main elements of his solution come together: the 'being in becoming' and the temporal 'leaps'. That Damascius is thinking of *parts* as what comes into being is suggested by his treatment above of a competition, which

can be present so long as it is accomplished a *part* at a time (*kata meros*).[27]

The same expression (*kata meros*) is used of a dance. And Aristotle was probably thinking of the *parts* of a competition, such as the wrestling and the discus, when he said that a competition has its being in coming to be,[28] and connected this with the idea of one thing coming into being after another. Damascius keeps repeating that time does not progress by *instants*, but by leaps,[29] and these leaps are treated as *parts* of time, *inter alia* in the passage[30] which immediately follows the insistence on being in coming to be.

But if that is Damascius' solution, the existence of time depends on the *coming* into existence of temporal leaps, and hence, it would seem, on their *actual* existence, however brief. And then the question can be raised all over again: if a leap is infinitely divisible, how does it, even briefly, exist? We shall have embarked on a regress, if we answer, 'by the coming into existence of *its* parts', and we know that Damascius would not want to say, 'by the coming into existence of instants'. What he needs once again is a suitable sense in which a temporal leap can be said to exist briefly as an *indivisible* whole. We have already seen that this is what he is looking for, at least some of the time. And once again the passage which immediately follows the insistence on being in coming to be is relevant. For it maintains not only that the leap is a *part* of time, but also that it has *being* as opposed to *coming* to be.[31]

[26] ap. Simpl. *in Phys.* 799,8-18 (SP 84,12-24).
[27] ap. Simpl. *in Phys.* 798,1-3 (SP 80, 34-7).
[28] *Phys.* 3.6, 206a32.
[29] ap. Simpl. *in Phys.* 796,34-797,5; *in Prm.* = *Dub. et Sol.* 2.236,15-18; 2.242, 5-6 (SP 78,29-36; 88,12-16; 92,4-5).
[30] ap. Simpl. *in Phys.* 799,20 (SP 84,26).
[31] ibid.

These are the two main components of Damascius' solution, but there are some embellishments. For one thing, he sometimes offers a stronger formulation of his view. Not only does the whole of time exist, but

the whole of time exists (*einai*) simultaneously (*hama*) in reality (*en hupostasei*).[32]

I imagine that in this stronger formulation Damascius is again inspired by Aristotle. For in discussing the infinity of an infinite process, Aristotle allows us to say that it exists '*in actuality*' (*entelecheiāi*), so long as one bit of the process is actually occurring, just as a competition can be said to exist *in actuality*, so long as one bit of it is actually occurring.[33]

There is another embellishment which proved a bit more problematic. In the middle of his discussion, Damascius introduces a diagnosis of how the paradox of the parts of time arises.[34] It involves dividing time into past, present and future. But this threefold division, he says, is only relative to us; in itself time is single and continuous. Moreover, divisions and instants have only potential existence, and are inserted only *in thought*. Finally, it involves something of a falsification to divide time into past, present and future (or, for that matter, into days, months and years). For only in something stationary can you have segments genuinely distinguished and laid out side by side. Time, however (the inferior time), is essentially flowing. It is by treating it as stationary that we overlook the crucial fact that it flows and so has its being in becoming.

The passage is closely modelled on some ideas of Alexander of Aphrodisias, already encountered in Chapter 2. For Alexander insists that time is unitary, that instants divide it only potentially, and that this should be taken to mean that they divide it only in thought. Moreover, part of the evidence for this last claim is that time flows, and so instants cannot display temporal parts standing still side by side.[35] Alexander is himself inspired by a passage in Aristotle's *Physics*,[36] and is in turn acknowledged by Simplicius as a source of several of the ideas.[37]

There is another resonance in Damascius' passage. For the idea that the division into past, present and future is merely relative to *us* and that in *itself* time is single and continuous gives us a foretaste, although only a foretaste, of Russell's idea that the division into past, present and future is egocentric, and of the associated solution of Chapter 4 which treats the paradoxes as concerning only our relation to time, not time itself.

But although Damascius introduces these further ideas simply as a supplement to his solution, Simplicius argues that they are fatal to the idea of the leap. For if time is a flowing thing, then, on Damascius' own principles, it must be a falsification to view it as segmented into leaps.[38] Damascius seems to play into Simplicius' hands. For he says that you destroy the flowing form

---

[32] Simpl. *in Phys.* 775,33-4 (SP 66,8-9).   [33] *Phys.* 3.6, 206b13-14.

[34] ap. Simpl. *in Phys.* 798,5-799,8 (SP 82,1-84,12).

[35] *de Temp.* (Théry) p.94, ll.22-42; p.97, ll.4-9 (translated in Chapter 2); and ap. Simpl. *in Phys.* 748,21-30.

[36] *Phys.* 4.13, 222a12-14; cf. 4.11, 220a11-14.   [37] *in Phys.* loc. cit.

[38] *in Phys.* 797,27-36 (SP 80,22-31).

of time, if you take a present stretch as being in actuality (*energeiāi*) bounded (*peratoumenon*) by nows on either side.[39] Yet this is almost exactly the description he gives elsewhere of his demiurgic leaps: they are bounded (*peperastai*) by two nows in actuality (*kat' energeian*).[40] Again, in a passage already quoted, Damascius says:

> Of the former [things which have their being in coming to be] one could say the reverse, that if part of them had not already happened, while some was yet to be, then some or all of their parts would have being (*einai*), [which is absurd] because being is quite inappropriate to these things; it is coming to be which belongs.[41]

This makes it sound as if it is absurd to suppose that *any* part of time ever has being. In fact, this cannot be Damascius' intention, for he goes straight on to make an exception of the leaps and to imply that they do have being.[42] The terms which he uses for being (*huphestanai, einai*) are undoubtedly meant to be contrasted with the preceding denial of being (*einai*) to *other* parts of time, such as past and future. Simplicius recognises that Damascius would insist on his leaps. But he is right that their existence is threatened by the supplementary remarks about the impropriety of dividing a flow.

Damascius on the ceasing instant

I have concentrated so far on Damascius' solution to Aristotle's first paradox. But he also tackles the paradox of the ceasing instant. His answer is that there is *no* time at which an instant ceases to exist. To say that there is would be to imply that time stands in need of a second time. And then you ought to think that a measure stands in need of a measure. But to think like that is to embark on a fatal regress. The words are Damascius' own:

> Next [Aristotle] maintains, that if the now ceases to exist, it does so either during itself or in another now, since what ceases to exist does so in time, just as what comes into existence does so in time. Clearly this argument calls for a time of time; yet he himself denounced the idea that there is motion of motion. And in general if we try to take measures of measures, we will go on to infinity taking one cubit measure to measure another cubit measure, as if that needed measuring (reading: *metrētou*), and positing one set of numbers as prior to another. But if that is out of place, and if each of the things mentioned is perfectly capable of giving its own peculiar character to whatever needs it, without itself needing to participate in the very thing which it itself is, then the suggestion about time is also out of place. If someone said that there was such a need, I think he would be speaking under pressure from someone else, but it would be his own decision to give a measure to a measure. Thus there is no need for time to cease existing in time, nor a now in a now. On the other hand, it is not possible for several nows to exist simultaneously. For the reality (*hupostasis*) of *now* is seen in the flow (*rhusis*) of time against some presupposed

[39] ap. Simpl. *in Phys.* 798,24-6 (SP 82,24-6).
[40] *in Prm. = Dub.et Sol.* 2.242, 24-5 (SP 92,27-8).
[41] ap. Simpl. *in Phys.* 799,15-17 (SP 84,20-3).
[42] ap. Simpl. *in Phys.* 799,18-30 (SP 84,24-86,2).

rest (*stasis*) whatever it may be. How then, if time, which itself moves, has its being in becoming, can it avoid needing a further time to measure and order its parts, and prevent them from telescoping with each other? Or is the answer that time moves in such a way as to accompany motion and measure it, just as the cubit extends alongside the thing which is measured and preserves the peculiar character of a measure without needing something to measure it.[43]

Damascius' solution is clever, but not as clever as Aristotle's. For his main point, that there is no time of an instant's ceasing to exist, leaves it something of a mystery how we are to escape the second horn of Aristotle's dilemma: that we are stuck at the same instant as the people of ten thousand years ago. His answer also threatens to prove too much. For if, on pain of regress, there can be no time of an instant's ceasing to exist, can there also be no time of its *having* ceased? Aristotle's solution affirms that there is a time by which the instants of ten thousand years ago *have* ceased to exist. It thereby shows us clearly how to escape the second horn of the dilemma. And it does so without postulating a second time series; for the instants at which a given instant has ceased to exist belong to the only time series there is.

Damascius has a more general comment, that the paradox-mongers make the mistake of treating their now as if it enjoyed actuality and as if it were a part of time, when in fact it is a mere boundary without reality.[44]

Simplicius' solution

Simplicius rejects Damascius' stronger formulation, according to which the whole of time exists simultaneously.[45] He wants to maintain only the more modest thesis that time exists. He also disagrees, we have seen, with the idea of leaps. What he seeks to retain is just the idea that time has its being in coming to be:

> However, neither the whole stride all in one go nor the leap over the whole of a part seems to me to make sense, when applied to motion or time. For place, whose parts stay put, I think we can contemplate bunching things together in this way. But for things which have their being (*einai*) in coming to be, we cannot take anything all in one go, unless we do so merely in our own thoughts. For what was taken all in one go would have to be taken not as flowing (*rheon*), but as static, and not as coming to be, but as being (*on*). Yet what has this character among the things which have their being in coming to be? One should rather listen to the following remarks of the philosopher Damascius, and to what he said in his very own words.[46]

There follows the passage of Damascius about things which have their being in coming to be. But unfortunately Simplicius is wrong to think that, having dropped the leaps, he can without further explanation hang on to Damascius' idea that time has its being in something's coming to *be*. For

[43] ap. Simpl. *in Phys.* 799,35-800,16 (SP 86,8-29).
[44] ap. Simpl. *in Phys.* 799,30-5 (SP 86,2-8).
[45] *in Phys.* 775,33-4 and 777,11 (SP 66,8-9 and 68,27).
[46] *in Phys.* 797,27-36 (SP 80,22-31).

what is there left which can be said ever to *be*? Not *parts* evidently; and Simplicius would not, any more than Damascius, want to allow time to exist through the coming to be of *instants*. Simplicius seems committed to a theory of coming to be in which nothing, even briefly, has *being*. And that is not readily intelligible.

However, there is another part of Damascius' answer that Simplicius intends to endorse. For in the passage quoted, he points forward to Damascius' idea that in something flowing segments can exist only in thought, and he accepts it. It subsequently becomes clear, from the final sentence of his discussion of time, that he thinks this provides a solution. For if the 'nows' which punctuate time and segment it do not exist 'in actuality' (sc. but only in thought),[47] then, as Damascius himself indicated, the paradoxes, which turn on the punctuation and segmentation of time, cannot arise. This, according to Simplicius, is why Aristotle did not in the appropriate place state the solution of his paradoxes. It was because the non-actuality of the 'now' was not explained until a later book:

> Why, then, did Aristotle not solve the difficulties which he brought forward? Or is it because his argument called for the indivisibility of now and its not existing in actuality (*energeiāi*), which is demonstrated in the sixth book of the work in his discussion of motion.[48]

Although Simplicius did his writing after A.D. 529, Justinian put an end to teaching in the Athenian school in that year. In Chapter 13, I shall outline the controversy about how final or damaging that prohibition was. But what can be said straight away is that we know of no successor to Simplicius in Athens, and as far as discussion of the paradoxes is concerned, his can be taken as the last contribution of antiquity.

Retrospect over the contributions of antiquity

I described in the Introduction how late Neoplatonism, especially that in Athens, is almost universally thought to have been dead. This is not at all how Iamblichus and Simplicius saw the situation. They speak at one point as if it was Aristotle and the Stoics who had declined (*paratrepein*) from a Pythagorean philosophy of time, which they erroneously suppose to be earlier, while they themselves are returning to the superior tradition.[49] Moreover, elsewhere Simplicius boasts that Aristotle and Alexander had not managed to solve the paradoxes of time, which had found a solution only in his own day with Damascius and himself,[50] although he is aware that they

---

[47] Simplicius records the equation of 'not actually' with 'in thought' for Alexander at *in Phys.* 748,24, and for Damascius in the passage under discussion at *in Phys.* 798,10-11.

[48] *in Phys.* 800, 21-4.

[49] Simpl. *in Cat.* 351. See the very interesting note 52 in Philippe Hoffmann, 'Jamblique exégète du Pythagoricien Archytas: trois originalités d'une doctrine du temps', *EP* 1980, 307-23.

[50] *in Phys.* 795,33-5. In connexion with theories of space too, Simplicius claims originality (*kainoprepēs*) for the theories of Proclus (*in Phys.* 611,12) and Damascius (625,2), although he is also anxious to disclaim too much originality, since he wants the support of a long tradition (639,12). See Philippe Hoffmann's other interesting discussion in 'Simplicius: *Corollarium de*

supplied the *materials* for that solution.[51] I think we can reasonably feel, looking back over the theories of time which we have encountered, that the most interesting ones came from the Neoplatonist period. There was Iamblichus' static and flowing time, Augustine's time in the mind and Damascius' leaps. Moreover, this period seems to have had the best insight into the characteristics of McTaggart's A-series: its flow, the corresponding shift in judgments and the egocentricity of the present. This is not to forget that I have presented Aristotle as much the sharpest thinker in his solution of the paradox of the ceasing instant. Even so, on the other paradox, the 'parts of time', it was a Neoplatonist exploitation of an Aristotelian insight which came closest to the solution which I myself favour. For Damascius and Simplicius applied to the paradoxes the Aristotelian idea that the verb 'to be' has a different sense when it is applied to time. Admittedly, the sense described is very different from the one which I recommended as affording a solution to the paradoxes. None the less, this may be enough to restore some credit to the despised Athenians. In Chapter 14 I shall discuss the far more spectacular successes of their contemporary in Alexandria, Philoponus, along with their own replies to Philoponus.

Infinitely divisible leaps of motion recur in early Islamic thought, as will become clear in Chapter 25, and I hope that the foregoing discussion of Damascius will help to throw light on the Islamic texts. I believe that there has not been a full appreciation of the difference between atom-length leaps and infinitely divisible leaps, nor of the purpose of arguments involving the leap, nor of their Greek provenance.

It was not only infinitely divisible leaps, but also the Aristotelian paradoxes, which continued to live after the end of antiquity. But solutions did not improve in mediaeval Europe, if we may take as typical the treatment apparently given by Peter of Spain (died *c.* A.D. 1277) to the ceasing instant. According to a recent report, he held that an instant exists, begins to exist and ceases to exist simultaneously.[52] As regards the other paradox, the 'parts of time', we have already encountered it recurring in Leibniz.

Loco', in *Astronomie dans l'antiquité Grecque*, Paris 1979, Actes du colloque tenu à l'Université de Toulouse-le-Mirail, 21-3 Octobre 1977.

[51] Simplicius mentions Alexander's use of the idea of having being in coming to be, at *in Phys.* 735, 28-30. For other acknowledgments by Simplicius, see above.

[52] Norman Kretzmann, 'Aristotle on the instant of change', *PAS*, supp. vol. 50, 1976, 105.

PART II

TIME AND ETERNITY

CHAPTER SIX

Does Time Require Change?

Many philosophers have held that time cannot exist without change. Two reasons for thinking this appear already in a passage of Plato's *Timaeus* (37E1-38C6), where he is describing the creation.

> For there were no days, nights, months and years before the heavens came into being, but he devised that they should come into being then just when the heavens were assembled. These are all parts of time, and 'was' and 'shall be' are forms of time which have come into being. We are wrong when we apply them, without realising it, to the everlasting being. For we say that it was, is and shall be. But really only 'is' belongs properly to it. 'Was' and 'shall be' are appropriately said of becoming (*genesis*) which proceeds in time, for they are processes (*kinēseis*). But it is not appropriate for what is always immovably (*akinētōs*) in the same state to be becoming (*gignesthai*) older or younger through time, nor once to have become <so>, nor now to have completed becoming <so>, nor to be <so> in the future, nor in general to receive any of the attributes which becoming (*genesis*) attaches to the moving objects of sense perception. These attributes have come into being as forms of time which imitates eternity and circles round according to number, ...

> Be that as it may, time has come into being with the heavens, in order that, having been created together, they may be dissolved together, if ever their dissolution should come about. And <the heavens> have come into being in accordance with the model of eternal nature, in order that they should be as like it as possible. For the model has being for the whole of eternity, whereas the other has come into being and is and shall be right through the whole of time. So from design and thinking of this kind on the part of God, with a view to time's coming into being, and in order that it might start to be, the sun and moon came into being and the five other stars which are called wanderers, in order to define and preserve the numbers of time.

First argument: growing older

One reason for supposing that time cannot exist without change is the idea expressed here that the things to which we apply the words 'was' and 'shall be' undergo the special change of becoming older. (That is probably why Plato calls 'was' and 'shall be' processes. There is no need to worry about becoming younger, which can be explained in a footnote.[1] I shall also

---

[1] The idea is that, as something becomes older than its former self, its former self becomes younger than it. For further detail, see Chapter 8.

postpone until Chapter 9 the question whether Plato is right to treat his ideal Forms as exempt from becoming older.) By 'becoming older' Plato does not mean becoming more decrepit – the universe does not do that (33A-B); all it does is gather more years.

It might be objected by those who believe that the universe has existed an infinite number of years already, that it cannot grow older because an infinity cannot be increased. But, first, Plato is saying that there has been only a finite number of years; and, secondly, it will in any case be explained in Chapter 14 that there is a perfectly good sense in which an infinite collection can be made to grow, even though there is also a sense in which it does not get any larger.

Second argument: passage

Gathering more years is a peculiar sort of change, and there is another, discussed in Chapters 3 and 4, namely the process whereby what is future becomes present and then past. It was argued that Aristotle did not manage to separate this dynamic or flowing way of thinking about time from the static way. None the less, his account presupposed the existence of flow. Can we say, then, that time necessarily involves these two kinds of change: growing older and the passage from future to past? I believe that by and large we can, but one or two objections must be considered, and the truth extracted from them.

First, some people believe that, even as things are, there is no passage of time, and that there is no flowing series of past, present and future. Hugh Mellor, for example, has revived McTaggart's difficulty that every occurrence would have to possess all three attributes, pastness, presentness and futurity, and that this is a contradiction.[2] The answer might seem to be obvious, that an occurrence will not possess the three attributes at the *same time*, so contradiction is avoided; for a present occurrence is future only in the past, past only in the future and present in the present. But McTaggart's reply is that shifting to these more complex attributes, such as present in the present and past in the present, leaves us with the same kind of contradiction as before, since an occurrence will have to have all of them. And if we try to show that it will have them only at different times by switching to yet more complex attributes, McTaggart's reply is that it will have to have all of these more complex attributes as well, so that contradiction is still not removed. There is, however, an answer available, I believe, to someone like myself who regards as equally real and equally important the static relations of earlier, simultaneous, later, and the flowing relations of past, present, future. For when asked to specify the times at which an occurrence will possess the three flowing attributes of past, present and future, we need not use flowing language. We can content ourselves with saying that the times will be *different*, or, if asked for an illustration, we can give it in the static language of *dates*: present in 1982, future in 1981, past in 1983. There can be no objection to falling back on static language in this context, for I am not engaged in

2 Hugh Mellor, *Real Time*, Cambridge 1981, ch.6.

explaining the *meaning* of the flowing language. Explaining the meaning is something I undertook in Chapter 3 in terms of the word 'this', but I am now engaged in a quite *separate* task of rebutting a charge of contradiction, and so have left myself free to fall back on the static terms in order to do so.

A further reason which has been given for denying that there is a flowing series of past, present and future involves an appeal to Relativity Theory. It has been said that the theory discredits the notion of presentness, because it relativises to an observer's frame of reference the question of which distant events are now present. But the answer to this is that it relativises the flowing and non-flowing time conceptions alike.[3]

Another kind of objection is more modest: it would admit that, as things are, there is a flowing past, present and future, but it would argue that in other circumstances there would not have been. After all, I have admitted that in some sense the notions of past, present and future are egocentric, in that I have connected them with the word '*this*': to judge that something is future is to judge, 'it is later than *this* time'. What, then, if there had been no conscious beings? There would still have been time, but would it have contained a past, present and future? I believe that this doubt is mistaken in at least two ways. For one thing, we can relate the imagined situation in which there are no conscious beings to the time of our conscious imagining. We can then view the person-less situation as past, present, or future in relation to *ourselves*. We can do so even if we imagine the person-less situation to be very different from the one we know, since we can, for example, imagine that the universe might have been nothing but a mass of flames up to *now*. By talking of 'now', we do not insert ourselves illegitimately into the supposedly person-less situation. On the contrary, to imagine the situation is not to imagine ourselves as part of it. Admittedly, we may not be *obliged* to think of the person-less situation as past, present, or future in relation to ourselves; but there is a second way in which we can hardly avoid allowing that the concepts of past and future apply to a situation which contains no conscious life. For I see no plausible way of denying that events in that situation would come to be over and past. When we think of them as doing so, we need not think of them as becoming past in relation to *ourselves*. We simply reflect that one event would be over by the time some other event occurred.

But how can this be, it may be objected, if the concept of pastness contains a reference to conscious beings? What matters here, I think, is the point made in Chapter 3 that the reference to conscious beings can be very *indirect*. Little is involved in imagining an event in the person-less situation becoming past. We need only imagine the universe bereft of conscious life, while further imagining that, if there *had* after all been an intelligent being present, he *would* have been able to say: 'so and so is earlier (or later) than this time.' Once again, this is not illegitimately to imagine conscious beings as both present and absent. It is rather to imagine a counterfactual situation within a counterfactual situation: having imagined that conscious beings might have been absent, we further imagine what could have been said by one, if one had after all been present.

[3] For this answer, see William Godfrey-Smith, 'Special relativity and the present', *PS* 36, 1979, 233-44, esp. 233-5.

There is, however, one circumstance in which I think the passage of time would have to be thought of, not indeed as altogether absent, but as interrupted. And that is if there are indivisible times such as the time-atoms of Diodorus, discussed in Chapter 2, or the times both divisible and indivisible of Damascius, discussed in Chapter 5. Some modern physicists have revived at least the first of these two ideas. A time-atom has a positive length, and yet, surprisingly, there is meant to be no sense, or no *physical* sense, to the idea of its being part way through. It therefore admits no *internal* flow, for there is no stage at which some of the time-atom is 'over' and some 'to come'. None the less, the time-atom is subject to flow in *another* way, for a future time-atom becomes present and then past. The same may apply to the 'freezes' which I shall discuss later in this chapter. These are imaginary periods in which no changes occur. If nothing occurs to mark intermediate stages within these freezes, then there may be no sense to the idea of an intermediate stage at which some of the freeze is 'over' and some 'to come', and so no sense to the idea of internal passage. If so, this will yield the interesting result that passage depends on other kinds of change more closely than time depends upon passage. Time depends upon passage to this extent, that although a time-atom or a freeze may involve no *internal* passage, it will itself be subject to passing away.

It is desirable now to turn from mere passage and growing older, which are rather peculiar sorts of change, to the more ordinary kinds of change which have already begun to enter the discussion. Can it be argued that the existence of time depends on these?

Third argument: clock processes

One argument would be that time requires change because it requires *clock* processes. This argument would probably appeal to Plato, for in the quoted passage he says that time came into being with the heavens. It is hard to see any reason other than the fact which he mentions at the beginning of the passage, that there would be no days, nights, months or years without celestial revolutions. The idea that time requires celestial motion is reinforced in the following paragraphs. He describes the planets (38E4-5) as 'the bodies which had to help in producing time', and he says (39D1) that the wandering motions of the planets actually are time. These passages, incidentally, seem to rule out the view of those who believe that Plato allows time without motion.[4] What will need to be investigated, however, is the different issue of why Plato subsequently allows after all a disorderly motion '*before*' the heavens were created. Does this mean that he oscillates on the question of whether time requires clock processes? For there are no clock processes in that period of chaotic motion. I shall discuss this question in Chapter 17.

[4] I.M. Crombie, *An Examination of Plato's Doctrines*, London 1962, vol. 2, p. 208; Galen, if we can believe the tenth-century A.D. writer Ibn Abī Saʿīd, translated by S. Pines, 'A tenth-century philosophical correspondence,' *PAAJR* 24, 1955, 111f., reprinted in A. Hyman (ed.), *Essays in Medieval Jewish and Islamic Philosophy*, New York 1977. But I shall suggest below that Galen only ascribed to Plato time without *orderly, celestial* motion.

In relation to our own concept of time, I believe it cannot be maintained that time requires clock processes for its *existence*. Admittedly, our concept of time would be impoverished if we did not treat some processes or other as supplying a kind of clock, not even such processes as the generations of man, or the return of seasons. We could not then assign particular lengths to periods. But we would still have the concepts of before and after, and many other temporal concepts. We could even compare two processes as longer or shorter, either by simply remembering them as such, or when they overlapped in convenient ways, one extending beyond the other at one end, without being exceeded at the other. I conclude that if the existence of time requires that of change, this will not be because it requires the occurrence of clock processes.

In so far as Plato makes time depend on clock processes, I would distinguish Aristotle as being more cautious. There is a rival interpretation according to which Aristotle is like Plato, and makes time involve change because he thinks of it as essentially *measured*.[5] It has even been said that he would have no reason to think of mere duration, in the absence of measurement, as involving change.[6] But we shall see in this chapter that he has a reason, based on the example of the 'sleepers in Sardinia', for thinking that *any* duration would involve change. And I shall argue in the next chapter that it is a misconception to suppose that Aristotle makes *measure* a defining characteristic of time: number and measure should not be equated. If we had to construct, out of the materials which Aristotle provides, an argument that time requires *celestial* motion, that argument would not involve the notion of *measurement* at all.[7]

One author has drawn an even more far-reaching conclusion from Aristotle's supposed definition of time as a measure: 'to the Greek mind time = *chronos* was essentially measured time'.[8] If that were so, then we might expect nearly all ancient philosophers to agree with Plato that time requires clock processes. Another consequence would be that Plato would immediately escape several criticisms: his insistence on clock processes could not be faulted for failing to conform to *our* concept of time, if he was concerned instead with a distinct *Greek* concept of *measurable* time. Again, he could not be faulted for his subsequent talk of disorderly motion '*before*' the heavens were created, for this use of 'before' would not contradict the claim that *time*, that is, *measurable* time, began only with the heavens.

In fact, however, the situation is not so simple. For one thing, in the quoted passage Plato makes not only measurable time, but even 'was' and 'shall be' begin with the creation of the heavens. I shall consider the controversy surrounding this in Chapter 17. For another thing, I am not persuaded by

---

[5] Gregory Vlastos, 'The disorderly motion in the *Timaeus*', in R.E. Allen (ed.), *Studies in Plato's Metaphysics*, London 1965, 386-8, reprinted from *CQ* 1939; and 'Creation in the *Timaeus*: is it a fiction?', in Allen, 413, reprinted from *PR* 1964; John Whittaker, *God, Time, Being*, Symbolae Osloenses, supp. vol. 23, 1971, 25, n.5.

[6] Whittaker, op. cit., 17.

[7] Instead we might refer to Aristotle's view that infinite time requires infinite motion (*Phys.* 8.1, 251b10-28), and that only *circular* motion can be infinite, since it alone can be continuous (*Phys* 8.8).

[8] Whittaker, op. cit. 25, n.5.

the general thesis that to the Greek mind time (= *chronos*) was essentially *measured* time. We might already wonder about Anaxagoras and his followers, or at least about the sources who report them. For these sources, Aristotle and Simplicius, are willing to use the word 'time' of their theory, and to describe them as thinking that there was rest in the universe for an infinite *time* before there was motion.[9] Yet it is not clear how this infinite time without motion could be *measurable*. I have already raised a doubt about the central contention that Aristotle defines time as measure, and the further contention that this is his basis, indeed his only basis, for thinking that time requires change. I would now add that in *Phys.*8.1 Aristotle argues against Plato's creation of time in a way that is appropriate only if 'time' means what it means for us. It does not occur to Aristotle that Plato could have been confining his attention to the restricted concept of *measured* time. Similarly, Aristotle's pupil Eudemus of Rhodes believes that, since there was disorderly motion in Plato's account *before* the heavens were created, Plato ought to have admitted that there was *time* then too.[10] There is more: in Chapter 17 we shall see that Plutarch and Atticus interpret Plato as meaning that there was indeed a (disorderly) *time* before the creation of the heavens, although as regards Plutarch, I shall have to ask whether he went quite as far as calling it *time*. These are only examples: another comes from Philo, who describes the Stoics as postulating time in the absence of days and nights during a universal conflagration.[11] Basil and Augustine too, as I explained in Chapter 2, make the connexion between time and measure a rather unexpected one.[12]

My general conclusion is that time does not involve clock processes *essentially*, and that ancient thinkers would not all have supposed that it did.

'Regularity' of the fundamental clock

A caution needs to be introduced before we go further. It concerns Aristotle's claim that regular circular motion, that is, celestial motion, is above all the measure of time, because its number is the best known (*Phys.*4.14, 223b18-20). One problem is why we should think of it as above all the measure, or even as a regular one. As a child I believed that my afternoon rest was over when I could hear an aeroplane crossing the sky. We can imagine a tribe which selects not celestial movement, but aeroplane movement, or, better, cloud movement as their measure of time. Where we talk of hours, they talk of 'clours', a clour being the passage of a cloud from one end of the sky to the other. To them our choice of hours might seem a most unsatisfactory arrangement; for sometimes our hours race by, at the rate of a score to the clour, while sometimes they dawdle, so that only a fraction of an hour gets

---

[9] Arist. *Phys.* 8.1, 250b23-251a5; Simpl. *in Phys.* 1121,21-6.

[10] ap. Simpl. *in Phys.* 702,24 (= Eudemus fr.82b Wehrli).

[11] Philo *Aet.* 1.4; 10.54.

[12] In Chapter 2, I explained how they would allow time to continue, while the heavenly clock stood still (Basil *adv. Eunom.* 1.21; Aug. *Conf.* XI.23). Time, though then unmeasured, would still for Augustine be a measure with which we could time the movements of a potter's wheel. Augustine's time of the six days of creation also needs no ordinary clock, but then it is not ordinary time.

fitted into a clour. These people would not agree that celestial motion was regular, let alone 'above all the measure'. How could we persuade them of the superiority of our system?

It is partly a matter of convenience. For among our tribesmen breakfast on a windy day will be protracted over a period of many clours, while on a still day a single clour may suffice for breakfast, dinner, lunch and tea. There is no predictability here, because cloud movements do not correlate systematically, as celestial movements do, with scores and scores of other processes. This bears on both questions, on regularity and on being above all the measure. It bears on regularity, because a claim that celestial motion is regular can be supported by showing its correlation with other processes, whereas cloud movement can at best be *deemed* regular. It bears on being above all the measure, in so far as it shows that celestial movement is at least a better measure than cloud movement.

But the argument about regularity may be thought only to create a *fresh* problem. For if celestial motion is shown to be regular by reference to *other* processes taken as a standard, how can it be declared in the same breath to be *itself* above all the standard? But the answer is that there is no inconsistency in bringing other standards to bear on it, while still thinking it the best standard.

There are other reasons for thinking celestial movement to be a particularly good measure. For one thing, distant tribes cannot see each other's clouds, nor get them synchronised with each other. By contrast, the heavens are visible to all men and this is probably part of what Aristotle means, when he locates the superiority of celestial motion in the fact that the number of this motion is the best known. For another thing, clouds may catch each other up, which is not a hazard at least with the *fixed* stars, and clouds may be wholly missing from the sky or may stay motionless, so that cloud movement is interrupted.

I shall consider some other possible difficulties for clours in Chapter 14. But the last problem, that cloud movement may suffer interruption, is one which Aristotle argues in another context will befall any change other than circular motion.[13] A main argument is that rectilinear motion in a finite universe involves changes of direction and that such changes involve pausing. For, it seems to be assumed, there will be a first instant of having reached the point of reversal and a first instant of having left it. These cannot be the *same* instant, nor can they be adjacent, since no instants are adjacent. They must then be separated by a period of rest.[14] Here Aristotle seems temporarily to forget the principle that I shall ascribe to him in Chapter 26, that there is a first instant of having reached a terminus, but *not* a first instant of having left it.

Despite this oversight, I think it right to agree with the claims that celestial

[13] *Phys.* 8.7, 261a27-8.8, 265a12.

[14] *Phys.* 8.8, 262a31-b3; b21-263a3; cf. 8.7, 261b5-6 and 9, to be discussed in Chapters 21 and 26. For subsequent controversy, see A. Koyré, *Études d'histoire de la pensée philosophique*, Paris 1961, 63; H.A. Wolfson, *Crescas' Critique of Aristotle*, Cambridge Mass., 1929, 623-5; William A. Wallace, 'Galileo and Scholastic theories of impetus', in A. Maierù and A. Paravicini Bagliani (eds) *Studi sul XIV secolo in memoria di Anneliese Maier*, Rome 1981, 278

motion is regular and a good measure. But I would hesitate to accept anything as *above all* the measure. For one thing, we may find that there are several measures interchangeable with each other. For another, we are likely to want different measures for different purposes. Further, for some purposes we may hope in the future to discover better measures than any available so far.

Fourth argument: the verification of freezes

Aristotle offers a quite different argument in *Phys*.4.11, 218b21-219a1 for saying that time requires change:

> But time is not without change. For when we do not ourselves change at all in our thoughts, or fail to recognise that we are changing, we do not think that time has elapsed, any more than do those who sleep, as the story goes, alongside the heroes in Sardinia, upon being woken up. For they fit the later now on to the earlier and make them one, removing what intervenes because they were unconscious. Thus just as, if the now was not a different one, but was one and the same, there would have been no time, so also when it is not recognised as being different, the interval is not believed to be time. So if the belief that there is no time comes to us when we mark no change but our mind seems to stay at a single, indivisible <stage>, whereas we say that time has elapsed when we perceive and mark a change, it is clear that time does not exist (*ouk estin*) without (*aneu*) change and alteration.[15]

Aristotle here moves from an epistemological premise (we *notice* time, when and only when we *notice* change) to an ontological conclusion (time does not *exist* without change). That the conclusion is ontological is made clear by the *ouk estin* in 218b33, which is decisive enough on its own. But there is in addition further evidence, first in the general structure of Aristotle's argument. He starts by arguing that time is not identical with change (4.10, 218b9-20). Then comes our passage in which he argues that it depends on change (4.11, 218b21-219a2), and from these two premises he concludes (219 a2-10) that time is something belonging to change (*tēs kinēseōs ti*). The next question is what its exact relationship is to change, and this leads to the definition of time in 219b2. What is significant is that the interim conclusion, that time is something belonging to change, is taken up in the last book of the *Physics* at 8.1, 251b10-28. Here Aristotle is prepared to argue not only from the definition of time (251b12), but also from the interim conclusion that time is something belonging to change (*pathos ti kinēseōs*), for the existential conclusion, that if there is time, there is change, and hence that if time has no beginning or ending, neither does change. Evidently, then, the interim conclusion, that time is something belonging to change, has existential import. It implies that the existence of time involves the existence of change.

The existential claim is also made explicitly in *GC* 2.10, 337a23-4:

---

[15] *kinēsis* and *metabolē*: Aristotle has said in 4.10, 218b19-20 that he will not distinguish between these. Normally *metabolē* differs by including creation and destruction, as well as motion, growth and change of quality. But now *kinēsis* will also include creation and destruction, as is made explicit at 4.14, 223a31.

If the time is continuous, so must the motion be, if it is impossible for time to exist (*einai*) apart from (*chōris*) change.

Some commentators have been unwilling to allow that in 218b33 Aristotle intends the ontological conclusion that time cannot *exist* without change,[16] because they have not seen how an ontological conclusion can be derived from epistemological premises. But in fact the argument can be made very plausible, even one of the most plausible of the arguments for saying that time cannot exist without change. Aristotle needs to add only two supplementary premises. One premise, which is closely related to what he actually does say, is that changeless times would be in principle undetectable. The other is some sort of verificationist premise, either that it is meaningless, or that it is false, to postulate undetectable times. This is brought out in a highly original paper by Sydney Shoemaker, who represents the verificationist as arguing for falsity.[17]

Whether he argues for falsity or for meaninglessness, his premise can be made to seem appealing. Let us take first the charge of meaninglessness. If I claim to have an extra pair of hands, but it emerges that you cannot see them, or feel them, or shake them, and I cannot do anything with them, you could justifiably say that you would not call those *hands*. Are undetectable times so very different from undetectable hands? If someone does postulate that undetectable times occur, then presumably a large number might intervene every day between breakfast and lunch without our being any the wiser. But does this suggestion have any content or meaning?

There is a particularly close connexion between *length* of time and verification, which might be admitted even by those who do not wish in general to be verificationist. If there were such undetectable periods, they could not meaningfully be called long or short. Could they even be said to have any length, and if not, could they still be called *periods*?

The other verificationist tack questions the *truth* of the idea of undetectable times. If we assume that periods must by definition have some length, then the intervention of undetectable periods between breakfast and lunch would lead to complete scepticism about the length of time between the two meals. And that scepticism might itself be taken as a sign that the postulate of undetectable times was false. This is the version of verificationism which would be most likely to appeal to Aristotle. For he was prone to assume the falsity of radically sceptical theses, because of his whole conception of how to do philosophy. He thought that the right method was to start from what had been accepted (*endoxa*) either by people in general, or by the great philosophers. If one then modified the elements which led to perplexities, what remained standing would be the truth.[18] Moreover, a view accepted by everyone is unchallengeable,[19] and such would be the view that time can be detected, and that we can know, at least roughly, how much time has

---

[16] E.g. J. Moreau, *L'Espace et le temps selon Aristote*, Padua 1965, 101-8; endorsed by Julia Annas, 'Aristotle, Number and Time', *PQ* 25, 1975, 101.

[17] Sydney Shoemaker, 'Time without change', *JP* 66, 1969, 363-81, see p. 366.

[18] E.g. *NE* 7.1-2, 1145b2-7; 1146b6-8.

[19] *NE* 10.2, 1172b35-1173a2; cf. *Phys.* 2.5, 196b14.

elapsed. The claim that there are undetectable times will, on this view, not be meaningless, but rather false.

So much for the verificationist premise, which rejects undetectable times. But what now of the other premise, according to which time without change could not be detected? Shoemaker attacks this premise, by arguing that, in bizarre circumstances, changeless times could be detected. His story can be retold as follows. Imagine that the universe is divided into three parts, and that each part periodically undergoes a freeze centred on New Year's Day, in which all change is arrested. The freezes are detectable from other parts of the universe through the use of extra-sensory perception, not involving the travel of light. In one part, the freeze occurs every third year, and lasts three minutes, being preceded and followed by a three-minute period of sluggishness. In a second part of the universe, the freezes occur every fourth year, and the period both of the freeze and of the sluggishness is four minutes. In the third and last part of the universe, the frequency of the freeze is five years, and the period of freeze and of sluggishness is five minutes. A simple calculation suggests that there will be a simultaneous, total freeze of the whole universe after sixty years, in other words, a three-minute period in which no change occurs anywhere. Such a changeless period cannot be called undetectable. It is detectable because the preceding regularities provide evidence for its occurrence. Moreover, its occurrence would be most beautifully and elegantly confirmed, if the expected periods of freeze and of sluggishness were observed, and even appeared to be shortened by just the predicted amounts. They would appear to be shortened by amounts corresponding to the periods when the observers were themselves immobilised and unable to observe.

When it is said that the changeless three minutes would be detectable, it is not meant that there would be conclusive proof of its occurrence. Proof is not what is needed. To show that three-minute freezes are not a meaningless postulate, like undetectable hands, we need only show that they leave behind traces of themselves and of their duration, not conclusive proof. Similarly, to show that the hypothesis of three-minute freezes does not have to be false by Aristotle's criterion, we need only show that it does not plunge us into a scepticism which violates what all mankind believes. And this can be shown, for there would be a straightforward and sensible hypothesis about when total freezes did and did not occur. Hence we should have a good idea, even though not a conclusive proof, about how much time had elapsed since breakfast.

I want to suggest, however, that there may be a simpler way to challenge the premise that times without change would be undetectable. It is a familiar experience that everything within a narrow field of vision may be seen to be frozen still, as when I watch a tree through a narrow window on a windless day. What if everyone reported that they had seen everything frozen for a while, clocks included, and people in mid-stride? Moreover, they noticed no changes within themselves: none of them grew bored, or tired or hungry. Of course, science tells us that a freeze on all movement would be impossible, and catastrophic even if it were possible. But, as in Shoemaker's example, we are to imagine a situation governed by different laws of nature from our own.

Would this not be a case of a time without change being detected? Galen is said to have argued against Aristotle along comparable lines. Surely, he maintained, we can contemplate changeless things without our thought being involved in change.[20]

Two lines of objection may be raised: first, if the freeze is to be experienced as having any duration, will not the thoughts of observers have to change? Will they not have to think, 'it's frozen ... it's still frozen ... it's *still* frozen', each of these being a *new* thought? Or if the psychological state of the thinker does not have to change, will not at least the proposition which he thinks have to change, as a matter of logic, as the reference of 'now' shifts in 'it is now frozen'? In reply, I doubt if the thinker's psychological state has to change. For everyone may retrospectively estimate the experience as having been very short, and an experience which is perceived as being so short hardly leaves room for the ideas in the thinker's mind to change in the way suggested. But what about the objection that the *reference* of 'now' will keep shifting? One difficulty with this is that I do not know how many 'nows' there would be, for the context has to determine the length of a 'now' (which will not necessarily be instantaneous), and in the present context there is nothing to motivate talk, even retrospective talk, of a *plurality* of 'nows', much less talk of an infinity of instantaneous ones.

The other line of objection is that no one can observe that his own mind has just become frozen, for such a thought would be self-refuting, since it would be a *new* one. But the answer is that when the observers' minds become frozen, they are not to be imagined as observing this change, but simply as observing the subsequent changelessness.

I am inclined to think, then, that this does provide a simpler way of detecting a period of time without change. A third possibility of detection arises from the theory which I shall discuss further in Chapter 24; that time comes along in atomic chunks. A time-atom would be an indivisible period with a positive length, and being indivisible, it would leave no room for change to occur within it. Philosophers must, however, be careful about the order in which they argue. It would be no good simply alleging that there may be time-atoms and pointing out that these would be times without change. For in order to meet the present objection that time without change would be undetectable, it would be necessary first to show how there could be empirical evidence for changeless time-atoms. In fact, those physicists, a minority, who have postulated time-atoms have usually been guided by empirical evidence that there may be tiny periods within which no change occurs.

This last argument is the only one of the three to suggest that the world, as it is, may possibly contain evidence of times without change. The other two arguments suggest only that in radically different circumstances there would be evidence. Moreover, none of the arguments suggests that time would be detectable, if there were no change at all. They all envisage changes alternating with pockets of freeze. None the less, the general upshot goes contrary to Aristotle's argument. For although I would join him in rejecting

---

[20] ap. Simpl. *in Phys.* 708, 27-32; ap. Themist. *in Phys.* 145,2.

undetectable times, I am not convinced that pockets of time without change would have to be undetectable.

The discussion has concerned pockets of time *within* the history of the universe, but I must also consider a further question: could there be time unaccompanied by change *outside* the history of the universe, supposing that the universe, or the change in it, has an ultimate beginning or end?

Fifth argument: the impossibility of relocating the history of the universe

One objection to this last suggestion is less than conclusive. It might be protested that, if the universe has a finite history, and is surrounded at either end by changeless time, then it ought to make sense to imagine the whole history of the universe occurring earlier or later, instead of occurring when it does. Yet the objection would be that this hypothesis has no meaningful content, for there is no real difference corresponding to the talk of different locations for the history of the universe.

I agree that the hypothesis has no meaningful content, but none the less I doubt if the objection succeeds. The reason can be brought out by considering space as opposed to time. For we can imagine a spatially finite physical universe situated within infinite space. And somebody might produce a parallel argument to show that it would be meaningless to talk of the entire physical universe having all along been located elsewhere within infinite space. What interests me is that no relevant conclusion follows from the argument. For we can still talk of there being empty space outside the physical universe, even if we cannot imagine the physical universe having all along been located elsewhere within that space. Why, then, cannot we talk in a parallel way of there being time without change before and after the history of the universe, even though we cannot imagine the history of the universe differently located within that time?

Sixth argument: temporal distance, unlike spatial distance, requires markers

If this is to be shown impossible, it will be necessary to trade on some of the differences between time and space, and this brings me to a sixth argument. Why is it that we cannot make sense of the idea of five years before the beginning of the universe, nor yet of a long or a short time before, whereas we can make sense of the idea of millions of miles beyond the outermost star?

We can define a mile as the distance which *would* be covered, *if* a standard measuring rod were placed end to end so many times. The emphasis is on what *would* happen, *if* something were done. It would be impractical to require that every mile-long route through space should actually be paved with measuring rods placed end to end, or with other markers. A year is different: it is *actually* marked out by a revolution of the earth around the sun, or by equivalent events. We are therefore fairly safe in defining a year as the period in which these events *actually* take place. A small qualification is needed, however, because our discussion of periods of detectable freeze shows that, in special circumstances, we should allow that a year had passed even in the absence of these events.

The situation can be summarised by saying that temporal distance, unlike spatial distance, normally requires to be marked out with markers. And the reasons for the difference can be stated quite simply.[21] First, if a period of time had not initially been provided with markers, we could not subsequently go back to that time and provide it with markers. Secondly, if *per impossibile* we should go back and do so, we would be violating the terms of discussion, for we would no longer be talking about a period of time without change. Contrast the spatial case: we are in principle free at any time to apply measuring rods to a previously unmeasured distance. And in doing so, we would not be violating the conditions of discussion. For it would remain true that the distance in question had been unmarked at the time we first talked about it.

I think this is enough to explain why the absence of actual markers deters us from talking about five years before the earliest change, when it does not deter us from talking about a million miles beyond the furthest matter. It would even deter us from talking about a *long* or a *short* time before the earliest change. But I am not so clear that it should stop us from talking about *time* before change began. For it needs to be considered whether such time would be undetectable, if once again there were an observer, say, an angel, who, at least at first, did not change. Could he not be startled to notice change beginning, and be impelled by the contrast to notice that he had previously been aware only of changelessness for as long as he could remember? To give sense to the idea of time before change, we would need only the *possibility* of such observation. Certainly, the observer could assign no *length* to the period of rest – for that, he would need actual changes. But could he not be aware of its existence? I leave this question unanswered, but while it remains unanswered, I must regard the present argument as inconclusive. It convinces me that we cannot make sense of the idea of long or short periods before the earliest change, but it leaves me unclear that we cannot make sense of the idea of earlier *time*.

Seventh argument: why not sooner?

It can now be seen that one of the earlier arguments, the fifth, is closely related to an argument offered by Leibniz.[22] In 1715-16, in a famous exchange of five papers with Clarke, Leibniz attacked Newton's absolutist view, that time exists independently of change. One major argument, to which I shall return in Chapter 15, started from the shared Christian assumption that God gave the universe a beginning. The question why God did not shift the entire history of the universe back to an earlier time is unanswerable, Leibniz says, if it makes sense. And, according to Leibniz, it will make sense, if, and only if, the absolutist view is right. For then, and only

---

[21] For a different diagnosis, see W.H. Newton-Smith, *The Structure of Time*, London 1980, 42-3.

[22] Leibniz, *Letters to Clarke* 3 §6; 4§§6, 13, 15, 5§§52; 55-60, ed. H.G. Alexander, *The Leibniz-Clarke Correspondence*, Manchester 1956, pp.26-7; 37-8; 75-7.

then, there will exist a time, independently of the history of the universe, against which that history could be shifted backwards. The Newtonians then make God act without sufficient reason. Leibniz's rival view is that time is simply a relationship between events, a view which naturally entails that the existence of time requires that of change.

Leibniz's argument goes beyond the fifth one, for that urged that the absolutist has to give sense to the senseless question why the history of the universe was not relocated. Leibniz's argument adds that, if the question has sense, as the absolutist must suppose, then he cannot answer it, but must make God act without sufficient reason. In either case, the objection is effective against a strong form of absolutism, which believes that a time scale can exist, whether or not there are any events at all. But I have argued that it is not effective against a more modest version of absolutism, which tries to give sense to the idea of time before change, but in strong dependence upon the subsequent occurrence of change.

Eighth argument: the identity of indiscernible times

There is another argument which draws on the ideas of Leibniz. It has been described by G.E.L. Owen, and attributed as an unspoken assumption to Parmenides.[23] I shall not myself interpret Parmenides this way, when I come to the subject, but none the less the argument is certainly one of the better ones. It is a principle commonly attributed to Leibniz that if A and B are indiscernible (i.e. have all their properties in common), then A is identical to B. As applied to times, this would mean that if any time A has all the same characteristics as any other time B, then there are no distinct times. That in turn means that, if there are to be distinct times, some change must occur, so that times A and B may be characterised by different states of affairs.

A full discussion of the principle of identity of indiscernibles would require us to consider what kinds of property are to be counted as relevant to discernibility. But the principle can be given some initial plausibility in more than one way. It might be made to rest on principles of verification; for the claim that there were two non-identical times or other entities would be unverifiable, if there were no properties by which they were distinguishable as two. But the principle can also be made to seem plausible, for example, by considering how else the notion of identity can be defined, if not in terms of having all properties in common.

Ninth argument: the definition of time

Another way to convince oneself that the existence of time does, or does not, require the existence of change is to challenge oneself to find a definition not of *identity*, nor of temporal *distance*, as in the previous arguments, but of *time itself*. It turns out to be very hard to define time without making some reference to change. But we must be careful to consider whether the reference implies that time requires actual change or only the possibility of change. For example, if we define time as what enables a subject to assume contradictory predicates, or as what enables bodies to occupy the same place as each other, this will imply only that time makes change *possible*, not that it requires change as something *actual*.

Aristotle does not at first base his thesis, that the existence of time requires that of change, on the definition of time. Rather, as shown earlier in the chapter, he argues in the opposite direction from that thesis (218b21-219a2) to the definition of time (219b2). However, in 8.1, 251b10-28, he turns round and argues back the other way. His definition, which I will try to explain in the next chapter, says that time is a number, or that which is countable in change, in respect of before and after. Here he infers from the definition (251b11-13) that if time exists and exists always without a beginning, then so also does change.

I shall anticipate discussion of what Aristotle's definition means by saying that it brings in change in two different ways. I believe Aristotle envisages someone observing that the present stage of some change is different from a past stage, and counting the stages. Noticing the difference between present and past involves change, in the sense of the *passage* of time, and the subject matter in which the difference is noticed is itself a change.

With this reference to the definition of time, I have finished introducing the arguments for the idea that time requires change. There have been nine in all, more than are recognised in certain recent treatments of time. Indeed, one contemporary writer draws attention to the paucity of current arguments on either side of the case.[24] I shall conclude by considering which philosophers in antiquity allowed time to exist without change.

Time without change in antiquity

Not all ancient philosophers accepted that time must be accompanied by change. Believers in time-atoms were committed to denying that change could occur within a time-atom. Further, it was seen in Chapter 3 that Iamblichus spoke of his higher kind of time as not moving; what moved was the lower kind of time which depended on the higher. This would not yet mean that the higher time existed independently of change, if it were open to Iamblichus to follow Plotinus' idea that time is somehow dependent on changes in the soul. But Iamblichus differs from Plotinus on this point, and declares, in his commentary on Plato's *Timaeus*, that time is:

not created along with the motion and life proceeding from the soul.[25]

Again, in expounding pseudo-Archytas, Iamblichus explains with approval that Archytas makes the higher time to be the number and the cause of the

---

[24] Newton-Smith, op. cit., 17 and 46, comments that there are only three types of argument for time requiring change, only one of them, the verificationist one, being of much significance in his view, and that there are only two types of reply to it. Compare also John Lucas, *A Treatise of Time and Space*, London 1980, 17-28.

[25] ap. Simpl. *in Phys.* 794, 12-15 (SP 42, 16-19).

change which is the cause of all other changes. This number

> is ordered in front of that change in the order of causation.[26]

This implies that change depends on the higher time, and not vice versa, although this evidence for (higher) time without change in Iamblichus is made slightly less clear cut by another report of his comments on Archytas.[27]

Acceptance of time without motion seems to have been more widespread among Presocratic philosophers, if we can believe Aristotle, who reports that Empedocles made the universe undergo alternating periods of rest, and that Anaxagoras subjected it to an infinite period of rest before Mind introduced motion.[28] To Anaxagoras another source adds his pupil Archelaus and Metrodorus of Chios.[29]

Sambur

sky cites three ancient philosophers as having taken an absolutist view of time, but the evidence is not very convincing.[30] Two of them are Aristotelians: Strato (head of the Lyceum from 288 to *c.* 269 B.C.) and Boethus (fl. 50 B.C.). Both of them want to assign time to the category of *quantity*. Strato defines time as the *quantity* in which there takes place travelling, campaigning, fighting, sitting, sleeping, etc. This seems quite inconclusive. Boethus merely distinguishes between time, for example, a month or year, and what is *in* time, for example, what *lasts* a month or year, with a view to justifying his assignment of time to the category of quantity. I do not think that this establishes anything either.

The evidence on Galen comes from a tenth-century Islamic source:

> Let me know whether you consider that time is necessarily consequent upon motion, so that there cannot be time otherwise than through motion, [which] is the cause of its existence – this being [the view of] Aristotle – or whether you consider that time has an existent nature, that it is a substance which subsists *per se*, and that motion only measures it as a surveyor measures land with a cubit. Galen – according to what Alexander states about him in the treatise in which he opposes [Galen's views] on time and place – held this opinion, [which] is refuted by Alexander. Galen's view was that time is eternal (*a parte ante*), and that it does not need motion [in order] to exist; and he states that Plato was of a like opinion on this point. That is to say that he considered that time is a substance, meaning by that duration, and that motion measures it. Galen states accordingly that motion does not produce time for us; it only produces for us days, months and years. Time, on the other hand, exists [according to him] *per se*, and is [not] an accident consequent upon motion.[31]

[26] ap. Simpl. *in Phys.* 786, 20 (SP 32, 28).

[27] Simpl. *in Cat.* 351, 13-16: in this report, Archytas is represented as giving the opposite verdict and as making the higher time depend on change in the soul. A solution which has been suggested is that this last represents the interpretation which Simplicius, rather than Iamblichus, places on Archytas' words. So, Philippe Hoffmann, 'Jamblique exégète du Pythagoricien Archytas: trois originalités d'une doctrine du temps', *EP*, 1980, 307-23, esp. 313-14.   [28] *Phys.* 8.1, 250b23-251a5.   [29] Simpl. *in Phys.* 1121, 21-6.

[30] S. Sambursky, *Physics in the Stoics*, London 1959, 100-2; *The Physical World of Late Antiquity*, London 1962, 9-14 cites the following: Strato, ap. Simpl. *in Phys.* 790, 1-15; Boethus ap. Simpl. *in Cat.* 348, 2; Galen in the report of Ibn Abī Saʿīd.

[31] Ibn Abī Saʿīd, translated by Pines, loc. cit.

I suggest that this evidence is suspicious. The view which we can reasonably take Galen to have ascribed to Plato, and which I shall discuss in Chapter 17, is not the view that time is independent of *motion*, but that it is independent of *celestial* motion. For Galen finds in Plato an eternal motion of matter before the creation of the *heavens*.[32] The treatise in which Alexander is said to have interpreted Galen otherwise may well be Alexander's *de Tempore*. But if so, the Islamic report is misrepresenting Alexander. For Alexander there takes his unnamed opponent (who may, indeed, be Galen) to think that there could be time, even if there were no motion in the *celestial* sphere.[33] This falls short of the view that there could be time, even if there were no motion *at all*.

What we do find is people arguing in a way which may unconsciously presuppose that there could be time in the absence of change. Such may be the position of those whom Augustine attacks for asking what God was doing before he created the universe.[34] Were they assuming that there would have been an earlier period in the absence of any changes, or would they have assumed changes to be going on in God?

The result of this survey is that it was the exception in ancient Greek thought to allow time without change. As for the arguments that time does require change, they will prove useful when, in the following chapters, I consider the possibility of timeless entities, and the idea that the universe might have had an absolute beginning in time.

[32] Galen *Compendium Timaei* 4, ll.1-13, in Plato Arabus, ed. Kraus-Walzer.
[33] Alexander *de Temp*. (Théry) 93,22-8.
[34] Aug. *Conf.* XI.12; 13; 30.

CHAPTER SEVEN

Time, Number and Consciousness

The definition of time as number

I have discussed Aristotle's treatment of time in Chapters 1, 4 and 6. But I want now to concentrate on his definition of time as *number*, which has proved very hard to understand. The definition is given, along with the rationale for it, in the following passage (*Phys*.4.11, 219a14-b8):

> So the before and after in place come first, and here they depend on position. But since there is before and after in magnitude, they must exist also in motion, and be analogous to the before and after in magnitude. But there is before and after in time as well, because time and motion always follow each other. (The before and after in motion is motion as regards its substratum, but its essence is different, and is not motion.)

> We recognise time when we set boundaries to motion, bounding it by before and after. And we say that time has elapsed when we take notice of the before and after in motion. We set boundaries by taking the before and after as different and as having something distinct between them. For when we notice that the ends are different from what is in the middle, and the mind says that the nows are two, one before and one after, we say that there is time then and that this is time. For what is bounded by the now is considered to be time: let that be taken as given. When, therefore, we perceive the now as one, and not (reading: *toi*) as one before and one after in a motion, or when we perceive it as a single terminus, albeit a terminus of what comes before and after, then no time is thought to have occurred, because no motion. But when we perceive a before and after, then we say there is time. For this is what time is: *the number of motion in respect of before and after*. So time is not motion, but exists in so far as motion contains number. (A sign of this: we discriminate more and less by number, but more and less motion by time.) So time is a kind of number. But since number is spoken of in two ways, for we call 'number' both what is counted or is countable and that with which we count, time is what is counted and not that with which we count.

To summarise, time, for Aristotle, is that which is countable in motion in respect of before and after. The word for motion is *kinēsis*, and it can apply to other kinds of change. Thus in 4.14, 223a29-b1 Aristotle extends its meaning and says:

> Someone might be puzzled as to the sort of change (*kinēsis*) for which time is the number. Is it not change of any sort? For things come into being and perish

and grow and change quality and move, all in time. So time is the number of each change in so far as it is change; which is why it is the number of continuous change quite generally, and not of a particular kind of change.

The revised statement, then, is that time is that which is countable in *change* in respect of before and after.

What is being counted here? One thing that is being counted is 'nows' or instants. Thus in the quoted passage at 219a27-8, he talks of how 'the mind says that the nows are two'. Later in the chapter, at 219a25-8, the very essence of now is said to reside in being countable, and I shall argue that he is thinking of counting 'nows' again at 220a10-21. However, there is a second thing which he thinks of as being counted. For his definition of time speaks of counting some *aspect of motion*, and the quoted lines 219a22-6 suggest that he means we count *instantaneous stages* of motion.[1] As a ball rolls across the floor, we count the stage at which it is at an earlier position and the stage at which it is at a later position.

We recognise time when we set boundaries to a motion, bounding it by before and after. And we say that time has elapsed when we take notice of the before and after in motion. We set boundaries by taking the before and after as different and as having something distinct between them.

If I say that the reference is to counting instantaneous stages in a motion, this is because the before and after in motion has just been distinguished (a14-19) from that in place and time, the reference being presumably to *stages* in a motion. That the stages are *instantaneous* is suggested by their being boundaries. That they are both instantaneous and *counted* is suggested by Aristotle's going on in a26-30 to associate the distinguishing of them with the distinguishing of two instantaneous nows as two:

For when we notice that the ends are different from what is in the middle and the mind says that the nows are two, one before and one after, we say that there is time then and that this is time. For what is bounded by the now is considered to be time: let that be taken as given.

The passage harks back to an earlier one, which I discussed in Chapter 6, about those who sleep alongside the heroes in Sardinia (218b21-219a1). If, upon waking, the mind seems to have stayed at a single indivisible stage (*en heni kai adiairetōi*, 218b31), then these people fit the later now on to the earlier and make them one (218b25-6), and so do not think that time has elapsed (218b23). In each of the two passages, the recognition of time depends upon distinguishing two instants as two, and that in turn upon distinguishing two stages in a change as two.

So instants and instantaneous stages are among the things that get counted. And when the definition of time refers to counting '*in respect of before and after*', the italicised phrase will naturally refer to earlier and later

---

[1] Although Aristotle objects (see Chapter 26) to saying that something is moving at an instant, he allows us (*Phys.* 6.8, 239a35-b3; 8.8, 262a30; b20) to say that it is at a point or level with something at an instant.

instantaneous stages, as the ball rolls across the floor. But the definition implies that yet a third thing gets counted, namely, time itself; for time is said to be countable. Perhaps, then, the idea is that we count periods of time *through* counting instantaneous 'before' and 'after' stages. To sum up: time is a countable aspect of change, because we count periods, and we do that when we count the instantaneous stages which bound the periods before and after.

If this is what the definition means, it calls for further completion, because time has turned out to be only *one* of the things that get counted. Stages and instants get counted as well, not to mention spatial positions and spaces. To complete the definition of time, Aristotle should add a clause to rule out these rival candidates as not the things intended.

Another difficulty in the definition is anticipated by Aristotle himself. For he foresees that someone might complain that the words 'before' and 'after' are temporal words, and so make the definition circular. At any rate, this is presumably why, in the quoted passage, he explains that before and after should be understood primarily as spatial notions: we perceive the moving body at different spatial positions, one spatially before, one spatially after (219a14-19). The defence is ingenious, but regrettably unsound. For we distinguish two positions as before and after (as opposed, say, to here and there), only if we have in mind an imagined movement such that the moving object is temporally before at the position we call 'before' and temporally after at the position we call 'after'.[2]

The relation of number to measure

What is the relation between counting and measuring? So far Aristotle has focused only on the former. But I have explained that in counting instantaneous stages or nows we are in effect counting the intervening stretches. I can now add that, if the counting is done in a *regular* way, the resulting stretches of time may constitute units of measurement. Thus if we count the stages of celestial motion at regular spatial intervals, we shall in effect be counting periods of time which can serve as units in measuring other motions. There is a link, then, between the initial idea of time being counted through the counting of stages in motion and the further idea of time serving as the measure of motion.

In this new context, where Aristotle considers time as the measure of motion, and in other contexts too, he occasionally ignores his claim that the sense in which time is a number is that it is something countable not something *with* which we count. This claim is made twice,[3] and yet it seems to be neglected at least as often.[4] In a sense, both ways of looking at time are acceptable, because if we make a regular count of the periods occupied by a

[2] See Joseph Moreau 'Le temps selon Aristote', *Rev. Phil. de Louvain*, 46, 1948, 75; *L'Espace et le temps selon Aristote*, Padua, 1965, 117-21; G.E.L. Owen, 'Aristotle on time', in Peter Machamer and Robert Turnbull (eds), *Motion and Time, Space and Matter: interrelations in the history and philosophy of science*, Ohio State University, 1976, 24-5, reprinted in *AA* 3; I am indebted to a very clear statement of the point by Norman Kretzmann in unpublished work.
[3] *Phys.* 4.11, 219b5-9; 4.12, 220b8-9.
[4] E.g. 4.11, 219b4-5; 4.12, 220b3-5; 220b16-17.

uniform motion, we obtain time units that can be used for counting. Why, then, does Aristotle twice reject one of these viewpoints? I take his idea to be that, since there could have been time even if there had not been regularity, it cannot be made part of the *definition* of time that it is that *with* which we count.

Many writers make the link between number and measure much closer. They believe that Aristotle defines time not only as the number, but also as the measure, of change, using the terms 'interchangeably' or 'without distinction'. Others divide on whether Aristotle defines time as what measures change or as a measurable aspect of change.[5] Moreover, large conclusions are drawn from the equation of number with measure: Plotinus devotes a chapter to attacking the supposedly Aristotelian definition of time as the measure of change. Gregory Vlastos concludes that Aristotle must accept Plato's consistency, when the latter talks of irregular motion '*before*' the beginning of time. As noted in Chapter 6, the word 'before' will not reimport reference to time, if time requires *measurable* motion. John Whittaker, in a sentence already quoted, infers from the equation of number with measure that 'to the Greek mind time = *chronos* was essentially measured time, ... and therefore not precisely identical with duration'. In the light of this, it is worth dwelling for a moment on the supposed equation.

Aristotle certainly calls time the measure of motion, but he does not offer this as part of the *definition* of time. Measuring receives only the most indirect mention[6] in the section where he introduces his definition. It works its way into the discussion later, but at this early stage there is no sign that the counting which he has in mind will always be evenly spaced, so as to provide units of measurement. To obtain even spacing, one needs access to a clock, ultimately to the celestial clock, in Aristotle's view, and he makes it no part of the definition of time that one has such access. Nor do I see any passage in which Aristotle uses the terms 'measure' and 'number' without distinction. Each seems to me to carry its own distinctive meaning, even when they are brought into close relation.

Not only does measuring receive only scant mention in the passage but the one mention which occurs seems to confirm my claim that 'number' and 'measure' are not interchangeable terms. The reference comes in a brief parenthesis, where Aristotle seeks to corroborate the definition of time as number.

A sign of this: we discriminate more and less by number, but more and less motion by time.[7]

The corroboration is already suspect, because it seems to support the idea

---

[5] Plotinus 3.7.9 (1); A.H. Armstrong, note ad loc. in the Loeb edition, Cambridge Mass., 1967; Cornford Loeb translation of 219b1-2, Cambridge Mass., 1963; Moreau, *L'Espace et le temps selon Aristote* 125; 129-30 (who prefers measurable aspect); Julia Annas, 'Aristotle, Number and Time', *PQ* 25, 1975, 97-113, esp. 100; Gregory Vlastos, 'Creation in the *Timaeus*: is it a fiction?', in R.E. Allen (ed.), *Studies in Plato's Metaphysics*, London 1965, 413; and 'The disorderly motion in the *Timaeus*' (reprinted from *CQ* 1939), ibid., 386-8; Whittaker, *God, Time, Being*, 25, n.5.   [6] 4.11, 219b3-5.   [7] *Phys.* 4.11, 219b3-5.

that time is number in the irrelevant sense of that with which we count. Aristotle appears to sense this, for he goes on immediately to say:

> So time is a kind of (*tis*) number. But since number is spoken of in two ways, for we call 'number' both what is counted and that with which we count, time is what is counted and not that with which we count.

In other words, the oblique reference to time as a measure establishes one sense in which time is a number, but this sense is viewed as not strictly relevant to the definition. This is strong confirmation that 'number' is not interchangeable in the definition with 'measure'.

The best and most interesting attempt I know to connect number more closely with measure is made by Julia Annas.[8] She draws attention to *Metaph.* 10.1-3, where Aristotle says that the unit in counting should be viewed like the unit in measuring. Just as a one-foot rule is used for measuring feet, so a horse is the unit for counting horses. She succeeds in showing that much of what Aristotle says about time and measurement, especially in *Phys.*4.12, is written in the same spirit. Thus the use of a horse as a unit for counting horses is twice compared with the use of a unit for measuring time or motion.[9] Elsewhere measuring is connected with numbering in other ways,[10] and, in the following passage, with the claim that time itself is a number:

> Not only do we measure motion by time, but also time by motion, because each is marked off by the other. For time marks motion, being its number.[11]

This is one of the passages I mentioned in which time looks like a number more in the sense of that *with* which we count.

None the less, Annas scrupulously admits that, in *Phys.* 4.11, 219a10-b2, Aristotle's focus is on counting instantaneous nows, and that these, being sizeless, are not at all like units of measurement. What she concludes from this is that Aristotle has not fully unified his treatment of time. But as an alternative, she makes the subsidiary suggestion that what we are really counting, when we pick out instantaneous nows, are the units which fall between them, and not the nows themselves at all.[12]

I think it is possible to restore some unity to Aristotle's discussion, while accepting Annas' helpful comparison with *Metaph.*10.1-3, just so long as we take Aristotle to be comparing counting with measuring in a way that falls short of *identifying* them. For his concern with counting instantaneous nows is by no means incompatible with such a comparison. First, in counting nows we shall, in *favourable* circumstances, be counting intervening units of measure, and we can think of ourselves as doing *both*, not one rather than the other. Secondly, even when our interest is *only* in counting nows and *not* in counting intervening units, a major part of Aristotle's analogy with measuring survives.

---

[8] op. cit.   [9] *Phys.* 4.12, 220b18-24; 4.14, 223b13-15.
[10] *Phys.* 4.12, 221b14-16: to be in time is to have one's being measured by time, as to be in number is to have one's being measured by number.
[11] *Phys.* 4.12, 220b14-17.   [12] Annas, op. cit., 108-9.

For what he wants to say is that the unit we use for counting is not a mysterious Platonic Unit, but something of the same kind as what is being counted, in this instance a *now*. The analogy is that in measuring feet we use not a Platonic Unit, but a one-foot rule.

I should devote a moment to Annas' subsidiary alternative, according to which Aristotle does not really think of us as counting instantaneous entities at all. The main evidence against this has been given already in the reference to counting nows, and counting instantaneous stages, and in the reference in the definition to counting 'in respect of before and after'. But there is one additional piece of evidence in *Phys.*4.11, 220a10-21, that Aristotle is concerned with counting instantaneous entities. He here considers three views: time may be countable in the way in which a single mid-point is countable, or in the way in which two end-points are countable, or in the way in which *parts* are countable. This is the opportunity for him to plump for *parts*, if he believes that we do not really count instantaneous entities. But in fact he plumps for end-points. The passage runs as follows:

> For the point both holds the line together and divides it. For it is the beginning of one part and the end of the other. But when you take it in such a way as to treat the one point as two [sc. as beginning and end], you have to pause, if one and the same point is to be beginning and end; whereas the now is ever different, because of the motion of the moving body.

> Thus time is number not in the way that a single point can be numbered [sc. as two] because it is a beginning and an end, but rather in the way that the ends of a line can be numbered. Nor is it a number in the way that *parts* are numbered. One reason is that already stated: you will treat the mid-point as two, so that a pause will occur. But further it is clear that the now is not a *part* of time, any more than the dividing point is a part of motion, or the point a part of the line: it is two lines which are parts of a line.

Time as continuous, number as discrete

A by-product of this discussion is that we can lay to rest a bogey which has troubled commentators perhaps more than any other about Aristotle's definition of time as number. The criticism is made by Plotinus among others.[13] Time is continuous, but number (whole number) is discrete. How, then, can time be number? The interpretation just offered is that time is a countable aspect of change because we count periods, when we count instantaneous stages in change. What is discrete here is the stages which we choose to count. But time is not itself a discontinuous thing merely because we can divide it by picking out instantaneous stages. On the contrary, it is infinitely divisible, in the sense that we can divide it at stages as close together as we please, and its infinite divisibility is precisely a mark of its continuity.

Time dependent on consciousness in Aristotle

The connexion of time with *counting* leads Aristotle to ask whether there can

[13] Plotinus 3.7.9 (1-2); also Strato, ap. Simpl. *in Phys.* 789, 2-4. Alexander's solution, which will be discussed below, is that number is inserted into time only *in thought*.

be time without consciousness, and this brings me to the second topic for this chapter. Aristotle states his position in *Phys*.4.14, 223a21-9:

> One may be puzzled whether or not time would exist, if soul did not. For if there cannot exist someone to do the counting, then there cannot be anything countable, so that clearly there will not be number. But if nothing else is of a nature to engage in counting other than soul, and indeed the intellectual part of soul, then time cannot exist, if there is not soul, but at best the substratum of time (as, for example, if change can exist without soul, and before and after exist in change). But time is this before and after only in so far as they are *countable*.

Has Aristotle made a silly mistake here? Not by any means a silly one. The claim that number requires countability is certainly justified, when we remember that by 'number' he has said that in this context he will mean the countable aspect of something. Even if this had not been his meaning, we should have to recall that it is part of Aristotle's anti-Platonist programme to make number a function of our counting operations, as Julia Annas has brought out.[14] Where there may seem to be a simple error is in Aristotle's use of certain other modal expressions. For example, why, at the beginning, does he envisage that there '*cannot*' exist someone to do the counting, and that there *cannot* be anything countable? But I presume he only means it *cannot* be the case that, if soul does not exist, there exists someone to do the counting. And again it *cannot* be the case that, if no one exists to do the counting, there is anything countable. The really interesting modal term will then not be the 'cannot', but the 'countable' (*arithmēton*, countable, is distinguished from *arithmoumenon*, counted, at 4.11, 219b6-7). And the interesting thing about it is the claim that countability requires the existence of beings to do the counting.

Aristotle's view explained by reference to controversies on possibility

Aristotle's claim that without anyone to count, there is no countability and no time, has seemed too surprising to many commentators. And they have tried to reinterpret the text.[15] But I think that Aristotle's conclusion, although mistaken, is not surprising, if we turn our attention away from time to the notion of *possibility* and to such related modal motions as countability. These were the subject of very real difficulties which were of keen interest to Aristotle and his successors. He kept returning to the difficulties in a series of works, which I shall take in their probable chronological order.

First, in the *Categories*, if as is probable that is a genuine work of his, he argues on the opposite side of the case. If there were no animals in existence, there would still all the same be knowables (*epistēta*) and perceptibles (*aisthēta*), such as body, hot, sweet, or bitter.[16] Next in order probably comes

---

[14] Annas, op. cit.
[15] A list of interpretations is supplied by P.F. Conen, *Die Zeittheorie des Aristoteles*, Munich 1964, 156-69. In a clear-headed discussion, he cites Simplicius, H. Carteron, W. Bröcker, A.-J. Festugière and J. Moreau.   [16] *Cat*. 7, 7b33-8a6.

our passage in the *Physics*, in which he reverses position and says that nothing would be countable, if there were no souls. Hence only the substratum (*touto ho pote on estin ho chronos*) of time would exist. In agreement with that, he argues in the *Metaphysics* (at least on one interpretation[17]) that if there were no beings with soul, then perhaps (*isōs*) there would be no perceptibles (*aisthēta*), but there would still exist the substrata (*hupokeimena*) which give rise to perception. Finally, however, there comes a much more nuanced passage in the *de Anima*:[18] sound, flavour, colour and in general perceptibles in their active state exist only so long as perception is actually occurring. But in so far as they are only *potentially* active, they exist when perception is not actually going on.

Aristotle's successors continued the controversy on possibility and were divided among themselves. Thus the Aristotelian Boethus rejects Aristotle's treatment of countability. The countable can exist without someone to count, just as the perceptible can exist without a perceiver.[19] Alexander disagrees: a thing is not countable in the absence of soul.[20] Simplicius echoes this: without a soul that can count, we should lose time as the countable aspect of movement.[21] He would also like to correct Aristotle's statement in the *Categories*, in order to obtain a parallel verdict about perception and knowledge. At least, he seems to be saying that Aristotle goes astray in supposing that there might be perceptibles and knowables without any living beings. And he himself favours the opposite view that nothing is potentially perceptible, if there is no potential perception.[22]

The controversy becomes more intelligible in the light of a related one which began in the generation after Aristotle. According to Philo, the dialectician, it is possible for a piece of wood at the bottom of the ocean to be burned, and similarly a shell at the bottom of the sea is perceptible, in virtue of the 'bare fitness' (*psilē epitēdeiotēs*)[23] of the subject. Most Stoics disagreed, and said that a second condition must be fulfilled before there is a possibility: the absence of external obstacles.[24] If this Stoic view is applied to the case of counting, the absence of living beings to do the counting might well be considered an external obstacle, and there would not then be a possibility of counting.

---

[17] *Metaph.* 4.5, 1010b30-1011a2, if *aisthēta* means perceptibles rather than things perceived.

[18] *DA* 3.2, 426a15-26.

[19] Boethus ap. Simpl. *in Phys.* 766, 17-19 (= Themistius *in Phys.* 163, 5-7), and 759, 18-20.

[20] ap. Simpl. *in Phys.* 759,20-760,3; *de Temp.* (Théry) 95,11-12.

[21] *in Phys.* 760,33-761,5. He thinks it more serious, however, that without soul as the source of becoming, we should lose '*time itself*', since its being consists in a process of becoming.

[22] *in Cat.* 196,12; and 27-33. I owe this reference to Robert Sharples, who offers 'goes astray' as the translation of *apopheretai*.

[23] G.E.L. Owen interprets *epitēdeiotēs* as having an almost opposite sense in later antiquity, and as standing for the absence of interfering factors (comments on Sambur123 in A.C. Crombie (ed.), *Scientific Change*, N.Y. 1963, 97). I am not sure, however, that it cannot be interpreted as still having the Philonian sense.

[24] I have cited the evidence, and argued for the Stoics intending a second condition for possibility in *Necessity, Cause and Blame*, 78-9. Philonian examples of things in the sea being perceptible or combustible are supplied by Philoponus *in An. Pr.* 169,20; Alexander *in An. Pr.* 184,12ff.; Simpl. *in Cat.* 196,1.

The ancient sources do not distinguish clearly between countability and the possibility of counting. But these probably need to be distinguished. For we ought to ask whether countability requires a possibility of counting. And we may get different results with different predicates, for example, with combustible and visible. A piece of wood hundreds of feet underground might still be thought of as combustible, even though the intervening earth removed the possibility of burning it. On the other hand, it would not in most contexts be thought of as visible: the intervening earth would be an external obstacle which removed *both* the possibility of seeing *and* the visibility. What now about the non-existence of any living viewers? This might be thought of as removing the possibility of viewing, but not surely the visibility.

However, it is not only with different predicates, but even with different contexts, that we get different results. This emerges from a discussion in the Middle Platonist, Taurus.[25] He says that one can (after all) call a body at the centre of the earth visible, since it is of the same genus as things that are visible. And he is surely right; for if the *context* were concerned with a contrast between things perceptible to the senses and things apprehended only by the mind, then it would be legitimate to class a rock at the centre of the earth as visible. This is not to deny that it would be illegitimate, if the context were instead concerned with what to see on a sight-seeing tour.

Taurus is recognising the Philonian ('bare fitness') account of visibility, although he does not suggest that it is the only usage. Alexander, however, takes the other side. It is not possible for a pebble to be seen, he says, when it is in the depths of the sea, or for chaff to be burned, if it has been ground into atoms.[26] It may have been easier for Alexander to reach this conclusion, just because he expresses it in terms of the possibility of being seen, rather than, like Taurus, in terms of visibility.

There is another controversy perhaps closer to the one we started with. According to Aristotle (*Cael* 1.9, 279a11ff), there cannot come to be (*out' enchōrei genesthai*) any bodily mass beyond the edge of the heaven, and he infers that there is not anything out there in which it is possible (*dunaton*) for there to be body, and so there is not any extra-cosmic place or void. In English, we might distinguish the question whether there is anything out there in which it is *possible* for there to be body from the question whether there is anything out there *capable* of receiving body. And so we can sympathise with the Stoic retort expressed by Cleomedes (*de Motu Circulari Corporum Caelestium*, ch. 1): Aristotle might as well say that because water cannot (*oukh hoion te*) exist in the desert, there cannot be a vessel there capable (*dunamenon*) of receiving water. Aristotle is obliged to treat the capacity to receive body like countability as something which is missing, when the corresponding possibility (of body being received or of a count taking place) is missing.

My point has been that it is by no means a simple and obvious question when something is countable, visible, combustible or receptive. I believe that Aristotle has made a mistake in supposing that, where there exists no one to

---

[25] ap. Philoponum *Aet.* (Rabe), 146, 10-13.

[26] Alexander (?) *Quaestiones* 1.18, p.31,11; *in An. Pr.* 184,12ff. I am grateful to Robert Sharples for the first reference.

do the counting, there is no countability. But it is not an obvious or a silly mistake, and not unexpected enough, to require us to look for a different interpretation of his words. It is an interesting result, if I am right, that an idealist view of time, or one which, if not strictly idealist, at least makes time depend on consciousness, should arise out of something so different as considerations about *possibility*.

I shall not comment on the attempts to reinterpret Aristotle's words, except to say one thing. No escape from the obvious interpretation is provided by the concession which he entertains at the end of the quotation:

> time cannot exist, if there is not soul, but at best the substratum of time.

For even if soul were not required for the existence of the substratum (the before and after in change), he still would not concede that time could exist without a soul to engage in counting. All that could exist would be the *substratum*, that is, the subject matter to which counting is applied; and this substratum is not itself a kind of time.

Admittedly, there is something odd about Aristotle's entertaining even this possibility. For although the before and after in change is not a kind of time, it might reasonably be objected that it implies the existence of time. For one thing, 'before' and 'after' would, on most views, be taken as temporal terms; but Aristotle, it will be recalled, has insisted that they are primarily spatial. For another thing, the idea of change implies *time*, as Aristotle himself would normally agree. Since he holds that there cannot be *time* without soul, he ought not then to entertain it, even as a hypothesis, that there might be *change* without soul.

In fact, Aristotle's remark about the substratum of time concedes almost nothing. For he expresses doubt about the substratum existing without soul, when he goes on to say: 'as, for example, if change can exist without soul.' The 'if' is not fulfilled in Aristotle's view. For change in the universe requires motion, which in turn requires an unmoved mover or movers, who must act as final causes, inspiring the *souls* of the spheres which carry round the stars.[27] Alexander and Simplicius offer this as an independent reason for saying that time (since it requires change) requires soul.[28] And no doubt Plato would agree in spirit, since he makes time depend on celestial motion[29] and motion on soul.[30]

The philosophical issue

Although Aristotle's reason is not likely to appeal to us, are there any reasons why we should accept the dependence of time on consciousness? I do not know of any good ones. Certainly, there cannot be a straightforward connexion if we believe that there was time *before* any consciousness emerged.

---

[27] *Phys.* 8. elaborated in *Metaph.* 12.7-10.
[28] Alexander *de Temp.* (Théry) 95,12-15; Simpl. *in Phys.* 760,11-26.
[29] *Tim.* 37E1-39D1.
[30] *Phdr.* 248C5-246A2; *Laws* 10.891E4-899D3.

But I will just mention two reasons which might make someone believe that there is a dependence, since these two reasons emerge out of issues that have already been discussed.

One suggestion would be that time-words like 'past', 'present' and 'future' are all *egocentric* in a way which I have explained. If, as I have maintained, the existence of time involves that of past, present and future, must it not then involve the existence of conscious beings? I have already (in Chapter 6) given my answer to this, which is 'no'. In spite of the egocentricity, we can intelligibly imagine that there might have been no conscious beings throughout the whole of the past, present or future.

Another argument might be constructed out of what I said in that chapter about measuring. For this implies that there can be no one answer to the question whether the battle of Britain was longer than the battle of El Alamein. Perhaps in hours it was, but in 'clours' it was not. In other words, judgments of comparative length, if expressed in terms of units of measurement (hours, clours), all presuppose the selection by conscious beings of a standard of measurement (celestial movements, cloud passages). Does this not mean that time requires the existence of conscious beings to do the selecting?

It does not, for at least two reasons. First, judgments of comparative length, couched in terms of units of measurement, constitute only a small fraction of our talk about time. The existence of such units is not essential to time's existence. Secondly, we can in any case imagine that there might have been no conscious beings, and still describe the imaginary situation in terms of hours and days. We should simply and legitimately be applying to the imaginary situation the units of measurement which in the actual situation we have obtained by conscious selection.

Other thinkers

What other ancient philosophers made time dependent on consciousness? I have already commented that Plato did so in effect, by making time depend on motion, and motion ultimately on soul, although this dependency of time on soul was not a conclusion to which he himself drew attention. Plotinus made the connexion explicit when (as we shall see in Chapter 10) he defined time as the life of the soul. The soul here is the soul of the universe, so he is not making time depend on the conscious activity of individual persons. Admittedly, he says (3.7.13 (65-9)) that time is in every soul, and hence in us, but he is not thinking of us as individuals, since he adds that all souls are one. And in 3.7.9 (78-84), he explicitly insists that, although the origin of time lies in soul, this is not because time depends on our measuring operations.

The view which connects time to human consciousness even more closely than does Aristotle's is the view discussed in Chapter 2 by which Augustine was tempted. He thought, it will be recalled, that to escape the Aristotelian paradox, we might have to say that past and future were the mental states of memory and expectation. J.F. Callahan has located an anticipation of Augustine's idea in the Cappadocian Church Father, Gregory of Nyssa, who died five years before Augustine wrote the *Confessions*.[31] But the evidence does not strike me as convincing. In Callahan's best passage, Gregory is talking of the passage of time, of which he says:

> These effects (*pathē*) are peculiar to created things, whose life is divided in the direction of expectation and memory, according to the division of time.

But neither this passage, nor the others cited, appear to me to be *identifying* past and future with memory and expectation.

There are other philosophers, however, who might be interpreted as making time depend on consciousness. For example, Epicurus, in certain fragments of his lost work *On Nature*,[32] says that time is an appearance (*phantasia*), and a concomitant (*sumbebēkos*) of an appearance. For first, we have an appearance of days and nights, and then, in accordance with that, we conceive some length connected with days and nights which is capable of measuring all motion. Sextus Empiricus similarly attributes to the followers of Epicurus and Democritus the view that time is a *phantasma* (a mental image, perhaps), resembling day and night.[33]

I have already remarked that Aristotelians were very much divided on the relation of time to consciousness. Two of them, Critolaus (*c*. 190-155 B.C.) and Antiphon (or Antiphanes), are said to have held that time is not a reality (*hupostasis*), but a concept (*noēma*) or a measure (*metron*).[34] Two others, however, disagreed. Themistius (*c*. A.D. 317-388) criticises Aristotle for apparently conceding that time is merely a concept in our minds (*ennoia monon tēs hēmeteras psuchēs*), having no reality of its own (*phusis oikeia*).[35] And Boethus (first century B.C.), complains against Aristotle that a measure has no existence in reality (*hupo tēs phuseōs*), since measuring and counting are our own activities. What Aristotle should have acknowledged, he says, is that the countable can exist even without someone to count, just as the perceptible can exist without a perceiver. Time, on this view, is not a measure, although it can be counted and measured.[36]

Alexander (fl. *c*. A.D. 205) has the most interesting of the views put forward within the Aristotelian school. First of all, he endorses Aristotle's claim that there would be no time without souls to engage in counting. Then he adds the extra argument that time requires souls equally in order to cause

---

[31] J.F. Callahan, 'Gregory of Nyssa and the psychological view of time', *Atti del XII Congresso Internazional di Filosofia*, vol. 11, Florence 1960, 59-66. The translated passage is from *contra Eunom*. 1.370-2 (PG45, col. 368A) Other passages cited are *contra Eunom*. 12.459, PG45, 1064C-D; *de Beatitudinibus* 4, PG44, 1245B; *de Mortuis*, PG46, 520D-521B.

[32] Herculanean Papyrus 1413, edited by R. Cantarella and G. Arrighetti, in *CE* 2, 1972, 5-46 (5 I; 9 I; 9 VII).

[33] *M* 10.181-8.

[34] Stobaeus *Eclogae* 1.8, Diels, p. 318.

[35] Simpl. *in Phys.* 766,13-17 (= Themist. *in Phys.* 163, 1-5).

[36] Simpl. *in Phys.* 766,17-19 (= Themist. *in Phys.* 163,5-7), and 759,18-20. Boethus' lost discussion of time has been rediscovered by Pamela Huby, who presented her findings at the Institute of Classical Studies in January 1980, and has now published them in 'An excerpt from Boethus of Sidon's commentary on the *Categories*?' *CQ* 31, 1981, 398-409.

motion.[37] Finally, he further offers some reasons for claiming that the divisions which we ascribe to time, when we talk of instants, years, months, days, 'is', 'will be', 'before' or 'after', are simply added by us *in thought*. For time is really unitary, and this is shown by two considerations adduced in the extract from Alexander's *de Tempore* translated above in Chapter 2. First, we need to explain how time can be a number, and yet continuous. The answer is that numerosity is introduced only in thought, and so does not interfere with continuity.[38] Secondly, motion can be divided by rests, but time cannot in the same way be divided by intervals which are *not* times. Hence time is not divisible in actuality, but only in thought.[39] Simplicius reveals yet a third reason which motivates Alexander. It crops up in the interpretation of Aristotle's claim, already discussed, that the 'now' divides time only potentially, not in the way that a point, which stands still, divides a line.[40] Alexander understands Aristotle to mean that because of the *flow* of time, an instant cannot divide time in the sense of displaying its parts on either side standing still in separate positions. He infers that the instant divides only in thought. Indeed, in the *de Tempore* his claim is that it *exists* only in thought.

It has already been noticed, especially in Chapter 5, how much these claims of Alexander influenced the late Neoplatonists, Damascius and Simplicius. Simplicius repeats that there would be no time without souls, and he gives both Alexander's reasons, that souls are needed for counting and for causing motion.[41] He also finds in his master Damascius, and himself endorses, the claims that time in itself is unitary, that instants divide time only in thought, and that, because of the flow, they cannot display temporal parts side by side.[42] Damascius adds in the same passage the further consideration that the division into past, present and future is relative to us, different people having a different present.

I have given Alexander's own reasons for saying that the divisions of time exist only in thought, but in case it is hard to muster imaginative sympathy, let me quote a somewhat analogous idea from a modern author, Thomas Mann:

> Time has no divisions to mark its passage, there is never a thunderstorm or blare of trumpets to announce the beginning of a new month or year. Even when a new century begins, it is only we mortals who ring bells and fire off pistols.[43]

There were other claims in antiquity that the divisions of time were

---

[37] *de Temp.* (Théry) 95,11-15; and ap. Simpl. *in Phys.* 759,20-760,3. I am indebted for most of the references in this paragraph to the paper of Robert Sharples, written in collaboration with F. Zimmermann, 'Alexander of Aphrodisias, *On Time*'.

[38] *de Temp.* 94,23-4.

[39] *de Temp.* 94,37-43.

[40] *in Phys.* 748,21-30, reporting Alexander's interpretation of Aristotle, *Phys.* 4.13, 222a12-14. What I translated and discussed above was the *related* passage, *Phys.* 4.11, 220a11-14.

[41] *in Phys.* 760,11-26; 760,33-761,5.

[42] *in Phys.* 797, 29-34; 798,6-26.

[43] Thomas Mann, *The Magic Mountain*, translated by Lowe-Porter 1928, from the German of 1924.

subjective, but they had quite different motivations from Alexander's. Philo, for example, describes the division into months, years and other periods as 'opinions of men' (*dogmata anthrōpōn*), but his point, according to one persuasive interpretation, is that these divisions are geocentric, rather than that they are in the mind. For he holds that the shadow which constitues day and night is not found in the supralunary region.[44] Exactly the same point is made later by Plotinus.[45] Tatian, in his attack on the Greek philosophers, denies that they are entitled to divide (*merizein*) things up into a past, present and future (the passage is translated in Chapter 2 above). For how can the future become past? The illusion that it does is due to the movement of men.[46]

Returning from the divisions of time to time itself, we find that its subjectivity continued to be discussed in Islam, where certain theologians argued, like Damascius, that the words 'was' and 'will be' involved a relation to us. A report of this is translated above in Chapter 3. Averroes took them to mean that the past and future are not real things in themselves, and do not possess existence outside the soul, but are only constructs of the soul. His own response is to concede that time is something which the soul constructs within change. In this he is perhaps influenced by Aristotle's account of time as the countable aspect of movement.[47]

In modern philosophy, the view that time is somehow dependent on consciousness still reappears in very diverse forms, for example, in Berkeley, in Kant, and in Bergson.[48]

[44] Philo *de Fuga et Inventione* 56 f., as interpreted by Whittaker, op. cit. second study, in the light of *de Iosepho* 146.

[45] 4.4.7 (9-12).

[46] Tat. *ad Gr.* 26 (PG6, 861).

[47] Averroes *Tahāfut al-Tahāfut* (Bouyges) pp. 72-4, tr. van den Bergh pp. 41-2.

[48] Berkeley, *Principles of Human Knowledge* (1710), esp. 1 §94; Kant, *Critique of Pure Reason* (ed. A 1781; ed. B 1787) esp. ch. on the antinomies; H. Bergson, *Essai sur les données immédiates de la conscience*, Paris 1889, translated as *Time and Free Will*, London, 1910, esp. pp. 98-112; 224; 227.

CHAPTER EIGHT

Is Eternity Timelessness?

In the autumn of 1853, a professor of my college, F.D. Maurice, was dismissed for taking a view of eternity not unlike that which I shall try to defend here. The full scholarly controversy with the Principal of the College is available in print, and it makes interesting and impressive reading.[1] The Principal understood eternity to mean an endless duration distinguished from time through involving no 'measures'. Maurice protested that duration is a temporal notion, and that eternity involves no duration. So far my view will seem close to his. But there are differences, in that he was primarily concerned to deny that a loving God could be taken to have threatened punishment that was eternal in the sense of unending duration. He therefore understood eternal punishment as punishment with a certain quality, namely, that of cutting us off from knowledge of God, and he based this interpretation on John 17:3. This in turn involves a further difference, that he was discussing the eternity not only of God, but also of creatures, and not only of the saved, but also of the damned. These, according to Aquinas, would be three different kinds of eternity.[2] A final difference is that, unlike myself, he was concerned with the New Testament, and thereafter almost exclusively with Protestant authors. He recognised that the view of the Apostles and the Fathers was not uniform, although he insisted that the Creeds, Prayers and Articles were compatible with his interpretation.

I shall be talking only of a tradition within Greek philosophy and of its influence on one strand in the Judaeo-Christian conception of God. I shall find this influence as early as Philo Judaeus in the first century A.D., but I shall not attempt to decide whether it can also be found in the New Testament.

The concept of eternity appears very early in Western thought in one of the first Presocratic philosophers, Parmenides of Elea (born c. 515 B.C.). It is taken up by Plato and the Platonists and this is the route by which it comes to influence Christian thought. Eternity is standardly contrasted with time

---

[1] F.D. Maurice, *Theological Essays*, 2nd ed., Cambridge 1853 (the concluding essay was the object of criticism); R.W. Jelf, *Grounds for Laying Before the Council of King's College, London, Certain Statements Contained in a Recent Publication Entitled 'Theological Essays, by The Rev. F.D. Maurice, M.A., Professor of Divinity in King's College'* (a set of nine letters), 2nd ed., Oxford and London 1853; F.D. Maurice, *The Word 'Eternal' And The Punishment of the Wicked*, Cambridge 1853 (a final letter). Restitution was made to Maurice subsequently by a whole series of commemorative honours.

[2] *ST* 1a, q.10, a.3.

and is said by the Christians I shall be discussing to be a characteristic of God. To the question raised in the chapter heading, whether eternity is timelessness, I shall answer with a qualified 'yes', after explaining what I mean. But the case will need arguing, for there are plenty of rival interpretations which have been ably supported.

Parmenides

In his poem *The Way of Truth*, Parmenides discusses an unspecified subject 'it'. I favour the suggestion that the subject is whatever can be spoken and thought of, or alternatively whatever we inquire into.[3] The crucial sentence for our purposes comes in fr.8 DK, 1.5 and the first half of 6:

> Nor was it ever (*pot'*), nor will it be, since it now is,
> all together, one, continuous.

It is the denial of 'was' and 'will be' which expresses some concept of eternity – but what concept?

Eight interpretations

I shall distinguish eight main interpretations. One is that Parmenides means only to deny creation and destruction,[4] and that his aim in doing so is to show that his subject is *everlasting* and is not already over or still to come.[5] A second interpretation is that Parmenides objects to the flowing concepts of McTaggart's A-series, discussed in Chapter 3 above. On this view, he would not mind talking of earlier and later but rejects the use of the tenses 'was' and 'will be', because these belong to the flowing A-series. This interpretation converts eternity in effect into McTaggart's static B-series.[6] Some modern philosophers would certainly reject the flowing A-series as representing some kind of illusion. A common feature of these first two interpretations is that Parmenides would be allowing *duration* to his subject. A third view, the one I shall advocate myself, is that Parmenides is groping towards the idea that his subject exists, but not at any time, neither at any point, nor over any period of time. I shall express this by saying that the subject is *timeless*. Such an idea is sometimes applied by modern philosophers to numbers, or to truth, which are thought of as existing, but not as existing in time. But it has been

---

[3] The first is the suggestion of G.E.L. Owen, the second that of Jonathan Barnes. G.E.L. Owen, 'Eleatic questions', *CQ* n.s.10, 1960, 84-102 (repr. in D.J. Furley and R.E. Allen, *Studies in Presocratic Philosophy* vol.2, London, 1975), and 'Plato and Parmenides on the timeless present', *Monist* 50, 1966, 317-40 (repr. in A.P.D. Mourelatos (ed.) *The Pre-Socratics*, Garden City N.Y., 1974). Jonathan Barnes, *The Presocratic Philosophers*, London 1979, vol.1, 163.

[4] H. Fränkel, *Wege und Formen frühgriechischen Denkens*,[2] Munich 1970, p.191, n.1; L. Taran, *Parmenides*, Princeton 1965, pp. 175-88.

[5] Malcolm Schofield, 'Did Parmenides discover eternity?' *AGP* 52, 1970, 113-35 (see p. 120); Denis O'Brien, 'Temps et intemporalité chez Parménide', *EP*, 1980, 257-72.

[6] I am grateful to Norman Kretzmann for proposing this interpretation to me for consideration, although it is not the one favoured by himself and Eleonore Stump in their current work on the subject ('Eternity', *JP* 88, 1981, 429-58).

objected that Parmenides uses the words 'now is' and that these imply some kind of existence in time.[7] A fourth view, therefore, denies that Parmenides is making his subject timeless but accepts that he is doing more than making it everlasting, or freeing it from the possibility of creation and destruction. He is postulating a very special sort of time, an enduring present, unaccompanied by past or future.[8] This is one of several interpretations which ascribe to Parmenides a very special sort of temporal now. A fifth takes it that the reference is to a single instant – that is all the time there is.[9] A sixth assimilates Parmenides' claim to that of the later Aristotelian paradox, discussed in Chapters 1 to 5 above. According to this, only a sizeless present exists, since the past exists no more and the future does not exist yet.[10] A seventh interpretation by John Whittaker cuts across the 'timeless' and the 'special present' interpretations,[11] while an eighth differs from them all and takes Parmenides to be referring to *cycles* of time.[12]

The 'special present' interpretations

The three interpretations which stress 'now is', and refer it to a special temporal present, must face grave objections. Two of them, the 'enduring present' and the 'single instant' interpretations, have the disadvantage that they give to Parmenides a view which looks incoherent from the start. For it is hard to see how a *temporal* present can exist without past or future or how an instant can exist which does not limit a preceding or following period. The fourth, or enduring present, interpretation faces extra problems, if the notion of *endurance* is taken seriously. For if the present *endures*, how can eternity be 'all together', as Parmenides says in the quoted passage, fr.8, l.5? And how are we to understand *endurance* without a past or future? It may be retorted that the rival 'timeless' interpretation also suffers from incoherence, but I believe the case is entirely different. For I shall argue in the next chapter that, for certain entities, for example, for certain universals, a 'timeless' interpretation is defensible, and that for others it is seductive, even though it can be shown by argument to be wrong. The charge against the enduring present is the opposite: it appears so baffling that argument would be needed to establish not its incoherence, but its coherence.

There is another weakness which especially affects the fifth, or 'single

[7] Barnes, op. cit., vol.1, 192; and Myles Burnyeat in unpublished work. The point stems from Owen, 'Plato and Parmenides on the timeless present'. Owen's view is that Parmenides does not succeed in detensing the verb 'is', as the third view would like him to, but incoherently wishes to allow it its familiar use by connecting it with 'now' in l.5 and with 'remaining fixed' in ll.29-30, while none the less cutting it off from its connexions with 'was' and 'will be'.

[8] This is in effect the incoherent position in which Parmenides ends up, on Owen's interpretation, after having tried to take the *different* line which abolishes temporal distinctions. It is also the interpretation taken by Myles Burnyeat in unpublished work which he has been kind enough to show me.

[9] Barnes, op. cit., 193-4.

[10] This idea is suggested by Ammonius *in Int.* 133, 16-21 (translated in Chapter 16), and hinted at by Stump and Kretzmann, op. cit., 444.

[11] Whittaker, *God, Time, Being*, 1-66.

[12] William Kneale, 'Time and eternity in theology', *PAS* 61, 1960-1, 90-2.

instant', interpretation. It would be more logical for a thinker who had whittled time down as far as an instant to take the next step and abolish time altogether. This emerges particularly clearly from Jonathan Barnes' account of why a Parmenidean might be tempted by the 'single instant' view of eternity. He might reason that if time requires change, then, since the universe of Parmenides contains no change, it should contain no time periods.[13] But the right conclusion to draw, in my view, would be that, lacking time periods, it would also lack any instant. Philosophical considerations here would actually favour the 'timeless' interpretation.

There are overwhelming objections to the sixth, or 'past no more, future not yet', interpretation. For the paradox (which is, in any case, not mentioned before Aristotle over a century later), does not deny that the past *was* and that the future *will be* (1.5). It denies only that they *are*. Moreover, the paradox is far from claiming that everything exists 'all together' (1.5), and that there are no divisions (1.22). On the contrary, it thrives on the division between past and future and relegates everything to one or the other.

When we come to Plato and Plotinus, it will turn out that there are yet further obstacles to the 'special present' interpretations. But I must now consider the main case on the other side, in favour of these interpretations. It is that Parmenides uses the words 'now is', and it is alleged that these must be construed temporally. It do not see that this is so. On the contrary, by divorcing his 'now is' from 'was' and 'will be', Parmenides makes it very hard to construe the phrase temporally. How can the 'is' be tensed or the 'now' be temporal after that?

It will in any case be unwise to put too great a stress on the words 'now is', when we come to the later tradition. For not all of Parmenides' successors borrow the notion of 'now' or of 'present' from him. Plato does not; Thomas Aquinas does not. It so happens that Plotinus does, and consequently so do many of his successors. But he does not make a great point of this divergence from Plato, and his talk of presentness does not feature in the actual definition of eternity, but only in the general discussion of it.

But if I say that the 'now is' is not to be construed temporally, how is it to be construed?[14] That is a fair question. I would suggest that the 'now is' is to be taken rather closely with the 'all together' which follows. Parmenides is trying to express the idea that his subject exists not stage after stage, but all together. On this interpretation the function of 'now is' is not to keep a last toe-hold on the idea of time, but to express the idea of not being extended. And the alternative to being extended is not, of course, occupying a point. Once this non-temporal use of 'now' is understood, we can no longer allow significance to be attached to the mere phrase 'eternal now', as if it must refer to some instant or enduring present. We shall have to inspect the individual occurrences of any such phrase, in order to see how best to construe them.

I would cite in support of my way of taking 'now is' some of the later

[13] op. cit., 194.

[14] The interpretations of Barnes and myself provide two fairly clear cut alternatives, but I am not sure how 'now is' is construed in certain other contributions to the subject. I do not understand, for example, the notion of presentness introduced by Stump and Kretzmann, op. cit., 434.

thinkers, although, admittedly, they are as often commenting on Plato as on Parmenides. Many of them treat the 'now' or the 'is' as non-temporal. One example comes in a passage of Boethius (*c.* A.D. 480-525), who more than almost anyone shaped the Christian concept of eternity.[15] In it he insists that the 'now' of eternity is *different* from the 'now' of time. The same treatment is given to the verb 'is' by Boethius' older contemporary Ammonius (born 435-45, died 517-26), who was the leading Neoplatonist in Alexandria.[16] The following is the relevant extract from his commentary on Aristotle's *de Interpretatione*:

> For we shall not allow anyone to say that the knowledge of the gods runs along with the flow of things, nor that anything with them is past or future. Nor shall we allow that 'was' and 'will be' are used among them, words which, as we have heard from Plato's *Timaeus*, signify some change. Only 'is' is used, and that not the 'is' which is counted in with 'was' and 'will be' and contrasted with them, but the 'is' which is conceived before the level of temporal images, and which signifies the gods' undeviating unchangeability. This is the 'is' which the great Parmenides also says belongs to all that is thinkable, when he says 'for it was not, nor will it be. All together, it *is* alone'.

Also striking are the slightly earlier comments on Plato's use of 'is' by Proclus (A.D. 412-85), who had been head of the Athenian Neoplatonist School, and who taught Ammonius and at least indirectly influenced Boethius:

> You see that he [Plato] here applies to the world the three tenses and does not withhold 'is'. So it is clear that when he above applied 'is' to eternal (*diaiōnios*) being and not to creation, it was the 'is' which is removed from all temporal extension (*chronikē paratasis*), and which stands still in the eternal (*aiōnios*) now, that he allotted to the intelligible beings. For he allows that the 'is' which is akin to 'was' and 'will be' belongs also to sensible beings thanks to their participation in the being which really is.[17]

I can thus claim that the 'now' and 'is' of eternity were construed often enough (and with explicit reference to Parmenides, in the case of Ammonius) as *non*-temporal, contrary to the fourth, fifth and sixth interpretations of Parmenides.

There is one last factor which may be thought to favour the 'special present' interpretations. I shall refer later to a peculiar psychological experience of absorption in the present. Might not such an experience have commended the idea that there is only a temporal present, without any future or past? I would say first that it *ought* not to commend this idea, for in fact the experience of absorption in the present presupposes the existence of time in the ordinary sense, with its past and future, even if the past and future are not attended to during the experience. To obtain a denial of past

---

[15] *de Trinitate*, 4, ll.64-77.

[16] *in Int.* 136,17-25: on whether Boethius learnt from Ammonius, or whether they were independently indebted to Proclus, see below.

[17] *in Tim.* (Diehl) 3, p.51, ll.15-21, commenting on Pl. *Tim.* 38 B6-C3.

and future, one would need as a supplementary premise some strong metaphysical thesis such as the thesis that only the experienced is real. And even then, one would not have overcome the charges of incoherence, levelled above, against the idea of a present without past or future. In any case, there is no appeal to this kind of psychological experience in the texts of Parmenides and Plato which introduce eternity. Later authors probably were influenced by psychological experiences, but as often by an experience of timelessness as by an absorption in the present.

The 'everlasting' interpretation

I can now turn to the first two interpretations, which (like the fourth) make Parmenides allow *duration*. Their idea that eternal things exist at *all* times is almost the opposite of the idea that they are timeless, i.e. that they exist, but at *no* times. It might seem that the two ideas share the assumption that eternal things lack beginning or end. But even this need not be so, for some authors, later than Parmenides, maintain that time itself has a beginning or end, in which case so also would everlasting things.

Malcolm Schofield has made the best attempt, I believe, to portray Parmenides as ascribing everlasting duration to his subject, although the case has also been ably argued by others.[18] In effect, Schofield understands Parmenides' words in ll.5 to 6, 'nor was it ever (*pot'*), nor will it be, since it now is, all together, one, continuous', as meaning that it is not past nor future, since it is present. He starts by reporting the attempt of Fraenkel and Taran to construe *pot'* as meaning not 'ever', but 'once', and this is also the interpretation of O'Brien. Parmenides' lines would then mean: 'it is not the case that it was *once* (*pot'*) in existence (but is no longer), or that it will one day exist (but does not yet), since ... ' But Schofield declines to take this quick route to his conclusion, since, like Owen before him, he holds that the word *pote* could not ordinarily be understood this way in Greek, and he provides strong linguistic evidence to prove the point. So instead of taking *pot'* to *imply* the extra parenthetical words ('but is no longer', etc.), he simply adds them alongside as *tacitly* understood, while allowing *pot'* to retain its conventional meaning.[19] The resulting paraphrase runs: 'it is not the case that it ever (*pot'*) was in existence (but is no longer) nor that it ever will be (but is not yet), ...'[20] Parmenides' point will be that his subject has not already perished, and is not waiting to be created, but is, on the contrary, everlasting.

An advantage of this interpretation is that it immediately explains why the following lines (6 to 21) concentrate so heavily on disproving creation and destruction. A major difficulty, however, is that in asking us to understand and supply the parenthetical words Parmenides would be asking far more than could reasonably be expected. It is not encouraging to consider the proferred parallel, 'What do you mean, Jones *was* a good vicar? He is *now*.'

---

[18] Schofield, op. cit; cf. O'Brien, op. cit.

[19] Owen 'Plato and Parmenides on the timeless present', 320-1; Schofield, op. cit., 122-3.

[20] op. cit., 127.

For this is not a genuine parallel so long as it omits the 'ever'. I find it hard to escape the conclusion that, 'nor was it ever, nor will it be' means what it seems to mean, namely, that the subject has no past or future.

An even more serious obstacle to Schofield's line of interpretation has been suggested by Alexander Mourelatos and Myles Burnyeat.[21] It is that the crucial l.5, on Schofield's rendering, denies only *past*, not *future*, ceasing, and only *future*, not *past*, beginning. It therefore fails in the very task which Schofield wanted it to perform – that of asserting everlastingness. O'Brien considers something like this objection, and admits that, on his construction, Parmenides does not initially rule out future ceasing or past beginning. The initial statement therefore rules out less than the ensuing argument. Even in the ensuing argument, O'Brien acknowledges, future ceasing is ruled out at best *implicitly*. There are yet other obstacles to the representation of Parmenides' subject as everlasting. How can he say that it exists 'all together' (l.5) or 'without divisions' (l.22), if it is spread out through the whole of time? Schofield candidly acknowledges that a denial of past and future would be historically appropriate for Parmenides, since it would place him, with some neat verbal echoes, in a continuous tradition of controversy: he would be answering Heraclitus, answered by Melissus and Anaxagoras, supported by Plato.[22]

Why, then, should the 'timeless' interpretation any longer be resisted? Schofield draws attention to two obstacles. First, developing a point of Taran's,[23] he asks whether the following lines, with their emphasis on excluding creation and destruction, can be shown to be relevant to a claim of timelessness, as the words 'since' (*epei*, l.5), and 'for' (*gar*, l.6) imply they are meant to be. This challenge is a fair one, and I shall offer an answer at the end of this chapter.

The second difficulty, which is also strongly emphasised by Fraenkel, Taran and O'Brien, is that various of Parmenides' phrases may seem to imply duration.[24] Thus he says that his subject 'remains' and 'lies' 'fixed' (ll.29-30). However, when these expressions are put beside the ones which I shall cite in a moment, and which seem to *deny* duration, one set or the other must be reinterpreted, and there is an obvious reinterpretation for the words 'remains', 'lies', 'fixed'. For they provide a natural way of expressing the merely *negative* point, just announced in l.26, that there is no motion or change. There is no particular need to take them as implying duration. Much the same goes for other phrases which superficially suggest duration. Parmenides says that it 'exists without beginning or ceasing' (l.27), and that 'what is crowds on what is' (l.25). The latter could suggest distinct temporal stages; but, in fact, it looks again like part of an attempt to express the *negative* point, made in l.22, that there are no divisions (and hence no

---

[21] Alexander Mourelatos, *The Route of Parmenides*, New Haven Connecticut 1970, 106, published simultaneously with, not in response to, Schofield. Similarly Burnyeat in unpublished work.

[22] op. cit., 134-5. Schofield does not question that Plato was denying a past and future to his ideal Forms, but others do, as will be seen below.

[23] Schofield, op. cit., 119; 121; 125; 127; Taran, op. cit., 178.

[24] Schofield, op. cit., 128-9; Fraenkel, loc. cit.; Taran, op. cit., 177; 180-1; O'Brien, op. cit.

divisions into periods). As for the denial of beginning and ceasing, this is as appropriate to the timeless as to the everlasting. The word 'now' is stressed by Taran and O'Brien, but I have already stated the case for construing this non-temporally. Finally, we can in charity supply a suitably atemporal paraphrase, when Parmenides slips into temporal language at ll.36-7, and says, 'for nothing else is, *or will be*, besides what is'.

The 'timeless' interpretation

The case for the 'timeless' interpretation, at least in connexion with the later tradition, will not be completed until the next chapter, where I shall try to show why in certain contexts the notion of timelessness has a positive appeal. But I can say now that the case also rests on the many expressions in Parmenides and his successors which appear to deny duration and temporal distinctions. The main ones in Parmenides are 'nor was it ever, nor will it be', 'all together' (both in l.5) and 'without divisions' (l.22). Between them, these phrases count against the first or 'everlasting', the fourth or 'enduring present', and the second or 'B-series' interpretations.

The 'B-series' interpretation

This last interpretation, according to which Parmenides is referring to the static B-series, runs into the additional difficulty that there is not, so far as I know, any conscious formulation of a distinction between static and flowing time, before that of Iamblichus in the fourth century B.C. And even Iamblichus' distinction was argued in Chapter 3 to differ from that of McTaggart.

Whittaker's interpretation

John Whittaker's interpretation deserves separate discussion, both because it is so different from the others, and because it is part of an extensive and informative survey of the history of the concept of eternity up to the time of Plutarch.[25] He is interested in the idea of *non-durational* eternity, and so far it might sound as if his concept is exactly the same as the 'timeless' one which interests me. If I am not sure whether it is the same, this is because Whittaker does not stress the point which I shall be stressing in this chapter, that 'now' and 'always' are often given non-temporal senses. He simply reports that his concept of non-durational eternity 'involves the conception of the "now" as an indivisible point',[26] while the concept of the 'now' is combined with the idea of 'always'.[27] An example of this combination with 'always' is provided by Plutarch, who says of God: 'He has completed always in a single now.'[28] My own view of this is that Plutarch means something

---

[25] Besides his *God, Time, Being*, cited above, see his 'The eternity of the Platonic Forms', *Phronesis* 13, 1968, 131-44; and his 'Ammonius on the Delphic E', *CQ* n.s.19, 1969, 185-92.

[26] *God, Time, Being*, 25, n.3.

[27] ibid., 45-6.

[28] Plut. *On the E at Delphi* 393A-B.

like: the whole of God's life is telescoped together rather than spread out. Whether Whittaker takes it the same way I am not sure.

Whether or not Whittaker has in mind the same concept of eternity as myself, his thesis about its history is radically different. He suggests that Plutarch was the first person to embrace it, and, in direct opposition to what I should expect, he proposes that it would be easier to make sense of it when, as with Plutarch, it was viewed as the attribute of a conscious personal God, rather than of a Platonic Form. I shall argue in Chapters 9 and 11 that, on the contrary, the idea of timeless eternity can be made quite plausible in connexion with some of the Platonic Forms, whereas it requires some very special metaphysics in Plotinus to make it at all palatable in connexion with a conscious being.

Whittaker argues his case fully, and offers three reasons which, in his view, 'render it obvious that ... Parmenides cannot possibly have propounded the doctrine of non-durational eternity'.[29] This places an onus on me to explain, if I am not persuaded. The first reason, that the concept is abstruse, and so would have required Parmenides to offer his contemporaries more explanation, does not allow for the possibility that Parmenides himself was groping. The second reason is that the views of another Eleatic, Melissus, exclude non-durational eternity, and that the ancients do not record Melissus as disagreeing with Parmenides on this point, even though Aristotle picked out certain *other* disagreements between them. But I doubt if this silence is significant: some of the Neoplatonists do ascribe the idea of timeless eternity to Parmenides, and they would have been motivated to bring out agreements rather than disagreements among their forebears on this important anticipation of their own views. Thirdly, Whittaker argues, following Taran,[30] that Parmenides could have had only one reason for introducing the idea of non-durational eternity – the belief that mere duration (as opposed to measured time) implies change; yet Greek philosophers down to and including Aristotle had no inkling of such an idea. I would disagree with this reason on more than one count. First, I shall analyse Parmenides' argument at the end of this chapter, and will argue that he had *several* reasons for introducing the idea of non-durational eternity, all of them different from the one envisaged by Whittaker. Secondly, Myles Burnyeat has suggested to me an interpretation according to which Parmenides would be articulating the argument that Whittaker desiderates in fr.8, 1.19.[31] Although I have not made use of Burnyeat's interpretation, I do think that it is perfectly possible. Thirdly, I believe the idea that mere duration involves change is present in both Plato and Aristotle. For Plato says that 'was' and 'shall be' are processes (*kinēseis*), and that they involve the change of growing older (*Tim.*38A), while Aristotle offers the argument based on the 'sleepers in Sardinia', which I analysed in Chapter 6, in order to show that duration implies change.

---

[29] *God, Time, Being*, 16-17.

[30] Taran, op. cit., 181.

[31] Mere duration involves change, in that it involves the coming into being of a thing's *future* existence.

Those are the three reasons that Whittaker brings to bear on Parmenides, but there is a fourth prong in his general line of argument. For he makes the claim, noted in Chapters 6 and 7 above, that 'to the Greek mind time = *chronos* was essentially measured time',[32] and he uses this claim in order to argue that in authors before Plutarch denials of time should be taken as denials of *measured* time, rather than as denials of *duration*. I have already disagreed with the general claim about the Greek conception of *chronos* in Chapter 6. None the less, there is one aspect of what Whittaker says with which I would agree: we cannot judge, merely from a denial of time, which concept of eternity, if any, is being expressed. Plato and Philo in certain passages both treat time as a system of days, nights, months and years. So when they describe something as being outside time, it is one *possibility* that they mean no more than that it is outside this system of measurement. However, that is only *one* possibility; Whittaker would agree that we have to look at the details to decide. I shall look at some details for both Plato and Philo later in this chapter, and I shall argue that sometimes they do mean more than that.

I come now to Whittaker's positive thesis about Parmenides, which forms the main part of his study of that thinker.[33] It is based on a well documented account of disagreements in the Neoplatonist records of Parmenides' original words, although the disagreements focus on 1.6, rather than on the crucial 1.5, of fr.8. Whittaker thinks we cannot be content with any of the surviving readings. His warning is salutary for all interpreters. But the main reason for despair seems to me to have gone, once we reject Whittaker's argument that the Neoplatonists must have been wrong to ascribe to Parmenides a durationless conception.

The cyclical interpretation

I will finish with William Kneale's cyclical interpretation. Kneale points out that, according to some ancient sources, Parmenides was at one stage a member of the Pythagorean school. He draws attention to a view about time which will be considered in Chapter 12, and which is ascribed to certain Pythagoreans by Aristotle's pupil, Eudemus of Rhodes. On this view, everything will happen again in cycles, precisely as it is happening now. In that case, the universe, though finite, will have no beginning or end. Nor will it be appropriate to say that anything *was* so, for it is equally true that the same thing *will* be so. We therefore have a motive for banning the use of 'was' and 'will be', and even of 'is', if that is taken in a temporal sense. And there is also good reason, if this is how Parmenides is thinking, for him to describe his subject later in fr.8 (1.43) as 'like the bulk of a well-rounded sphere': it will indeed be cyclical. So far there is much that fits Parmenides' wording. But Kneale describes his idea as only a guess, and there is, I think, too much that does not fit. It would be hard, for example, to understand 1.5 as expressing the Pythagorean view, when it says 'since it now is, all together, one

---

[32] *God, Time, Being*, 25, n.5.
[33] ibid., first study, 1-32.

continuous'. This does not bring out the idea of a cycle, and I doubt if 'all together' is even compatible with what the Pythagoreans intend. Moreover, the supporting arguments which immediately follow, and which ban creation and destruction, do not offer the Pythagorean reason that a circle has no beginning or end.

I conclude provisionally that the 'timeless' interpretation fits Parmenides best, and I should now like to see what happened to the concept of eternity after Parmenides. To put it briefly, my suggestion will be that Plato clouded the issue by placing alongside the implications of timelessness more phrases implying everlasting duration than can conveniently be explained away. This made it necessary for Plotinus to make a decision and his decision was in favour of timelessness.

Plato

The best attempt I know to argue that Plato ascribes eternity to his Forms only in the sense of everlasting duration is that of John Whittaker, but he has others on his side.[34] On the other hand, Taran and Schofield who interpret Parmenides' conception of eternity in that way do not so interpret Plato. On the contrary, Taran regards Plato as the first atemporalist.[35]

In many dialogues, as Whittaker shows, Plato simply talks of the Forms and other entities as existing *always*, without any attempt to represent them as timeless. The two exceptional dialogues are the *Parmenides* and the *Timaeus*. The order of composition of these is still a matter of controversy, but merely for purposes of discussion, let us take the *Parmenides* first. In this dialogue Plato represents Parmenides as creating paradoxes about the One, that is, about the unitary subject to which alone he accords existence. In one passage (140E-142A), Parmenides is made to argue that his One is not in time (not *en chronōi*, 141A5; A6; C8; D5), and has no share of time (*oude chronou autōi metestin*, D5; *mēdamēi mēdenos metechei chronou* D1; E4), and that consequently it cannot exist. This conclusion is held to be wrong, but the error in the argument is not diagnosed. The reason for saying that the One is not in time is (141A-D) that whatever is in time is always becoming older than its (former) self, which means that its (former) self is becoming younger than it. At the same time it must be of the same age as itself. But none of these things can be true of the One.

These ideas are found also in the *Timaeus*, but in connexion with Plato's ideal Forms, not with Parmenides' One. The Forms are said to be incapable of growing older or younger. Yet this time the conclusion is not that they fail to exist but rather that being is precisely what we must ascribe to them,

---

[34] John Whittaker, 'The "eternity" of the Platonic Forms'. Earlier F.M. Cornford had interpreted eternity in Parmenides and Plato as a duration (*Plato's Cosmology*, London 1937, 98; 102), and W. von Leyden had given qualified support to this interpretation of Plato in 'Time, number and eternity in Plato and Aristotle', *PQ* 14, 1964, 35-52. A.H. Armstrong declares Whittaker's view possible in 'Eternity, life and movement in Plotinus' accounts of *nous*', *Le Néoplatonisme* (Report of the International Conference on Neoplatonism held at Royaumont, 9-13 June 1969), Paris 1971, 67-74.

[35] Taran, op. cit., 175.

rather than the various other appellations that we may be tempted to use. In the following passage (37C6-38C3), Plato, through the person of Timaeus, has just finished describing God's creation of the world as a living and moving body, modelled upon the ideal Forms.

> When the father who had generated it saw it moving and living (*zōn*), a shrine created for the everlasting (*aïdioi*) gods, he rejoiced and was well pleased, and planned to make it still more like its model. So as the model is a living organism (*zōon*) and everlasting (*aïdios*), he tried to make this universe an everlasting (*aïdios*) living organism as far as possible. Now the nature of the living organism is eternal (*aiōnios*), and it was not possible to confer this completely on the generated thing. But he planned to make a moving likeness of eternity (*aiōn*), as it were, and while arranging the heavens, he simultaneously made a likeness of eternity (*aiōn*) which stays still in unity, a copy which moves according to number and is eternal (*aiōnios*), namely, that which we have called time. For there were no days, nights, months and years before the heavens came into being, but he devised that they should come into being then just when the heavens were assembled. These are all parts of time, and 'was' and 'shall be' are forms of time which have come into being. We are wrong when we apply them, without realising it, to the everlasting (*aïdios*) being. For we say that it was, is and shall be. But really only 'is' belongs properly to it. 'Was' and 'shall be' are appropriately said of becoming (*genesis*) which proceeds in time, for they are processes (*kinēseis*). But it is not appropriate for what is always (*aei*) immovably (*akinētōs*) in the same state (*kata t'auta echon*) to be becoming older (*presbuteron*) or younger through time, nor once to have become <so>, nor now to have completed becoming <so>, nor to be <so> in the future, nor in general to receive any of the attributes which becoming (*genesis*) attaches to the moving objects of sense perception. These attributes have come into being as forms of time which imitates eternity (*aiōn*) and circles round according to number. (Other inaccurate things we say besides these are, for example, that what has become *is* such as to have become, and what is becoming *is* becoming, and what will become *is* such as will become, and the non-existent *is* non-existent; but perhaps it is not the right moment at present to spell all this out.) Be that as it may, time has come into being with the heavens, in order that, having been created together, they may be dissolved together, if ever their dissolution should come about. And (the heavens) have come into being in accordance with the model of eternal (*diaiōnios*) nature, in order that they should be as like it as possible. For the model has being for the whole of eternity (*aiōn*), whereas the other has come into being and is and shall be right through the whole of time.

What is Plato's conception of eternity? Much of what he says in the quoted passage would suggest that the ideal Forms are eternal in the sense of being timeless. Thus Plato differentiates eternity (*aiōn*) from time which he treats as a mere likeness. And he denies that 'was' or 'shall be' are applicable to the Forms, or that they grow older.

Perhaps the ban on growing older (though not the ban on 'was' and 'shall be') might be explained away, if growing older were simply a matter of gathering more *years*. For *years* do not exist independently of the created heavens, and so, in the absence of the heavens, the Forms might have a temporal existence without gathering *years*. But how could they avoid advancing in age, even if there were no unitary amounts by which they advanced? The obvious answer is that they must be timeless.

On the other hand, there are problems about the 'timeless' interpretation, which may suggest that after all the Forms are eternal only in the sense of having everlasting duration. Some of these problems are more serious than others. Thus when the ideal model of the living universe is itself called a *living* organism, we need not take this as implying duration. For, as Cornford and Whittaker point out,[36] the Form of a living organism is not necessarily a living organism in quite the same sense. There is a notorious problem about whether the Form of large things is large in the same sense as the things which partake in that Form.[37] And a corresponding doubt will apply to the Form of *living* things. Again, we need not be worried that at 37D7 Plato gives the epithet eternal (*aiōnios*) unexpectedly to *time*. This does not show eternity to involve duration, for it must be understood in the light of the heavy qualifications which have preceded it: it was not possible to confer eternity *completely* on the generated thing, but only a *likeness* of eternity. If, then, time is immediately said to be eternal, this must be in a different way from the model,[38] and the latter may still be eternal in the sense of being timeless. There is a third point over which it is needless to worry. Thus when Plato calls time an image of eternity (37D5 and 7), this does not imply likeness in all respects, and in particular does not imply that both alike have duration.

More disturbing, however, is Plato's use of the temporal word 'before', when he says that the Forms and space and processes of coming into being all existed *before* (*prin*, 52D) the creation of the heavens. This might be an isolated lapse. But another epithet, *aïdios*, is *repeatedly* applied to the Forms both in the quoted passage and at 29A. Perhaps I have over-interpreted in translating this 'everlasting'. But this is a common meaning, and the word cannot be reinterpreted as implying timelessness in all its occurrences. For the stars are called *aïdia* at 40B5 and probably 37C6.

Whatever may be thought about the word *aïdios*, it is still harder to explain away the fact that Plato often speaks in the *Timaeus* of the Forms as existing *always* (*aei*). Thus at 27D6-28B1 he says:

> What is it that always (*aei*) is, and has no coming into being, and what is it that always (*aei*: not present in all MSS) is coming into being, and never (*oudepote*) is? The one is to be grasped by the mind with reason, and is always (*aei*) in the same state. The other is opined by opinion combined with irrational sense perception, and keeps coming into being and going out of existence, but never (*oudepote*) has real being. Again, everything which comes into being necessarily does so through some cause, for nothing can have its genesis without a cause. Now a creator necessarily makes everything good, whenever he produces the

---

[36] Whittaker, op. cit. in n.34; Cornford, op. cit., 40. Correspondingly, in *Sophist* 248Eff., Cornford takes the point to be not that the 'friends of the Forms' should admit that the Forms are living, but that they should admit living things *as well as* Forms among the things they count as real (*Plato's Theory of Knowledge*, London 1935, 245).

[37] For recent developments in the controversy, see Gregory Vlastos, 'The Unity of the virtues in the *Protagoras*', in his *Platonic Studies*, Princeton 1973.

[38] Owen points out (op. cit. (in n.19), 333) that the difference is subsequently emphasised by Plato's using the compound word *diaiōnios* for the eternity of the Forms (38B3).

form and features of his work by looking towards what is always (*aei*) in the same state and using something of that kind as a model.

The passage goes on to describe the divine creator's model as in the same state (29A1; A7), as everlasting (*aïdios* 29A3; A5), and as steady and firm (*monimos, bebaios* 29B6). At 35A2, Plato depicts the highest kind of being as one which is always (*aei*) in the same state. The main account of eternity, which has already been quoted, includes the claim (38A3) that the Forms are always (*aei*) in the same state immovably, and not becoming older or younger through time. It is repeated at 48E6 that the ideal model is always (*aei*) in the same state. Finally, 50C5 and 51A1 both talk of the copies or likenesses of the things which exist always (*aei*).

This tattoo of occurrences of 'always' is all the more significant, because Parmenides avoids the word in his description of reality. His avoidance of it in *The Way of Truth* is to be contrasted with his use of it in reporting the *mistaken* views of mortals in *The Way of Opinion* (fr.15), and with his predecessor Heraclitus' use of it in the sentence, 'it was always and is and will be always-living fire' (fr.30).[39] It was Plato's use of 'always', I believe, which brought things to a head, and led Plotinus in the third century A.D. to his very influential distinction between two senses of 'always'. But before I come to this, I must set aside an alternative interpretation.

I have suggested that Plato allowed implications of timelessness and of duration to stand side by side in his account of eternity without offering a resolution. But Whittaker has a different interpretation. He understands time in Plato in a narrow way as merely a system of measurement, which began with the creation of the heavens. Eternity can then differ from time, in spite of possessing duration, because it is not a system of measurement. Whittaker adds that it also lacks a beginning and is partless. In fact, the last point (partlessness) is crucial. For the other two points are insufficient to distinguish eternity from the longer period of time (as we should naturally call it) which includes the stage before measurement was introduced. And one unwanted result of that would be that eternity would belong not only to the Forms, but also to space, matter and motion. For we are told that space existed before the introduction of order and measurement, and that it contained vestiges of earth, air, fire and water in disorderly movement (*Tim.*48B; 52D-53C; 69B). In order to show why eternity does not belong to space, matter and motion, Whittaker needs to rely on his point that eternity differs by being *partless*. But in a sense the duration of space, matter and motion is itself partless, in that it is not, or not at first, divided into the days, nights, months and years, which Plato calls *parts* of time (37E). If the partlessness of eternity is to differ *significantly*, it must be partlessness in some much more radical sense. Perhaps the sense is that eternal things cannot in any way have lives divided into earlier and later phases. But if that is so, it is hard to see how they can still have duration.

There is a further difficulty for Whittaker's suggestion. If eternity has duration, how can it exclude growing older and the use of 'was' and 'shall

---

[39] These points about Parmenides are made by Mourelatos, op. cit., 107.

be'? To the first question Whittaker suggests the answer that there would be nothing for the Forms to grow older than. But there would be their former selves.

Plotinus

Sticking to the view, then, that Plato did not decide between making eternity timeless and giving it everlasting duration, I shall now turn to the resolution offered by Plotinus (A.D. 205-269/70). The crucial passage is *Enneads* 3.7.6 (23-36). In the immediately preceding lines, Plotinus has been making one of his many attempts to explain that the eternal intellect has no duration. I shall start translating from l.15:

> So it does not contain any this, that and the other. Nor therefore will you separate it out, or unroll it, or extend it, or stretch it. Nor then can you find any earlier or later (*proteron, husteron*) in it. If then there is neither any earlier nor any later about it, but 'is' is the truest thing about it, and indeed is it, and this in the sense that it is by its essence and life, then again we have got the very thing we are talking about, namely, eternity (*aiōn*).

> But when we say 'always' and 'not alternately existent and non-existent', we must view this way of speaking as a concession to ourselves. For 'always' is perhaps not being used properly (*kuriōs*). Adopted for clarifying the imperishable, it may divert the soul into imagining a lengthening of something which is increasing and again of something which will not ever fail. Perhaps it would have been better only to describe it as 'being'. But although 'being' is an adequate word for substance, people thought that *coming* to be (*genesis*) was also substance, and therefore needed for their understanding to add in the word 'always'. It is not that being is one thing and *always* being another, any more than the philosopher is one thing and the *true* philosopher another. But because there was such a thing as pretending to do philosophy, the addition 'true' came into use. In the same way, always is added to what has being, and the word 'always' to the word 'being', so that we say 'always being'. And this is why we must take the 'always' as 'truly' being. The 'always' has to be included in that unextended (*adiastaton*) property which in no way needs anything beyond what it already possesses. And it possesses everything.

Plotinus' point then is that 'always' does not have its proper sense, but merely has the function of denoting *true* being, as opposed to coming to be. That 'always' has a non-temporal sense is stated more briefly at 3.7.2 (28-9):

> I mean not the 'always' in time (*en chronōi*), but the kind we think when speaking of the *aïdion*.

A similar treatment is given to 'first', 'before', 'after' and 'origin', to which Plotinus also gives non-temporal senses.[40]

This device of giving a non-temporal sense to 'always' and other such words is Plotinus' way of resolving the ambiguity left by Plato. Suggestions of

---

[40] 4.4.1 (26-31); 3.7.6 (52). Plotinus further reminds us, 3.7.6. (49-57), that Plato himself, at *Tim.* 37E, says that the application of 'was' to the eternal is an improper usage.

duration are henceforth to be discounted, by reference to the idea of a non-temporal sense. Just how influential this device was among Plotinus' pagan and Christian successors we shall see shortly.

But let us notice first how many protestations there are in Plotinus that the eternal should not be thought of as spread out in time. It is unextended (*adiastatos*),[41] partless (*ameres*),[42] all together (*homou pasa*),[43] as if in a point (*hoion en sēmeiōi*),[44] and is not of any size (*tososde*).[45] Its life must not be partitioned (*meristheisa*),[46] but has pure partlessness,[47] and its thinking is timeless (*achronos*).[48] It is not scattered (*skidnasthai*)[49] like time into an interval (*diastasis*)[50] but is like a point.[51] It does not run along with and is not stretched out with (*sumparathein, sumparateinein*) the movement of the soul.[52] These can be added to the protestations in the main passage just quoted.[53] We were there told that the eternal intellect does not contain any this, that and the other (*allo kai allo*). You will not separate it out (*diistanai*), nor unroll it (*exelissein*), nor extend it (*proagein*) nor stretch it (*parateinein*). You cannot find any earlier or later (*proteron, husteron*) in it. Finally, there is an emphatic denial of extension in the following passage:

> So <the life of that which has real being> must not be counted by time, but by eternity. And this is neither more nor less (*pleon, elatton*), nor of any length (*mēkos*), but a 'this', and unextended (*adiastatos*) and not temporal (*ou chronikon*). You must not then join what has being to what does not, nor time or the temporal always (*to chronikon de aei*) to eternity, nor must you stretch out (*parekteinein*) the unextended (*adiastatos*). But you must take it all as a whole (*pan holon*), if you take it at all, and take not the indivisibility which is found in time, but the life of eternity which is not made up of many times, but is all together (*pasa homou*) from the whole of time.[54]

These denials of duration, which are repeated again and again by the later Neoplatonists, seem to me to exclude most of the rival interpretations of eternity, including the idea of an *enduring* present. Indeed, the only surviving alternatives to the 'timeless' interpretation would seem to be those which treat eternity as a mere instant. But these too are ruled out by the last quotation's insistence that what eternity involves is 'not the indivisibility which is found in time'. In any case, all these interpretations, enduring present, instanthood and the rest, turned out above to have difficulties of their own. So far as I can see, all that is left in the field is the 'timeless' interpretation. Kretzmann and Stump have allowed in their work that an eternal life is atemporal in one way, but only in a way that knowledge of logic might be called atemporal in comparison with proving a theorem of logic.[55]

---

[41] 3.7.2 (31); 3.7.3 (15 and 38); 3.7.11 (54).
[42] 3.7.3 (18-19).   [43] 3.7.3 (18-19 and 38); 3.7.11 (3).
[44] 3.7.3 (18-19).   [45] 3.7.6 (47).   [46] 3.7.6 (47).   [47] 3.7.6 (47).   [48] 4.4.1 (12).
[49] 6.5.11 (14-21); in 1.5.7 (15-17) *skidnasthai* characterises time.
[50] 6.5.11 (14-21).   [51] 6.5.11 (14-21).
[52] 3.7.13 (44-5); compare 3.7.8 (55-6) where the same terms appear in a mooted definition of time as the extension alongside which motion runs (*sumparathein*) and stretches (*paratasis*).
[53] 3.7.6 (15-22).   [54] 1.5.7 (20-31).
[55] Stump and Kretzmann, op. cit., 446.

Plotinus' rejection of duration seems to go far beyond this.

Admittedly, Plotinus may not remain quite consistent. A.H. Armstrong has argued that he often slips back into ways of talking which suggest that the eternal enjoys duration after all and even succession and change.[56] But this is at most a view into which he *slips*. And meanwhile much of the evidence which might *seem* to connect him with such a view will be disqualified by his distinction of non-temporal senses.

It is worth adding that Plotinus would probably extend the idea of a non-temporal sense to the concept of *life*. We think of life as something spread out in time. But the life which constitutes eternity is the life of the intellect, and it consists of a very special kind of thinking. I do not myself understand how any thinking can fail to be temporal. But Plotinus is persuaded that it can, partly because it is a type of thinking which involves no *progress*, and partly because it involves a *sense* of timelessness. We would think the sense of timelessness a mere *illusion*, but it would be no good bringing against Plotinus the evidence of clocks and of other events, to show that in reality our intellectual thought had taken time. For in his view, the world of clocks and of physical things is *less* real than the world of the intellect, and so cannot be used to impugn the impression of timelessness. Nor could we fault Plotinus by alleging that he makes our intellectual thinking *alternate* in time with other events; for I shall explain in Chapter 11 how he believes the intellectual thought within us to be *uninterrupted*.

Non-temporal senses after Plotinus

The device of distinguishing non-temporal senses was not new with Plotinus, but only new in its application to the word 'always'. Plato had already described the soul as being 'prior' (*proteron*) to the body and 'older' (*presbuteron*), not only in birth, but also in excellence.[57] Aristotle had distinguished non-temporal senses of 'prior' and 'posterior'[58] and of 'origin' (*archē*),[59] and so, for 'origin' (or 'beginning') had Philo Judaeus.[60] Early and middle Platonists, in interpreting Plato's account of creation in the *Timaeus*, had given non-temporal senses to 'origin' and to 'created' (*genētos*).[61] The new idea is that 'always' has a non-temporal sense, and, interestingly enough, this idea appears simultaneously in Plotinus and in his fellow-student, the Christian Origen. For Origen distinguished non-temporal senses not only for Christ's 'descending',[62] for 'was'[63] and for 'origin' but also for 'always'.[64]

---

[56] A.H. Armstrong (ed.), *The Cambridge History of Later Greek and Early Mediaeval Philosophy*, Cambridge 1967, 246-7 and 455; 'Eternity, life and movement in Plotinus' account of *nous*', 67-74; 'Elements in the thought of Plotinus at variance with classical intellectualism', *JHS* 93, 1973, 13-22. The references are to Plotinus 5.8.4; 6.2.8 (26-41); 6.7.13.

[57] *Tim.* 34C4.

[58] E.g. *Cat.* 12, 14a27-b13; *Metaph.* 7.1, 1028a23, on which see the note of W.D. Ross, for many further references (*Aristotle's Metaphysics, A Revised Text with Introduction and Commentary*, Oxford 1924).   [59] E.g. *Metaph.* 1.3 and 5.1.

[60] Philo *de Opificio Mundi* 26.9.| [61] References in Chapter 17   [62] *On First Principles* 1.2.11.

[63] *Commentary on John*, fr. from Cod. Monac. 208, on John 1:1, Prussian Academy edition vol.4, ed. Preuschen, p.564; cf. *On First Principles* 1.3.4.   [64] *On First Principles* 1.2.9

The Holy Spirit would never have been included in the unity of the Trinity, that is, of God the unchangeable Father and His Son, if he was (*erat*) not always (*semper*) the Holy Spirit. Of course, the words we use, 'always' and 'was', and any other such word with a temporal meaning (*temporalis significatio*) that we appropriate, must be understood in an elastic way as an artless expression. For the meanings of these words are temporal, whereas the things of which we are speaking, although described in a temporal way for handling in our discussion, go by their nature beyond any understanding in a temporal sense.[65]

After Plotinus and Origen, we find non-temporal senses ascribed to innumerable words. What matters is that they are ascribed to the word 'always' and to other words associated with *eternity*. Thus the Athenian Neoplatonist Proclus (*c.* A.D. 411-485) finds non-temporal senses for 'always'[66] and 'everlasting' (*aïdios*),[67] for 'infinite'[68] and for 'is'.[69] Pseudo-Dionysius applies the idea to 'eternity' (*aiōn, aiōnios*).[70] The application to 'always' is repeated by the anonymous author of a commentary on Aristotle's *Categories*.[71] In Alexandria, Ammonius may be credited with non-temporal senses of 'is' and of 'everlasting'.[72] The Athenian Neoplatonist Simplicius, a pupil of Ammonius, distinguishes non-temporal senses for 'always', 'everlasting' and 'is'.[73] His fellow-pupil and opponent the Christian Philoponus does not, as we shall see, take the same view of eternity, but none the less finds a non-temporal sense for 'always',[74] as does Ammonius' successor Olympiodorus.[75] And thereafter, in the Islamic world, we find Avicenna (A.D. 980-1037) distinguishing a non-temporal sense for eternity.[76]

I have omitted the most important influence on Christian thought – Boethius (*c.* A.D. 480-525/6). In a classic passage, we find at least three things: first, the famous distinction between a temporal and eternal 'now'; secondly, a distinction, not always noticed,[77] between a temporal and an eternal 'always'; and thirdly, the use of two separate words, *sempiternitas* and *aeternitas* for everlastingness and eternity respectively. *Sempiternitas* is coined from the Latin *semper* (always).

---

[65] *On First Principles* 1.3.4 from Rufinus' Latin version. I am grateful to Graciela Gayoso for these references.

[66] *in Tim.* (Diehl) 1.238,15; 1.239,2ff; 1.278,10.

[67] *in Tim.* 1.278,9f. (cf. 1.239,12); *Elements of Theology* 55.

[68] *in Tim.* 1.278,10f; *in Prm.* 1230,12f.

[69] *in Tim.* 3.51,15-21 (translated above).

[70] *On Divine Names* 10,2-3.

[71] Anon. *in Cat.* CAG 23. 53, 20ff.

[72] ap. Zach. *Ammonius* or *de Mundi Opificio* PG 85, 1032A-1033A.

[73] *in Phys.* 1154,29ff.; 1155,15ff.; *in Cael.* 95,21.

[74] *Aet.* (Rabe) 104-7; 472; *in Phys.* 456,17-458,16; and ap. Simpl. *in Phys.* 1158,31-5.

[75] *in Phaed.* 13.2 (Westerink).

[76] *Tractatus de Diffinitionibus et Quaesitis,* tr. into Latin by Andrea Bellunensis, in *Avicennae Compendium de Anima etc.,* Venice 1546, f.137B, repr. in M. Gierens as text 30 in *Controversia De Aeternitate Mundi,* Textus et Documenta, series philosophica 6, Pontificia Universitas Gregoriana, Rome 1933. There is a French translation direct from the Arabic by A.-M. Goichon: *Livre des Définitions,* Cairo 1963.

[77] See Stump and Kretzmann, op. cit., 432-3.

But the 'is always' (*semper*) which is said of God signifies a single idea, as if he has been there in all the past, is there in some way in all the present, and will be there in all the future. According to philosophers, the same can be said about the heavens and the other deathless bodies, but it is not said of God in the same way. For he is always because 'always' in him belongs to the present, and there is a great difference between the present of our affairs, which is now, and the present of divine matters. For our now, as if running, creates time and sempiternity (*sempiternitas*), whereas the divine now stays not moving, but standing still, and creates eternity (*aeternitas*). If you add to that name the word 'always' (*semper*), you will create the continual, untiring, and hence perpetual race of that which is now, namely sempiternity.[78]

The use of the word 'always' in the last sentence of the quoted passage will only be confusing, if one fails to notice that Boethius has here reverted to the *temporal* sense of the word.[79] In a passage from *Consolation* 5.6, to be quoted below, we shall see that Boethius appeals to the device of non-temporal senses again, this time distinguishing a non-temporal sense in which God is *older* (*antiquior*). He also repeats the distinction between two kinds of present.

Distinct words versus distinct senses

It introduces new clarity when Boethius tries to separate the words *sempiternus* (elsewhere *perpetuus*)[80] and *aeternus*, and to apply them to the everlasting and the eternal respectively. For although Augustine had distinguished between time and eternity often enough,[81] he did not separate these words.[82] Nor did Rufinus in his Latin paraphrase of Origen's Greek, or Macrobius or Calcidius.[83] We found that Plato had made no conscientious attempt to separate the Greek words *aïdios* and *aiōnios*. Nor yet did Philo.[84] Nor did Boethius himself separate *perpetuus* and *aeternus* in earlier work,[85] so he is trying to make a new point here. It has been said that Boethius would have found the Greek words distinguished in Proclus, but I have not noticed the distinction in the passages suggested.[86] What is true is that the Greek words are distinguished very shortly afterwards in Simplicius.[87]

Once it is made, it introduces a new possibility of picking out eternity, not

---

[78] *de Trinitate*, 4, ll.64-77.

[79] *Pace* Stump and Kretzmann, loc. cit.

[80] In *Consolation* 5.6, in a passage to be translated below.

[81] E.g. *Conf.* VII. 10 and 17; IX.10; XI.6-8; 11; 13; 30; XII.9; 11; 12; XIII.15; *City* XI.4; XI.6; XI.21; XII.15-16; XII.18. *de Trin.* V.1.2; XII.23; *Enarratio in Psalmum* 102, sermo 11.

[82] Augustine describes God as 'sempiternal without time' (*sine tempore sempiternum*, *de Trin.* V.1.2; cf. *Conf.* VII.17; XI.7; XI.10; *City* XII.18 for 'sempiternal').

[83] Origen *On First Principles* 1.2.11; Macrobius *in Somnium Scipionis* 2.11.4; Calcidius *in Tim.*312.

[84] See H.A. Wolfson, *Philo*, Cambridge Mass. 1947, vol. 1, 234-5.

[85] P. Courcelle cites Boethius *in Isagogen* (Brandt) 257,6. (P. Courcelle, *Les Lettres Grecques en Occident de Macrobe à Cassiodore*, 2nd ed., Paris 1948, translated as *Late Latin Writers and Their Greek Sources*, Cambridge Mass., 1969, ch.6, 313; also in *La Consolation de philosophie dans la tradition littéraire*, Paris 1967, 266. The same interchanging of *perpetuus* and *aeternus* appears in an account of the resurrection at the end of the *de Fide Catholica*, to which I shall return in Chapter 13.

[86] Courcelle, *La Consolation de philosophie dans la tradition littéraire*, 225-6; P. Merlan, 'Zacharias Hermiae, Ammonius Scholasticus and Boethius', *GRBS* 9, 1968, 193-203, n.20.

[87] *in Phys.* 1155,13.

through distinguishing two senses of the *same* word, but by insisting on *different* words. We find this development a little later in Olympiodorus, who by A.D. 541 held the chair in philosophy at Alexandria. He insists that we may not use the word *aïdios* at all for the eternal, but only the word *aionios*:

> And we must not neglect either that the eternal (*aiōnion*) and the everlasting (*aïdios*) are different. For the eternal is what is whole in the manner of a whole now which is divorced from past and future, but exists in the present now. That is everlasting, however, which always (*aei*) exists and is viewed in all three times. Hence we call God eternal, because he does not have his being in time, but has all time, present, past and future as a now; for this is the nature of the eternal. We do not call him everlasting, because he does not have his being in time.[88]

Olympiodorus does allow temporal and non-temporal senses for the word 'always', but here too he gives things a radical turn. For he insists that the proper (*kuriōs*) use of 'always' is the *non*-temporal one:

> Immortality in the proper sense (*kuriōs*) is in God, seeing that he is eternally (*aiōniōs*), and does not have was, is and shall be. For everlasting things (*aïdia*) do not in the proper sense (*kuriōs*) exist always (*aei*), because the temporal designations 'was', 'is' and 'shall be' apply to them, and these do not hold always. Rather, the parts of time, was, is and shall be, perish.[89]

Philoponus' dissent

Many other Christians, as we shall see, accepted the Platonist idea of eternity as lacking duration. But before coming to them, I shall record one striking case of dissent. The Christian Philoponus, who is idiosyncratic in so many things, explicitly argues against Proclus that eternity has duration. He thus takes the opposite side to his contemporary Boethius, who had accepted the influence of Proclus' school. Philoponus even picks up the very words which Proclus (and earlier Plotinus, Basil and Gregory)[90] had withheld from eternity and associated with time, and applies them instead to eternity. Eternity must be an extension (*paratasis*), Philoponus says four times, and Proclus' own ideas prove it, because Proclus calls eternity a *measure* (*metron*) of the life of the living being, and a *measure* cannot be without parts (*ameres*). It must be an extension (*paratasis*), and a breadth (*platos*). It must, so to speak, be stretched out alongside with (*sumparateinesthai*) the being of eternal things. Philoponus suggests as a comparison that you might in your thoughts make the heavens stand still. There would then no longer be distinct segments of time, but there would still be a homogeneous extension (*paratasis*) running alongside with (*sumparathein*) the being of the *kosmos*. Eternity, on this conception, is a sort of unsegmented time, and Philoponus finishes by

---

[88] *in Meteor.* 146,15-23.
[89] *in Phaed.* 13.2 (Westerink)..
[90] For Plotinus and Proclus, see above; for Basil and Gregory below.

exploiting Plato's willingness to use the word 'always'. The passage runs as follows:

> Suppose eternity (*aiōn*) measures the being of eternal things (*aïdia*), as time measures the motion of the heavens, which is what Proclus himself says in the following passage: 'Time is the measure of the motion of the heavens, as eternity is the measure of life of the living being itself.' Well then, if eternity is the measure of the life of the living being itself, or at any rate of eternal (*aïdia*) things, it is absolutely necessary presumably that eternity (*aiōn*) should not be a single point (*sēmeion*), but a sort of breadth (*platos*) and extension (*paratasis*) stretched out, so to speak, alongside (*sumparateinomenon*) the being of eternal things (*aiōnia*). It should not be cut, like time, into distinct segments, I mean, years, months, nights and days. Rather, it is like this: if someone halted the heavens and the motion of the sun in his thoughts, there would no longer be distinct segments of time in his mind, and yet there would nevertheless be in his mind a sort of extension (*paratasis*) running along (*sumparatheousa*) evenly together with the being of the *kosmos*. In the same way surely one can say of eternal things (*aïdia*) that, even if there is not a motion of temporal length (*chronikon diastēma*) for the eternity (*aiōn*) which measures their being, nevertheless there is at any rate a sort of uniform extension (*paratasis*) thought of together with their being. For eternity (*aiōn*), as I said, is not some point without parts (*ameres sēmeion*), nor do eternal things (*aiōnia*) exist merely at some single point: which is indeed why Proclus said that eternity (*aiōn*) was a measure of the life of eternal things (*aïdia*), and before him Plato. For Plato says, 'time has come into being with the heavens, in order that, having been created together, they may be dissolved together, if ever their dissolution should come about, so that (the heavens) may be as like the model as possible. For the model has being for the whole of eternity (*aiōn*), but (the heavens) right through the whole of time.' So, if Plato says that the paradigm for the *kosmos* has being for all eternity, and is the measure of life of eternal things (*aïdia*), clearly he does not think that eternity (*aiōn*) is some single point, for all is not a point, nor does a measure lack parts (*ameres*). Rather, he thinks eternity is some single, uniform extension (*paratasis*), not cut by any differentiation, but staying always (*aei*) the same, and remaining without change in itself.[91]

One defect in Philoponus' account is that he considers that the only alternative to eternity being an extension is that it should be a point (*sēmeion*) and that eternal things should exist in a point. Plotinus put it better, when he said that in eternity it is *as if* everything were together in a point (*hoion en sēmeiōi homou pantōn ontōn*).[92] We are not to think of eternity as a single instant – that was one of the interpretations rejected earlier in this chapter.[93] It is merely *analogous* with a point in various ways, notably in being unextended. The analogy with a point is repeated by Plotinus elsewhere:

> Which is why it is not in time, but outside all time, for time is always spread out (*skidnasthai*) into an interval (*diastasis*), whereas eternity stays still in the

[91] *Aet.* (Rabe) 114,20-116,1.
[92] 3.7.3 (18-19).
[93] I would reject also the characterisation of the timeless view of eternity as making eternity into a frozen static, or isolated instant (Stump and Kretzmann, op. cit., 430; 432).

same, and dominates and is greater by its *aïdios* power than time, which seems to go into a plurality. It is as if a line, which seemed to be going to infinity, depended on a point, and ran around it, with the point reflected in it wherever it ran, while the point did not run, but the line circled around it.[94]

From this and the battery of other Plotinian passages, referred to above, it is clear that Philoponus is departing from Plotinus, when he gives extension to eternity.

Boethius

I hope that it will now be possible to see Boethius in a better light. It was he above all who transmitted to the Christian middle ages the Neoplatonist concept of eternity which I have been discussing. He did not side with Philoponus, although he was his contemporary, and there has even been speculation that, like Philoponus, he might have studied under Ammonius in Alexandria. But it has been very fully argued on the other side that he simply had access in Rome to writings from the school of Ammonius' teacher Proclus.[95] I have already quoted Boethius' distinction of non-temporal senses in *de Trinitate*. But he uses many additional devices for denying duration in the *Consolation of Philosophy* 5.6. This passage deserves to be quoted in full, all the more because it has sometimes been interpreted, wrongly I believe, as allowing eternity to involve duration after all. Boethius speaks repeatedly of the whole (*tota*) possessing and embracing of the whole (*totum*) extent, presence, or completeness of life all together (*simul*) and equally (*pariter*). The phrases echo the oldest description of eternity in Parmenides as 'all together' (*homou pasa*). The passage reminds us of Ammonius, who also echoes Parmenides, in a discussion which is quoted partly in this chapter (above) and partly in Chapter 16.

> Eternity therefore is the whole and perfect possession all together (*tota simul et perfecta possessio*) of a life which cannot end, which becomes clearer from a comparison with temporal things. For whatever lives in time progresses as something present from what is past to what is future, and there is nothing placed in time which could embrace the whole extent of its life equally (*totum vitae suae spatium pariter*). It does not yet grasp tomorrow, and it has already lost yesterday. Even in today's life you do not live more than in the moving and transitory moment. So what is subject to the condition of time is not yet such as rightly to be judged eternal (*aeternus*), even if (*licet*), as Aristotle believed of the world, it never began to exist, and does not cease, but has its life stretched

---

[94] 6.5.11 (14-21).

[95] It is the suggestion of Courcelle, op. cit. in n.86, ch. 6, that Boethius might have gone to Alexandria, to hear Ammonius. But James Shiel has argued that the commentaries of Boethius and Ammonius are instead influenced by a common source, Proclus' school in Athens ('Boethius' commentaries on Aristotle', *Mediaeval and Renaissance Studies* 4, 1958, 217-44); and Minio-Paluello has dismissed as unfounded the hypothesis that Boethius left Rome to study in Alexandria: 'Severino Boezio', in *Dizionario Biografico degli Italiani* 11, 1970, p.13 of offprint; 'Les traductions et les commentaires aristotéliciens de Boèce', *Studia Patristica* 2 = *Texte und Untersuchungen zur Geschichte der altchristlichen Literatur* 64, 1957, 358-65, repr. in Minio-Paluello's *Opuscula*, Amsterdam 1972, 328-35; *Encyclopaedia Britannica*, 1968 ed., 842-3.

out with the infinity of time. For even if its life is infinite, it does not include and embrace the whole extent of that life all together (*totum simul infinitae licet vitae spatium*), since it does not yet possess the future and it already lacks the past. So that which embraces and possesses equally the whole completeness (*plenitudinem totam pariter*) of a life which cannot end, and for which there is not some of the future missing nor some of the past elapsed – that is rightly held to be eternal (*aeternum*). And it must be in possession of itself and always present to itself, and must have present to itself the infinity of moving time. Hence those are not right who hear that Plato thought this world had no beginning in time and will have no end, and who conclude that the created world is in this way made co-eternal (*coaeternus*) with the creator. For it is one thing to be drawn through an endless life, which is what Plato attributed to the world, and another to have embraced equally the whole presence (*totam pariter praesentiam*) of a life which cannot end, which is clearly the special characteristic of the divine mind. Nor should God be thought older (*antiquior*) than created things by some amount of time, but rather by the peculiarity of his nature which is simple. For this status of his motionless life as present is <merely> imitated by that infinite movement of temporal things: and since the latter cannot represent or equal the former, it lapses from immobility into motion, and shrinks from the simplicity of presentness into the infinite dimensions of future and past. Since it cannot possess equally the whole completeness (*totam pariter plenitudinem*) of its life, it seems to some extent to emulate what it cannot completely express, in so far as it in some way never ceases to be. It does so by tying itself to whatever present it is which attaches to this tiny fleeting moment, a present which bears some likeness to the present which stays still, and so it bestows on what it touches an appearance of being. But since it could not stay still (*manere*), it seized on an infinite journey in time, and thus became such as to continue by travel a life whose completeness (*plenitudinem*) it could not embrace by staying still (*manere*). Thus if we want to apply names appropriate to the things, let us say, following Plato, that God indeed is eternal (*aeternus*), but the world perpetual (*perpetuus*).

The passage says that eternity is not spread out, as it would be if it had duration. Admittedly, there are phrases which might be taken, in the absence of our previous explanations, as implying duration after all. But the device of distinguishing non-temporal senses occurs twice in the passage as a reminder, and this device must certainly be applied to the solitary occurrence of the word 'always'. Duration has been read not only into the word 'always', but also into the talk of completeness, of life, and (here and in the passage from *de Trinitate*) into the talk of staying still (*manens, permanens, consistens*).[96] Someone might want to add the point that eternal life cannot end. But it will already be clear how these ideas are to be understood in the context of an eternal life. God can possess the *completeness* of life, precisely because his life is not spread out. Staying still is a negative idea excluding change rather than implying duration, and the same goes for unendingness.

Six other denials of duration

I have so far told a rather clear-cut story: Parmenides was **groping** for the

---

[96] Stump and Kretzmann, op. cit., 432-3; 446.

concept of timelessness. Plato clouded the issue with his talk of 'always'. Plotinus restored the idea of timelessness, thereby influencing the Neoplatonists and some prominent Christians. Philoponus dissented, but Boethius paid no attention, and transmitted the traditional concept to the Latin middle ages. In the sub-plot, Plotinus distinguished non-temporal senses of the words involved, but some of his successors preferred to use two distinct words for temporal and non-temporal eternity. The story appears neat, but there are four ways in which I want to elaborate it. First, I want to fill in some gaps in the history of the denial of duration. It will be seen that it appealed more widely than I have yet brought out in the Judaeo-Christian tradition. Secondly, I want to introduce some necessary qualifications into the story. Thirdly, I shall consider why Aristotle has so far been given no role. Lastly, I shall fulfil my promise to see if I can interpret the text of Parmenides.

Contrary to what is sometimes said,[97] the denial of duration goes back earlier than Plotinus although naturally before he introduced his clarifications, and also within the Judaeo-Christian, as opposed to the Platonist, tradition, these denials tend to be less clear-cut. The denial is found, however, in the Middle Platonist Plutarch (*c.* A.D. 45-*c.* 125) in the following passage.

> Hence it is irreverent, in the case of what is, to say even that it was or will be. For these are deviations and changes and alterations and belong to what is not of a nature to remain in being. God, however, if this needs to be said, is not in time, but in eternity (*kat' aiōna*), which is changeless and timeless (*achronos*) and undeviating, containing no earlier or later, no going to be or pastness, no older or younger. Single, he has completed 'always' in a single now, and that which really is in this manner only *is*, without having come into being, without being in the future, without having begun, and without being due to end. Thus we must reverence and love him, and address him by saying, 'Thou art', or, by Zeus, as some did long ago, by saying 'Thou art one'.[98]

Jews and Christians in Alexandria also picked up this Platonic tradition. Thus Philo Judaeus says in the first century A.D.:

> So with God there is no future, for he has put beneath himself the very boundaries of all times. Indeed, his life is not time, but eternity (*aiōn*), the archetype and model for time. And in eternity nothing is past or future, but simply has being.[99]

It has been disputed whether this passage expresses the idea of timeless eternity, and it must be admitted that Philo does not have an unwavering grasp of the idea, since he has just been talking of God's *fore*sight and *fore*thought (*promētheia, pronoia,* §29). Elsewhere too, I would agree, Philo falls

---

[97] While I agree with Bevan's caution about finding it in the New Testament, I shall be dissenting from his suggestion that it reached Christianity only after the establishment of Neoplatonism (S. Bevan, *Symbolism and Belief*, London 1938, 97).

[98] Plut., *On the E at Delphi* 393A-B.

[99] Philo *Quod Deus Immutabilis Sit* 6.32.

short of the timeless conception.[100] But I am not persuaded by Whittaker's ingenious attempt to reinterpret the present passage.[101] He takes Philo to mean not that God is timelessly eternal, but that past and future events have a certain stable and everlasting subsistence within God's life, or, better, in his mind, and are coextensive with it. I shall explain in a footnote why I am not convinced, and prefer to stick to the view that Philo does touch here on the idea of timeless eternity.[102] I would hesitate, however, to endorse the further suggestion that the author of St John's Gospel was influenced by Philo, when he used a tenseless present:

> Before Abraham was I am.[103]

The controversy between Maurice and the Principal of King's College will reveal how difficult it is to be sure that the timeless conception of eternity appears in that Gospel.

A person who does record the traditional conception is Clement of Alexandria (died A.D. 215) in his *Stromateis*:

> Eternity (*aiōn*) holds together the future, the present and indeed the past of time in a hair's breadth (*akariaiōs*).[104]

Clement's junior in Alexandria, Origen, says that no earlier or later can be understood in the Son or the Holy Spirit.[105] Moreover, in a statement to which I shall return, he places God not only above time, but also above eternity:

> It must be understood elastically, when we say that there was no time when he was not. For even these words bear the sense of a temporal description, I mean, 'when' and 'no time'. But what is said of the Father, Son, or Holy Spirit should be understood as being above all time, above all the ages (*saecula*) and above all eternity.[106]

---

[100] In *de Sacrificiis Abelis et Caini* 76, where Philo says that with God nothing is old or past, I would agree that Philo may mean only that God's precepts are never out of date. I would add that in *de Mutatione Nominum* 47, 267, in the same breath as making God's word eternal, not temporal, Philo describes it in relation to God as something which '*will* be begotten'.

[101] Whittaker, *God, Time, Being*, 36-40.

[102] Whittaker first claims it was a commonplace that, even for mortals, past and future events subsist with a certain stability, so as not to be really past or future. But I would interpret his references differently; one, the Stoic idea that past and future *subsist*, even though they do not *exist*, was given a different interpretation in Chapter 2. Another, Damascius' much later idea that the whole of time exists simultaneously in reality, was even interpreted in Chapter 5 as implying the *opposite* of stability: one temporal 'leap' after another becomes present, and then past, in a repeated flux – it is in this process of *coming* to be that gives being to the whole of time. Whittaker's second step is the suggestion that Philo is extending the supposed commonplace from mortals to God, as a way of explaining God's omniscience; but I should expect Philo to try, so far as possible, to *contrast* God with mortals, rather than to assimilate him. Thirdly, Whittaker suggests that the stably subsisting events would be located in God's *mind*, but Philo's immediate reference is to God's life, rather than to his mind, and from the passages discussed below in Chapters 13 and 15, I should have expected Philo to locate in God's mind not the events themselves, but, if anything, archetypal patterns of them.

[103] John 8:58. See W. Kneale, 'Time and eternity in theology', 94.   [104] *Stromateis* 1.13.

[105] *On First Principles* 2.2.1.   [106] *On First Principles* 4.4.1.

I have already warned that these authors do not always have a very firm grasp of the idea of timelessness. The point has been illustrated in connexion with Philo. Equally from many of Clement's accounts of God's eternity in the *Stromateis* one could well understand that he exists merely throughout all time without beginning or end. Origen, like Augustine after him, often speaks as if God had foreknowledge, not timeless knowledge.[107] Moreover, in escaping from a temporal beginning for the Son, Origen falls, in the following passage, into applying to God the idea of time and of 'stretching out along with', which his contemporary Plotinus and subsequent Neoplatonists associated with time, rather than eternity.

> But because of all this, the high birth of the Son is not clearly presented, when God, for whom it is always today, says to the Son, 'You are my son; today I have begotten you'. For there is no evening of God, I think, since there is no morning either, but the time (*chronos*) stretching out along with (*sumparekteinōn*) his uncreated and everlasting (*aïdios*) life, if I may so put it, is for him the today in which the Son has been begotten. In this way, no beginning is found for the Son's begetting, since no day is found either.[108]

After Plotinus, accounts of eternity in the Christian tradition tend to become fuller. I shall consider just two more. Gregory of Nyssa (died A.D. 394) denied that for God, or for Christ, or before the creation, or even for the 'waters above the heaven', there was any *diastēma*, where *diastēma* is a word used by his brother Basil for defining time,[109] and which denotes spatial or temporal extension with beginning and end. Such things are without *diastēma* (*adiastatos*), or quanitity (*aposos*). Nor is God's eternity *measured* by *diastēma*. Beginning and end in their turn imply *paratasis*, the word we have seen Proclus using for temporal extension, and this too is excluded by Gregory. Moreover, we are wrong to apply to superior beings the past and future of *diastēmatikē paratasis*. Some things escape the differences of older and younger (*presbuteron, neōteron*), or earlier and later (*proteron, husteron*), and are free of all temporal succession (*akolouthia*). God leaves no *diastēma* behind him and does not journey on in his life to anything lying in front of him. His nature does not run along with (*sumparatrechein*) times, and is not to be viewed in company with past and future. There is nothing which moves alongside him, and of which part is past and part future. This denial of duration is expressed chiefly in Gregory's treatise against the Arian heresy, *contra Eunomium*.[110]

It is problematic[111] which of these Greek Fathers was known, and when, to the Latin-speaking Augustine (A.D. 354-430). But at any rate, when

---

[107] E.g. *On Prayer* 5-6; *On First Principles* 1.2.2. But see Chapter 16 for the idea, some of which is in Augustine, that talk of providence and forethought can after all be understood as a timeless and changeless knowing of the future not *as* future.

[108] *Commentary on John* 1.29 (31), 204.

[109] Basil *adv. Eunom.* 1.20-1, defines time as a *diastēma* stretched out together with (*sumparekteinomenon*) the holding together of the *kosmos*.

[110] *contra Eunom.* 1.359-64 (PG45, col. 364); 1.370-1 (PG45, 368A); 1.685-9 (PG45, 461-4); 2.459 (PG45, 1064C-D = Book XII in Migne's numbering) 8.5 (PG45, 796A); 9.2 (PG45, 809B-C; also *in Eccl.* 7, PG44, 729C-D; *in Hex.* PG44, 84D. On Gregory's view, see H. von Balthasar, *Presence et pensée*, Paris 1942, 1-10.

[111] See on this Courcelle, *Late Latin Writers and Their Greek Sources*, ch.4.

Augustine wrote the *Confessions* about six years after Gregory's death, he too made eternity exclude duration. For in eternity *everything* is present (*totum praesens*, *Conf.* XI.11), and God's years stand *all together* (*anni tui omnes simul stant*, XI.13). There is no succession in God's Word, but all is spoken *together* (*simul*), and (Augustine adds rather confusingly) sempiternally (*sempiterne*, XI.7).[112]

Qualifications

It is time to introduce qualifications into the story and make it less clear-cut. Some of the necessary cautions have been issued already: we have seen that many of the authors cited, especially those from the Judaeo-Christian tradition, did not achieve consistency in their accounts of eternal being. We have seen that Greek and Latin writers did not attempt until a very late date to reserve different words for timeless eternity and everlastingness. We have seen, starting with Philo,[113] numerous uses of the word *aiōn* with a temporal meaning. But more needs to be said than that. Lampe's Patristic Lexicon gives 'eternity' as only one of many meanings of *aiōn*, and even then does not confine the word to *timeless* eternity. Sometimes the plural form *aiōnes* is used, the reference being to the ages. Another phenomenon is that we find God described by Christians as existing *before* or *above* the ages,[114] or again *before* or *above* eternity.[115] We encountered an example of this in Origen (above). In such passages, the meaning varies. For the idea may be that God is above the ages and everlasting time. But in some authors it can instead represent the idea that God cannot even be described as timelessly eternal, but is best described negatively, or not at all.

I have entirely ignored one conception of *aiōn* which is found in certain Gnostic writers. They speak of eternity (*aiōn*) and contrast it with time, and yet they seem to mean by eternity something thoroughly temporal, admitting all sorts of successive happenings. Their concept is so far removed from the tradition which I have been considering that I shall do no more than refer to some secondary literature on it[116] and set it on one side.

Even when reference is made to the *timeless* eternity with which I have been concerned, the absence of duration is by no means the sole point of emphasis. In different contexts we find stress placed on the lack of beginning or end, or the absence of change. The word 'present' or 'now' is treated as doing more than just expressing the absence of duration. Even in an author like Boethius,

---

[112] Cf. *Conf.* IX.10; *de Trin.* XII.23; *Enarratio in Psalmum* 102, sermo 11.

[113] A list is given by H.A. Wolfson, *Philo*, Cambridge Mass., 1947, vol.1. p.171, to which may be added Philo *de Mut. Nom.* 47, 267. See also Whittaker, *God, Time, Being*, second study.

[114] E.g. Origen *On First Principles* 4.4.1; Athanasius *contra Arianos* 1.12; Ps-Dionysius *On Divine Names* 5.4; 10.3.

[115] Origen, *On First Principles* 4.4.1; Basil *adv. Eunom.* 2. 17-18; Gregory *in Eccl.* 7 (PG 44, col. 729C-D); *contra Eunom.* 1, ch. 26 (PG45, col. 364); Ps-Dionysius, *On Divine Names* 2.10; 5.10; 10.2-3.

[116] Henry-Charles Puech, in *Eranos Jahrbuch* 20, 1951, translated as 'Gnosis and time', in Joseph Campbell (ed.), *Man and Time, Papers from Eranos Yearbooks* vol.3, London 1958, 38-84. The French version is reprinted in Puech's *En quête de la Gnose*, vol.1, Paris 1978. I am grateful to D. Kyrtatas for drawing my attention to relevant texts.

who strongly emphasises this absence of duration, we shall (in Chapter 16) find the notion of presentness put to imaginative new uses, to show how God's knowledge is compatible with freedom.

Then, inevitably, there are writers who, unlike Boethius, use the notion of presentness in a way that reimports time, and who are unaware of the original implication of durationlessness. Two passages from modern authors, Richard Jeffries and Thomas Mann, are quoted side by side in one recent article,[117] and they nicely illustrate two opposite tendencies. In the first, Jeffries seems to be treating the eternal now in a thoroughly temporal way. Even though his intention is to describe a transcendence of time, the effect is more like a concentration on the temporal present:

> I cannot understand time. It is eternity now. I am in the midst of it. It is about me in the sunshine; I am in it, as the butterfly floats in the light-laden air. Nothing has to come: it is now. Now is eternity; now is the immortal life. Here this moment, by this tumulus, on earth, now; I exist in it.

In the second passage, Mann comes much closer to articulating the denial of temporal duration:

> We walk, walk. How long, how far? Who knows? Nothing is changed by our pacing, there is the same as here, once on a time the same as now, or then; time is drowned in the measureless monotony of space, motion from point to point is no motion more, where uniformity rules; and where motion is no more motion, time is no longer time.

Jeffries' absorption in the temporal present provides a distinct motive for emphasising presentness in an account of eternity. I cannot exclude the possibility that some writers of late antiquity grafted this idea inconsistently on to the older idea of an unextended eternity.

Aristotle

So far I have entirely excluded Aristotle from the discussion. And it may seem that he has little to contribute to the development of the concept of eternity. In Chapter 10 I shall deny this, maintaining instead that his contribution comes from a quite unexpected quarter, that is, from his account of thinking. It is true, however, that if we look at his discussion of time, he seems to add little that proved influential. For he does not use the notion of timelessness, as we have understood it, even though he talks of things not being *in time*. This needs some explaining.

The passage where he discusses what it means for something to be, or not to be, in time is in the *Physics*.[118] The connexion of thought is not always

---

[117] Richard Jeffries, *The Story of My Heart*, London, new ed. 1912, 30; Thomas Mann, *The Magic Mountain*, 1924, tr. H.T. Lowe-Porter, from the section, 'By the ocean of time'. The first is paraphrased, the second quoted, in John J. Clarke, 'Mysticism and the paradox of survival', *Int. PQ* 11, 1971, 165-79 (repr. in John Donnelly, ed., *Language, Metaphysics and Death*, New York 1978).

[118] 4.12, 220b32-222a9; 4.13-14, 222b16-223a15.

clear, but I shall make some suggestions. First, time is defined as the *number* of motion, in that it is the *countable* aspect of motion.[119] And a thing will be *in time*, if it has the appropriate kind of number, and its being is measured by that number.[120] It emerges that only rest and motion and the things which undergo these can then be *in time*, because only these are subject to the appropriate kind of counting and measuring.[121] This is the main statement, but there are several corollaries.

From his close comparison between being *in time* and being *in number*, Aristotle infers (*ei de touto*: if this is so, 221a17; *epei*: since, 221a26) that in order to be *in time*, a thing must be *included within* (*periechetai*) time and cannot last for the whole of time. I assume the connexion of thought is that infinity is not a number. To be numerable, a thing must have a finite number assignable to it. The result, that things in time are included within time, suits Aristotle because it supplies an analogy with his account of being *in place*, which involves being surrounded (221a18; a29-30). At the same time, it rules out a rival interpretation of being in time as being *simultaneous* with time, which accordingly gets discussed at this point (221a19-26). Further, the fact that things in time cannot last for the whole of time fits with the common idea that time is a destroyer which makes things age (*gēraskein* 221a31) and perish, an idea which Aristotle therefore introduces with a *dē* (then) at 221a30.

There is one more corollary which may cause puzzlement. Aristotle says that things which begin or cease to exist, whether by a process or instantaneously, are in time (221b28-31). This is connected with the point that they do not last as long as time. It may be thought that he has forgotten to consider whether these things are subject to motion and rest. In fact, this addition can be taken as given, in the case of those things which begin or cease by a process, for Aristotle believes that all these are subject to motion (*Metaph.*8.1, 1042b3-6). Indeed, they undergo motion when they come into being (222b23), so only the status of those which begin or cease instantaneously is in question. These include coincidences, relations, processes, points, lines, surfaces, instants, units, in general indivisibles and wholes, activities such as perceiving or being pleased, and certain non-substantial forms. Perhaps he thinks that some or all of these have sufficient connexion with motion to qualify as being in time.

It can now be seen that Aristotle's conception of not being in time is quite unlike the conception of timelessness which we have been considering. For things like the stars which, on his view, exist at *all* times, and which are therefore not timeless in the sense that concerns us, will none the less be *not in time* according to Aristotle's criterion. Is there anywhere, it may be asked, where he comes closer to the concept of timelessness? The obvious place to search is his account of God's eternity (*aiōn*), which is sometimes construed as being a kind of timelessness.[122] There are two important passages.

One is *Cael.*1.9, 279a12-b3, where Aristotle says:

[119] 4.11, 219b1-2.   [120] 4.12, 221a13; b14-16; b21-2.   [121] 4.12, 221b20-3.

[122] W. von Leyden, 'Time, number and eternity in Plato and Aristotle', *PQ* 14, 1964, 35-52, goes as far as to say, after some *caveats*, that if *aiōn* is applied to what lies outside the first heaven, 'it must practically come to mean the same as timelessness'.

At the same time it is clear that there is neither place nor vacuum nor time outside the heavens.

The reason why there is no time outside the outermost spherical shell of the universe is that there is no motion there, because no body. He then goes on to talk of the things there (*ta'kei*), meaning either God or the outermost spherical shell, or both, and says:

> Which is why the things there are not of a nature to have a place, nor does time make them age (*gēraskein*), nor is there change in any respect in the things which are arranged outside the furthest motion. Unchanging and unaffected, they continue for the whole of *aiōn* (*diatelei ton hapanta aiōna*), with the best and most self-sufficient life (*zōē*). Indeed, the word (*aiōn*) was a divine utterance on the part of antiquity. For the completeness which embraces (*periechon*) the length of life of a thing, and which is not naturally exceeded, is called its *aiōn*. By analogy the completeness of all the heavens and the completeness which embraces (*periechon*) the whole of time and infinity is *aiōn*. It takes its name from *aei einai* (always being), and is deathless and divine.

If God exists where there is no time, this might seem to give him the timelessness we were looking for. But the sequel makes this doubtful. For *aiōn* gets connected etymologically with 'always' (a word not detemporalised until Plotinus). Things are said to 'continue (*diatelein*) for the whole of *aiōn*'. And *aiōn* is said to 'embrace the whole of time and infinity'. Although the last expression is hard to understand (and I shall return to it in Chapter 14), the general impression created is that possessors of this special sort of *aiōn* have everlasting duration rather than timelessness. And there are two pieces of evidence that Aristotle is only trying to say that in his special sense they are not *in time*. First, they do not age (*gēraskei*), that is, grow more decrepit, as it will be recalled that things in time do. Secondly, they are not included or embraced (*periechetai*) by time in the way that things in time are, for it is explained that *aiōn* is itself what does the embracing. I conclude that the eternity (*aiōn*) of Aristotle's God is not timelessness, although it does involve not being in time in Aristotle's sense.

That God has everlasting duration emerges still more clearly from the other important passage which comes in the *Metaphysics*.[123] Here we are told about God's life that:

> It is a way of life like the best we ever have for a short time. For he is *always* (*aei*) in that state, which for us is impossible. ...

> If, then, God is always (*aei*) in that good state in which we are sometimes (*pote*), that is wonderful. ...

> God's self-dependent actuality is a life most good and everlasting (*aïdios*). We say then that God is a living being, everlasting (*aïdios*) and most good, so that life and continuous, everlasting (*sunechēs, aïdios*) *aiōn* belong to him. For that is what God is.

[123] 12.7, 1072b13-1073a13.

Parmenides fr.8, ll.1-25

I shall return to Aristotle's influence on the concept of eternity in Chapter 10. But I want to complete this chapter by returning to an earlier challenge. The question was whether the lines which immediately follow Parmenides' fr.8, ll.5 and 6 can be understood in conformity with the idea that in l.5 Parmenides is groping for the idea of timelessness. I shall try to show that they can be so understood by sketching out one possible interpretation of the first 25 lines of the fragment. It will only be one possible interpretation, and my comments will concentrate on the general structure of the argument, while ignoring many points of detail, including many disputes about the correct MS readings. I follow the suggestion of Owen and Barnes, it will be recalled, that the unnamed subject is whatever can be spoken or thought of, or inquired into. Parmenides' dimly conceived idea in l.5, I maintain, is that whatever can be spoken or thought of is timeless. The difficulties of understanding are not surprising, given that we have here a poem, recording the words of a goddess, composed by a philosopher struggling with novel ideas, written at the dawn of Western philosophy, and transmitted via divergent manuscripts.

> 1-4 Only one story and one path is left, namely, (a) that it is; and on this path are very many signs that (b) it is without creation or destruction; (c) whole, unique, unmoved and perfect.

> 5-6 (d) Nor was it ever, nor will it be, since (e) it now is, all together, (f) one continuous. For ...

The last word in this quotation is 'For ... '. Which of the preceding points are supported in the following lines (6 to 25)? I would suggest that ll.6 to 11 support (f) 'continuous', by showing that there are no *gaps* before or after the existence of the subject. Lines 12 and 13, if not merely recapitulating, will be adding a new argument in support of (e) 'all together', by showing that there cannot be belated accretions to the subject. A third argument comes in ll.19-20, and supports both (d) 'nor will it be' and (e) 'all together' by showing that the subject is not spread out over time, and in particular does not have a future. These three arguments all depend in some way on (b) 'it is without creation or destruction', and this fact answers the very legitimate question raised by Schofield, why (b) features so prominently in these lines. They do so, I am suggesting, not because (b), the absence of creation and destruction, represents Parmenides' ultimate interest in ll.5 to 21, but because (b) is used in turn for supporting (d), (e) and (f).

Finally, we get an argument in ll.22 to 25 which supports (d), (e) and (f) simultaneously, by showing that there are no divisions. If there are no divisions, we will have (f) continuity (explicitly mentioned in ll.23 and 25); and (d) no divisions separating a past, a present and a future, which in turn means that (e) the subject exists all together, and is not spread out over time (I take the word 'now' to be glossed by 'all together'). This last argument also establishes the first of the items in (c), namely 'whole'. Indeed, 'whole' is probably just a way of saying (f) 'continuous'.

Let us now look at the next group of lines:

6-11 For what generation will you seek for it? How and whence did it grow? (i) I shall not allow you to say or think 'from non-being'. For it is not sayable or thinkable that it is not. (ii) And if it did begin from nothing, what need would have driven it to grow later rather than earlier? Thus it must either be completely or not at all.

The premise for the first argument, that you cannot speak or think of the non-existent has already been argued in earlier fragments which I have not recorded. The second, or 'why not sooner?', argument, here formulated for the first time in Western thought, came to be repeated again and again through the ages, as will be seen in Chapter 15. In ruling out creation, Parmenides expects us to see that at least the first argument will count against destruction as well. He has thus eliminated discontinuities before and after the existence of the subject.

12-13 Nor will the force of trust allow that something could ever come into being from non-being to be set alongside the [original] subject.

There are plenty of alternative ways of rendering these lines, but this way at least makes the point a new one: there can be no accretions to the original subject, and this removes at least one barrier to its existing 'all together'. Again the premise will be the unthinkability of anything, including an accretion, springing out of the *non-existent*.

13-18 Therefore Justice has not loosed her chains or allowed it to come into being or to perish, but rather holds it fast. And the decision about these things lies in this: it is, or it is not. But it has been decided, as is necessary, to leave aside the one path as unthinkable and unnameable, for it is not a true path, and that the other path, according to which it is, is actually the real one.

This portion sets aside an alternative hypothesis, namely, that the subject does not exist *at all*. It is rejected on the familiar ground that the non-existent is unthinkable.

19-20 How could what is be subsequently? How could it come into being? For if it came into being, it is not; nor is it, if it is at some time going to be.

Line 11 implied that, if the subject ever came into being, then it cannot be said to be '*completely*' (*pampan*). The new suggestion may well be that the same applies, if some of the subject's existence is stored up for the future. Then too the subject cannot be said to exist completely, for some of its career is still to come. By understanding the word 'completely' from l.11, we get a neat argument which supports (d) 'nor will it be' and (e) 'all together', by exploiting once again the ban on creation. This much-exploited ban can thus be relevantly recalled in the next line.

21 Thus coming into being is extinguished and ceasing to be is not to be heard of.

We now come to the last of the arguments which concern us.

> 22-5 Nor does it have divisions, since it is all alike, and there is not more here
> and less there, which would prevent it from being continuous; but it is all full of
> what is. Thus it is all continuous, for what is crowds on what is.

Parmenides here seems to construe divisions as (temporal) gaps in which
there would have to be *less* of what is. What is would have become thinner or
rarer, and the gaps would not be quite full of it. This would violate the
conclusion already reached that what can be spoken or thought of has to
exist *completely*. It would also involve trying to think about the non-existent
portion whose absence had created the thinning out. Even divisions between
past, present and future would involve gaps of this kind. Thus Parmenides
can insist not only on what he explicitly mentions, continuity, but also on the
absence of the familiar temporal distinctions, and on his subject's existing
'all together'.

I have been attempting no more than to show that the 'timeless'
interpretation of 1.5 can fit as easily as any other into the surrounding
context.

CHAPTER NINE

Is Anything Timeless?

I must now fulfil my promise, made in the last chapter, to consider whether there are any entities which can plausibly be regarded as timeless, and if so, which. Martha Kneale has argued boldly that the notion of timelessness is otiose and could be allowed to collapse into that of sempiternity (existence at *all* times), or that of necessity.[1] Let us then consider what things have been thought of as timeless.

Universals

As we saw in the last chapter, Plato sometimes spoke of universals as existing timelessly. If we take *justice* as an example of a universal, then whether he was right depends on what it means to say that justice *exists*. One view is that of Aristotle in the *Categories*, which implies that to say that justice exists is to say that just men (or acts, or institutions) exist. It is quite possible in that case that justice does not yet exist, or that it should not have existed for long. We could then, *pace* Plato, correctly, if quaintly, say that justice is growing older. And we could correctly use tenses and say that it did exist, or that it will.

In order to defend Plato, we should need to look for some other context of discussion. Somebody, for example, might say that there is such a thing as justice, meaning to contrast justice with something like a perfect society, which he views as incapable even of being coherently conceived. Or somebody might say that there is such a thing as a perfect answer to an elementary arithmetic examination – a hundred answers right out of a hundred – but no such thing as a perfect essay in philosophy. There may be a brilliant essay, or an essay of genius, but this is not the same thing as a *perfect* essay. The claim here, that there is such a thing as a perfect answer paper in arithmetic, has a sense which is shown by the contrast with philosophy, and which has nothing to do with there being *instances* of the perfect arithmetic answer. For all the speaker knows, no one has ever scored a hundred marks out of a hundred. His claim that there is such a thing as a perfect answer paper implies at most the *possibility* of instances. Similarly, the claim that there is such a thing as justice need not, in this sort of context, imply that there ever have been instances.

An analogous case is suggested by David Armstrong, but dismissed too

---

[1] M. Kneale, 'Eternity and sempiternity', *PAS* 69, 1968-9, 223-38.

quickly, I believe.[2] Following an example of Hume's, we might assemble samples of all the available shades of blue, arrange them in a steady graduation and notice that there was a missing shade of blue. We may say that there is such a shade, even if we are convinced, after search, that there are no instances at all. Once again, I think the sense of the claim can be brought out by a *contrast*: if someone suggests that there is also a missing shade beyond the darkest navy, we may reply that there is no such shade of blue, because anything darker than that would be black.

The suggestion is, then, that the question whether there is such a thing as justice can be understood in *one* way as independent of the question whether there are actual instances. But ought we to conclude that if justice exists, it exists *timelessly*, that is, at *no* times, or rather that it exists at *all* times? The trouble with saying that justice exists (or, more naturally, that there is such a thing) at *all* times, is that this makes it sound as if we are reverting to the claim that there are actual instances, only adding that there are instances at *all* times. This is the case for saying that, on *one* interpretation, the existence of justice can be viewed as timeless, except that, as remarked, it is more natural to put the point not in terms of existence, but of there being such a thing.

None the less, it is hard to carry the argument over to the case of other universals discussed by Plato, such as fire and snow. The claim that there is such a thing as snow almost irresistibly suggests that there are *actual* instances, even if only intermittently. Moreover, in that part of the *Timaeus* where Plato comes closest to making universals timeless, 37C-38C, the example which he has most in mind is that of a living being. And to say that there is such a thing as a living being does again strongly suggest that there is at least one actual instance, even though that is not the way that Plato wants to take it. I can therefore offer only a partial vindication of Plato. But even a partial vindication can be useful, if it suggests a philosophical motivation for him. Sometimes his theory of Forms is discussed without a search for philosophical motivation, in the manner of an uninterpreted algebraic system. Then we lose both one of the main reasons why his theory should interest us and one of the tools which (with due caution) we can use in assessing his meaning.

Numbers

Numbers are another thing which some people have thought to be timeless. But that will depend on what sort of thing numbers are. Some have suggested that two is the set of all twosomes in the universe, others that numbers are logical constructions out of our arithmetical operations. The settling of these questions will clearly affect whether numbers can be viewed as timeless.

Truth

A favourite candidate for a timeless entity is truth. This suggestion is more plausible, if it is confined to the truth of certain kinds of proposition, for

---

[2] David Armstrong, *Universals and Scientific Realism*, vol. 1, Cambridge 1978, ch. 7, pp. 64-5.

example, of the proposition that $2 + 2 = 4$, or that the universe has neither beginning nor end. If these propositions are true at all, it might be said, their truth exists timelessly. But why should we not rather say that they are true at *all* times? A child who misunderstood, and asked if it had been true in his grandfather's time that two and two were equal to four, could naturally be corrected by saying that it has *always* been true that two plus two equals four. By contrast, the corresponding correction to the inquiry whether there was such a thing as justice in his grandfather's time, namely, that there has *always* been such a thing, would carry an implication not necessarily wanted – that the interest was in there being *instances* of justice.

I do not know if there are any good arguments for viewing the truth of the proposition that $2 + 2 = 4$ as timeless rather than omnitemporal. One argument would be that, on a certain view, time would come to an end, if the created universe came to an end, but the truth of the proposition that $2 + 2 = 4$ should be treated as *independent* of the fate of the universe. On that view, the proposition's truth should not be thought of as existing at any times, lest it be ended by the ending of the universe.

Many logicians have not been content to argue about particular truths, but have maintained that *all* truth is timeless.[3] The most striking difficulty for this view is provided by those many examples in which what is said begins or ceases to be true. For example, if the commander says, 'The decisive battle lies in the future,' it would seem that his belief, the sentence he utters and the proposition he expresses may be true now, but cease to be true later. What makes this change of truth-value possible is the presence of the 'token-reflexive' or 'flowing' time-expressions, which were distinguished in Chapter 3. The verb 'lies' is tensed, and the tenses are an example of a token-reflexive device. Other examples are the terms 'now', 'soon', 'recently', 'today', 'tomorrow', 'yesterday', 'ago', 'hence', 'at present', 'in the future', 'in the past'. The point is that it may be true at one time, but not at another, to describe something as happening 'now', 'soon', or in other token-reflexive ways.

There are other ways as well in which we seem to connect truth closely with time. We say, 'Your prediction *was* true,' selecting the past tense, 'was', to mark the fact that the prediction is past. Again, in sorting out the historical situation at some past date, we may say, 'At the time of Queen Elizabeth's birth it *was* true that there was going to be a second world war (but not true that there had been one).' Here we use the past tense again, 'was', to show that the time we are thinking about, the time of Queen Elizabeth's birth, is earlier than the present discussion. Philosophers have argued about artificially tenseless sentences like, 'There is (tenseless) a battle at Hastings.' Could this be used to express a true proposition before there was a battle, or a proposition at all before there was a Hastings? If not, does the sentence *become* true at some time? All these ways of thinking seem to make truth exist at particular times.

Believers in the timelessness of truth have several replies. As regards the idioms involving the use of '*was* true', these might be said to call for

---

[3] I have discussed this view in an earlier book, *Necessity, Cause and Blame*, ch. 5.

paraphrase along the following lines: 'What you predicted is (no past tense here) timelessly true.' Or again: 'At the time of Queen Elizabeth's birth, there was going to be (no mention of truth) a second world war.'

The hardest case for believers in the timelessness of truth is provided by the examples of changing truth value. Some atemporalists take a modified position, and concede that a *sentence* or *belief* can cease to be true. But they insist that truth attaches primarily not to these, but to the *proposition* expressed by a given use of a sentence. If instead we ascribe truth to the *sentence*, 'The decisive battle lies in the future,' this must be understood as meaning that the sentence can be used to express a *proposition* which is timelessly true. And if we say that the sentence ceases to be true, this must be understood as meaning that it can no longer be used to express the timelessly true proposition which it could have expressed before, but can only be used to express a *different* proposition which is timelessly false. At no stage in this analysis is it conceded that a *proposition* ceases to be true. And if truth attaches primarily to propositions, then the theory needs only a little reformulation. It will now say that truth *in its primary application* attaches timelessly.

But why do the atemporalists believe that the proposition expressed by a given use of the sentence 'the decisive battle lies in the future' is true timelessly, if true at all? One argument is that the same proposition could be expressed, if the tenses and other token-reflexive expressions were replaced, and that once that replacement was made, the proposition expressed would not even appear to change its truth value. For example, if today is 1 January 1980, the commander could say, 'The decisive battle is (tenseless) later than 1 January 1980.' That sentence would *add* some extraneous information about dates, but, on the theory under consideration, it would not *subtract* anything, and the proposition expressed, if true, would not cease to be true.

One trouble with this idea has been brought out in Chapter 3. The point is that something *would* have been subtracted; for tenses, and other token-reflexive devices, have a unique action- and emotion-guiding force, and if the commander does not supply this, the soldiers will need to supply it for themselves. Changing the example, to make it more dramatic, let us suppose the commander says, 'A hand grenade is thrown (tenseless) into this room on 1 December 1978.' The soldiers will need to be able to judge whether 1 December 1978 is *today*, or years into the *past* or *future*. For without this information they will not know whether to take action or to feel any urgency. Of course, most soldiers could easily supply the action- and emotion-guiding information for themselves. But that does not alter the situation; for the point is that the commander's tenseless sentence does not include it.

The atemporalists therefore need a different argument, and one argument runs as follows. Suppose it was earlier true to say, 'The decisive battle lies in the future.' Then the proposition expressed will not change its truth value when the decisive battle is over. Rather, it will merely then need to be expressed by the *differently* tensed sentence, 'The decisive battle *lay* in the future.' And that sentence clearly expresses something true. The illusion of a change to falsity is created by the fact that the *original* sentence will by now express a false proposition. But that is an irrelevance, because that proposition will not be the *same* as the one originally expressed. This replaces the old move, according to which a tenseless and a tensed sentence express the same proposition. The new move is to say that two differently tensed sentences express the same proposition.

In order to cast doubt on this, let us first take an example in which we have a positive interest in whether something lies in the past or in the future. Thus if I am interested in a house coming on the market, it makes a difference whether I am told beforehand 'The house will be sold,' or told only afterwards, 'The house has been sold.' Some people may be persuaded straight off that I am not being told the same thing on the two occasions. For in one case I am told, 'The sale is after now' and in the other case, 'The sale is before now'. But, in case it is not clear that this constitutes different information, I would argue that what I am told on the later occasion guides my emotions and actions quite differently. It may disappoint me and lead me to desist from making a bid. If I earlier expressed my interest, I may be indignant with the man who has informed me too late. And he cannot plead that by telling me 'The house has been sold,' he is giving me the *same* information as would earlier have given by saying 'The house will be sold.' I do not mean that the different guidance is part of what is expressed by use of the two different sentences. Rather it is a *symptom* of the difference in what is expressed, when I am told that the sale lies in the future or alternatively that it lies in the past.

This is not to deny that what was expressed by saying, 'The house will be sold' could later be *confirmed* by saying, 'The house has been sold.' In order to *confirm* what has been said, we do not have to express an *identical* proposition, but only one which is *relevantly* related.

I conclude that the last strategy for avoiding change of truth value is unsuccessful: we cannot identify the true proposition expressed by 'The house will be sold' with the true proposition expressed by a later use of 'The house has been sold.' On the contrary, the first proposition will cease to be true in a certain respect after the sale. This fact may be concealed, if we switch to an example in which we have no *interest* in the respect in which the proposition ceases to be true. It would then be positively *misleading* to say that the proposition ceased to be true, because it would give people to understand that it ceased to be true in some *relevant* respect. None the less, even then it would be *true* that the proposition changed its truth value; the only misleading thing would be to *say* so in an ordinary conversational context.

The more general upshot is that, even if the truth of some things (for example of the proposition that $2 + 2 = 4$) could be viewed as timeless, this is not so for the truth of any and every proposition.

Is the idea of eternity inconsistent?

Martha Kneale attacked not only the idea of timelessness as otiose, but Boethius' idea of eternity as inconsistent. For in that idea Boethius combined with the notion of timelessness the notions of *all at once* (*simul*) and of *life*. Since both of these are temporal notions, she objected, there is a double inconsistency, and both points have worried other people too.[4] Eternity, according to Boethius, is the endless and perfect possession of life all at once (*Consolation* 5.6).

From the discussion in the last chapter, it will be clear how Boethius should be defended, at least on certain charges. 'All at once' is *not* a temporal notion, but expresses the idea that God's existence is *not* extended or spread out stage after stage.

The treatment of *life* proved more complicated. In Plato, it emerged, the eternal Form of living beings, though itself described as a living being, is not necessarily a living being in quite the ordinary sense. In Plotinus, the life in question consisted of a special kind of thinking, which he was led to envisage as timeless (although I did not follow him), partly because of its non-progressive character and the loss of any *sense* of time, and partly because of the metaphysics which made him discount as less real the world of clocks and of other physical events which we might use to impugn the sense of timelessness.

When Boethius speaks of timeless life, the kind of life which he has in mind is that of the Christian God. So we need to consider what kind of life that is before deciding whether it is compatible with timelessness. I shall postpone that task until Chapter 16, because it will be easier to carry out after I have considered (in Chapter 15) the idea of a *changeless* God acting as a creator. I will only say for the present that William Kneale has objected that God's life must at least involve *acting*, and that he and many others have complained that a timeless being cannot *act*.[5] That is one of the subjects which I shall consider in Chapter 16.

Conclusion

I have argued that the idea of a timeless entity can be made sense of in a few cases. In other cases it cannot be, but that takes some showing. Often the incoherence cannot be assumed straight off. Consequently, the idea of a timeless being has a certain plausibility which is missing from some of the other ideas considered in the last chapter, particularly from the idea of a being with an enduring temporal present, but no past or future, or from the idea that an instant is all the time there is.

---

[4] M. Kneale, op. cit., 227. On *simul*, cf. Richard Swinburne, *The Coherence of Theism*, Oxford 1977, 220-1. On *life*, W. Kneale had already complained of a contradiction ('Time and eternity in theology', 99), and Swinburne (p.218) cites several modern theologians who do so (Paul Tillich, Karl Barth, O. Cullman).

[5] W. Kneale, op. cit., 99; Robert Coburn, 'Professor Malcolm on God', *Australasian J. Phil.* 41, 1963, 155; J.R. Lucas, *A Treatise on Time and Space*, London 1973, 303; Nelson Pike, *God and Timelessness*, London 1970, 104-7; Richard Swinburne, op. cit., 221 (cf. p. 218 for references to modern theologians).

CHAPTER TEN

Myths about Non-Propositional Thought

Plotinus connected eternity with the second of two kinds of thinking. First there is *dianoia*, which is often called discursive thinking, and which is the activity of the soul (*psuchē*), and then there is the different activity of the intellect (*nous*), which is often called non-discursive thinking. It is commonly held that non-discursive thinking does not involve entertaining propositions. That is, it does not involve thinking *that* something is the case. Instead, one contemplates concepts in isolation from each other, and does not string them together in the way they are strung together in 'that'-clauses. It is further supposed that Plato and Aristotle anticipated Plotinus in postulating this non-propositional thinking.

I have four aims in this chapter. One is to show how Plotinus connects eternity with thinking, and so in turn to show that Aristotle does after all make a contribution to the concept of eternity. He does so by influencing Plotinus' conception of *thinking*.

My second aim is to deny that non-propositional thinking is to be found in Plato, Aristotle, or Plotinus, at any of the points where it has most commonly been detected. In order to show this for the case of Plotinus, I shall have to explain some of Aristotle's ideas about thinking and how Plotinus transformed them. He certainly believed that there is a mystical state in which we have contact with something much simpler than any proposition. But I shall maintain that he regards this mystical experience as *above* the level of thinking, while thinking in its highest form he treats as propositional.

I have a third aim, because among the many Aristotelian ideas about thinking that influenced Plotinus, one is especially interesting. It is that non-discursive thought does not involve seeking, and that in general contemplating the truth is more rewarding than seeking it. This was certainly a majority view in Greek philosophy. But I want to trace and endorse the minority view which takes 'perpetual progress' as an ideal, as opposed to static contemplation.

My fourth aim is connected with the need to describe some of Aristotle's views about thinking. They have proved very difficult to understand, and I do not believe that he is at his strongest on this subject. But I think it can be seen why he says what he says, so the remaining aim will be to explain his account of thinking, though not to justify it.

I shall start, then, by showing how Plotinus connects thinking with eternity.

Plotinus: eternity and thinking

There is a startling feature in Plotinus' concept of eternity, for he defines time and eternity as two kinds of *life*. This idea is not new with him: it appears already in Philo, in the first century A.D., who is influenced by Middle Platonism. Eternity, according to Philo, is the life of the intelligible world, time the life of the perceptible world.[1] How could this idea develop out of Plato?

Plato had already made time and eternity to be the attributes of living things. For he thought of the physical universe as a living organism, animated and moved by a soul. Moreover, the Forms on which this living organism was modelled had themselves to be, in some sense, a living organism; for Plato always makes model and copy resemble each other. To make the Forms living was even more tempting for his successors, once they had made Forms to be thoughts in the mind of the divine intellect, and had taken over the Aristotelian idea, to be discussed below, that this living intellect is identical with its objects. The remaining step is to make time and eternity not merely the attributes of living things, but their *lives*.

However, Plotinus' theory is not the same as Philo's, because in Plotinus we find not just two levels of reality, but three: soul, intellect and the One. Soul is treated, like Plato's realm of perceptible things, as being temporal, while intellect is treated, like Plato's ideal Forms, as eternal. Above these Plotinus postulates the One, which cannot properly be described by either of these terms,[2] or by any terms at all.

Although soul and intellect are both psychological entities, the difference from Plato's two levels is not so great as may at first appear. Thus the intellect corresponds to Plato's Forms, in that it apprehends forms; not only that, but, in a sense to be explained, it is identical with the forms which it apprehends. Admittedly, soul, at the level below, does not sound very like Plato's perceptible realm. But Plotinus has a definite reason for diverging from Plato here: matter has so little hold on reality that he does not think material things deserve a separate slot in his scheme. He pays more attention to the soul which animates Plato's physical universe. Finally, at the summit, Plotinus' One represents a higher level than anything recognised by Plato, but it too is modelled on things which Plato says. In the *Republic*, Plato says that there is one supreme Form, the Form of the Good, which is 'not being, but beyond being' (*epekeina tēs ousias*, 509B8-9). Again, in his dialogue the *Parmenides*, Plato discusses the One of Parmenides, and Plato's negative description of Parmenides' One has been shown to have had its influence on Plotinus.[3]

When Plotinus takes up the idea that time and eternity are *lives*, he makes time to be the life of the soul and eternity the life of the intellect. This is the point at which eternity and thinking come to be connected. For the life which constitutes eternity, being the life of the intellect, is a life of *thinking*. But the

---

[1] Philo *de Mut. Nom.* 47,267; cf. *Quod Deus ...* 6.32.

[2] Despite the use of the words *aïdion* (everlasting, 5.4.2 (19)) and *aei* (always, 6.8.16 (33)).

[3] E.R. Dodds, 'The *Parmenides* of Plato and the origin of the Neoplatonic One', *CQ* 22, 1928, 129-42.

thinking must be of a very special kind, if it is to be non-temporal. Before trying to understand it further, let us record Plotinus' definitions of time and eternity. Eternity (*aiōn*) is

> a life which stays in the same state, always having everything present to it, and not one thing after another but everything together; again not some things at one time others at another, but a completeness without parts, and with everything together as if in a point before flowing out into a line.[4]

Again, it is

> a life concerned with being, residing in being, all together, not extended in any direction.[5]

Time, on the other hand, is

> a life of the soul subject to movement which progresses from one mode to another.[6]

Time is a mere copy of eternity, the latter being

> a life which stands still and alike in the same state already without boundaries.[7]

The soul and intellect of which Plotinus speaks are a sort of world-soul and world-intellect. But we can also find them within ourselves. And he believes that humans can ascend from the kind of thinking characteristic of the soul to that characteristic of the intellect, and finally to a mystical experience of the One. In order to throw light on Plotinus' account of these different kinds of thinking, I shall now turn to the ideas of Aristotle which influenced him.

Aristotle: non-discursive thought is propositional

I shall start with Aristotle's account of non-discursive thinking, which comes in two related chapters.[8] The usual interpretation is most clearly articulated by A.C. Lloyd, though he is concerned with Plotinus rather than with Aristotle. It is that non-discursive thought involves contemplating things in isolation without thinking anything *about* them.[9] In thinking that beauty is truth, my mind passes from beauty to truth. That is offered as an example of ordinary discursive thinking. But the suggestion is that this passage from

---

[4] 3.7.3(16-20). [5] 3.7.3(36-8). [6] 3.7.11(43-5). [7] 3.7.11(45-7).

[8] *Metaph.* 9.10, 1051b27-1052a4 and *DA* 3.6, 430b26-31.

[9] A.C. Lloyd, 'Non-discursive thought – an enigma of Greek philosophy', *PAS* 70, 1969-70, 261-74. For the explication in terms of knocking unconscious, see p. 270. Lloyd agrees that in the end the idea of non-discursive thought, so understood, is incoherent. For a valuable survey of interpretations of Aristotle, see E. Berti, 'The intellection of indivisibles according to Aristotle *de Anima* 3.6', in G.E.L. Owen and G.E.R. Lloyd (eds), *Aristotle on Mind and the Senses, Proc. 7th Symposium Aristotelicum*, Cambridge 1978. Berti expresses agreement with the kind of interpretation which I offered in discussion on that occasion, and which I later put into print in *Necessity, Cause and Blame*, and in E. Berti (ed.), *Aristotle on Science: the 'Posterior Analytics', Proc. 8th Symposium Aristotelicum*, Padua 1981.

concept to concept already implies the possibility of contemplating something in isolation. For will there not be a stage at which my mind is contemplating beauty without yet having passed to truth? Admittedly, I do not allow the concept of beauty to remain in isolation, for I promptly link it up with the concept of truth. But (it is suggested) if I were knocked unconscious by a no. 68 bus before I had done so, I should then have thought of beauty in isolation.

There are several objections to this suggestion. One is that I could hardly be said to have thought of beauty, if I did not go on to think something about it, even if only that I wanted to know what its attributes were. It must also be doubtful that Aristotle can be considering the kind of interrupted thinking just described. For one thing, he seems to regard the kind of thinking in question as the loftiest achievement of man in his happiest moments and the permanent activity of God.[10] That is suggested, at any rate, by the fact that these kinds of thinking are all compared with touching.[11] It is hard to see what is so lofty about interrupted thinking. It is also hard to see how contemplating something in isolation, without thinking anything about it, could lead to *truth*, as Aristotle says that non-discursive thinking does.[12] For we should expect there to be neither truth nor falsehood, unless we are in some sense *combining* concepts, and indeed Aristotle himself sometimes expresses this view.[13]

It may be thought easier to defend the non-propositional interpretation, if we drop from the explication the notion of being knocked unconscious. Perhaps the idea should rather be that, although the non-discursive thought of beauty is involved in the discursive thought that beauty is truth, none the less it cannot ever occur on its own. But this new suggestion will not do when we come to Plotinus, since he clearly does believe that non-discursive thought can occur without discursive, and Aristotle probably believes the same, at least for the case of God. In any case, the new suggestion is little improvement on the old, since it is still not clear why merely having beauty in mind should be thought of as a very lofty achievement or as involving truth. Loftiness and truth might be relevant if, instead of merely having beauty in mind, we had a full understanding of it. But then understanding is *propositional*, since it involves appreciating *that* so-and-so is the case.

Instead of trying to defend a non-propositional interpretation, it may be better to look for an entirely different one. And I shall start by asking why it was ever supposed that Aristotle had in mind the non-propositional contemplation of isolated concepts. The most important reason is his saying that in this kind of thinking we do not predicate anything of anything (*ti kata tinos*, *DA* 3.6, 430b28), nor is there any assertion (*kataphasis*, *Metaph.*9.10, 1051b24). But I think there is a better interpretation available.

One of the loftiest achievements for a human being, according to Aristotle, is to engage in theoretical science, and this involves knowing the essences,

---

[10] *Metaph.* 12.7, 1072b14-26; *NE* 10.8.
[11] *Metaph.* 9.10, 1051b24-5; 12.7, 1072b21.
[12] *DA* 3.6, 430b28; *Metaph.* 9.10, 1051b24.
[13] *DA* 3.6, 430a27-b6; *Cat.* 4, 2a7-10; *Int.*1, 16a9-18.

that is, roughly speaking, the defining characteristics,[14] of the various subject matters. For Aristotle's account of a science is that one knows the definitions of the basic entities in that science, and by reference to these definitions can explain the further characteristics of all the entities concerned. In our two chapters, Aristotle is talking about subjects which are incomposite (*asuntheta* 1051b17; *adiaireta* 430a26), in the sense, I believe, that they do not involve matter as well as form. He is further discussing, I believe, *definitions* of these incomposite subjects, which state what their essences are. Hence the reference to 'what it is' (*ti esti*, 1051b26; b32), and to 'what it is in respect of essence' (*ti esti kata to ti ēn einai*, 430b28). Aristotle's non-discursive thinking will then involve contemplating the definitions of incomposite subjects. But in that case, the thinking must be *propositional*; for it will involve thinking *that* such-and-such an essence belongs to such-and-such a subject. How can this be squared with the claim that there is no asserting, nor predicating something of something?

I think the answer is that Aristotle often views definitions as being statements of *identity*. They do not therefore require us to predicate one thing of another, but involve simply referring to the same thing twice. This is not assertion or predication as Aristotle usually understands it. That Aristotle sometimes thinks of statements which give the essence of something, or part of its essence, as *identity* statements has been argued by G.E.L. Owen and Christopher Kirwan.[15] The evidence is that he says that it is by being something *other* than a pale thing (viz. a man) that a man is pale, but it is not by being something *other* than an animal that he is an animal.[16] Again, pale is predicated of an individual man as one thing of *another*, whereas man is not predicated of him as one thing of *another*.[17] There is a further statement even closer to our interests. In *Metaph*.7.11, 1037a33-b7, Aristotle is talking of a subject which is not a composite involving matter as well as form (*suneil-ēmmenon tēi hulēi*). Here at least, he says, the subject is *identical* with its essence.

We are now in a position to understand what is perhaps the most surprising statement of all. Aristotle says that in this kind of thinking you cannot be mistaken, but can only touch or not touch (*thigein, thinganein*, 1051b24-33). The idea is, perhaps, that if you try to state the essence of an incomposite subject and fail, you are not in error, because you have not succeeded in talking about the subject at all. You have not made contact with it. The contact metaphor is more useful than the seeing metaphor here,

---

[14] It is a slight oversimplification to identify form or essence with defining characteristics. That is the picture given by early works such as the *Posterior Analytics*: form or essence consists of genus and differentia, and the form or essence of lunar eclipse (to take one example) would be the moon's loss of light due to screening by the earth. In later works, however, the form is restricted to something less than the full defining characteristics, for certain material characteristics are excluded. Thus (*DA* 1.1, 403a25-b9) the form of a house is a shelter protecting from wind, rain and heat, whereas the full defining characterics would include being made of stones, bricks and timbers.

[15] G.E.L. Owen, 'The Platonism of Aristotle', *PBA* 50, 1965, 125-50, esp. 136-9 (repr. in *AA* 1). Christopher Kirwan, *Aristotle's Metaphysics Books gamma, delta and epsilon*, Clarendon Aristotle Series, Oxford 1971, 100.

[16] *An. Post.* 1.4, 73b5-10; 1.22, 83a32; *Phys.* 1.4, 188a8; *Metaph.* 14.1, 1087a35; 1088a28.

[17] *Metaph.* 7.4, 1030a2-6; 10-14.

because there are degrees of clarity in seeing, but contact is an all-or-nothing affair. Plato had also maintained that one could not be mistaken about certain identity statements. No one, mad or sane, has ever said to himself that a horse was an ox.[18] And Plato like Aristotle uses a tactual metaphor, namely, that of grasping (*ephaptesthai, Tht.*190C6).

I think that this interpretation meets the requirements that any interpretation must meet, since it explains the loftiness and the possibility of truth, while at the same time explaining the absence of predication, the impossibility of error, and the insistence that we can only touch or not touch.

Plato's Republic: knowledge of the Forms is propositional

There is another kind of thinking which I believe to be propositional. In Plato's *Republic* 509D-541B, there is a discussion, which strongly influenced Plotinus, of how philosophers can ascend through dialectical training to knowledge of the ideal Forms. It is very commonly taken that the knowledge they acquire is some kind of 'knowledge by acquaintance'.[19] By that is meant a knowledge like that involved in knowing a person, and it is usually supposed to be non-propositional. There is controversy as to whether Plato later renounced this conception of knowledge in the *Theaetetus*,[20] but on the *Republic* there is fairly widespread agreement. I do not think, however, that the common interpretation is right.

The thinking described in this passage of the *Republic* seems to me, on the contrary, to be propositional. Thus progress towards knowledge of the Forms is said to start from questions like 'What is largeness?', 'What is smallness?'.[21] For many years, there will be an intensive course in dialectical argument, which involves[22] question and answer. The questions are designed to trap the answerer into a contradiction and so to refute him (*elenchein*).[23] The method is meant to enable one to grasp what (or that which?) each thing is (*ho estin hekaston*),[24] and eventually what (or that which?) goodness itself is (*auto ho estin agathon*).[25] In 534B3-534C5, Plato concludes:

---

[18] *Tht.* 190B-C; cf. 188B and *Phd.* 74C1-2.

[19] I think I can fairly ascribe this view to Gilbert Ryle, commenting on *Tht.* 184B-186E, in 'Plato's Parmenides', *Mind* 48, 1939, 129-51 and 302-25 (repr. in R.E. Allen (ed.), *Studies in Plato's Metaphysics*, London and New York 1965, see pp. 136-41); D.W. Hamlyn, 'The communion of forms and the development of Plato's logic', *PQ* 5, 1955, 289-302; R.S. Bluck, 'Logos and forms in Plato: a reply to Professor Cross', *Mind* 65, 1956, 522-9; and ' "Knowledge by acquaintance" in Plato's *Theaetetus*', *Mind* 72, 1963, 259-63; W.G. Runciman, *Plato's Later Epistemology*, Cambridge 1962, 40-5; J.H. Lesher, *Gnōsis* and *Epistēmē* in Socrates' dream in the *Theaetetus*', *JHS* 89, 1969, 72-8; John McDowell, *Plato, Theaetetus*, Clarendon Plato Series, Oxford 1973, 115-16.
In disagreeing with the common interpretation, I have been anticipated by Myles Burnyeat in an unpublished paper delivered at Princeton in 1970, 'The simple and the complex in the *Theaetetus*'. Burnyeat also draws attention to the wording in *Rep.* 534B-C. Other dissenters are Gail Fine, 'Knowledge and logos in the *Theaetetus*', *Phronesis* 24, 1979, 70-80; Julia Annas, *An Introduction to Plato's Republic*, Oxford 1981, 280-4.

[20] Ryle postulated a renunciation, and was followed by Hamlyn, but Bluck (1963) and McDowell (p.193) disagree.

[21] *Rep.* 524C11.   [22] *Rep.* 534D9.   [23] *Rep.* 534C1; C3.   [24] *Rep.* 532A7; 533B2.   [25] *Rep.* 532B1.

Do you not call a man a dialectician, if he gets an account (*logos*) of the being (*ousia*) of each thing? And will you not deny that a man understands something, if he does not have such an account, and in so far as he cannot give an account of the thing to himself or others? ... And is it not, then, similar with goodness? If someone cannot define (*diorisasthai*) the Form of the Good with an account, separating it from all other things; if he cannot come through all refutations (*elenchoi*) as if in battle; if he does not desire to produce real refutations rather than merely seeming ones; if he does not in all these things journey through with an unfaltering account; will you not deny that such a man knows goodness itself, or anything else that is good?

Propositions are involved throughout this account. For the questions, answers and refutations all bear on propositions, and what is being sought is definitions. The Form of the Good is not itself a proposition. But to know it is to know the proposition that goodness is so-and-so.

But I must face a series of objections to the claim that the thinking here is propositional. First, Plato is insistent that the kind of knowledge involved cannot be conveyed in writing.[26] Why not, if it is propositional? I would answer that these comments of Plato's are entirely appropriate to definitional knowledge in philosophy, and do not at all imply a non-propositional knowledge. Definitional knowledge cannot be conveyed in writing, because one cannot be said really to know that goodness is so-and-so, until one has gone through the dialectical process. One must try one definition after another, seeing how the others fail, and how the successful one exactly surmounts all previous difficulties, and achieves all that the others could not.

A second objection to the propositional interpretation would be that apprehending the Form of the Good is described as if it were a kind of vision and is compared with coming to see the sun. Such experiences are certainly non-propositional. But I would reply that the analogy with vision is a very appropriate one, provided it is understood in the right respect. For it describes what it is like to realise that the new definitional formula does at last achieve what all the others could not.

A third objection is that I have not taken into account what Vlastos so rightly stresses: the almost religious significance which Plato attaches to apprehending the Forms.[27] But I think the religious significance is not out of place, when so much importance has been attached to the ascent, and when so much of life has been devoted to it. The dialectical training is not completed until the age of thirty-five, and it is not expected that the supreme Form will be understood until a further fifteen years of practical experience has been gained in public service (535A-541B).

This last point helps to answer a possible fourth objection. For I may be said not to have taken into account that the apprehension of Forms is supposed to have practical consequences, in the understanding of mundane questions of justice and injustice in the city state. I think the requirement of public service makes it more plausible than it would otherwise have been that

[26] *Prt.* 329A; 347E; *Phdr.* 274B-277A; and (if genuine) *7th Letter* 341C-344D.
[27] Gregory Vlastos, 'A metaphysical paradox', *PAPA* 39, 1966, 5-19 reprinted in his *Platonic Studies*.

the understanding gained will be of a kind to have appropriate practical consequences.

The propositional interpretation has some positive advantages. For one thing, it gets rid of what Vlastos calls 'the gravest flaw in [Plato's] theory'.[28] For, on Vlastos' view, it is the experience of having a vision of the Forms which is supposed to provide infallibility, and this, I agree, would be an unsupportable theory. But on the foregoing propositional interpretation, what provides infallibility is the years of dialectical thought and of public life which prepare the ground for appreciating at last, in a sudden insight, the correctness of the correct account of goodness. There is a second advantage of the propositional interpretation. For the other makes the logical, argumentative side of dialectic quite unconnected with the religious mystical side, one merely following the other.[29] The foregoing interpretation connects the two sides together. I agree with Julia Annas, who makes both these criticisms of the non-propositional interpretation, though she gives an interestingly different reply.[30] A final advantage is that what is going on is quite intelligible, if the experience to which Plato is referring is the joy of finally coming to understand what goodness, justice, or beauty really are; whereas I must confess that I do not understand what Plato can plausibly be referring to, if he has in mind some *non*-propositional acquaintance with such things – they seem to be things of the wrong kind. A god might reasonably be thought to admit of such acquaintance in mystical experience, but even if goodness and justice are thought of as divine, I do not quite understand how they can.

I have attempted to discuss only the thesis that knowledge of the Forms in the *Republic* is non-propositional. Some commentators have argued for a *different* thesis, namely, that when Plato later goes on to *analyse* what knowledge is, he is *confused* between propositions and more ordinary objects of acquaintance. On this larger and more complex issue I shall not comment. The evidence on it is drawn from a *later* work, the *Theaetetus*.[31]

I must now return to Aristotle and to some of the other ideas which he bequeathed to Plotinus.

Aristotle: the act of thinking is identical with its object

One difficult saying of Aristotle is that the act of thinking is identical with the object of thought. The basis of this idea can safely be traced (although this is not always recognised) to a discussion in *Phys*.3.3. When an agent acts on a patient, the activity of the agent is in a certain sense identical with the activity of the patient, and both are located in the patient. For example, the activity of some teacher and the activity of his pupil can be called a single activity, and can be located in the pupil. Aristotle wants this result, because

---

[28] Vlastos, op. cit., *Platonic Studies*, 57.

[29] Admittedly, this is what happens in Plotinus.

[30] Julia Annas, loc. cit., says that Plato's central interest is more in understanding than in infallibility.

[31] Runciman, op. cit., 45; McDowell, op. cit., 115-16 and see index under '*connaître* and *savoir*'.

at the end of the *Physics* he will make his God an unmoved mover, and he wants no activity of causing motion to go on within the deity. But he spells out very carefully what kind of identity is to be found here. It is not, he says five times, an identity of essence,[32] for the essence of teaching and the essence of learning are quite different things. He might have put his point by saying that it is a merely numerical identity: if you are counting activities on a particular occasion, there are not two different activities to be counted. In fact, he tries out various other formulations. Properly speaking (*kuriōs*), teaching is not the same as learning; it is rather that teaching and learning are predicated of a single process.[33] It is not like the identity of cloak and mantle, but more like that of the road from Thebes to Athens and the road from Athens to Thebes. Nor should you expect, since the identity is not one of essence, that the activities we are identifying will have all their predicates in common.[34]

The idea is re-applied in the *de Anima*. The activity of the man who hears and the activity of a resounding object in arousing his hearing can be viewed as a single activity.[35] Again, and by analogy, the activity of thinking is identical with the actively working object of thought; not, admittedly, with a stone, if you are thinking of a stone (for there is no stone in the soul), but with the intelligible *form* of the stone,[36] that is, roughly speaking, with its defining characteristics.[37]

What does this idea mean, when it is applied to the case of thinking? Aristotle maintains that, when we think of something, its intelligible form is in the soul,[38] and that the thinking part of the soul must *receive* the form,[39] and is the *place* of forms.[40] We might initially understand this by saying that the defining characteristics of the thing will be in one's mind. Aristotle's idea will then be that, if we are counting, we should not count the act of thinking and the defining characteristics at work in our minds as if they were two distinct things.

G.E.M. Anscombe has defended the idea that we should not speak of two distinct things here.[41] If we want to know whether a person understands a theorem, it is the theorem which we ask him to expound. There is not a second thing, the understanding of the theorem, which we must ask him to expound as well.

Aristotle: the intelligible form is in the soul

The idea that the defining characteristics of a thing are in the soul can be given a more concrete sense, if we consider some further remarks of Aristotle's. One way for them to be in the soul would be for them to be embodied in a mental image. And Aristotle does say that the object of

[32] *Phys.* 3.3, 202a20; b9; b12; b16; b22.   [33] *Phys.* 3.3, 202b19-21.
[34] *Phys.* 3.3, 202b14-16.   [35] *DA* 3.2, 425b26-426a26.
[36] Esp. *DA* 3.8, 431b20-432a1; also 3.4, 429b6; b30-1; 430a3-7; 3.5, 430a14-15; a19-20; 3.7, 431a1-2; *Metaph.* 12.7, 1072b21; 12.9, 1074b38-1075a5.
[37] See above on the identification of intelligible form with defining characteristics.
[38] *DA* 3.8, 431b28-432a1.   [39]*DA* 3.4, 429a15.   [40] *DA* 3.4, 429a27-8.
[41] In G.E.M. Anscombe and P.T. Geach, *Three Philosophers*, Oxford 1961, 60.

thought, or intelligible form, is in, or is thought within, an image.[42] As to how it can be within an image, there is a revealing passage in the *de Memoria*.[43] If you want to think of a triangle, you will place before your mind's eye a triangular image, but will attend to its features selectively, ignoring the irrelevant ones. You will ignore its exact size, for example, since this is irrelevant to its triangularity. Aristotle points out that the same treatment is given to physically drawn diagrams in geometry. In the example which he chooses, that of a triangle, it is easy to understand the idea that the defining characteristics are in the image. For the image can simply be a plane figure with three straight sides.[44]

Aristotle further distinguishes between the intelligible form in its potential state and in its actual state.[45] We can perhaps speculate that the defining characteristics of the triangle are considered to be present only potentially, until they are separated out from other characteristics by the act of attending to which he refers. Be that as it may, it is the *actualised* form which Aristotle declares identical with the act of thinking.

The idea that the defining characteristics are within the image will be harder to understand for some examples. When I think of man, the form of man (rationality) can hardly be embodied in an image in the same way as the form of triangle. None the less, Aristotle clearly thinks that his account will apply to all cases. You may, he says, want to think of something altogether sizeless. In that case, you will still put before your mental gaze an image which has a size, but you will ignore the fact that it has a size.

Aristotle: the intellect is identical with the object of thought

Since the act of thinking is numerically identical with the object of thought, Aristotle is equally willing to say, in some of the passages cited above, that the *intellect* is identical with that object. This would not mean, in normal cases, that the *thinker* was identical with it. For a human thinker is more than an intellect. But God constitutes a special case, because Aristotle conceives his God as being *nothing but* an intellect. Accordingly, God is identical with the object of his thought.

The intellect thinks itself

On one persuasive interpretation, which is followed among others by Plotinus, this last point explains Aristotle's further claim that intellects, including God, think of themselves:[46] naturally so, if they are identical with

---

[42] *DA* 3.7, 431b2; cf. 432a4-5.

[43] 449b30-450a7. See Sorabji, *Aristotle on Memory*, London 1972, 6-8: the *de Memoria* is an important, though under-used, source for Aristotle's theory of thinking.

[44] Many modern philosophers have maintained that having a mental image is never like seeing a picture. I assembled evidence in *Aristotle on Memory*, ch. 2, which shows, I believe, that for many people it is extremely like that.

[45] *DA* 3.8, 431b24-6.

[46] This interpretation is most fully defended by Richard Norman 'Aristotle's Philosopher-God', *Phronesis* 14, 1969, 63-74 (repr. in *AA* 4). The connexion of thought is also made by

the objects of their thought. Self-thinking is guaranteed, for in thinking of the objects, they will be thinking of themselves. There need be nothing narcissistic in the claim that God thinks of himself, or regressive in the claim that he thinks of his own thinking.

Aristotle: human thought requires images

Aristotle believes that all human thinking requires images.[47] He is indeed committed to believing this, if the thought process is one of attending to the right features, in the way described above. But he also has a more metaphysical reason for thinking images required.[48] Thus he accepts Plato's view that forms are objects of thought, but rejects his view that intelligible forms can exist separately from the sensible world. Rather, they need a sensible vehicle, and a convenient vehicle for *intelligible* forms is provided by the so-called *sensible* forms. An example of a sensible form would be the colours of external objects, which during perception are taken on by one's eye-jelly.[49] Subsequently, these colours in the eye-jelly can leave behind an imprint in the central sense organ, which in turn gives rise to images. First the colours in the eye-jelly, and subsequently the images, can provide a vehicle for the intelligible forms. There is a further disagreement here with Plato, who explicitly maintained that dialectical thought rises above the need for images.[50] We shall see that Plotinus makes a claim parallel to Plato's.

God's thought, in Aristotle's view, is evidently different. For images depend on physiological organs; whereas God is immaterial, so that his thinking must be imageless. Aristotle never explains, however, how God escapes the need for images, or how the disagreement with Plato can be maintained once imageless thought has been allowed.

Aristotle: the agent intellect and the real self

In *DA* 3.5, Aristotle briefly introduces the agent intellect. His account of the intellect so far has made it seem analogous to a material cause, because it passively receives forms. But it depends for being activated on there also being an active efficient cause to bring it from potentiality to actuality. To serve this purpose, Aristotle postulates that there is a second intellect which thinks incessantly and for ever. It can reside both in us and separately from us, and it involves no memory. He makes little more of this 'agent' intellect, but the commentators made a great deal of it. The Aristotelian Alexander of

Plotinus *Enn.* 5.3.5 (21-48), and perhaps by ps-Alexander, *in Metaph.* CAG 671,8-18. Among modern commentators, G.E.M. Anscombe has a related interpretation in *Three Philosophers*, 60. For Aristotle's claim that intellects think of themselves, see *DA* 3.4, 49; *Metaph.* 12.7, 1072b19-21; 12.9, 1074b33-5; 1074b38-1075a5.

[47] *Mem.* 449b31; *DA* 3.7, 431a16; 431b2; 3.8, 432a8; a13.

[48] *DA* 3.8, 432a3-9. I have discussed these points in *Aristotle on Memory*, 6-8.

[49] That Aristotle thinks our eye-jelly takes on colour patches when we see I have argued in more than one place, most fully in the revised version of 'Body and soul in Aristotle' in *AA* 4, 49-53, with nn.22 and 28, which expands the earlier version in *Philosophy* 49, 1974, 72-6, with nn.30 and 35.

[50] *Rep.* 510B; 511C; 532A.

Aphrodisias maintained that it was God,[51] and also that what he called our 'material' intellect could somehow become this 'agent' intellect, since when it thinks of the agent intellect it becomes the object of its thought.[52] No doubt, there is yet another Aristotelian discussion behind Alexander's idea. For, in his account of the happy life in *NE* 10.7, Aristotle says that the real self is thought to be, or to be above all, the intellectual part of us,[53] and that we should so far as possible act the immortal (*athanatizein*).[54]

Aristotle: thinking, unlike processes, is complete at any moment

There are two last points to be made about Aristotle's theory of thinking. He classes thinking as an *energeia* (activity) rather than a *kinēsis* (process).[55] The English renderings do not properly bring out the distinction. Aristotle's idea is that as soon as you can use the present tense 'is thinking', you can use the perfect 'has thought'. For thinking is not, like building a temple, a process which has to *wait* before it is complete.[56] It might be protested that this ought to be said only of certain kinds of thinking. Proving a theorem surely does remain incomplete until the end, even if contemplating a premise does not.

Aristotle: contemplating truth is superior to seeking it

It may be a connected fact that Aristotle describes the happiest and most pleasant possible life as one of *contemplating* philosophical truths rather than *seeking* them (*zētein*).[57] For seeking is defined by reference to the goal of finding, and is in a certain sense (admittedly, a different sense) incomplete until it gets there. Moreover, it was a view which appealed at least to some members of Plato's Academy that the goal must always be better than the process of reaching it.[58] Even if this view is not plausible, when taken so generally, it is at least fairly natural to suppose that the whole point of seeking is to possess the object sought.

Personally, I think this natural supposition overlooks the fact that part of the pleasure of philosophical activity is emerging from the state of perplexity which Aristotle describes in *Metaph.*1.2 (982b11-983a21). Aristotle's God, who has always known and contemplated the truth, has missed this peculiar philosophical excitement. If we too had been so born, or so educated, that we never got into a state of perplexity, we should, I think, have missed

[51] *de An.* CAG 80,16-92,11.

[52] *de An.* CAG 89,21-2.

[53] *NE* 10.7, 1178a2; a7; similarly 9.4, 1166a16-17; a22-3; 9.8, 1168b28-35; 1169a2; *Protrepticus* fr.6 (Walzer). The view is implied in Plato's *Phaedo* and *Republic* 611C. For references in Plotinus to becoming the intellect, see the end of this chapter.

[54] *NE* 10.7, 1177b33, inspired by Plato *Tim.* 90B-C. For reference to becoming godlike in Plotinus, Porphyry (*homoiōsis theōi*) and Augustine (*deificari*), see Georges Folliet, ' "Deificari in otio" Augustin Epistula 10.2', *Recherches Augustiniennes* 2, 1962, esp. 226; 234.

[55] *Metaph.* 9.6, 1048b24; b34; 9.8, 1050a36.

[56] *Metaph.* 9.6, 1048b18-35; *Sens.* 6, 446b2-3; *NE* 10.4, 1174a14-29.

[57] *NE* 10.7, 1177a25-7.

[58] *NE* 7.12, 1153a8-9, with reference to Speusippus.

something of value. Perhaps the point would be clearer if we distinguished three stages rather than two: seeking the truth, winning it and contemplating it. Philosophers differ on whether they enjoy the search: some find it exciting, others agonising. But for many of them, winning the truth, if they reach that stage, provides the greatest pleasure of all. After that, they typically want not to stay contemplating it, but to tackle a new perplexity. To remain in contemplation, so far from being the most rewarding activity, would soon become tedious. And this difficulty would only be obviated, if with Plotinus we could think of contemplation as *timeless*, or as excluding any sense of time, so that a sense of tedium would also be excluded.

Gregory of Nyssa's perpetual progress: two rival traditions

There are still disputes, however, about the value of philosophical perplexity. Wittgenstein compared the person caught in philosophical perplexity with a fly trapped in a fly bottle, or with a man scratching an itch. And after him, John Wisdom compared philosophical perplexity with mental illness which calls for therapy. These similes would suggest that emerging from perplexity is a relief rather than an exhilaration. Indeed, they invite the question whether it would not have been better to avoid entering the fly bottle in the first place. Wittgenstein may have been ambivalent on this question; Peter Winch has shown me a passage which he has translated for publication, in which Wittgenstein says:

> I am by no means sure that I should prefer a continuation of my work by others to a change in the way people live which would make all these questions superfluous. (For this reason I could never found a school.)[59]

Wittgenstein speaks here as if it would have been better, like Aristotle's God, never to have suffered perplexity at all.

In antiquity, Plato was on the same side as Aristotle. He did not, like Wittgenstein, speak of search as disagreeable, but he did speak of contemplation as the superior state. Thus he records with approval the claim that the gods are not eager for wisdom, as philosophers are, because they already have it,[60] and that philosophers after death may hope to be rewarded with full knowledge.[61]

Another view was that of the ancient Pyrrhonian sceptics, represented by Sextus Empiricus.[62] For them, as for Wittgenstein, philosophical perplexity was a source of *disturbance*. Unlike Wittgenstein, however, they expected that the effort of seeking the truth would eventually lead to equanimity, since it would lead them to suspend judgment on every philosophical issue. This would come about, because they would find that the arguments were equally strong on either side of every case, so that no conclusions could be drawn.

---

[59] *Vermischte Bemerkungen*, translated by Peter Winch as *Culture and Value*, Oxford 1980. The passage cited was written in 1947.

[60] *Symp.* 204A; cf. *Phdr.* 278D.

[61] *Phd.* 64A-69E.

[62] E.g. Sextus *PH* 1.1-30.

And this on one interpretation would free them from disturbance about matters of everyday life,[63] since the most elementary assumptions about daily life would also call for suspense of judgment.

There was, however, a minority tradition. Augustine records one version of it,[64] putting it into the mouth of Licentius, who represents another group of sceptics, those of Plato's Academy, and ascribing it to Cicero. It was presumably expounded in Cicero's lost work, *Hortensius*. The view is that for man, as opposed to God, happiness consists in *seeking* the truth. Some sceptics must have felt forced to say this, when they reflected that their scepticism denied to man all hope of knowing the truth. Augustine's own preferred answer, however, is that the truth has been revealed to us in Scripture.

At a more trivial level, Plutarch, who is also in the Platonist tradition, records two relevant anecdotes. One concerns a man who did not want his uncertainty resolved, because he wanted the pleasure of seeking. Another concerns Democritus, who wanted to know how, on atomist principles, his cucumber could taste of honey. When he was told that it had been stored in a honey jar, he was angry, and insisted that he would go on looking for a scientific explanation.[65]

But the most influential expression of the minority tradition is found in the Christian philosopher Gregory of Nyssa (*c.* A.D. 331-396), who wrote after Plotinus but before Augustine. He viewed mystical experience of God, not as something static, but as a perpetual discovery.[66] Since the distance between the soul and God is infinite, there will always be more to understand, and the more we understand, the more we recognise that God is incomprehensible.[67] But we will never feel satiety, because we can always progress. Thus he describes the soul as:

conforming itself to that which is always being apprehended and discovered.

Again, he describes the beatific vision as follows:

Then, when the soul has partaken of as many beautiful things as it has room for, the Word draws it again afresh, as if it had not yet partaken in beautiful things, drawing it to share in the supreme beauty. Thus its desire is increased in proportion as it progresses towards that which is always shining forth, and because of the excess of good things which are all the time being discovered in that which is supreme, the soul seems to be touching the ascent for the first

[63] See Myles Burnyeat, 'Can the sceptic live his scepticism?', in M. Schofield, M. Burnyeat, J. Barnes (eds), *Doubt and Dogmatism*, Oxford 1980.

[64] Augustine *contra Acad.* I.7 (= Cic. *Hortensius* fr. 101, Müller); 9; 23; III.1. I am indebted for this reference and the next to Myles Burnyeat.

[65] Plut. ap. Montaigne, *Apologie de Raimond Sebond* (= *Essays* 2.12), somewhat less than half way through. The Democritus story comes from Plut. *Quaest. Conv.* 1.10 (*Mor.* 628B-D), but I have not been able to track down the other.

[66] The following two quotations are from Gregory of Nyssa, *On the Soul and Resurrection* (*de Anima et Resurrectione*) PG46, col. 93C; and *On the Song of Songs* (*in Canticum Canticorum*)5, PG 44, col. 876B-C; see also 12, col. 1037; and *de Vita Moysis*, PG44, cols 403D-404D.

[67] *de Vita Moysis* PG44, cols. 372C; 376C-377B; 380A; 404A-409B.

time. For this reason the Word says again to the awakened soul 'arise', and to the soul which has come 'come'. For to him who really arises there will be no end of always arising. And for him who runs towards the Lord, the space for this divine race will never be used up. For we must always be aroused, and never cease from coming closer by running.

Gregory was influenced by his brother, Basil of Caesarea, who had expressed a somewhat similar view: we shall not be able to understand God even in eternity, or he would be finite; we shall only know him more perfectly.[68] Other thinkers, such as Philo and Origen, had allowed room for progress in mystical-experience, perhaps in Philo's case inexhaustible room for progress.[69] But there is none of Gregory's suggestion that progression is something of value in itself. Indeed, Gregory provides something of an *answer* to Origen, who fears that we may feel *satiety* when we stand on the highest rung in our progress.[70] Surprisingly enough, Augustine comes somewhat closer to Gregory in one of his later works, the *de Trinitate*, XV.2.2. He is worried that the psalmist's exhortation *always* to seek God's face implies that we shall *never* find, and that such a prospect should sadden, not delight us. His answer is that incomprehensible things make the intellect engage in an alternating sequence of finding and seeking, each finding being sweeter, and each seeking more ardent.

Arthur Lovejoy has described how the idea of a perpetual progress after death suddenly became popular in the eighteenth century and recurred in thinker after thinker.[71] It was embraced even by thinkers whose systems made it difficult to accommodate. Leibniz, for example, held ours to be the best of all possible worlds. But the best possible world turns out to be one in which there is room for everyone getting better. This is in spite of the fact that each individual is defined by its position on the great scale of being, and every rung on the ladder is filled, but filled by only one specimen. In our day, Louis MacNeice has vividly expressed his dislike of the static quality of Parmenides' eternity, and has opted for change, even if progress is accompanied by retrogression: 'action along with error, growth along with gaps'. The following are the opening lines of his poem 'Plurality':

> It is patent to the eye that cannot face the sun
> The smug philosophers lie who say the world is one;
> World is other and other, world is here and there,
> Parmenides would smother life for lack of air
> Precluding birth and death; his crystal never breaks −
> No movement and no breath, no progress nor mistakes,
> Nothing begins or ends, no one loves or fights,
> All your foes are friends and all your days are nights.

---

[68] *Letters* 233-5; *contra Eunom.* 1.5.11.

[69] Philo *de Post. Caini* 5.14-6.21; 44.142; *Quod Deterius* 24.86-90; *Legum Allegoriae* 2.21; 3.15; cf. *de Somniis* 1.8-11; Origen *On First Principles* 1.3.8.

[70] *On First Principles* 1.3.8. Philo also speaks of souls falling because of satiety: *Quis Rerum Divinarum Heres Sit* 239-40.

[71] Arthur Lovejoy, *The Great Chain of Being*, Cambridge Mass. 1936, ch.9.

Plotinus: three levels of experience

I come now to Plotinus' three levels of experience. He thinks that with suitable discipline we can progress from one kind to another. At the bottom comes discursive thinking which he calls *dianoia*. Above that is the non-discursive thinking of the intellect. He believes that we can 'become' this intellect and engage in the same kind of thinking. Finally, there is union with the One, a union which is above thinking altogether. I shall argue that Plotinus transforms Aristotle's conception of non-discursive thinking by treating quite differently the idea of contact with the incomposite. On the other hand, he is like Aristotle in making non-discursive thought propositional. It will not be easy to establish this last point, since in order to do so, I shall have to combat a rival interpretation, which has been persuasively articulated by A.C. Lloyd. I would not dissent lightly from someone to whom I owe so much of my understanding of Plotinus.

Lloyd ascribes to non-discursive thought four attributes which I shall try to argue belong only to the higher level of union with the One.[72] He maintains that non-discursive thought involves no complexity, and hence (secondly) is not directed to propositions, since these are complex. He believes, thirdly, that it involves no self-consciousness, and fourthly that it is typically described in terms of contact. To make up our minds what Plotinus' view is, let us consider his three levels of experience in turn.

Discursive thought

An important mark of discursive thought for Plotinus is that it takes one thing after another progressively, and is consequently spread out in time.[73] Indeed, the discursive thought in which the soul engages actually constitutes time.[74] It also depends, unlike non-discursive thought, on contemplating imprinted images.[75]

Non-discursive thought is propositional

In contrast, non-discursive thought, in which the intellect engages, is not spread out, but timeless.[76] Indeed, this kind of thinking constitutes the timeless eternity which Plotinus describes as neither extended nor progressing.[77] He further declares that it does not involve seeking (*zētein*), but possessing knowledge.[78] He maintains that the intellect in action is identical with its objects.[79] And from this several consequences flow. First, the intellect does not depend on mere images of its objects, since it can actually be identical with them.[80] Secondly, Plotinus is able to represent his theory as more akin to Plato's than it would otherwise have been. For it is no longer so big a divergence that the eternal realm for Plato is the realm of Forms, while

[72] Lloyd, op. cit, 263; 266; 268.
[73] E.g. 3.7.11(36-40); 5.3.17(23-5); 6.9.5(7-12).   [74] 3.7.11.
[75] 5.3.2; 5.3.5(23-5).   [76] 4.4.1.   [77] 3.7.3.   [78] 5.1.4 (16).
[79] 1.8.2(16); 5.1.4(21); 5.3.5 (21-48); 5.4.2(46-51); 5.9.5(7-48); 5.9.8(3-4); 6.9.5(14-15).
[80] 3.9.1(8-9); 5.3.5(21-5); 5.5.1(50-65); 5.9.7(1-8).

for Plotinus it is the intellect, once it can be maintained that the intellect is *identical* with the forms which are its objects.[81] Thirdly, Plotinus maintains that the intellect thinks of itself.[82] And he supports this idea in just the way I took it to be supported in Aristotle. For if the intellect is identical with its object, then, in thinking of its object, it will be thinking of itself.[83]

Now why should it be supposed that Plotinus makes non-discursive thought non-propositional? Lloyd gives as one reason that this kind of thinking involves no transition from concept to concept. This is true, if by transition is meant a chronological passage which occupies time. But if there is no chronological transition, it does not follow that there is no complexity in the thought. Indeed, Plotinus repeatedly maintains that the intellect and its thinking are complex,[84] and that the object of thought is complex.[85] As I understand it, this actually excludes the idea that non-discursive thought is directed to concepts taken in isolation.

I see no barrier, then, to supposing that non-discursive thought is directed to propositions. And there is actually evidence in favour of this. For one thing, Plotinus agrees with Aristotle in making the intellect capable of truth (which suggests that it grasps propositions), but not of falsehood.[86] For another thing, as will be seen, the route by which we attain to non-discursive thought is through discovering the definitions of things, in terms of genus and differentia.[87] The stage of non-discursive thought seems to involve contemplating these definitions arranged into a unified network.[88] And definitions are propositional in form, since they tell us *that* so-and-so is so-and-so.

There is a further question on which I am inclined to understand Plotinus differently from Lloyd, namely, that of self-thought. Lloyd takes it that, since the intellect is identical with its object, it cannot think of itself, at least not in the primary sense.[89] But Plotinus' view, so far as I can tell, is the opposite. We have seen him arguing that the identity of intellect and object actually *guarantees* self-thought.[90] Moreover, he adds that this is self-thinking in the proper sense (*kuriōs*), whereas discursive thought involves self-thinking only in a secondary sense.[91]

From this I conclude that the four descriptions considered by Lloyd (non-complex, non-propositional, not self-directed, tactual) do not belong primarily to non-discursive thought. They belong rather to the higher level of union with the One, although I would not deny that occasionally[92] the metaphor of touch is applied to the lower level as well. Let me now consider the higher level.

Non-propositional contact with the One is not a kind of thinking

At the higher level, I believe, we find a significant departure from Aristotle. The object of thought is always complex for Plotinus, as we have seen.

---

[81] 3.9.1(1-20). [82] 2.9.1(33); 5.3.2-6. [83] 5.3.5(21-48).
[84] E.g. 5.3.10-13; 6.4.4(23-6); 6.7.39(10-19); 6.9.4(3-6); 6.9.5(14-16).
[85] E.g. 4.4.1(16-38); 5.3.10; 5.3.13; 6.4.4(23-6); 6.9.5(14-16).
[86] 5.5.1(1-6 and 65-7). [87] 1.3.1-4. [88] 4.4.1(16-38).
[89] Lloyd, op. cit., 266. [90] 5.3.5(21-48). [91] 5.3.6(1-5). [92] E.g. 1.1.9(12).

Plotinus diverges from Aristotle in thinking that this is so, even in the case of an identity statement, like 'I am this'. Even here, as will emerge in a passage to be quoted shortly, the I and the this would be two things. To avoid this duality, one would have (absurdly) to say 'am am', or 'I I'.[93] Much the same view is found in 6.7.38: the identity statement, 'I am the good' would import too much duality into the One. On the other hand, Plotinus would, for his own reasons, agree with Aristotle that we do have dealings with what is non-complex. For Plotinus' One lacks complexity altogether. Moreover, he would agree again that *contact* is an appropriate metaphor for these dealings. He uses the same words for touching as Plato and Aristotle had used, or cognate ones, namely, *thixis, thingein, thinganein, haphē, ephapsasthai, epaphē, sunaphē, sunaptein, prosaptesthai*.[94] But just because this touching is not directed towards a complex object, Plotinus would depart from Aristotle, by denying that the touching should be classed as thinking. He repeatedly says that, because the One is simple, it does not think,[95] and when we achieve contact or union with it, we are not thinking of it.[96] Our contact is rather a pre-thinking (*pronoousa*), 5.3.10(43)). And the One engages not in thinking, but in a super-thinking (*hupernoēsis*, 6.8.16(33)). It exists before thought (*pro tou noēsai; pro noēseōs*, 5.3.10(48); 6.9.6.(43)). Despite the fluidity of Plotinus' thought, there is not much wavering on this view:[97] I shall return to it in the next chapter.

A passage which incorporates some of Plotinus' most important ideas is 5.3.10 (28-52). This seems to say that thinking requires a complex object, and that the contact which one might make with a non-complex object is not a kind of thinking at all. I think this shows that the contact cannot be non-discursive thinking, as it is in Aristotle, and as Lloyd takes it to be in Plotinus:

> What can you think of which does not contain diversity (*allo kai allo*)? For if every object of thought is a verbal formula (*logos*), it will be multiple (*polla*). A thing is conscious of itself by being a diversified (*poikilon*) eye, or an eye of diverse colours. For if it encountered an object that was one and indivisible, it would be rendered speechless (*alogeisthai*). For what would it have to say or know about itself? For if a wholly indivisible thing had to describe itself, it would have first to say what it was not; so that in this way too it would be multiple (*polla*) in order to be one. Then when it says 'I am this', it will speak

---

[93] 5.3.10(34-7).

[94] 5.3.10(41-4); 5.3.17(25-34); 6.7.36(4); 6.7.39(15-19); 6.7.40(2); 6.9.4(27); 6.9.7(4); 6.9.8(19-29); 6.9.9(19); 6.9.10(27); 6.9.11(24).

[95] E.g. 3.9.9(1); 5.3.13(10); 5.4.2(18); 5.6.4-5; 6.7.35 and 37-42; 6.8.16(31-6); 6.9.6(42-5).

[96] E.g. 5.3.10(41-4); 5.3.13(37); 5.3.14(3); 6.7.35(30 and 44-5); 6.7.39(18-19); 6.7.40(1); 6.9.4(1-6); 6.9.10(7-21); 6.9.11(11). cf. Porph. *Aphormai* ch.25, p.11,4: the One is contemplated by an unthinking (*anoēsia*) superior to thinking.

[97] The nearest Plotinus comes to deviating in is 5.4.2(18), where he says that the One engages in a downright thinking (*katanoēsis*), and in a thinking different from the thinking of the intellect. But even this qualified ascription of thinking cannot be taken too seriously. For one thing, the treatise is an early one, and for another, a contemporary work, 3.9.9(22), denies *katanoein* to the One after all. When Plotinus wants to ascribe to us any apprehension of the One (3.8.9(20); 6.8.11(23)), or to ascribe to the One itself any apprehension (6.7.38(26); 6.7.39(2)), he uses instead the Epicurean term *epibolē, epiballein*, which is meant to convey something different from any kind of thinking.

falsely if the this of which it speaks is other than itself; while if the this is an accident of itself, it will be describing a multiplicity (*polla*). Otherwise it will simply say 'am am' or 'I I'. But what if it were only two things and were to say 'I and this'? Must it not rather be many things (*polla*)? For it is diverse (*hetera*) in kind and manner, and is a plurality (*arithmos*) and many other things. Hence a thinking thing must take a diversity of objects (*heteron kai heteron*), and what is being thought, while it is being thought, must be diverse (*poikilon*). Without this [diversity], there will be no thinking (*noēsis*) of it, but only a contact (*thixis*) and, as it were, a grasping (*epaphē*), which is unsayable and unthinkable, but which pre-thinks (*pronoousa*). The intellect (*nous*) will not yet have come into existence, and that which is touching (*thinganon*) will not be thinking (*nooun*). In contrast, what is thinking must not remain simple, especially if it is thinking of itself, for it will divide itself in two, even if it does not speak its thoughts.

Thus [the absolutely indivisible] will not need to make a fuss about itself. For what would it discover, if it thought? Its essence belongs to it before all thinking (*pro tou noēsai*). For consciousness (*gnōsis*) is a sort of desire and a sort of discovery after search (*zētein*). Hence that which contains absolutely no differences remains by itself and searches for nothing about itself; whereas a thing that unrolled itself would have to be multiple (*polla*).

What I am suggesting is a curious turn-around of Aristotle's ideas. Aristotle is prepared to assimilate non-discursive thought to contact, only because he has a way of representing the proposition thought as non-complex, provided it is an identity proposition. Plotinus rejects this possibility in the passage quoted. Hence for him thinking and contact with the simple have to be two separate things.

It might be objected to my interpretation of the foregoing passage that Plotinus is confining his remarks to *discursive* thought, when he says that thinking is directed to a complex object. Admittedly, the notion of *unrolling itself* in l.52 and of *search* in ll.50-1 are elsewhere declared inapplicable to the non-discursive thinking of the intellect.[98] But everything else suggests that it is discursive and non-discursive thinking alike for which Plotinus demands a complex object.

I shall conclude by outlining the method of ascent, which confirms the idea that non-discursive thought is propositional.

Plotinus: our ascent to the One

In 1.3.1-4, Plotinus describes the method of ascent from discursive to non-discursive thinking and finally to union with the One. Musical people and lovers are to be led up to the intelligible realm, by having their thoughts directed away from particular beautiful things to beauty itself, in the manner of Plato's *Symposium*. Philosophers, however, have by their nature already made a start upwards. So they should be trained in the style of Plato's *Republic*, first in mathematics and then in dialectic. The account of dialectic in 1.3.4 is derived from Plato's *Republic*, *Phaedrus* and *Sophist*. It involves finding the definitions of all intelligible things, by using the method of

[98] 3.7.6(16); 5.1.4(16).

division, which divides genera by their differentiae into species. It also involves seeing the interconnexions between all these things, until the whole intelligible realm has been analysed. Finally, the dialectician rests from labour and quietly contemplates. The intellectual process here is perfectly familiar to philosophers today, or at least to those philosophers, the majority, who concern themselves with definitions; and it involves propositional thinking, at least until we reach the One.

The intellectual training is only one side of the preparation; there is also a moral training. The flesh is to be renounced. Porphyry's *Life* gives us a picture of how Plotinus put this renunciation into practice, and how he influenced others. One thing he has in common with many later mystics is taking little food and little sleep (§8).

There are two stages of the intellectual ascent, the first up to the intelligible realm, the second within it up to its highest point (1.3.1(13-18)). On the way up to union with the One, you must first (to echo a phrase from Alexander of Aphrodisias) 'become' intellect.[99] For the final stage of union with the One, you must simply wait as for the sunrise.[100]

[99] 5.3.4(28); 6.7.34(31); 6.7.35(4); 6.9.3(22-3).
[100] 5.5.8(3-7).

CHAPTER ELEVEN

Mystical Experience in Plotinus and Augustine

We cannot understand the summit of Plotinus' ascent if we do not consider the character of mystical experience. Plotinus is not merely describing an abstract possibility. He claims to be speaking from experience, and he offers an autobiographical account for example at 4.8.1. His pupil Porphyry supplies more information in the *Life of Plotinus*, which is prefixed to many editions, surely one of the pithiest biographies ever written. What other biography tells us, within the first three paragraphs, such details as that its subject refused enemas, and that he declined to be weaned until the age of eight? The biography gains importance from telling us (not quite reliably for the earliest writings) the order of composition of Plotinus' works, and the rationale for the present arrangement, which is Porphyry's own. What matters for present purposes is Porphyry's report that Plotinus achieved the mystical state a number of times. Whereas Porphyry did so only once, Plotinus achieved it four times when Porphyry was with him (§23). This may mean four times in his presence, or four times during their six years together in Rome.

Features of Plotinus' mysticism

Plotinus is a source of inspiration for a long tradition of mystical experience. From earliest times, people have had extraordinary experiences; but that described by Plotinus has quite distinctive features, of which several deserve to be enumerated.[1] (i) First, there is a sense of timelessness. Indeed, at the stage of union with the intellect Plotinus claims you have become one with a timeless being. And the happy life (1.5.7) would be a timeless one. (ii) The experience involves union with, or at least contact with and closeness to, a superior being, first the intellect, and then the One. Plotinus sometimes calls the One 'God'. (iii) While the union lasts, you leave the physical world behind and do not[2] even experience mental imagery. (iv) There is a temporary loss of any sense of self as a distinct being. Indeed, Plotinus considers it an actual loss of self. (v) Although the earlier stages of ascent involve, for Plotinus, strenuous philosophical activity, you have simply to

---

[1] Many are set out by R. Arnou, *Le Désir de Dieu dans la philosophie de Plotin*, Paris 1921.
[2] 5.8.11(4); 6.5.7(3-6).

wait for union with the One.[3] It is like calling a god into the house who may or may not come.[4] When the One comes, Plotinus says that you are seized and as if possessed, and you may be held aloft by it.[5] (vi) The last part of the experience, union with the One, is not cognitive. It is not an experience of thinking; for the object of the experience does not have the complexity of a proposition. Nor is it an experience of knowing.[6] (vii) The experience has no special connexion with death. Plotinus enjoyed it a number of times in life, while at death the soul will be in the intelligible realm, only if philosophical discipline has freed it.[7] Even then, we are liable to reincarnation, although there may be[8] souls which escape rebirth. (viii) To detect the soul, the intellect and the One residing within you, you need to be very quiet, and to listen for one voice to the exclusion of others.[9] (ix) The procedure involves turning inwards into yourself.[10] (x) If you see the One, you are filled with indescribable love and longing.[11]

In the Christian mystical tradition which followed Plotinus, many of these features occur again and again. That tradition is too complex to permit any single account. But great efforts have been made during the twentieth century to reach a decision on what experiences should be counted as mystical. One subject on which there has been some consensus is the role of *imagery*. From the thirteenth century onwards, many Christian mystics, especially, it has been said, women, did experience visual imagery. None the less, there is now some agreement that visual imagery should be viewed as an extraneous concomitant.[12] This is in effect an attempt to return to the earlier Christian tradition, and to a tradition closer to Plotinus.

As regards the non-cognitive character of the highest mystical experience, the idea that it excludes knowing is less radical than the idea that it excludes thinking, and is already found as early as Philo in the first century A.D.: the mystic starts with some idea (*ennoia*) of God in his intellect, and may make some progress in finding him, but ultimately he must comprehend that God is incomprehensible.[13] A still more modest idea is present in the second century A.D. in the Middle Platonist Albinus: one of the four ways of comprehending God by the intellect is the negative one of knowing what he is *not*.[14] But these ideas do not yet imply that mystical experience will be non-cognitive, and, as Pierre Hadot has pointed out,[15] Plotinus goes a giant stride

---

[3] 5.5.8(3-7).   [4] 5.3.17(28-32).   [5] 5.8.10(8); 5.8.10(43); 5.8.11(2).

[6] There is no *gnōsis*, 5.3.14(2); nor *epistēmē*, 6.7.41(38); 6.9.4(1-2).

[7] 6.4.16(36-47).

[8] 3.2.4(8-11); 3.4.6(30-5); 4.3.24(21-8); for doubts on whether permanent escape is allowed, see Chapter 12.

[9] 5.1.2; 5.1.12.

[10] 1.6.9(8); 5.8.10(31-43); 6.9.7.

[11] 1.6.7(12-28); 6.9.4(16-23); (6.7.34-5).

[12] See, e.g., C. Butler, *Western Mysticism*[2], London 1926; C. Baumgartner, *Dictionnaire de spiritualité* ed. Marcel Viller, II.2, Paris 1950, cols 2171-2194.

[13] Philo *de Post. Caini* 5.14-6.21; 48. 168-9 *Quod Det.* 24.86-90; *Leg. Alleg.* 1.36-8; 2.21; 3.15; *de Mut. Nom.* 7-15.

[14] Albinus *Didaskalikos* 10, in *Platonis Opera*, ed. Hermann, vol. 6, p. 164, 6ff.

[15] P. Hadot, 'Apophatisme et théologie négative', in his *Exercices spirituels et philosophie antique*, Paris 1981, 185-93.

further when he says that our contact with the One is not even a *thinking*. Plotinus does, however, argue that in spite of this we can express something about (*legein ti peri*) the One, because we can say what it is *not*, and so can possess it in a way, like people who are themselves possessed or inspired, although we cannot express it itself (*legein auto*).[16] Damascius retracts even this concession, and holds that we cannot say anything about the One.[17] I do not know how far these more extreme ideas, which make the highest experience non-cognitive, entered into Christian thought. But certainly Gregory of Nyssa, alongside his idea of perpetual progress, took up the idea that God remains incomprehensible at least as regards his essence (*ousia*).[18] And pseudo-Dionysius popularised among Christians the idea of mystical experience as an unknowing in which one knows with supra-intellectual knowledge,[19] a tradition which is still reflected in the title of the anonymous fourteenth-century mystical work, *The Cloud of Unknowing*.

Returning to Plotinus, it is instructive to notice the choice of metaphors, besides that of contact, which he considers appropriate for union with the One. He talks of union or of two becoming one;[20] of interpenetration such as bodies cannot enjoy;[21] of being present;[22] of being together;[23] of there being nothing in between, a state imitated by lovers;[24] of wanting to be mingled;[25] of there being no difference;[26] of two centres coinciding;[27] of regarding the One as an internal possession, rather than external;[28] of regarding it as being oneself, and oneself as being it;[29] of not being able to see where one's boundaries stop and it begins.[30] Often enough, Plotinus also uses visual metaphors, but he does not regard these as quite right, because they fail to emphasise that you no longer regard the One as something distinct from yourself.[31] Talk of seeing merely illustrates how difficult it is to avoid speaking in dualities.[32] And he even adapts it, so as to talk of the seeing light and the seen light becoming one.[33]

Plotinus' descriptions of *joy* and *passion* overlap with those just recorded. He talks of loving with a true or intense love (*eran alēthē erōta; erōs suntonos*), of terrible and piercing longings (*deinoi, drimeis pothoi*), of wanting (*boulesthai*) to be mingled, of desire (*ephiesthai*), of erotic passion (*erōtikon pathēma*), of being astonished at beauty (*epi kalōi agasthai*), of amazement (*thambos*), of being shocked with pleasure but without hurt (*ekplēttesthai hedonēi, ablabōs*), of being united in a state imitated by lovers and of satiety (*koros*). The intellect has left its senses (*aphrōn genētai*), and become drunk with nectar (*methustheis tou nektaros*), such drunkenness being better for it than sobriety.[34]

---

[16] 5.3.14.

[17] *Dub. et Sol.* §7 (Ruelle vol. 1, p.11, 20).

[18] *de Vita Moysis* PG44, 372C; 376C-377B; 380A; 404A-409B.

[19] *de Myst. Theol.* PG3, 1001.

[20] 6.7.34(13); 6.9.3(12); 6.9.10(12-21); 6.9.11(4-6).

[21] *koinōnia*, 6.9.8(30).

[22] 5.3.17(30); 6.7.34(14); 6.9.4(3); 6.9.7(4); 6.9.8(28-35).

[23] 6.9.3(11-12); 6.9.7(21-3); 6.9.10(10).   [24] 6.7.34(13).   [25] 1.6.7(13).

[26] 6.9.3(13); 6.9.8(31-3); 6.9.10(20-1).   [27] 6.9.8(19-31); 6.9.10(7-21).

[28] 5.5.7(23-35); 5.8.10(36-45); 5.8.11(13-22).   [29] 5.8.10(42); 6.9.9(59).   [30] 6.5.7(14-17).

[31] 5.5.7(31); 5.8.10(33-45); 5.8.11(11-22); 6.9.10(11-21); 6.9.11(5 and 22).

[32] 6.9.10(11-21).   [33] 5.6.1(17-22).   [34] 1.6.7(13-18); 6.7.34-5; 6.9.4(16-35).

Such descriptions may be unexpected for at least two reasons. First, Plotinus is associated with the Stoic ideal of *apatheia*: absence of passion. This is what is required for the higher virtues,[35] even if the Aristotelian idea of *moderate* emotion will do for the lower ones. But I think it must be significant that, four times in one of the passages just referred to, Plotinus describes the joy involved in union with the One as *eupatheia*: a good state of feeling.[36] For the Stoics did after all approve three kinds of rational *eupatheia*, one of which was joy (*chara*, *gaudium*), which Cicero describes as tranquil (*placide atque constanter*) and rational.[37] The passions which they rejected were defined as disobedient to reason and contrary to nature,[38] and similarly the passions which Plotinus rejects are evidently bodily ones, since *apatheia* is enjoined in 1.2.6, in order that the soul may not suffer along with the body.

The second reason why union with the One might be expected to be passionless is that the One itself experiences no passions. Nor then should the human soul, one would think, when it becomes united with the One. Part of the answer is that Plotinus is describing the state of someone who has caught sight of the One but not yet reached it, or alternatively of one who has returned from a state of union. None the less, the experience itself is described as one of *eupatheia*. I do not know how to reconcile this with the passionlessness of the One, but I suspect one answer is that, in the case of the One, union involves something less than identity.

Plotinus' mysticism and the self

Problems are raised by Plotinus' descriptions of the *loss of self* in mystical experience, some of which have already been cited. He speaks of not being able to see where one's boundaries stop and the One begins.[39] He claims that in union with the One, your soul is no longer a soul, because it is not living then, but beyond living; it is not even an intellect, because it has to be like the One.[40] There are not two things, seer and seen, but the seer is unified with the One, and is, one might say, no longer altogether himself.[41] It is as if he has become another, and not himself, and as if two centres have coincided, although they can become two again.[42] If your soul is in the intelligible realm at the time of death, then it does not exist in actuality, although it has not perished either.[43] Again, when your soul is united with the intellect, you are only potentially yourself,[44] even though your soul has not perished.[45] In the intelligible realm, it has no memory.[46] At most, the soul could say after returning, 'I have thought', but that would be after changing back.[47]

This creates a philosophical problem, because Plotinus must want it to be *ourselves* who enjoy mystical experience, for temporary union with the Intellect or the One gives us our happiest moments, and ideally we should never descend from such union: that would be[48] the happiest possible life for

---

[35] 1.2.1(22); 1.2.6.   [36] 6.7.34(34-8); 6.7.35(24-6); cf. 6.9.9(39).

[37] Cic. *Tusc.* 4.6.12-14; Diog. Laert. *Lives of Eminent Philosophers* 7, 115-16; Lactantius *Inst. Div.* 6.14-15.

[38] Stobaeus *Ecl.* 2.88, Wachsmuth (= SVF 3.378).   [40] 6.7.35(42-5).

[41] 6.9.11(4-6 and 11-12).   [42] 6.9.8(19-31); 6.9.10(15-21).   [43] 6.4.16(41-7).   [44] 4.4.2(8).

[45] 4.4.2(29).   [46] 4.3.32; 4.4   [47] 4.4.1(8-9).   [48] 1.5.7(23-5).

us. How could such moments, or such a life, be the happiest possible for us, if it is not *we* who enjoy it? In order to see how Plotinus would answer, we need to notice some of the qualifications that he introduces into his account of the loss of self. In leaving behind the world of matter, we increase our stature, since what accrues to us in bodily life comes from non-being and therefore diminishes us.[49] Again, God is in those who can receive him like another self.[50] It is, in any case, too simple to speak as if we had only one self. Sometimes Plotinus speaks as if we had two selves, an intellect and a lower soul.[51] Sometimes he adds a third intermediate self, a higher soul, which engages in discursive thinking, above the level of our lower soul, but below that of the intellect.[52] This is most truly our self.[53] Evidently, then, it is not altogether true that the self is lost, for we have several selves.

Plotinus' mysticism and time

I shall now turn to some of the problems about mysticism and *time*. The intellect is a timeless being. If it is right to think of ourselves as temporal, how can it be *we* who become the timeless intellect? Again, how can *anything* temporal '*become*' something timeless? And if the thinking in which the intellect engages is a *timeless* activity, how can we sandwich that activity between stretches of our *temporal* life? These initial questions all receive a kind of answer, if the timeless intellect already is one of our selves, so that it is not right to call us merely temporal beings. They receive a further answer, when Plotinus adds that part of our soul is never dragged down, but remains in the intelligible world, so that we are 'always' thinking.[54] This may seem contrary to experience: are we not sometimes asleep or unconscious? But Plotinus maintains that even then the activity of thinking can go on, for, unlike Descartes,[55] he believes that there is no need for us to be conscious of it.[56] The situation is again presented as an *unchanging* one, when Plotinus is talking of the intellect not in individuals but in its own right. It is not *successively* in the two states of thinking thoughts and of unthinking union with the One; rather it is in both states 'always'.[57] 'Always' in these contexts should presumably be construed in some non-temporal sense.

These passages seem to obviate the need for a temporal being to *become* timeless, or to *sandwich* timeless activities between temporal ones. For one of our selves *already* is a timeless being, and its timeless activity goes on *uninterruptedly*. But if that puts the earlier questions to rest, it raises the question why Plotinus ever talked of our *becoming* intellect or *becoming* united to the One. What room is left for such temporal becomings? One thing which he says can happen is that our intermediate self is *directed* upwards towards

[49] 6.5.12(15-26).   [50] 5.1.11(10).   [51] 6.4.14.

[52] 1.1.7(13)-1.1.8(8); 1.1.11(1-8); 2.9.2(4-10); 5.3.3(34-9); 6.7.6.

[53] 1.1.7(17-24); 1.1.10; 5.3.3(34-9).

[54] 2.9.2(4-10); 4.3.30; 4.8.8.   [55] 1.4.9-10; 4.3.30; 4.8.8.

[56] In contrast to Descartes, he thinks self-consciousness quite unimportant, and complains that it even impairs the activities of reading or of heroism, if we are self-consciously aware of our activity (1.4.10).

[57] 6.7.35(19-33).

the higher activity of thinking or downwards towards our bodily activities.[58] We may act according to any of the three men who are within us, and each of us *is* that man according to whom he acts.[59] On this view, the thinking of the intellect within us is uninterrupted, but we *become* intellect when we are suitably directed.

But now it may be wondered how we can become one with the intellect, so long as we still have bodies anchored in the material world. To this Plotinus would presumably give an answer in terms of the unreality of body. A hint of his view has already been given in the remark that what accrues to you in bodily life comes from non-being.[60] He is influenced by a conception of matter which is outlined by Aristotle. It is arguable that that account of matter does not represent Aristotle's own view.[61] None the less Aristotle expounds an account according to which matter is a subject of properties, which can be imagined by mentally stripping away all properties whatever, even length, breadth and depth. Aristotle justifiably comments that it is hard to see what is left. It is this stripped conception which Plotinus takes over,[62] and so naturally he does not think of matter as having much reality. Not only does the stripping conception give matter a low degree of reality, but matter is in any case dependent for its continuing existence on the activity of soul from which it is considered to emanate.[63] And what goes for matter goes also for body, which is a *combination* of matter and form, but which inherits the unreality of matter.[64] Indeed, the human body is only the outside shadow of the man.[65]

There is still much that I do not understand in Plotinus' treatment of time, and one important thing is how anything which deserves to be called 'thinking' can be timeless. But the thinking in question is non-progressive and involves a sense of timelessness; and I have suggested in Chapter 8 that it would be no good bringing against Plotinus the evidence of clocks and of other physical events, to show the sense of timelessness to be an illusion. For Plotinus would think of that evidence as belonging to a less real world, and hence as unable to impugn the sense of timelessness associated with the intellect's thinking. We can also now see more clearly how, by making intellectual thought *uninterrupted*, he avoids the charge of having made it *alternate* with temporal events.

I have been assuming in this discussion that Plotinus' conception of mystical experience is heavily dependent on his metaphysics. But I would finish by saying that there is likely to have been influence in the opposite direction too: his metaphysics may have been influenced by his experience. Thus he found in Plato the idea that there is a realm of being to which you cannot apply the words 'was' or 'will be'. But he must have been all the readier to accept this because of the character of his mystical experience

---

[58] 1.1.11(1-8); 2.9.2(4-18).   [59] 6.7.6(15-18).   [60] 6.5.12(15-26).

[61] I agree with Malcolm Schofield, '*Metaph.* Z3: some suggestions', *Phronesis* 17, 1972, 97-101.

[62] 2.4.6; 8; 10.

[63] This is the natural interpretation of 1.8.7(17-22); 4.3.9; and the idea is located in three further texts (2.9.3(18); 2.9.10(40-2); 4.8.6(18-23)) by J.M. Rist, *Plotinus: the Road to Reality*, Cambridge 1967, 117-19.

[64] 3.6.6(31-77).   [65] 3.2.15(49).

which involved a sense of timelessness. His experience may even have encouraged him to resolve the ambiguities we earlier found in Plato's conception of eternity in the direction of making it into a concept of timelessness.

Features of Augustine's mysticism

I shall now turn to Augustine, in whose *Confessions* mysticism and time are central subjects. For in Books VII-X he gives a personal account of mystical experience, an experience which involves a relation to a being who is timelessly eternal. In IX.10 he wonders whether the saints enjoy without interruption the experience which he enjoyed so briefly. In Book XII he describes the realm in which the saints and others dwell, the heaven of heaven, and gives a fuller account of the way in which this heaven is divorced from ordinary time. He concludes Book XIII with the hope of repose there. I shall want to argue that the intervening account of time in Book XI helps us to see how it would be possible, in Augustine's view, for the saints to escape to a certain extent from temporality.

When the *Confessions* are read as containing this theme, then it is possible, as R.J. O'Connell has argued,[66] to see them as forming a unity. For it has not been clear what the first nine 'autobiographical' books have to do with the last four 'philosophical' ones, and the former have even been printed without the latter. But in fact the philosophical exploration of creation, time and eternity in X-XIII completes Augustine's reflections on the soul's journey in the autobiography.

Neoplatonist influence on Augustine's account of mystical experience

Augustine was bowled head over heels by his encounter with Platonism, as he subsequently describes in *Conf.* VII to VIII. While residing in Milan in A.D. 386, he came across some Platonist books in the Latin translation of Marius Victorinus, and this impelled him to one or more experiences which many would describe as mystical.[67] In three earlier writings, he names Plotinus.[68] The most famous of Augustine's experiences occurred at Ostia in A.D. 387. It is described in *Conf.* IX.10. There are further descriptions in VII.10, VII.17, VII.20 and X.40, which suggest a certain disappointment. Courcelle conjectures that these may represent some less satisfactory experiences as early as 386, but that has proved controversial.[69] On the occasion at Ostia, Augustine was in the company of his mother a few days before her death. They were discussing what the eternal (*aeterna*) life of the saints might be like; and in the course of their discussion, he felt that they had themselves

---

[66] R.J. O'Connell, *St Augustine's Confessions, the Odyssey of Soul*, Cambridge Mass. 1969, esp. p.11. See also the preceding volume, *St Augustine's Early Theory of Man, A.D. 386-391*, Cambridge Mass. 1968, and the subsequent article, 'Augustine's rejection of the fall of the soul', *Augustinian Studies* 4, 1973, 1-32.

[67] *Conf.* VII. 9-10; VII.20-1; VIII.2.

[68] *contra Acad.* III.18.41; *Soliloquies* I.4.9; and (according to five MSS) *de Beata Vita* IV.

[69] P. Courcelle, *Recherches sur les Confessions de Saint Augustin*, Paris 1968, 157-67.

touched the eternal. He wonders if there could be a whole everlasting (*sempiterna*) life which was like that moment of understanding; and he asks if such a life awaits the saints at the time of resurrection. The passage deserves to be quoted.

Augustine and his mother start by thinking in wonder of material things. Then their thoughts turn inwards (*interius*) and they come to their own minds (*mentes*), until finally they reach what Augustine calls, echoing Plotinus, 'the life which is wisdom'.[70]

> That wisdom is not made, but is as it was, and will always be so. Or rather 'was' and 'will be' are not in it, but only 'is', since it is eternal (*aeterna*), and 'was' and 'will be' are not eternal. While we were talking and gaping at it, we just (*modice*) touched it (*attingere*) with one whole shock (*ictus*) of our hearts. Then we sighed deeply, left behind the first fruits of our spirits fastened to it, and returned to the disturbance of our own voices, where words have both beginning and end. What comparison is there with your word, our Lord, who remains in himself without ageing, yet gives new life to all things?

Augustine and his mother envisage that everything else might fall silent (*silere*, *tacere*), including the images (*phantasiae*) of material things, along with dreams and visions of the imagination (*imaginariae revelationes*). Only one thing would be heard: the word of him who abides in eternity (in *aeternum*).

> Suppose we heard him himself, without these other things, as my mother and I then stretched out and with our fleeting thought (*cogitatio*) touched (*attingere*) the eternal (*aeterna*) wisdom which remains above all things. Suppose this were continued (*continuetur*), and other visions (*visiones*) of a different order were far removed, while this one vision were to seize and absorb and bury the beholder in inward joys (*gaudia*), so that his everlasting (*sempiterna*) life were like that moment (*momentum*) of understanding (*intelligentia*) for which we sighed. Is not this the meaning of 'Enter into the joy of your Lord'? When is it to be? Is it when we all rise again, but will not all be changed?

Augustine's descriptions of his experience in *Conf.* VII, IX and X have been shown to contain innumerable echoes of Plotinus.[71] There is the intellectual upward journey, which starts like those in Plato's *Symposium*, and in Plotinus 1.3.1-2, by admiring the beauty of material things, and goes on through the levels of soul (*anima*) and reason (*ratiocinans potentia*) to that which changelessly is.[72] The necessity of looking inwards into oneself to which Augustine refers more than once[73] is explicitly said in VII.10 to have been taught him by the Platonist books. The description of the highest level as the life which is wisdom has already been cited as Plotinian, although Augustine does not follow Plotinus' further description of God as *beyond* being. The idea in IX.10. of an *everlasting* life for the saints like Augustine's

---

[70] This is Plotinus' description of the life of the intellect at 5.8.4(35).

[71] Besides the more recent work of R.J. O'Connell, see P. Henry, *Plotin et l'occident*, Louvain 1934, chs. 3-4; *La Vision d'Ostie*, Paris 1938, 15-26; Courcelle, op. cit.

[72] VII.17; IX.10.

[73] VII.10; IX.10.

moment of understanding might well be suggested by Plotinus' talk in 1.6.7(27) of *remaining* in contemplation (*menōn en tēi theāi*), even though Plotinus ought to mean by 'remaining' not a *temporal* prolongation, but a *timeless* state of changelessness. Augustine himself introduces the Plotinian denial of a temporal 'was' and 'will be', if not for the saints, at least for God.[74] The metaphor of *touch* appears in the quoted passage (*Conf.*IX.10), though in Augustine it no longer has the Aristotelian connotation of an all-or-nothing affair. Augustine probably means, as elsewhere,[75] to contrast merely touching with actually *grasping*, and this is probably why he talks of 'just' (*modice*) touching, with *fleeting* (*rapida*) thought, in contrast to the *permanence* (*manens*) of the thing touched.

The similarities, however, continue, for Augustine echoes the Plotinian references to passion, when he talks of love (*amare*), of trembling with love and awe (*contremui amore et horrore*), of joy (*gaudium*), of sweetness (*dulcedo*) and of a loving and desiring (*desiderans*) memory.[76] One expression deserves special mention. For not only does Augustine twice describe mystical experience in terms of a shock (*ictus*), which Courcelle compares with Plotinus' verb *ekplēttesthai*,[77] but he applies in a different connexion the very phrase which Plotinus uses to describe mystical experience in 1.6.7 (17): shocked without harm (*ekplēttesthai ablabōs*; *percutit sine laesione*).[78] The Plotinian comparison of mystical experience with drunkenness also recurs in Augustine,[79] although H. Lewy has shown how this would have come to Augustine via Ambrose through a tradition starting with Philo, and need not be particularly Plotinian.[80] Augustine's experience does, however, relate to Plotinus in being passively received. Courcelle draws attention not only to the shock (*ictus*), but also to Augustine's expressions: 'sometimes you introduce (*intromittis*) me into a quite extraordinary experience (*affectus*)'[81] and 'you lifted me up (*assumisti*)'.[82] The account of the experience at Ostia, quoted from IX.10, exploits Plotinus' idea of silence, and of listening for one sound, and, along with other passages,[83] it agrees with Plotinus that mental images are a hindrance, and must be dispensed with. Again, as in Plotinus, the experience is not especially connected with death. Only the chosen may be expected to have it after the resurrection, while conversely some have it briefly during life.

Neoplatonist influence on Augustine's account of time

O'Connell has argued that Augustine's account of *time*, as well as his account of mystical experience, is influenced by Plotinus.[84] Thus Augustine's famous remark in *Conf.* XI.14 that he knows what time is, if no one asks him, but not

[74] IX.10; XI.11.   [75] *Sermon* 117, PL 663-4.

[76] *Conf.* VII.10; VII.17; IX.10; X.40.

[77] VII.17; IX.10; cf. Plot. 1.6.7(14 and 17); 6.7.31(8).   [78] *Conf.* XI.9.

[79] *contra Faustum* XII.42; *Enarratio in Psalmum 103*, 3.13.

[80] Hans Lewy, *Sobria Ebrietas*, Giessen 1929.

[81] *Conf.* X.40.   [82] VII.10.

[83] See *Conf.* VII.17; IX.10; *de Trin.* VIII.3.

[84] O'Connell, *St Augustine's Confessions*, 139-42.

if he has to explain it, has its parallel in Plotinus 3.7.1 (4-9). O'Connell has also argued for analogies between the *definitions* of time given by the two authors, but here the connexion must be viewed as more remote.[85]

Controversies

What I have said thus far may sound uncontroversial, but in fact controversies have raged, and there have been three main ones, which John-J. O'Meara has conveniently summarised.[86] One idea has been that what Augustine was converted to in 386 was not Christianity, but Neoplatonism, and that the *Confessions*, written more than a decade later in 397-401, represents the conversion as having been more Christian than it really was. Most commentators would now agree that this hypothesis creates a false antithesis between the Platonism and Christianity of this time and place. Admittedly, Platonism and Christianity were at times violently opposed; admittedly too Plotinus and Porphyry had written against Christians and Gnostics (whom they lumped together);[87] and Augustine himself later made the case against Platonism sharper. But it only gradually became apparent to him on what issues the two positions conflicted. In the early *contra Academicos* of November 386, Augustine says he is confident of finding in the Platonists what does not conflict with Scripture.[88] And in the *de Vera Religione* of 389-91, he says that the Platonists with the change of very few words and opinions would become Christians, as very many contemporary Platonists had done.[89] In fact, some of those who influenced Augustine in Milan, such as Theodorus, are commonly regarded as being both Platonists and Christians.

A second controversy has turned on whether it was Porphyry rather than Plotinus whose translated works first influenced Augustine. This was the view of Willy Theiler. But although Porphyry's influence is extensive in the period from A.D. 400 onwards, the work of P. Henry, P. Courcelle and R.J. O'Connell constitutes a fairly persuasive answer to the idea that Porphyry was the main influence at first.[90] Courcelle's discussion includes the exciting

---

[85] Augustine's account of time as an extension in the *mind* (*distentio animi, Conf.* XI.26) is compared with Plotinus' account of it as the life of the soul (3.7.11(43-5)). And Augustine's reference to *distentio*, literally a spreading out, is compared with Plotinus' insistence that that life is spread out into a length (*mēkos*, 3.7.11(10-29)). But, as was pointed out in Chapter 7, the soul of which Plotinus is talking is the *World* Soul, and he is not currently emphasising his point that the World Soul is in each one of us. Nor is Augustine referring to the idea which O'Connell detects in some of his earlier works (*St Augustine's Early Theory of Man*, 122-4), that our souls are identical with a World Soul.

[86] The most useful summary is in 'Augustine and Neoplatonism', *Recherches Augustiniennes* 1 (suppl. to *Rev. des Ét. Augustiniennes*), 1958, 91-111; but see also his book, *The Young Augustine*, London 1954.

[87] See the details in P. Hadot, 'Citations de Porphyre chez Augustin', *Rev. des Ét. Aug.* 6, 1960, 205-44.

[88]*contra Acad.* III.20.43.   [89] *der Ver. Rel.* VII.

[90] Willy Theiler, *Porphyrios und Augustin*, Halle 1933. For the replies, see references above. The introduction to O'Connell's *St Augustine's Early Theory of Man* contains a particularly full answer. O'Meara argues for the more plausible view that there is Porphyrian, alongside Plotinian, influence in the early Augustine, but see Hadot, op. cit., for a critical review of his ideas, as well as O'Connell.

claim that we can date to the very day when Augustine first heard various Plotinian ideas in the sermons of Ambrose in Milan, but this thesis has not been universally accepted.

The third controversy is over whether the early experiences described in the *Confessions* can be described as mystical. On one view, they occurred too early in Augustine's career for it to be plausible that they were mystical in the sense of involving a genuine contact with God.[91] But if such contact is given by grace, not merit, it is hard to speculate on what would count as too early. Without wishing to comment on whether any cases involve a genuine contact, I would only say that, from a psychological point of view, Augustine's early experiences seem to have had many of the features of accredited cases of mystical experience.

Augustine's mysticism and time

So much for the current state of scholarship. I should now like to ask for Augustine, as I did for Plotinus, whether his metaphysics throws any light on the connexion he sees between mystical experience and *time*. The problem is less acute than it was for Plotinus, because Augustine does not claim that one actually becomes a timeless being. Moreover, he uses temporal language, in the passage quoted from *Confessions* IX. 10, to describe the possible life of the saints. They will continue (*continuari*) in an everlasting (*sempiterna*) life of contact with the divine wisdom. On the other hand, he once uses the stronger metaphor of being changed into God in the manner of food (VII. 10). What temporal status would this give to the saints?

There is a possible clue to Augustine's view, if we recall from Chapter 2 the special temporal status which he assigns to the heaven of heaven in *Conf.* XII.[92] It has been argued that his heaven of heaven contains as members not only the angels, but also the souls of the just – his statements are sufficiently imprecise to allow this.[93] The heaven of heaven collectively is wrapped in a beatific contemplation of God, and so, according to *Conf.* XII, does not change, or have a past to remember, or a future to anticipate. Thanks to this, it is a participant (*particeps*) of God's eternity, which it can be said to enjoy (*perfrui*). It is thus lacking (*carens*) in all times, and is not spread out (*distenditur*) into times (recall that in Book XI time is an extension, *distentio*, in the mind). On the other hand, it is still *capable* of change, and this means that it is not coeternal (*coaeternus*) with God.

But how can it be supposed that the angels, or the saints, are lacking in times and not spread out into times? I suggest that it may again be relevant to refer to metaphysical theory. I explained in Chapter 2 that Augustine has several different suggestions about time. The one canvassed in *Conf.* XI.20; 26; 27, is that past, present and future are simply three states of mind. That

---

[91] O'Meara, *The Young Augustine*, 139; 200; 202-3.

[92] XII.9; 11; 12; 19; cf. XIII.15.

[93] Jean Pépin, 'Recherches sur le sens et les origines de l'expression *"caelum caeli"* dans les *Confessions* de saint Augustin', *Bull. du Cange*, 23, 1953, 272-3; quoted with approval by R.P.A. Solignac, *Oeuvres de Saint Augustin*, Bibliothèque Augustinienne, vol. 14, *Les Confessions, Livres VIII-XIII*, Paris, 1962, 592.

canvassed in *Conf.* XII.8; 11; 12 is that times require change. Both suggestions should make it easier for Augustine to imagine how some human souls might come to be lacking in times. They might do so if they could but be rid of the three mental states of memory, awareness and expectation, or if they could be sufficiently separated from the change which characterises the sensible world.

This suggestion would not downgrade the mystical experience either of angels or of saints. For it is, for one thing, extremely hard for a human to get rid of the familiar mental states. The flesh and the images associated with the material world keep dragging us down. In any case, to be lacking in times is at most only one aspect of mystical experience.

The suggestion needs finally to be qualified by recalling from Chapter 2 that Augustine elsewhere assigns to the angels a kind of change and hence a kind of time, but one of an unfamiliar sort.

It can be confirmed that Augustine intends us to connect his specu-lations in Book IX about the life to come with his description of the heaven of heaven in Book XII. For not only does he intermingle talk of the heaven of heaven in XII with talk of the place where he hopes to dwell himself (XII.11; 15; 16). But, further, he ties the last of these passages in XII particularly closely to the description of mystical experience in IX.10; for there he says that he and his mother left behind the first fruits of their spirits fastened to the eternal wisdom; while here in XII.16, he says that he wants to be gathered for eternity into the peace where the first fruits of his spirit are.

Augustine's mysticism and the self

There arises for Augustine, just as for Plotinus, the question how the self can survive translation into a non-temporal state of contemplation, a question which Augustine does not really solve. He thinks in *Conf.* IX.3 that his dead friend Nebridius will not be too drunk with the fountain of wisdom to remember him. On the other hand, he is prepared in his earlier work *de Vera Religione* XLVI.88-9 to treat some aspects of the self as unimportant. No one is fit for the kingdom of God unless he hates genetic relationships, because they are based on the flesh. To love a man because he is your son, rather than because he is a man, is to love what belongs to yourself rather than what belongs to God. But if we love eternity, we will hate such relationships. We should instead love the real self in another, and our real selves are not bodies. O'Connell has pointed out that even after his affectionate account of his mother at the time of her death, Augustine goes on in *Conf.* IX.13 to describe his parents as his *brethren* under God and his *fellow citizens* in the eternal Jerusalem, as if the earthly relationship were in that context unimportant.[94]

Caveats about Neoplatonist influence

I should like now to return to the question of Neoplatonist influence on Augustine and to introduce some *caveats*, drawing upon what has already been said about Plotinus. I have rejected, as do most

---

[94] O'Connell, *St Augustine's Confessions*, 110.

people nowadays, the crude view that it was Neoplatonism to which Augustine was converted in A.D. 386. But there is a more subtle interpretation which also calls for caution, and that is that in the first works which Augustine wrote after his conversion, he had an idea of mystical experience as austere, as intellectual and as offering an immediate escape. This idea, it may be said, was derived from Platonism, and it was an idea from which Augustine separated himself only later, when he came to write the *Confessions*.[95] Such an interpretation seems to me mistaken, but it can be focused in a short passage of the *de Quantitate Animae*, written as early as A.D. 388, two years after the conversion:

> We are now in the very vision and contemplation of truth which is the seventh and last (*ultimus*) step for the soul. Nor is it any longer a step (*gradus*), but a sort of abiding place (*mansio*) which is reached by the other steps. What its joys (*gaudia*) are, what the enjoyment of the highest and true good, what serenity (*serenitas*) and eternity are breathed on one there, how am I to say? It has been told, in so far as they judged it should be, by certain great and incomparable souls which we believe have seen and are seeing those things.[96]

There are at least three relevant features in this passage. First, Augustine is interested in *serenity*. It has been argued that at this period, under Platonist influence, Augustine's goal was an austere tranquillity, and that he was less aware of the importance of delight as a motive.[97] Moreover, the rejection of emotion has been seen as Plotinian.[98] Certainly Plotinus and Augustine both rejected *bodily* passion, and in that sense favoured *apatheia*. But it would be wrong to suggest[99] that Plotinus rejected emotion quite generally and influenced Augustine to do so. For both philosophers thought of a non-bodily kind of joy as perfectly *compatible* with serenity and indeed as vital to mystical experience. Thus Augustine, in the passage just quoted, combines the notion of *serenity* with that of *joy*, and a little later of *pleasure* (*voluptas*). And Plotinus equally speaks of the *eupatheia* which, under the Stoics, had been defined as a *tranquil* joy, and which he too sees[100] as going with tranquillity. Hence in bequeathing to Augustine a hope of serenity, he is not bequeathing a rejection of joy.

A second feature of the passage quoted is that Augustine sees the possibility in this life of reaching an *abiding* place, which is the *last* level for the soul. It is natural to connect this hope of permanent elevation with the influence of Neoplatonists who are surely included among the 'great and

[95] I am grateful to Martha Nussbaum for drawing my attention to this interpretation. She was describing a possible extrapolation from the work of Peter Brown and John Burnaby. English translations of the *de Quantitate Animae* and of Augustine's other works are listed in the very helpful tables provided by Michael Walsh in Peter Brown's classic biography, *Augustine of Hippo, a Biography*, London 1967.

[96] *de Quant. An.* 33,76.

[97] See Brown, op. cit., 145-6; 155; John Burnaby, *Amor Dei*, London 1938, 60-1; 68; 95.

[98] Dom Cuthbert Butler, to whose book I am indebted for bringing out the passionate character of Augustine's experience, modestly disclaims knowledge of Plotinus, but assumes that 'mere Platonism' is intellectual rather than emotional (*Western Mysticism*, 2nd ed. London 1926, 19-62; 232-6; esp, 41-2).

[99] I am not sure how far, if at all, this is suggested.

[100] E.g. 6.9.11(9-16 and 24).

incomparable souls' whom Augustine mentions.[101] Their influence seems all the more likely seeing that the hope was later abandoned.[102] But what I want to say is that the Neoplatonist writings contain plenty of warnings against this hope. If Augustine chose to concentrate on other passages, this may reflect his own inclinations, rather than representing a balanced assessment of the Neoplatonist position. Admittedly some works of Plotinus, to which he had access early, could have encouraged an expectation of a lasting elevation in this life.[103] But Augustine himself admits later in the *City of God* that in Porphyry's view, man cannot reach wisdom in this life.[104] Moreover, in the *Life of Plotinus*, Porphyry records the infrequency of Plotinus', and much more of his own, mystical experience.[105] As for Plotinus, he thinks that souls after death are liable to fall again and be reincarnated. And although he speaks of some souls as escaping,[106] he treats that as exceptional, and we shall see in Chapter 12 that it has been doubted whether he thinks that such escape can be permanent.

The third relevant feature in the passage quoted is that Augustine sees the highest mystical stage in *intellectual* terms as involving a vision and contemplation of *truth*. It has been said that Augustine's early experiences and Plotinus' were both intellectual, indeed, too intellectual to count as genuinely mystical.[107] But quite apart from the question whether mysticism excludes intellectualism, the suggestion is mistaken in at least two ways. First, for Plotinus the highest stage of mystical experience was never intellectual, since, as has been seen, it involved no thinking at all. Secondly, for Augustine the highest stage always remained intellectual, since it always involved a vision of the truth.[108] What has happened here can be seen more clearly from Augustine's much later work *de Genesi ad Litteram* XII. For there it becomes evident that he has taken as his model for mystical experience Plotinus' description of union at the *lower* level with the *intellect*, not at the higher level with the 'One. Augustine is talking in *de Gen.ad Lit.* XII of the exceptional visions enjoyed by Moses and by St Paul.[109] These visions are classified as intellectual, rather than sensory or spiritual, and this means that, as in Plotinus' account of intellectual thinking, the object of the vision is itself present, not merely an image of it,[110] and that the mind is capable of truth, but not of falsehood.[111] The conclusion to be drawn is that Augustine's

---

[101] Augustine uses a similar phrase of the Neoplatonists at *de Ordine* 2.28: 'great men almost divine'. Later, in *de Trin.* IV.15.20 he allows that certain pagans have been able by their own power to touch the light of unchangeable truth to a small extent.

[102] That it was later abandoned is strongly emphasised by Burnaby, op. cit., 52; 60-1; 63-6; 98; 316; and Brown, op. cit., 105; 146-7; 150; 152.

[103] In the treatise *On Happiness*, which was known to Augustine, Plotinus insists that men as well as gods can have a perfect and happy life (1.4.4; cf. 6.9.11(49). In *On the Beautiful*, one of the first treatises to influence Augustine (there are echoes of it in *de Quant. An.* 33,75), Plotinus talks of *remaining* in contemplation 1.6.7(27).

[104] X.29.   [105] §23.   [106] 3.2.4(8-11); 3.4.6(30-5); 4.3.24(21-8).

[107] Gerald Bonner, *St Augustine of Hippo*, London 1963, 84.

[108] This has been shown by Butler, op. cit., 38-42; 52; 61.

[109] *de Gen. ad. Lit.* XII.26.54-XII.28.56.

[110] *de Gen. ad Lit.* XII.6.15; cf. *Conf.* X.9-12; X 20; Plot. 3.9.1(8-9); 5.3.5(21-5); 5.5.1(50-65); 5.9.7(1-8).

[111] *de Gen. ad Lit.* XII.14.29; XII.25.52; cf. Plot. 5.5.1(1-6; 65-7).

intellectualist conception of mystical experience corresponds to *one* part of what Plotinus says, but depends on ignoring *another* part, namely, the *unintellectual* character of union at the higher level with the One. We must, then, reject the simple picture of an intellectual conception of mystical experience in Plotinus and of the gradual abandonment of such a conception by Augustine.

I do not want to say that there is no trace in Augustine of the theme of unknowing, for he does doubt that God is ultimately comprehensible.[112] But he does not describe the mystical state as an unknowing, much less as an unthinking, experience.

I have two more *caveats* to enter about Augustine's relation to Neoplatonism. For there are two ways in which he is *unlike* the Neoplatonists in his conception of mystical experience. But in each case what I want to say is that the difference is not clear cut. Augustine himself locates a major contrast, in *Conf.* VII.9, VII.20 and especially VII.21, in the Platonists' failure to recognise that mystical experience is a gift of *grace*. In contrast with this, Plotinus describes in 1.6.9 how you can advance in virtue by self-discipline to the point where you no longer need a guide (*deiknus*); you can just concentrate your gaze and see God and Beauty. Courcelle has shown how in *Conf.* VII.10 Augustine adapts the passage by inserting 'with you as guide' (*te duce*), and by quoting from the Psalms, 'since you are my help'.[113] There *is* a contrast here; if I say it is not clear cut, this is partly because in *City* X.29 Augustine admits that Porphyry recognises the need for grace in a way. For Porphyry talks about concession, saying that it is *conceded* (*concessum*) only to a few to reach God by virtue of their intelligence, and he further says that by the providence and grace (*gratia*) of God, those who live in accordance with intellect may reach perfect wisdom, although not during this life. As for Plotinus, in passages already referred to, he says that, although there is strenuous activity at earlier stages, for union with the One you simply have to wait (*menein*) as for the sunrise, not to pursue it (*diōkein*, 5.5.8(3-7)). It is like calling for a god (5.3.17(28-32)), who may, or may not come. When he does come, you are seized and, as it were, possessed (5.8.10(43); 5.8.11(2)). You may be held aloft by the One (5.8.10(8)), just as Augustine was lifted up by God.[114] Even in Plato, individual effort is not on its own enough to guarantee a good view of the Forms. He who has chosen the philosophical life for three incarnations running may win a speedier opportunity to view them. But whenever you gain the opportunity to view, there is an element of chance (*suntuchia*, *Phdr.* 248C6) in whether you succeed.

---

[112] The subject has been discussed by V. Lossky, 'Les éléments de "théologie negative" chez Saint Augustin', *Augustinus Magister* 1, Paris 1954, 575-81. Augustine says that God is better known by not knowing (*nesciendo*, *de Ordine* II.16.44; II.18.47). He speaks of a 'learned' ignorance', thereby supplying the title for Nicholas of Cusa's work in the fifteenth century, and explains in the same context that we cannot think of God as he is (*sicuti est*), but that we know what he is *like* (*quale sit*), and can *reject* descriptions of him (*Ep.* 130, 14.27). This last idea that we can know what he is *not* is found also in *de Trin.* VIII.2.3, while the idea that we cannot comprehend him appears, for example, in *de Trin.* XV.2.2 and *Sermo* 117 (PL663-4).

[113] Courcelle, op. cit., 128-9.

[114] Arnou, op. cit., 224-9, adds some further passages, although he rightly warns that Plotinus' recognition of the need for grace is not very extensive.

On the other hand, there is no suggestion that the gods will help you in your viewing.

The remaining point of contrast is that Augustine is less ready than Plotinus to talk of being *united* with God. It was the Manichaeans' view, from which Augustine had struggled to free himself, that the soul was of the same substance as God.[115] In Christian thought, there was always a danger of blasphemy in the suggestion that a human might *become* God. Thus in A.D. 1329, Eckhart was to be condemned by the Pope for having implied that in mystical experience he actually was God. On the other hand, the Middle English *Cloud of Unknowing*, later in the same century, talks freely of being 'oned' with God. What we find in Augustine is a reliance on ideas other than total union, such as the metaphors of touch, sight and light. None the less, the contrast with Plotinus is once again not clear cut. On the one hand, Augustine does sometimes come closer to the idea of unity: the closest turn of phrase that I have noticed is in *Conf.* VII.10, where Augustine speaks of being changed into (*mutari in*) God in the way that food is changed into a body. On the other hand, I have already doubted whether Plotinus means to speak of a complete identity with the One.

Identification of experiences

I have talked in this chapter about whether different people were having a similar experience. There are two rather different kinds of factor relevant to the question, and one is what is sometimes called the *intentional* object of the experience. It would, of course, be an extraordinarily important difference if one person were really in contact with God and another were not. But in talking of the *intentional* object of the experience, I am not talking of the being, if any, with whom they have *actually* made contact, but rather of how they *conceive* the being towards whom their efforts are directed. In this respect, there are close connexions between Plotinus and Augustine, because Augustine's conception of God is so much influenced by Plotinus, and by the earlier Platonist tradition coming through Philo.[116] The other prerequisite for similarity of experience comes at what might be called the phenomenological level: there is an absence of imagery, a loss of the sense of self and of time, a passivity, a quiet and calmness, and yet a love and joy.

Drugs and mania

Similarity of experience is relevant to another issue. Those who have listened to the accounts of modern drug-takers may be startled, as I have been, if they look at Plotinus and notice the very same descriptions of a loss of the sense of time and of one's own boundaries. The similarity has led to a challenge laid down, for example, by Aldous Huxley.[117] May not the taking of an

---

[115] Aug. *de Ordine* 2.46, and O'Connell, *St. Augustine's Early Theory of Man*, 124-31.

[116] This will become clearer in later portions of this book, e.g. in Chapter 15, where I discuss Philo's and Augustine's quasi-Platonist interpretation of the Biblical account of God as a creator.   [117] Aldous Huxley, *The Doors of Perception*, London 1954.

impregnated sugar cube give one the *same* experience as that enjoyed by the greatest mystics? According to one etymology, the 'beatniks', who dropped out of society in the 1950s and 1960s, and who in some cases took drugs, were so called because of their access to the beatific vision. This has seemed threatening, because it would then appear that the mystics and the drugtakers had as much or as little right as each other to claim that their experience put them in contact with God. The case of Plotinus in particular may seem worrying, since it is recorded that he often went without food and sleep[118] – a sure way of affecting the body's chemical balance. Nor could he claim that the intellectual upward journey gave the final experience a distinctive character, since we have seen that the final experience was quite unintellectual.

This kind of challenge has been wittily met by R.C. Zaehner.[119] He argued that Huxley had not gone far enough: why should not the manic phase of manic depression be counted as an equally valid route? But then, without claiming to know who might have made contact with God, he responded to the argument at the level of its claim to similarity of experience. Whatever the similarities, we cannot overlook the difference that Huxley saw himself united to the legs of a bamboo table, whereas Plotinus and the Christian mystics saw themselves as in contact with a supreme being, or, in Plotinus' case, with something beyond being. This difference of what I called 'intentional' object will have wholly transformed the experience. This is quite apart from the fact that with Huxley's mescalin, the visual imagery and the heightened awareness of physical objects is a central part of the experience, whereas for Plotinus and Augustine it was important that they left the physical world and visual imagery behind.

---

[118] Porph. *Life of Plotinus* §8.
[119] R.C. Zaehner, *Mysticism Sacred and Profane*, Oxford 1957.

CHAPTER TWELVE

Fear of Death and Endless Recurrence

I want next to talk about the fear of death, but in a selective way. I shall be concerned only with those arguments against fear which turn on considerations about *time*.

Almost any group of people will divide between those who say they are afraid of death and those who say they are not. Moreover, the fear of death is itself not a simple thing. For there are many different aspects of death which horrify people, and many corresponding fears. Arguments relevant to one fear will not be relevant to another.

The diversity of fears

Some people fear the *process* of dying: the pain, the saying goodbye, the helplessness.[1] Some fear the unknown either in the process of dying or in being dead. Others fear judgment or punishment, or both, after death. Others again fear the curtailment of existing projects and attachments, or the loss of opportunities which could have been expected. These last fears should be reduced for those who live a long and rounded life, who see their main projects (if they have projects) completed, who have enjoyed the experiences which life normally has to offer, and who have seen many of their friends go before them. Different again is horror at the rotting or burning of the body, if this horrifies not merely because of its association with the other things feared.

None of these fears is my concern, however. Rather, I want to talk about the horror which some people feel just at the thought that they may exist no more. Even here it is necessary to distinguish some deceptively similar fears. For some people accept that they will not continue to be in the future, but wish to avoid its being as if they had never been in the past. They may then want either to be remembered or to have a continuing effect on the world. A monument can work in either way. A plantation of oak trees (for example) may continue to give pleasure to others, even if the planter is not remembered. And wanting to be remembered can take different forms, for wanting to be remembered affectionately by those who knew you is different from wanting to be remembered by future generations.

People do not always keep these different fears distinct. I think an example is provided by Plato in the *Symposium*, when he makes Diotima suggest that

[1] Here the creation of hospices for the dying has helped.

people seek vicarious immortality through the children, or works of art, or institutions which they leave behind.[2] If this is really what they are doing, then they are guilty of the confusion of which others have complained.[3] For, in leaving these things behind, they are certainly not securing for themselves a *future*. I would suggest, however, that some of them have a more realistic aim, namely, avoiding the obliteration of their *past*.

Admittedly, there is something a little surprising even about this form of consolation. It is that the trace or the memories left behind are themselves so short-lived, in comparison with the future history of the universe, that the consolation provided seems out of all proportion to their duration.

Those who feel the horror which concerns me would accept the loss of *existing* projects and attachments, if only they could have more experience of some kind. So the horror must be distinguished from fear of that loss. Some would like more conscious experience in the future, even if it were only intermittent. For what they dislike is not non-existence as such, but a *permanent* end to existence. Some would like to have further consciousness, even if it were of a thoroughly unpleasant sort.

It is easy to misunderstand this attitude as being too irrational to require further discussion. For surely anyone, given a sufficiently unpleasant life, would eventually prefer extinction. This can safely be admitted. Two points need to be made: first, it does not prevent extinction from being a horrifying evil that in certain circumstances something else would seem even worse. Secondly, even if someone foresees that, given the pains of hell, he would come to prefer extinction, it does not follow that he can yet form that preference. It may be only with the stimulus of hell that he can form it.

Many will not share the attitude which I have been describing. But it is the one on which I shall concentrate, and it is certainly common enough. Bernard Williams quotes a statement of the kind from Unamuno:

> For myself I can say that as a youth and even as a child I remained unmoved when shown the most moving pictures of hell, for even then nothing appeared quite so horrible to me as nothingness.[4]

There is a counterpart in antiquity in Plutarch:

> ... The hope of immortality, a hope on account of which I can almost say that all men and women would be eager to have a biting match with Cerberus and to carry water to the leaky jar, if only they might stay in being and not be annihilated.[5]

In this chapter I want to consider a series of arguments intended to counteract this fear, and all having to do with time in one way or another.

[2] 207C-209E.
[3] Michael A. Slote, 'Existentialism and the fear of dying', *PQ* 12, 1975, 17-28 (repr. in John Donnelly (ed.), *Language, Metaphysics and Death*, New York, 1978).
[4] Unamuno, *Del sentimiento trágico de la vida*, tr. J.E. Crawford Flitch, London 1921, Fontana ed. 1962, p.28, cited by Bernard Williams, 'The Makropulos case: reflections on the tedium of immortality', in *Problems of the Self*, Cambridge 1973 (repr. in *Language, Metaphysics and Death*).
[5] *Epicurus Makes a Pleasant Life Impossible* 1105A.

Lucretius: past non-existence is no evil

One such argument is suggested by Lucretius' exposition of Epicurus:

> Again, look back and see how the ancient past of everlasting time before we are
> born has been nothing to us. Nature then shows us this as a mirror of future
> time after our final death. Does anything appear horrible there, does anything
> seem sad? Does it not stay steadier then any sleep?[6]

Lucretius may well be repeating a point which he has already made in Book
3, ll.832-42. It had been made still earlier in the pseudo-Platonic *Axiochus*,
and is reiterated by Cicero and subsequently by Plutarch:[7] just as we did not
suffer from past evils before our birth, so we will not suffer from future evils
after our death. This comparison between past and future is not found in the
extant writings of Epicurus, but the insistence on our immunity from
disturbance is.

However, Lucretius is sometimes interpreted as making a more interesting
point: we are not now horrified at the thought of our *past* non-existence. Why
then should we be horrified at the thought of our *future* non-existence?[8] If
Epicurus ever said this, it would give the lie to Plutarch's complaint that the
horror of non-existence was ignored by him.[9] I am not at all sure that
Lucretius does intend this argument, but it is none the less an argument well
worth discussing, and it clearly bears on the concept of time.

There may be exceptions to the claim that people are not horrified at the
thought of past non-existence. Nabokov, in his autobiographical work *Speak,
Memory*, describes someone who was horrified to see a film taken a few weeks
before he was born:

> He saw a world that was practically unchanged – the same house, the same
> people – and then realised that he did not exist there at all and that nobody
> mourned his absence. He caught a glimpse of his mother waving from an
> upstairs window, and that unfamiliar gesture disturbed him, as if it were some
> mysterious farewell. But what particularly frightened him was the sight of a
> brand-new baby carriage standing there on the porch, with the smug,
> encroaching air of a coffin; even that was empty, as if, in the reverse course of
> events, his very bones had disintegrated.[10]

Admittedly, this person's retrospective horror is not quite analogous to the
prospective horror we may feel at our future non-existence. For his horror
was at his not existing and not being missed *in that particular environment* – one

---

[6] *de Rerum Natura*, bk 3, ll.972-7.

[7] Ps.-Plato *Axiochus* 365D; Cic. *Tusc.* 1.37,90; Plut. (?) *Cons. ad Ap.* 109F; (adapted to a case
of bereavement) Plut. *Cons. ad Ux.* 610D.

[8] See the excellent paper by Thomas Nagel, 'Death', *Nous* 1970, 73-80. But I think he is right
to withdraw his proposed solution in the revised version printed in James Rachels (ed.), *Moral
Problems*, New York 1971; in Nagel's *Mortal Questions*, Cambridge 1979; and in *Language,
Metaphysics and Death*.

[9] Plut. *Epicurus Makes a Pleasant Life Impossible*, chs 25-30.

[10] Vladimir Nabokov, *Speak, Memory*, London 1967, p.19.

where he felt he should have belonged. There is no sign that he would have felt horror at his absence in the more remote past. For something closer to this, we might consider the horror that some people feel when they think of the infinitude of time, and feel their own brief existence dwarfed by it. Here the infinity of past time before their existence is just as effective a source of horror as the infinity of future time. But here too we do not seem to have an exact retrospective counterpart of our prospective horror.

Actual examples of a retrospective counterpart may be hard to find. But it seems conceptually possible that someone should be horrified by his past non-existence, while paying no attention to his future nonexistence, or even that he might attend to his future non-existence, but think it less bad. For before he existed, a small turn in fortune could have brought it about that he never came into existence at all, whereas at least this horrifying possibility is excluded, when he considers his future non-existence.

These examples should eliminate one mistaken reply to the 'Lucretian' argument. For it might have been alleged that it is conceptually impossible to feel horror at our past non-existence, so that our differential dislike of *future* non-existence is logically inevitable. But this is not so. Admittedly, we will not describe horror about the past as *fear*, since we can only *fear* that something *has* happened if its discovery lies in the future. But this is merely a point about nomenclature.

Another attempt to meet the 'Lucretian' argument also fails. Thus it might be urged that our greater dislike of *future* non-existence is not surprising, because it is simply one example of a much more general preference. We would rather a bad thing was over than to come.

This suggestion is not quite right as it stands. For with a bereavement or a disgrace we would typically prefer it to be future rather than past. Indeed, we feel this preference in relation to death itself. But perhaps the thought is that when, as with toothache, the bad effects of a thing can be wholly over, then we prefer them to be over rather than to come.

But the analogy of toothache is inadequate, because we think of pain as an evil regardless of its timing, however happy we may be when it is over. Our attitude to non-existence looks more irrational than this. For we normally view our non-existence as an evil, *only* in so far as it is future. Why should mere futurity make it an evil?

In any case, our attitude to toothache may itself need explanation, for it is not logically inevitable. After all, past pains can be recollected in all their concrete particularity, whereas, with future pains, we do not know what form they will take. Why does this not result in our minding past pains more?

A better analogy than pain to present to the 'Lucretian' would be separation from a friend. For in the case of friendship, we do not think it an evil, unless there are special factors, that there was a time before we had our friend's company. But we do think a permanent separation in the future to be an evil. The analogy is still closer. For we do not, except in a very intimate friendship, think it an evil if our future associations are intermittent. And in the same way, many people would not mind an intermittent future existence, interrupted by periods of suspended animation, just so long as their existence did not come to a permanent end. As with friendship, so here, it is not future absence as such that matters, but *permanent* future absence.

This robs the 'Lucretian' of his complaint that there is no parallel for our attitude to non-existence. But he may still not be satisfied. I shall return to the question whether he ought to be. But what he may be looking for is a *further* reason for treating past and future differently. Without this, he may think, our attitude is as irrational in the one example as in the other.

One suggested reason, in the case of friendship, for treating past and future differently, would be that we will *mind* a future separation at the time. But surely this cannot be the whole reason why we think the future separation an evil. We mind it because we think it an evil; it is not an evil merely because we mind it. The question then remains why we do not think of the past separation as an evil. The suggestion would not in any case help with the case of non-existence, since we shall not mind our future non-existence at the time.

A different kind of explanation, available in some instances, would invoke natural selection. Animals which minded past pain or danger more than future would not fare well in the evolutionary process. But although this sort of consideration may sometimes explain why we have a particular preference, it does not yet represent it as *rationally* based.

One response to the question of rationality would be to admit that we have no good reason for our preference about non-existence, but to insist that that does not make the preference *irrational*. After all, on pain of regress, we must just see some things as preferable to others *without* further reason, and to do so is neither rational nor irrational. But this plea overlooks an important part of the 'Lucretian' complaint, which is that our preference for past over future non-existence is a preference felt between things which look in many ways extraordinarily alike, almost mirror images of each other. In most cases in which we prefer one thing to another without further reason, the choice is at least between things which are not alike.

The best I can do is to return to the question of the *generality* of the phenomenon. We certainly do have a very general tendency to give a preference to one direction or the other in time. Thus we are not interested merely in the *quantity* of good things in our lives, but also in the *order* in which they come. Any sensible person would prefer a life of promotion to a life of demotion, even if the life of demotion started at the top, and included the same total quantity of advantages. Gregory of Nyssa, to take a more sublime example, was attracted by a future of perpetual progress, not by one of perpetual retrogression, albeit containing the same quantity of understanding. To make these comparisons fair, we must discount such differentiating factors as the thought that the progress is due to one's own efforts and the retrogression to one's lack of effort. When such allowances have been made, I think that preferences concerning temporal order and direction will turn out to be so widespread, that we would hardly be human without them.

The 'Lucretian', then, cannot represent our preference concerning non-existence as a stray anomaly. But he can claim that it is in a certain sense irrational, and we should congratulate him for bringing the fact to light. None the less, I doubt if his argument can be expected to alleviate fear. For one thing, we cannot change our preferences over temporal direction without shedding many of the attitudes which are characteristic of us as humans. For another thing, if we did change, we might become horrified at our past non-existence as easily as indifferent to our future. Although, on his death bed, Hume cited the 'Lucretian' argument to Boswell,[11] he did not present it as the source of his unconcern – he *never* felt uneasy at the thought of annihilation.

The reality of past and future

I now turn to a different kind of consideration. It has been seen that our horror is directed to the future rather than to the past. Sometimes it also takes the form of worrying that our death is *soon*. That our non-existence falls within the twenty-first century A.D. only concerns us in so far as the twenty-first century is *future* and *soon*. But *future* and *soon* are concepts which belong on the *flowing* side of the distinction which was drawn in Chapter 3 between static and flowing time conceptions. And some philosophers have tried to represent flow as a kind of illusion, while Iamblichus, we saw, represented the flowing time as not fully real. If 'future' and 'soon' express illusory ideas, then our fears too will represent some sort of illusion. I outlined this argument already at the beginning of Chapter 1, though I do not know a case of its actually having been used.

I cannot agree with the argument, because I have not agreed that the flowing time conceptions represent any illusion, although a remarkable sense of the unreality of the difference between past and future is generated by the film *Last Year in Marienbad*. The technique is to confuse us as to which sequences represent the past, the present, the future, or the imagined. But it is itself an illusion to feel that the difference between past and future is illusory. The best we can do is to try to think of the fact of our existence as more important than its position in the temporal A-series.

It may be wondered whether *all* the fears in this area involve the *flow* of time in the same way. There is one kind of horror that does not. For somebody might think in static terms and still feel the horror of being dwarfed, as he contemplated the shortness of his own life within the infinite length of McTaggart's static B-series.

Marcus Aurelius

The *Meditations* of the Stoic Emperor Marcus Aurelius are not impressive for the strength of the philosophical arguments. But they are impressive, if it is true that these are the private exhortations to himself of the ruler of the Roman world, and were not intended for publication.[12] He dwells on death, not only to rid himself of fear, but to impress on himself the vanity of life. One

---

[11] *Private Papers of James Boswell*, ed. Geoffrey Scott and Frederick A. Pottle, vol. 12, 1931, 227-32, repr. pp. 76-9 of Norman Kemp Smith's edition of *Hume's Dialogues Concerning Natural Religion*, 2nd ed., London 1947. I am grateful to Myles Burnyeat for the reference.

[12] So P.A. Brunt, 'Marcus Aurelius in his Meditations', *JRS* 64 1974, 1-20.

argument to do with time is that death, however prematurely it comes, can only deprive us of the momentary present. For this is all we possess, and a man cannot be deprived of what he does not possess.[13] The premise is surely mistaken, for a man can be deprived of what he *might* have had. Moreover, an unappealing feature of the argument for many people would be that it reduces the fear of death only by downgrading the value of life.

Stoic and Epicurean arguments against prolonging life

Marcus has another argument to do with time, which is found earlier in Lucretius and Cicero.[14] This argument does not deny that it is bad to be dead, but it claims that nothing is to be gained by *prolonging* life, since the everlasting period of death will be just as long whenever you die.

There was another tradition of arguments against prolongation which goes back as far as Aristotle. He had argued against Plato's Form of the Good that it would not be any more good through being everlasting, any more than a white thing is whiter for lasting longer.[15] An analogous idea, already hinted at by Aristotle,[16] was soon made explicit by Epicurus and the Stoics: life can be made perfect in a finite time, even if it is short. Perfection does not require prolongation.[17]

Plotinus against prolongation

But an entirely new twist was given to the arguments against prolongation by Plotinus, thanks to his views on time. Happiness attends the life which

> is not to be reckoned in time, but in eternity (*aiōn*). And this is neither more nor less, nor of any length, but is a 'this', and unextended (*adiastaton*) and not temporal (*ou chronikon*).[18]

In other words, happiness is not gained by prolonging life, because the happy life is a timeless one, to which the idea of prolongation cannot apply.

I have described in Chapters 10 and 11 Plotinus' conception of the life of intellectual contemplation. I did not agree (Chapters 8 and 11) that it could really be timeless, but I tried to explain, as far as I could, why he thinks of it as such. I have also expressed a preference for a rival tradition, which finds pleasure in winning truth rather than in contemplating it, and in making perpetual progress rather than in remaining fixed. Even if it involves only an *illusion* of timelessness, the life of static intellectual contemplation would be incompatible with most of the things which people hold dear in ordinary life. It would, for example, exclude association with friends. In Chapter 11, I

[13] Marcus Aurelius, *Med.* 2.14.
[14] Lucr. *de Rerum Natura* bk 3, ll. 1087-94; Cic. *Tusc.* 1.39.94; Marcus Aurelius *Med.*, e.g. 4.50.
[15] *NE* 1.6, 1096b3-5; *EE* 1.8,1218a10-15.
[16] *EE* 1.5, 1215b29.
[17] Epicurus *Principal Sayings* 19 and 20; *Letter to Menoeceus* 126 (= Diog. Laert. 10.126 and 145); Sen. *Ep. Mor.* 32; 93,7; 101,8-9.
[18] 1.5.7(23-5).

described Augustine's ambivalent attitude to this. On the one hand, he hoped his dead friend Nebridius would remember him; on the other, he imagined the saints rapt in changeless contemplation, and he rebuked himself for grief at his mother's death, when the ties of kinship ought not to be valued in that context. I shall have more to say shortly about the appeal to some people of this more Platonist conception.

The Makropulos case against prolongation

Bernard Williams has produced a different and a strikingly original argument against prolongation.[19] He maintains that an unending life would inevitably be intolerable. He explains his case partly by reference to the fictitious history of Elina Makropulos. In Williams' version she remained with the physical attributes and mental outlook of a woman of forty-two, and by the age of three hundred and forty-two was so bored, that she decided to die. I am not surprised; but the mistake, as it seems to me, lay not in the continuation of life, but in the retention of the outlook of a forty-two-year-old. That ruled out the perpetual progress of which I spoke in an earlier chapter, and which seems to me to be vital. Need we suppose that perpetual progress is conceptually impossible? Is there only a finite number of things which it is worth coming to understand, and a finite number of interactions between persons in which it is worth engaging? If not, need we accept that *any* human life, indefinitely prolonged, would have to become as tedious as that of Elina Makropulos?

There is another point which has already been made in another connexion. Even if an indefinitely prolonged life had to be an intolerable evil, it would not follow that extinction was not a horrifying evil as well. Certainly, as Williams makes clear, extinction is not something that many of us are yet ready to prefer.

Elina Makropulos seems to have missed *both* of the alternative strategies which might be used for trying to avoid boredom. One would be to seek a life which included friendships and activities not too utterly far removed from those which we know, but which also included the possibility of *progress*. The other would be to seek the kind of blessed life envisaged by many Christian mystics. This, whatever the *other* problems about it (and several discussed by Williams are relevant), would *logically* exclude a sense of tedium just because of its sense of timelessness. Indeed, it would be so different from the life we ordinarily know as logically to exclude most of the ills with which we are familiar, and it has recently been commended precisely on this ground.[20] Makropulos, however, pursued neither of these alternatives.

I do not deny that an unending life could not be very similar to the one we know, if it was to be tolerable, or if it was to be realised at all. But I do not feel sure enough of my powers of imagination to conclude that it could not be realised, and realised in a form that was worth living. An immortal person in

---

[19] Williams, op. cit.

[20] John J. Clarke, 'Mysticism and the paradox of survival', *Int. PQ* 11, 1971, 165-79 (repr. in *Language, Metaphysics and Death*).

communication with others would want those others to be immortal too, if every relationship was not to be fleeting. He could not then expect there to be careers, or marriages, or families as we know them. People would diverge from their past selves and from each other infinitely more than they do in a finite life. But I cannot feel confident that no arrangement of these differences would make an unending life possible and desirable.

Endless recurrence

For the remainder of the chapter I want to take up a theory which is sometimes said[21] to have comforted people in the face of death, since it denies our permanent extinction. I shall try to show, however, that, on the contrary, although the theory was widely believed, it was not treated as a source of consolation. Instead, it gave rise to a wide range of attitudes, most of them adverse.

I am referring to the theory that everything will recur again and again in great cycles. Not just individual persons, but whole worlds will be repeated. From an early date, there were theories of successive worlds being created and destroyed within fixed time periods, without these successive worlds necessarily being like each other. Such views are attested for Anaximenes, Anaximander, Heraclitus and Empedocles.[22] Heraclitus added the idea of a Great Year consisting of many ordinary years, although it is not recorded how he thought its length was determined.[23]

Plato does not have a theory of endless recurrence, but many of his myths contributed to subsequent theories. In the *Timaeus* he develops the idea of the Great Year, so that it represents the period within which all the heavenly bodies, including the wandering planets, return to their earlier alignments.[24] He speaks in the same work of periodic floods and conflagrations on the earth,[25] an idea that was to be exploited later. In the *Phaedrus*, he tells a myth in which some souls climb on to the back of the universe and are carried round by its thousand-year revolution, to view the ideal Forms. But others fall to earth and must be reincarnated ten times at thousand-year intervals before they have another chance to view the Forms, unless they choose the life of philosophy for three incarnations running, in which case they will gain an earlier opportunity.[26] The *Statesman* contains a quite different myth; the universe rotates in opposite directions in alternate periods. Anyone caught at the moment of the next reversal, will begin to shrink back towards infancy, but after that, procreated humans will be replaced by humans who are

[21] Jonathan Barnes, *The Presocratic Philosophers* London 1979, vol. 2, p. 203; Richard Gale, 'Some metaphysical statements about time', *JP* 60, 1963, 236-7.

[22] Anaximander ap. Euseb. *Praep. Evang.* 1.8.1; Anaximenes, Heraclitus, Empedocles ap. Simpl. *in Phys.* 1121; *in Cael.* 293-4; 307; Heraclitus and Empedocles ap. Arist. *Cael.* 1.10, 279b13-17. It is to be noted, incidentally, that Empedocles, fr. 17, ll.11-13, influenced Plato's account in the *Timaeus* of how time falls short of eternity (*aiōn*), but is a moving likeness of it. Empedocles says that in so far as things are always coming into being, they lack fixed *aiōn*; but in so far as they never give up exchanging places continually, they are always unchangingly in a circle. There is a connexion here with the later idea that rotation is not a change of place.

[23] Aëtius 2.32.3; Censorinus *de Die Natali* 18,10-11.

[24] *Tim.* 39D.   [25] *Tim.* 22B-23C.   [26] *Phdr.* 246A-249C.

resurrected out of the earth, for an appointed number of times.[27]

Aristotle takes up the idea of flooding and (not of conflagration but) of drying out, and suggests that the floods are due to a Great Winter, which forms part of the Great Year.[28] However, he recognises only *partial* destructions of the earth's surface. Within Aristotle's school, the author of *Problems* 17 envisages a much more ambitious theory, according to which numerically the same people are born again, so that we cannot say we are after, rather than before, the people of Troy.[29] But he too takes the Aristotelian view that people born later are the same as ourselves only in their form, that is, in being men.

Aristotle's pupil Eudemus of Rhodes ascribes a very extreme theory to certain Pythagoreans:

> One might be puzzled whether or not, as some people say, the same time recurs. For sameness is spoken of in different ways, and it does seem that a time the same in kind recurs, e.g. summer, winter and the other seasons and periods. Similarly, motions (*kinēseis*) the same in kind recur, for the sun will accomplish its solstices, equinoxes and other journeys. But if one were to believe the Pythagoreans' view that numerically the same things come again, and I will talk, staff in hand, to you sitting like this, and everything else will be alike, then it is plausible that the time too will be the same. For when the motion (*kinēsis*) is one and the same, and similarly there are many things which are the same, their before and after is one and the same, and hence so is their number. So everything is the same, which means that the time is as well.[30]

It seems to be Eudemus' *inference* that the *time* will be the same. Although he does not say so, the inference would destroy the Pythagorean theory. For if the time is the same, then, after all, things happen only once and there is no recurrence. The argument that the time will be the same does not seem to rely, as is sometimes thought,[31] on a Leibnizian principle of the identity of indiscernible times. I suggest it depends rather on the Aristotelian definition, discussed in Chapter 7, of time as the number of motion (*kinēsis*) in respect of before and after. The idea is that if the countable stages in what is going on are supposed to be the same, that simply means that the number of change in respect of before and after, in other words, time itself, is the same.

Zeno and Chrysippus, the first and third heads of the Stoic school, are supposed to have said that the same people and numerically the same things would all recur after the conflagration with which each cycle ends.[32] None the less, there was evidently some debate about whether in that case I am numerically one or diverse.[33] Some even said that a person exhibiting no changes, who was not, however, the same person, would appear in the next cycle.[34] Others said the opposite: that it would be the same person, but that small changes would be permissible, for example, in respect of freckles.[35]

---

[27] *Statesman* 269B-274E.   [28] *Meteor.* 1.14, esp. 352a28-31.   [29] Ps-Arist. *Probl.* 17.916a18-39.
[30] ap. Simpl. *in Phys.* 732,26-733,1.   [31] Barnes, op. cit., vol. 2, pp. 203-4.
[32] Zeno ap. Tat. *ad Gr.* ch.5 (= SVF 1.109); Chrys. ap. Alex. *in An. Pr.* 180, 31 (= SVF 2.624).
[33] Simpl. *in Phys.* 886,11 (= SVF 2.627).   [34] Origen *contra Celsum* 4.68 (= SVF 2.626).
[35] Origen *contra Celsum* 4.20 (= SVF 2.626); Alex. *in An. Pr.* 181,25 (= SVF 2.624).

The rival school of Epicurus did not hold a theory of eternal recurrence, but was willing to concede that for an individual person the same atoms might by chance meet in the same order, so that he lived again. This may often have happened in the past, in view of the immeasurable extent of past time. But that only shows that such recurrence would be nothing to us, seeing that the thread of memory is snapped.[36]

Finally, as we shall see, a doctrine of recurrent worlds was adopted by two great contemporaries, Plotinus, the founder of Neoplatonism, and Origen, one of the first Christian theologians.

Cyclical time versus cyclical events

It is interesting that not even the Pythagoreans and Stoics make it an explicit part of their theory that the *time* will be the same. On the contrary, it is the *events* which recur in cycles, whereas the idea that the *time* will be the same is put forward by Eudemus as a *reductio ad absurdum* of the theory. For if the time is the same, there is after all no recurrence. The point is worth making because it is often said that among the Greeks, in Cornford's words, 'time was conceived, not as a straight line, but as a circle'.[37] Cornford is able to quote not only Empedocles and Plato talking of time coming round or circling (*periplomenos*, *kukloumenos*), but also Aristotle, saying with possible reference to them:

> Even time itself is thought of as a sort of circle. And this opinion again is held because time is the measure of a circular motion, and is itself measured by a circular motion. So to say that the things which happen form a circle is to say that there is a sort of circle of time – and that is because time is measured by the circular motion [of the heavens], while there is nothing else to be seen in what is measured except for the measure, or the several measures which constitute the whole.[38]

Cornford further cites Proclus, who says:

> The advance of time is not, as it were, a line that is single, straight and infinite in both directions, but finite and circumscribed ...

> The movement of time progresses in accordance with the measures of the temporal monad, fitting end to beginning, and that an infinite number of times ...

> The whole of time is contained in a single revolution of the whole universe ...

> The revolution of the whole has as its measure the entire extent and development of time, and no extension is greater – unless it be by repetition, for in that way time is infinite ...[39]

[36] Justin *de Resurrectione* 6; it is not impossible. Lucr. 3.847-61 (856: you could easily believe it has happened before).

[37] Cornford, *Plato's Cosmology*, 103-5.

[38] Arist. *Phys.* 4.14, 223b28-224a2: Empedocles fr. 17, l.29DK; Plato *Tim.* 38A.

[39] *in Tim.* (Diehl) 3.29,3-5; 30, 30-2; 2.289,14 and 20-3.

This certainly provides one sense in which time is cyclical. But the idea is only that the seasons come round again, and so do the longer periods, including the longest period of all in which the heavenly bodies return to their original alignments. If, then, we view time as a measure, and that is the most relevant way of viewing it in this context, we must admit that it comes to an end and starts again. None the less, that is not the only way of viewing time, and Proclus has to admit that in another sense time continues infinitely. We do not therefore find the radical thesis, which has been put forward in the twentieth century by the great mathematician Kurt Gödel,[40] that time itself, as we might say, and not just time as a measure, or the events in time, may be cyclical. I have explained elsewhere[41] why I cannot view his hypothesis as coherent. The Greeks would not find it coherent either: Eudemus saw that it would destroy, not support, the theory of endless recurrence.

Reasons for the theory

What reasons were there for the belief in recurrence? At first we find only empirical evidence offered. As evidence for partial, not total, destructions on the earth, Plato and Aristotle cite records of floods or of the sea receding.[42] A set of four arguments for the Stoic view of periodic total destruction was assembled by Aristotle's successor, Theophrastus, and is preserved by Philo Judaeus.[43] The earth cannot have existed very long, since it has not yet been flattened by rain. At least the element of water is beginning to perish, since the sea has lowered round Delos. Each of the four elements is perishable, so the whole world order must be. The human race has not existed for long, as shown by the comparatively recent invention of various arts. Theophrastus holds to Aristotle's view, and concedes only that there have been *partial* destructions. Another argument of Zeno's is that the sun *must* burn things up, so long as there is fuel, and it does so by always using up slightly more vapour from the sea than it relinquishes.[44]

Jonathan Barnes has imaginatively reconstructed a more *conceptual* argument for recurrence, which he suggests might have appealed to the Stoics, based on the idea that there is only a finite number of possible states of the universe, and R.T. Wallis has claimed as Stoic an argument based on the idea that only a finite number of states is *knowable*.[45] I do not myself find

[40] Kurt Gödel, 'A remark about the relationship between Relativity Theory and Idealistic philosophy', in P.A. Schilpp (ed.), *Albert Einstein, Philosopher-Scientist*, New York 1951, 555-62.

[41] Sorabji, *Necessity, Cause and Blame*, 116-19.

[42] Plato *Tim*. 22B-23C; Arist. *Meteor*. 1.14.   [43] Aet. 118-131.

[44] This fragment of Zeno is preserved in a work of Alexander of Lycopolis, *contra Manichaeorum Opiniones Disputatio*, ed. Brinkmann, Leipzig, 1895, ch. 12, p.19, 2f., and was detected by J. Mansfeld. See his 'Providence and the destruction of the universe in early Stoic thought', in M.J. Vermaseren (ed.), *Studies in Hellenistic Religions*, Leiden 1979.

[45] Jonathan Barnes, 'La doctrine du retour éternel', in J. Brunschwig (ed.), *Les Stoiciens et Leur Logique*, Paris 1978, 3-20. Wallis cites Nemesius *On the Nature of Man* ch.38 (SVF 2.625), which says that the gods can know all the infinitely recurring events there are, by knowing the events of one cycle. But their need to know is not presented as an *argument for* recurrence. Rather, infinite recurrence having been postulated, it is explained how they can have knowledge of it: R.T. Wallis, 'Divine omniscience in Plotinus, Proclus and Aquinas', in H.J. Blumenthal and R.A. Markus (eds), *Neoplatonism and Early Christian Thought*, London 1981.

these arguments in the Stoic literature, but what is interesting is that they do
actually become explicit at a much later date in Plotinus and Origen, who
were both students in Alexandria of the mysterious Ammonius nicknamed
Saccas.[46]

Plotinus argues for recurrence in *Enn.* 5.7.1-3, by saying that there is only a
finite number of seminal reasons (*logoi*). So when as many creatures have
been produced as there are seminal reasons, a new period and a new *kosmos*
will have to start, containing the same creatures.[47]

It is because Origen is so early a Christian that he is free to entertain the
idea of successive worlds. His reason emerges gradually in *On First Principles*.
First, God cannot create an infinity of intelligent creatures, because infinity
is incomprehensible and uncontrollable (2.9.1). That infinity is unknowable
was an often repeated doctrine of Aristotle.[48] Equally, according to Origen,
what lacks a beginning is incomprehensible, and so our *kosmos* must have
had a beginning (3.5.2). But that raises a fresh problem: what was God
doing *before* he created our *kosmos*? How did he avoid idleness? It is to
answer this question that in 3.5.3 Origen suggests he earlier created a
succession of other worlds. He rejects the idea of exact repetition (2.3.4-5): a
world cannot be created for a second time with the same order of events, and
the same circumstances of birth, death and action. Rather, different worlds
must be created with more than minimal differences, or there would be no
freedom of the will. Nor did the crucifixion occur in earlier worlds. On the
other hand, Origen does envisage that the same people can appear in
successive worlds.

Origen's solution is not quite satisfactory. He confesses himself ignorant
how many worlds there may have been (2.3.4-5). But there seems to be a
dilemma: if there was a first one with a beginning, the question remains what
God was doing before that. If, however, there was an infinity of differing
successive worlds, or a first one without a beginning, the question remains
how on Origen's view God can comprehend an infinity of worlds or of days.
The question how God can comprehend an infinity of worlds received a great
variety of answers in antiquity.[49] Some Stoics said that in knowing one world,
the gods know all, since all are alike.[50] Plotinus, who allowed an infinite
succession of world-cycles, even though he denied an infinite number of
seminal reasons, said that the World Soul could know this infinity by
knowing its own single and unitary action upon the world.[51] The later

---

[46] This third-century Ammonius (not the Neoplatonist of the fifth to sixth centuries) is
mysterious because he had no official teaching post in Alexandria, wrote nothing, and had
pupils who agreed not to reveal his doctrines (Porph. *Life of Plotinus* §3). Yet, despite his
obscurity, he had among his pupils two or three of the most outstanding writers in Western
thought, Plotinus, Origen and Longinus, if (as is now disputed) this Longinus is the author of
On the Sublime. (The Origen mentioned as a pupil in Porph. *Life of Plotinus* §§3,14 and 20 may not
be the Christian, but the one so mentioned by Porph. ap. Euseb. *HE* 6.19.5-6, is.)

[47] 5.7.1(23-5); 5.7.3(14-19).

[48] *Metaph.* 2.2, 994b22; 3.4, 999a27; *An. Post.* 1.24,86a6; *Phys.* 1.4,187b7; 3.6,207a26; *Rhet.*
3.8,1408b28; 3.9, 1409b31.

[49] See the valuable discussion by Wallis, 'Divine omniscience in Plotinus, Proclus and
Aquinas'.

[50] ap. Nemesium *On the Nature of Man* ch. 38 (= SVF 2.625).   [51] Plot. 4.4.9.

Athenian Neoplatonists said that inferior entities will be seen as finite by superior ones, and so can be known by them.[52] Augustine, we shall see, simply denied that God was unable to know infinity.[53] But Origen in the end avoids this horn of the dilemma, by plumping for a *finite* succession of worlds, not an *endless* recurrence. Thus the idea of an end to the series (*perfectus finis*) seems already to be implied in 2.3.5, although others have said that he did not accept an end to the series until his later writings.[54] Meanwhile a beginning to the series is also implied, for created intelligences are said to have had a beginning,[55] and so is matter.[56] Since the succession of worlds is finite, it is not surprising to find Origen looking for a *different* solution to the problem of how God avoided idleness. In Chapter 15, we shall see him giving the different answer that God is the creator of a beginningless world of patterns, apprehensible by the mind, not the senses.

Augustine reformulated Origen's ideas, quoting the same scriptural passage from Ecclesiastes which Origen had cited in support of the idea of a succession of worlds, and which declares that there is nothing new under the sun.[57] Augustine's aim is to attack the whole idea of recurrence, but the version he attacks in *City* differs from Origen's, as we have it, in that it involves exact and endless repetition. Elsewhere, he attacks Origen by name for his theory of recurrence, and attributes an infinite repetition to him.[58] Among other arguments for endless recurrence, Augustine refers to the contention that God cannot know infinity, yet also cannot recently have begun the work of creation after repenting of earlier idleness, and equally cannot have created man without knowledge of what he was doing, just as the idea came into his mind. The solution proposed by Augustine's opponents is that God has recreated the world an infinite number of times, recycling a *finite* number of creatures. Thus God escapes earlier idleness, while the collection of creatures remains finite and knowable.[59] Part of Augustine's reply is that God *can* know an infinity. It must after all be admitted that he knows the infinity of numbers.[60] Perhaps this is made possible by the fact that he counts without succession in his thought.[61] But further God does not for present purposes *need* to know an infinity, since the universe had a beginning. And he avoids repenting of earlier idleness, because there is no change in him. Rather he has an unchanging plan and will that things should exist at one time without existing before, but these distinctions of before and after apply only to the existence of the creatures, not to the Creator.[62]

---

[52] Syrianus *in Metaph.* 147, 13-22; Procl. *Elements of Theology* 93; *de Decem Dubitationibus* ch. 11.

[53] *City* XII.18-19.

[54] Einar Molland, *The Conception of the Gospel in the Alexandrian Theology (Skrifter utgitt av det Norske Videnskaps-Akademi i Oslo*, Hist.-Philos. Klasse, 1938, no. 2, pp. 161-4), citing Origen's *Commentary on the Song of Songs* and his *Commentary on Romans* 6.8-10 (vol.6, pp.407ff. Lommatzsch).

[55] *On First Principles* 2.9.1-2; 4.4.8.

[56] *On First Principles* 1.3.3; 2.1.4; *Homilies on Genesis* 14.3.

[57] Aug. *City* XII.14; Origen *On First Principles* 3.5.3.

[58] *de Haer.* 43.

[59] *City* XII.18.

[60] *City* XII.19.   [61] *City* XII.18.

[62] *City* XII.18. For other passages in Augustine expressing a similar view, see Chapter 15.

Parallel arguments for reincarnation

There is a parallel between the foregoing arguments for the repetition of worlds and certain arguments for the reincarnation of souls.[63] The Neoplatonists denied a beginning to the human race, and concluded that there must be a recycling or reincarnation of a finite number of souls. Otherwise there would have to be an actual infinity of human souls awaiting or following embodiment, and an actual infinity, on Aristotelian principles, is impossible. While some Neoplatonists embraced the argument, Augustine used it to prevent them from hoping that all human souls might eventually be freed from reincarnation. Avicenna, we shall see,[64] preferred to argue that an infinity of souls was possible after all.

Attitudes to endless recurrence

I have already remarked that the idea of recurrent worlds provided solace to almost no one. Lucretius mentioned the recurrence of individuals, we saw, as a possible source of solace, but objected that, without memory, it would not help us. Plato believed that people were reincarnated up to ten times per great cycle, but he held that the better course is to escape reincarnation as much as possible, and to remain in the region of the Forms, as did Plotinus, Porphyry and Proclus.[65] It was a matter of controversy whether anyone could escape permanently: Porphyry thought so, and so, I believe, did Plato,[66] Proclus argued against,[67] while the view of Plotinus is debatable.[68] Origen, as already remarked, appears uncertain how many future worlds there may be, but he expects an end of the series, and he entertains, without explicitly rejecting it, the hypothesis that only those who need *correction* will be reborn in successive worlds.[69] None of these thinkers found recurrence a matter of solace.

---

[63] Aug. *City* XII.21; Sallustius *de Diis et Mundo* ch.20; Olympiodorus *in Phd.* 10.1.2-5 (Westerink).

[64] Chapter 14.

[65] Plato, see *Gorgias* 524A; 526C; *Phaedo* 82B-84B; 114C; *Rep.* 611B-612A; *Phdr.* 247B-249D; *Tim.* 42B. Plot. 3.2.4(8-11); 3.4.6(30-5); 4.3.24(21-8). For Porphyry, see Aug. *City* X.30; XXII.27. Procl. *in Crat.* ch. 117.

[66] Porphyry ap. Aug. *City* X.30; XXII.27. Of the Plato references in note 65 the clearest are *Phaedo* 114C and *Phdr.* 248C 3-8.

[67] *in Tim.* (Diehl) 3.278.10-27; cf. *Elements of Theology* prop. 206.

[68] See on the one hand the second and third of the Plotinus passages in note 65. But Aug. *City* X.30; XXII.27, credits the view only to Porphyry, and criticises Plato and Plotinus for not adopting it. I agree with R.T. Wallis and E.R. Dodds that there is no definite statement in Plotinus, but I am not quite persuaded by their arguments that most Neoplatonists must have denied all possibility of permanent escape. For (i) the need to avoid an actual infinity of souls by having some recycling would not exclude a *small* number of permanent escapes. (ii) Plotinus' theory of the soul as a 'frontier-principle between time and eternity' would not be compromised by that small number. (iii) I have not taken it that Plotinus intends an absolutely identical repetition of worlds, 4.3.12(12-19); 5.7.1(11-13); 5.7.3(13-18), although if he did, that would indeed exclude permanent escape. See R.T. Wallis 'Divine omniscience' p.235, n.71; and *Neoplatonism*, London 1972, 77 and 94; E.R. Dodds, edition of Procl. *Elements of Theology*, Oxford 1933, 304-5.

[69] *On First Principles* 2.3.1.

To some people, the idea of endless recurrence was a source of *dismay*. In the Stoic version, each world ends with a conflagration, and this led Romans at the time of the civil wars to wonder whether the cycle was about to bring their city to an end. There was speculation as to whether one could pass to the new age without suffering the conflagration.[70] Augustine was appalled by the theory, and gave the most eloquent explanation of his horror. The theory would make nonsense of the crucifixion and resurrection.[71] After our struggle from misery to the vision of God(notice how important this is to him), there would be a return to misery. Knowing this, we could not love God so much, nor would it help us to become blessed if we did. Or if we remained in ignorance, the happiness we felt would be based on delusion.[72]

In one context, Augustine speaks more favourably of a partially similar theory.[73] For some people believe that certain men will be reborn every 440 years. Those who think this cannot hold that there is anything logically impossible about the Christian belief in resurrection. In allowing an analogy here, however, Augustine does nothing to withdraw his criticisms of the theory of endless recurrence.

The range of reactions to endless recurrence is enlarged by Nietzsche and his commentators.[74] A first, idiosyncratic reaction of Nietzsche's is horror at the thought that 'the small man' and everything contemptible will recur. But a second reaction is that a person might be able to exult in the recurrence, if he could so live his life that he wanted no detail changed. Consequently, Nietzsche thinks that accepting endless recurrence as a mere possibility could give one's decisions much greater significance. A further reaction of Nietzsche's is to find endless recurrence terrifying, because it excludes any final aim for the universe.

Among commentators on Nietzsche, one follows an Epicurean line: we should be indifferent to the possibility of recurrence.[75] For we do not in the present life, and so will not in any life, remember or anticipate other ones. With memory and anticipation removed, we cannot feel any concern about future recurrences. This seems to me to be shown wrong by Augustine's reflections. A person might firmly believe in recurrence, and he might then foresee, like Augustine, that all his struggles were in vain, since he would relapse into the condition which he had left with such effort. This would rightly be a matter for concern.[76]

An attempt to cope with Augustine's kind of worry emerges from a different interpretation of Nietzsche.[77] On this interpretation, Nietzsche

[70] Mircea Eliade, *The Myth of Eternal Return*, translated from the French of 1949, New York 1954, 130-7.

[71] *City* XII.14. [72] *City* XII.21. [73] *City* XXII.28.

[74] The data are assembled in two very interesting articles: Ivan Soll, 'Reflections on recurrence', in Robert Solomon (ed.), *Nietzsche: A Collection of Critical Essays*, Garden City, N.Y., 1973; and Alexander Nehamas 'The eternal recurrence', *PR* 89, 1980, 331-56.

[75] Soll, op. cit., 339-42; endorsed by Nehamas, op. cit., 341.

[76] See the interesting study by Bernard Williams, which allows concern for a future self, despite knowledge that amnesia and other psychological changes will intervene first ('The Self and the Future', *PR* 79, 161-80, repr. in his *Problems of the Self*).

[77] Arthur Danto, *Nietzsche as Philosopher*, New York 1965, 210-13; but Nehamas disagrees, op. cit., 340.

recommends us to feel no concern for the *outcome* of what we do, but to take joy in the activity itself. This recommendation, though impossible for someone like Augustine to follow, is at least intended to overcome the frustration and the sense of pointlessness of which Augustine is so aware.

A different suggestion is that recurrence would imply fatalism,[78] which is usually understood as the view that action is pointless because outcomes are determined independently of how one acts. I do not see why fatalism should be implied, rather than determinism, the view that outcomes simply are determined, that is, rendered inevitable all along. Determinism differs from fatalism in allowing that certain outcomes may be due precisely to the efforts we make, or fail to make. There is a question, however, whether even determinism is implied. I have argued elsewhere that one abnormal version of recurrence does imply determinism.[79] But with more normal versions one might expect the situation rather to be that no one would have any reason to believe in recurrence, or at least in exact recurrence, unless he thought that something made such recurrence inevitable. This would fall short of strictly *implying* determinism.

Even this expectation, however, is shown to be too simple by Ivan Soll.[80] For Nietzsche was tempted to believe in eternal recurrence on grounds connected with *probability*, as well as on grounds connected with determinism. Suppose there is a finite number of possible arrangements of the universe. Then, if they are equally probable, considerations of probability would make us expect that in infinite time each would recur. Soll points out that this should have very different consequences from those anticipated by Nietzsche. For not only should my present life recur, but also a finite number of alternative versions of it. Things will presumably be even more bizarre, if we consider all *sequences* of these possible lives as equally probable.[81] What Soll further points out is that such a conception of recurrence should make our present decisions *less* significant, not more significant as Nietzsche anticipated. For whatever I decide now, I am likely to make different decisions in other cycles.

In illustrating how various are the possible attitudes to endless recurrence, I have reached the topic of infinite past time, which will be taken up in Chapter 14. My point has been that the theory was not a source of consolation, but often an additional source of horror.

---

[78] Nehamas, op. cit., 340-1.

[79] In *Necessity, Cause and Blame*, 119, I considered the version which is implied by Eudemus in the extract quoted above. According to this version the time of each future recurrence will (absurdly enough) be the same as the present time. If we try to take that self-contradictory idea seriously, we get the result that the events within each recurrent world will belong to the present. They ought then to be irrevocable, and so determinism will have been introduced by an unexpected route.

[80] Soll, op. cit.

[81] I could then expect long sequences in which my present life is the only one which recurs and long sequences in which only variants recur. Indeed, either kind of sequence ought already to have occurred with every possible length; and each of these sequences to have recurred an infinity of times.

PART III

TIME AND CREATION

CHAPTER THIRTEEN

Did the Universe Have a Beginning?
The Background

Did the universe have a beginning? With only a very few possible exceptions,[1] such a view was denied by everybody in European antiquity outside the Judaeo-Christian tradition. That tradition's belief that God could have given a beginning to the material universe would have seemed to most Greeks an absurdity. But some clarifications are needed as to just what the Greeks denied.

First, many Greeks allowed that the *present* orderly arrangement of the earth and heavens, what might be called the *kosmos*,[2] had a beginning; what they denied, as Aristotle makes very clear,[3] was that matter itself began. It is the latter question that I shall be concerned with, and with the former only as it relates to the latter.[4] Secondly, certain Greeks, notably Platonists, allowed, as we shall see,[5] that matter was created, because they took creation as not implying a *beginning*: its *beginningless* existence was due to God. Thirdly, we shall also see[6] that they allowed matter to be created *not out of anything*, in this sense, that the Creator needs no independently existent stuff in order to create matter. But once again this creation not out of anything does *not* imply a beginning. Moreover, we shall see that they preferred to call it creation from God, rather than creation out of nothing.[7]

Despite all these concessions, then, the Greeks still denied a beginning to the material universe. Now for some clarifications on what was said on the other side by a number of Christians. First, they allowed that *certain* things

[1] Listed in Chapter 15.

[2] Unfortunately Greek is not quite consistent in its use of *kosmos* and of *to pan* (the whole).

[3] Aristotle says that all his predecessors gave the (present) *kosmos* a beginning (*Cael*.1.10, 279b12); but that those among them who made the *kosmos* to be alternately assembled and dissolved were in effect making it eternal (*Cael*. 1.10, 280a11).

[4] There is a continuous literature on the former subject, and partly relevant to the latter, stemming from Aristotle's fragmentary *de Philosophia*. An important source, which preserves earlier arguments on either side by Aristotelians and Stoics, is Philo's *Aet.* (translated in Philo, Loeb vol. 9). This incorporates some fragments of Aristotle's *de Philosophia* and draws partly on a pseudo-Pythagorean treatise of the first or second century B.C., Ocellus Lucanus *On the Nature of the Universe* (translated by Thomas Taylor, London 1831). An Epicurean answer to the Aristotelians is given by Lucretius *de Rerum Natura* 5.90-508 (frequently translated). Many Platonist treatises took up the theme; some are lost, but a good specimen from the fourth century A.D. is Sallustius, *de Diis et Mundo* §§7; 13; 17 (translated, with commentary, by A.D. Nock, Cambridge 1926).

[5] Chapter 17. [6] Chapter 20. [7] Chapter 20.

did not begin: God, the Trinity, sometimes[8] various intelligible (as opposed to perceptible) creations. Secondly, some of them talked not merely of matter, but of a *formless* matter out of which the *kosmos* was made.[9] But so long as this formless matter began, the material universe still has a beginning. Thirdly, in Chapter 18 we shall note an innovation in Gregory of Nyssa: other Christians had reserved the expression 'from God' for the generation of Christ the Son from God the Father; but Gregory extends the phrase to the creation of the material universe. And it is, in fact, marginally more appropriate for his theory of creation, if that theory is that the created material objects are bundles of ideas from God's mind. None the less, since no independently existing matter was needed, he still feels free to call this a creation out of nothing, and, more important, since the ideas were not earlier collected into bundles he still thinks the material universe had a beginning.

Judaeo-Christian hesitancy

In the sense now explained, the orthodox Judaeo-Christian view has come to be that the material universe had a beginning. But what I should now add is that even among Jews and Christians this has been by no means a unanimous view. An early Biblical account of God's creative power in the Book of Job 28 and 38 has been taken to mean that God put order into pre-existing chaos, rather than creating out of nothing.[10] In post-scriptural Jewish literature, *The Wisdom of Solomon* 11:17 says that God 'created the *kosmos* out of formless matter', without saying whether that matter had a beginning; while the reference in 2 *Maccabees* 7:28 to creation 'out of things non-existent' can easily be reinterpreted.[11] This has led to the far-reaching view that there is no clear statement in the Bible, or in Jewish-Hellenistic literature, of creation out of nothing (in a sense which includes a beginning of the material universe). On the contrary, such a view was invented by Christians in the second century A.D., in controversy with the Gnostics.[12]

Personally, I should have thought that the opening of Genesis strongly suggests a beginning of the material universe. It makes no difference that it (naturally) does not specify whether it is formless matter or the ordered universe, including its matter, that begins: in either case we should in effect have an absolute beginning of the material universe. It is those who want to *avoid* this inference who have to go to lengths of reinterpretation. Thus one argument, for which Tertullian attacks the Gnostic Hermogenes, makes play with the word 'was' in the second verse: 'the earth was without form and void'. Not only is this formless earth taken to be formless matter, but the

---

[8] Chapter 15.

[9] Below in this chapter.

[10] See Robert M. Grant, *Miracle and Natural Law in Graeco-Roman and Early Christian Thought*, Amsterdam 1952, ch. 10, 'Creation'.

[11] H.A. Wolfson, *Philo*, vol.1, Cambridge Mass, 1948, 302-3; but there is a more convincing reinterpretation of Maccabees in Gerhard May, *Schöpfung aus dem Nichts*, Stuttgart 1980, 7.

[12] David Winston, 'The *Book of Wisdom*'s theory of cosmogony', *History of Religions* 11, 1971-2, 185-202; May op. cit., 1f. Winston cites as the earliest formulations Tat. *ad. Gr.* 5, and Theophilus *ad Autolycum* 2.4 and 10.

'was' is taken to show it pre-existent.[13] Tertullian also attacks the attempt to reinterpret the words which, in the first verse, are translated 'in the beginning' as meaning: in primordial matter.[14] In the English of the Revised Standard Version of 1952, the opening reads:

1. In the beginning God created the heavens and the earth.
2. The earth was without form and void.

I do not find it at all easy to reinterpret these words, not even when we turn to the versions of the text available in antiquity.[15]

None the less, it is certainly true that in philosophical discussions Jews and Christians were not unanimous. The viewpoint of Philo, the Alexandrian Jew of the first century A.D., is so controversial, and proved so seminal, that I shall return to it at the end of this chapter. But he issues at least one denial of creation out of nothing, along with many *apparent* affirmations.[16] Tertullian of Carthage (*c*. A.D. 160-220) attacks not only Hermogenes, but also the Gnostic Marcion, for accepting that matter has coexisted with God from everlasting.[17] In Alexandria, his contemporary Clement (died by A.D. 215) is said to have made matter timeless in one of his works (presumably, in the sense of having no beginning or end), and to have postulated many *kosmoi* before the *kosmos* of Adam,[18] although elsewhere he modifies these views.[19] Clement's younger contemporary in Alexandria, Origen (*c*. A.D. 185-253) certainly seems to have favoured the creation of a finite number of similar, though not identical, *kosmoi* in succession.[20] The eternity of matter, however, in the sense of its lacking beginning or end, is several times attacked by Origen as a view held by *others*.[21] It was accepted by Gnostics, Manichaeans

---

[13] Tertullian *adv. Hermogenem* 23.1; 27; similarly Basil *in Hex.* 2.2; Ambrose *Hex.* 1.7; Aug. *de Genesi contra Manichaeos* I.3.5.

[14] *adv. Hermogenem* 19.1.

[15] For these versions I would refer to Frederick Kenyon, *The Text of the Greek Bible*, 3rd ed., London 1975, revised by A.W. Adams.

[16] Of the two statements that creation out of non-being is impossible (*Aet*.2.5; *de Spec. Leg.* 1.266), the second may be concerned only with the creation of *compound* bodies. The numerous statements which have been taken to support the other side, and to postulate creation out of nothing, are all brought under a certain suspicion by May's point that talk of creation out of non-being and of bringing into existence is applicable to all plants and animals (see *Quod Deus ...* 25.119), and to parents and children (see *de Spec. Leg.* 2.38.225; May, op. cit., 7; 16; 21). That God brings non-existent things into existence not as a demiurge, but as a creator (*ktistēs, de Somniis* 1.13.76) has been variously explained away as meaning that he created even the four elements out of pre-existent matter, or that he created an intelligible world. The other suspected references are, 'out of non-being': *de Vita Mosis* 2.48.267; *de Deo* 7; *Leg. Alleg.* 3.10; 'the non-existent': *de Spec. Leg.* 4.187; *de Vita Mosis* 2.20.100; *Quis Rerum Divinarum Heres* 7.36; *de Migratione Abrahami* 32.183; *de Opificio Mundi* 31.81; *de Mut. Nom.* 5.46. I shall argue, however, that Philo does in effect endorse creation out of nothing, although not by name, in *Prov.* 1.6-7.

[17] *adv. Hermogenem* chs 1-4; *adv. Marcionem* 1.15.

[18] *Hupotuposeis*, ap. Photium, *Bibliotheca*, Cod. 109, printed in vol. 3 of the Prussian Academy edition of Clement, ed. Stählin, p.202.

[19] *Paidagogus*, 1.62, returns to the more orthodox view that God created everything, including matter.

[20] *On First Principles* 2.4-5; 3.5.3.

[21] *On First Principles* 1.3.3; 2.1.4; *Homilies on Genesis* 14.3.

and other heretics.[22] One argument used by Origen and by Gnostics and Manichaeans turned on a question to which I shall be returning: 'What was God doing before the creation?'[23]

A number of Christians seem to have accepted a beginningless or endless universe under *Platonist* influence. This is true of the bishop Synesius of Cyrene (A.D. 370/5-413/14), who studied under Hypatia in Alexandria,[24] of Elias, one of the sixth-century Christian heads of the Neoplatonist school in Alexandria,[25] and it is often thought to be true of Boethius, as regards his *Consolation of Philosophy*,[26] though not as regards his earlier *de Fide Catholica* (concluding lines).[27] The relevant passage of the *Consolation* was translated above in Chapter 8, but I am not quite convinced that he is endorsing the pagan view.[28]

Thomas Aquinas knew Boethius' discussion well.[29] In his thirteenth-century Paris, there was uncertainty again among Christians. This was due to the recent influence of Islamic and Jewish philosophy. Ghazālī (A.D. 1058-1111) had recorded arguments of the theologians for, and arguments of the philosophers against, a beginning of the universe, in his *Destruction of the Philosophers (Tahāfut al-Falāsifa)*, and Averroes (*c.* 1126-*c.* 1198), replying in his *Destruction of the Destruction (Tahāfut al-Tahāfut)*, comments that neither set of arguments achieves conclusive proof.[30] His Jewish contemporary

---

[22] For Gnostics, see perhaps Irenaeus *adv. Haereses* 2.28.3; Clement of Alexandria, *Extracts from Theodotus* 47.4; Tertullian *adv. Marcionem* 1.15; for Manichaeans, Gregory of Nyssa, *de Hominis Opificio* ch.23; Aug. *de Genesi contra Manichaeos* I.2.3; for unnamed heretics, Basil of Caesarea *in Hex.* 1.7 (referring to the *kosmos*); 2.2 (referring to matter); Ambrose *Hex.* 1.7.

[23] Irenaeus *adv. Haereses* 2.28.3; Origen *On First Principles* 3.5.3; Aug. *de Genesi contra Manichaeos* I.2.3.   [24] Synesius *Ep.* 105 makes the universe endless.

[25] Elias *in Cat.* 187, 6-7, treats the heavens as indestructible; 120. 16-17, reports it as Aristotle's view that the *kosmos* has the same duration as God.

[26] So P. Courcelle, *Late Latin Writers and Their Greek Sources*, Cambridge Mass., 1969, translated from the French of 1948, ch. 6, pp. 316; 320; 322; Philip Merlan, 'Ammonius Hermiae, Zacharias Scholasticus and Boethius', *GRBS* 9, 1968, 193-203; J. de Blic, 'Les arguments de Saint Augusutin contre l'éternité du monde', *Mélanges de Science Religieuse*, Lille, vol.2, 1945, 33-44.

[27] For argument that the very Christian *de Fide Catholica* is by Boethius, as well as the *Consolation*, see H. Chadwick, 'The authenticity of Boethius' Fourth Tractate, *de Fide Catholica*', *JTS* 31, 1980, 551-6. The closing lines envisage a new world of resurrected bodies after the end of this world.

[28] The lack of a beginning or end is put forward as a *hypothesis* of Aristotle and Plato: 'even if (*licet*) as Aristotle believed', 'even if (*licet*) its life is infinite', 'Plato thought (*visum Platoni*)', 'Plato attributed (*tribuit*)'. Boethius' interest is that *even* on this hypothesis, the status of God as eternal (*aeternus*) and timeless remains distinct, and the world can only rightly be called perpetual (*perpetuus*). Courcelle interestingly points out that this fact is not recognised in the anti-Platonist treatises of Boethius' Christian contemporaries Zacharias (*Ammonius* or *de Mundi Opificio*) and Philoponus (*de Aeternitate Mundi*). I am not quite convinced, however, by the further claim of Courcelle and others that Boethius is endorsing the view of Plato and Aristotle.

[29] Aquinas quotes the passage in Boethius, in order to defend Aristotle from the charge of having made the universe eternal in the same sense as God: *Aet.* §10, p.24 in the translation of Cyril Vollert, in Cyril Vollert, L.H. Kendzierski, P.M. Byrne (eds), *St Thomas Aquinas, Siger of Brabant, St Bonaventure, On the Eternity of the World*, Milwaukee Wisconsin, 1964.

[30] Nearly all of Ghazālī's *Destruction* is incorporated in Averroes' reply, which was edited by Bouyges, Beyrouth 1930, and is translated into English with an admirably informative commentary by S. van den Bergh, London 1954. For Averroes' remark, see p.22 (I shall use Bouyges' page numbers, which are recorded in van den Bergh's margin).

Maimonides (1135-1204) claims that neither side can be proved, so that he is free to accept a beginning of the universe on authority. In any case, the arguments for that side are somewhat better, in his view, even though not conclusive.[31] This modesty about the possibility of demonstration was to influence Thomas Aquinas (*c.* 1224-1274) and his teacher Albert the Great (d. 1280) in the next century.[32] Both concluded that it is revelation, not argument, which shows the universe to have had a beginning.

Two slightly younger Parisian contemporaries went a step further, indeed, a step too far. Boethius of Dacia (the Dane, not the sixth-century Roman) and Siger of Brabant maintained that philosophical argument showed the universe to be beginningless, but that none the less reason must bow to revelation.[33] They had to flee Paris in the condemnation of 1277, and there is a tradition that Siger was murdered. He had further added that the same species, opinions, laws and religions recur in cycles, although he claimed as a safeguard to be reporting, rather than endorsing, the view.[34] Clearly, the doctrine that the universe began was a source of difficulty. Even Bonaventure (*c.* 1217-1274), another Parisian of the period, but one who thought a beginning could be proved, confessed that he had felt uncertain, when he first encountered the rival arguments.[35]

Controversy did not end there. For among Jews in the next (the fourteenth) century, Gersonides accepted matter as having coexisted with God without beginning. In the seventeenth century, Spinoza's very different doctrine also excluded a beginning, while Schleiermacher, in the early nineteenth century, tried to introduce that denial into Christianity.[36]

Philoponus

I have so far illustrated the uncertainty felt, even within the Judaeo-Christian

[31] Maimonides *Guide for the Perplexed* 2.16 and 23 (English translations by M. Friedländer and S. Pines).

[32] Albertus Magnus explicitly refers to Maimonides, in *Summa Theologiae* 2, tr.1, q.4, a.5, partic. 3; *in VIII Physicorum*, tract.1, ch. 11. For Thomas Aquinas, see e.g. his *de Aeternitate Mundi*; *Scriptum Super Libros Sententiarum Magistri Petri Lombardi*, in 2, dist. 1, q.1, a.2 solutio; and *ST* 1, q.46, a.1 and 2. Article 1 rejects proofs that the universe is beginningless; while article 2 attacks proofs that there was a beginning, and under 'I answer', makes clear that we must rely on faith, not demonstration. John F. Wippel, 'Did Thomas Aquinas defend the possibility of an eternally created world?', *JHP* 19, 1981, 21-37, argues that in works up to the *Aet.* Thomas treats a beginningless universe as not disproven or disprovable, while in the *Aet.* he treats it as actually possible.

[33] Boethius of Dacia *Tractatus de Aeternitate Mundi* ed. G. Sajó, Berlin 1964, pp. 47-8; 60-2; Siger of Brabant *de Aeternitate Mundi*, ed. W.J. Dwyer, Louvain 1937. Translations of relevant works of Aquinas, Siger and Bonaventure, are conveniently assembled, with useful introductions in Vollert, Kendzierski and Byrne, op. cit. See also F. van Steenberghen, 'La controverse sur l'éternité du monde au XIIIe siècle', *Academie Royale de Belgique, Bulletin de la Classe des Lettres et des Sciences Morales et Politiques*, ser. 5, vol. 58, 1972, 267-87; Wolfson, op. cit., vol. 1, ch. 5.

[34] p. 42 of Dwyer; p. 93 of Kendzierski.   [35] *Collationes in Decem Praeceptis* 2.28.

[36] Gersonides, *Wars of the Lord* 4.2.1. On Gersonides and Spinoza, see Wolfson, op. cit., vol. 1, 302 and 323; 'The Platonic, Aristotelian and Stoic theories of creation in Hallevi and Maimonides', in *Essays Presented to J.H. Hertz*, London 1942, 427-42; *The Philosophy of Spinoza*, Cambridge, Mass. 1934, vol. 1, ch.10. On Schleiermacher, see Nelson Pike, *God and Timelessness*, London 1970, 108-10, with reference to Schleiermacher, *The Christian Faith* §§36-41.

tradition, about the idea of matter beginning out of nothing. But I now want to turn to the year A.D. 529 and to the Christian John Philoponus. Up to 529, Christians adopted a defensive position. They sought only to rebut the arguments that the universe *cannot* have had a beginning. In 529, however,[37] in Alexandria, Philoponus moved on to the attack: he sought to show that the universe *must* have had a beginning. The most striking and influential of his arguments had to do with the concept of infinity.

529 was an *annus mirabilis* for Christianity. The Emperor Justinian put a stop to teaching in the Neoplatonist school at Athens, where Philoponus' opponent Simplicius was working. St Benedict, it is commonly said,[38] founded the monastery at Monte Cassino. The Council of Orange settled outstanding matters on the Christian view of free will. And Philoponus published his arguments for the universe having a beginning in his *de Aeternitate Mundi contra Proclum*.[39]

The work is arranged as a reply to the eighteen arguments against a beginning by Proclus. Proclus (A.D. 411-485), the best known Neoplatonist after Plotinus, had been head of the late Neoplatonist school in Athens. His own treatise *de Aeternitate Mundi contra Christianos*, an attack on the Christians, is lost; but Philoponus summarises the eighteen arguments. And where the Greek manuscript is deficient, omitting Proclus' first argument, the want is supplied by an Arabic translation of Philoponus' summary, now available in French.[40]

Philoponus had already put forward the arguments which I shall be considering in 517, but he had tucked them into a commentary on Aristotle's *Physics*, as he was later to do in a commentary on Aristotle's *Meteorology*, and here the arguments are not very prominent.[41] But the *de Aeternitate Mundi contra Proclum* of 529 is the first of several (perhaps four) works which address the subject extensively.[42] The next is a work in six books against Aristotle. Book 6, against Aristotle's arguments that motion has no beginning, is summarised by Simplicius *in Phys.* 1129-82, while replies to Aristotle's arguments in the *de Caelo* are summarised by Simplicius *in Cael.* 25-201 (see esp. 118-201). A tiny fragment preserved in Arabic has been translated into English.[43] Evidence has been cited[44] for a third work, not directed

[37] The date 529 is given by Philoponus *Aet.* 579,14.

[38] Alan Cameron, however, in the article to be discussed below, reports a recent redating to A.D. 530.

[39] Philoponus *de Aeternitate Mundi contra Proclum*, ed. H. Rabe, Leipzig 1899.

[40] G.C. Anawati, 'Un fragment perdu du *de Aeternitate Mundi* de Proclus', in *Mélanges de Philosophie Grecque Offerts à Mgr. Diès*, Paris 1956, 21-5. The Arabic text has been published by A. Badawi, *Neoplatonici apud Arabes, Islamica* 16, Cairo 1954.

[41] The commentary on the *Physics* is shown to come from A.D. 517 by a passage at *in Phys.* 703,16. Philoponus had discussed the subject even earlier than that according to *in Phys* 55, 26, but the earlier discussion is not extant. The commentary on the *Meteorologica* is placed between the reply to Proclus and the reply to Aristotle by Étienne Evrard, 'Les convictions religieuses de Jean Philopon et la date de son commentaire aux *Météorologiques*', *Academie Royale de Belgique, Bulletin de la Classe des Lettres* 39, 1953, 299-357.

[42] The evidence is given in a magisterial article by H.A. Davidson, 'John Philoponus as a source of mediaeval Islamic and Jewish proofs of creation', *JAOS* 89, 1969, 357-91.

[43] Joel L. Kraemer, 'A lost passage from Philoponus' *contra Aristotelem* in Arabic translation', *JAOS* 85, 1965, 318-27.   [44] Davidson, op. cit.

polemically against any particular opponent, in which Philoponus tried to improve his arguments that the universe had a beginning. Some of the arguments may be summarised by Simplicius *in Phys.* 1326-36, and S. Pines has translated what he believes to be an Arabic summary,[45] although, we shall see, the summary is suspiciously close in some ways to the *contra Aristotelem*. Finally, there is a work more directed to the biblical account of creation: *de Opificio Mundi*. Surprisingly, little of this work has been translated,[46] but this situation will to a large extent be remedied by projected English translations of the *de Aeternitate Mundi contra Proclum* and of the fragments of the *contra Aristotelem*,[47] while the *de Opificio Mundi* is under consideration.

Late Athenian Neoplatonists

I should now like to say a little about Philoponus' Neoplatonist and Christian contemporaries, and first about the Athenian school where teaching was halted. There is surely room, in the present time of university closures, for a popular book on the treatment of philosophers in antiquity.[48] Simplicius and the Athenian school of seven men, including Damascius, its head, took refuge with King Khosroes at Ctesiphon in a part of ancient Persia which is now Iraq. They spent the year of 532 at his court, and there debated, among other subjects, whether the universe has an end.[49] But within a year they left. Understanding of this episode has been transformed through a brilliant, if controversial, article by Alan Cameron.[50] He claims that Justinian's interference was not particularly effective, and that Simplicius, who wrote his commentaries on Aristotle after this date, was able to write them in *Athens*. Cameron acknowledges that they are not written in the widely practised style as lecture notes and that this may be a sign that Simplicius was still not allowed to *teach*. But Simplicius none the less will have had the Athenian school's library for his *research*.

---

[45] S. Pines, 'An Arabic summary of a lost work of John Philoponus', *IOS2, 1972, 320-52*.

[46] On the eternity of the world the following brief extracts are available in translation, over and above the translations from Arabic already mentioned. W. Böhm, *Johannes Philoponus: Ausgewählte Schriften*, Munich 1967, is a valuable selection from Philoponus' work in German translation, but includes only 12 pages on the eternity of the world. The eighteen arguments by Proclus in favour of eternity, as summarised by Philoponus, were translated into English in 1825 in a version which is no longer in print, but are now available in German: Thomas Taylor, *The Fragments that Remain of the Lost Writings of Proclus, Surnamed the Platonic Successor*, London 1825; M. Baltes, *Die Weltentstehung des platonischen Timaios nach den antiken Interpreten*, vol.2, Leiden 1976. Finally, some of the arguments on eternity from Philoponus' commentaries on *Physics* and *Meteorologica* are translated into English by S. Sambursky, 'Note on John Philoponus' rejection of the infinite', in S.M. Stern, Albert Hourani, Vivian Brown (eds), *Islamic Philosophy and the Classical Tradition*, Essays Presented to Richard Walzer, Oxford 1972, 351-3, and by Robert B. Todd, 'Some concepts in physical theory in John Philoponus' Aristotelian commentaries', *ABG* 24, 1980, 151-70.

[47] To be published by Duckworth (see Introduction). Some assignments are still to be arranged.

[48] Exile was a stock philosophical theme, as was the correct behaviour for a philosopher under an adverse government.

[49] Agathias *History*, p.130,15.

[50] Alan Cameron, 'The last days of the Academy at Athens', *PCPS* 195(n.s.15), 1969, 7-29.

There is one reason for thinking Justinian ineffective which Cameron rightly eschews, and this is the idea that the Athenian school was in any case dying of its own accord, so that Justinian's moves made no difference. This very common view of a moribund school appears in Gibbon's *Decline and Fall of the Roman Empire* and is still repeated today.[51] I agree with Cameron that the school had plenty of life in it. He offers historical reasons for saying so, and I hope to have given concrete philosophical illustration of the fact in the arguments of Damascius and Simplicius recorded in Chapter 5 and in the arguments, to be considered below, of Simplicius against Philoponus. None the less, my own inclination is to think that what Cameron admits, the effectiveness of the ban on teaching, would have had a more severe effect on a philosophy school than he allows. Certainly Simplicius was able to compile his great scholarly commentaries. But one could wish for more live philosophical controversy, and for this a teaching situation can be a positive asset. Moreover, even scholarship requires for its continuation into the next generation either the training of new pupils or the prospect of importing ones trained elsewhere, either of which would have been difficult in the climate created by Justinian.[52]

Among the discussions aroused by Cameron's article, the most substantial opposition has come from Alison Frantz, who was criticised in the original piece, and who has returned to the fray, offering archaeological evidence that a pagan philosophy school in Athens, presumably the Neoplatonist one, was closed around 529 and never reoccupied by pagans.[53] I shall leave it to others to assess this evidence, and I will only say that it is not surprising if, after his ordeal, Simplicius thought of the Christian Philoponus with some bitterness. He refers to him as 'the grammarian', and was no doubt pleased that Philoponus' post in Alexandria was in grammar not philosophy. This was in spite of the fact that Philoponus had, like Simplicius, studied under the Neoplatonist philosopher Ammonius in Alexandria, and that most or all of his commentaries on Aristotle are adapted versions of Ammonius' lecture courses.[54]

Late Alexandrian Neoplatonists

Ammonius (born A.D. 435/45, died 517/26) had heard Proclus' lectures in Athens. On Proclus' death in 485, he went to take the chair in Alexandria,

---

[51] Gibbon, Ch. 40, §6; other references in the Introduction.

[52] As it was, the Slavs sacked Athens in 579, so that the demise of the school was doubly determined.

[53] Alison Frantz, 'Pagan philosophers in Christian Athens', *PAPS* 119, 1975, 29-38. For other contributions, see John Glucker, *Antiochus and the Late Academy, Hypomnemata* 56, Göttingen, 1978, 322-9; H.J. Blumenthal, '529 and its sequel', *Byzantion* 48, 1978, 369-85; Ilsetraut Hadot, *Le Problème du néoplatonisme Alexandrin: Hiéroclès et Simplicius*, Paris 1978, 33-4; Whittaker, *God, Time, Being*, 19-20.

[54] The manuscript titles for four of Philoponus' commentaries on Aristotle describe them as exegetical notes from the seminars of Ammonius, in three cases with the addition of some personal observations. See Étienne Evrard, 'Jean Philopon, son *Commentaire sur Nicomaque* et ses rapports avec Ammonius', *REG* 78, 1965, 596.

and his lectures there were heard at least by Philoponus, Simplicius and Damascius. Speculation that Boethius also went to listen to him has recently been discounted.[55] But another Christian who did hear him was Zacharias,[56] who later became bishop of Mitylene.

There is plenty of evidence that Ammonius, like Proclus before him, argued against a beginning or end of the universe. He did not write very much,[57] but his lecture courses were written up with additions by such pupils as Philoponus and Asclepius.[58] Asclepius ascribes to Ammonius, and Philoponus, following Ammonius, ascribes to Aristotle belief in a beginningless and endless universe.[59] Although there are signs that Philoponus may have played down Ammonius' acceptance of this belief,[60] his acceptance is none the less the subject of a whole dialogue by Zacharias.[61] Zacharias' treatise, called *Ammonius* or *de Mundi Opificio* (another work which has not been translated into English) purports to record a series of debates in which Zacharias took on Ammonius and some of his pupils. We get some picture of how discussions were conducted in Ammonius' circle.[62] Although we need not believe every detail of the account, and need not assign the same importance to Zacharias as to Philoponus, none the less we do find these two Christians sometimes using the same arguments as each other. At one point Zacharias (perhaps benefiting from discussions in Ammonius' classroom) seems to go into greater depth,[63] and we shall see in Chapter 19 that he preserves some suggestions which travel in the direction of occasionalism.

Relations between Neoplatonists and Christians, so bitter at Athens, were much closer elsewhere. In Alexandria, Ammonius is accused of reaching some unspecified agreement with the Christian authorities.[64] Not only did he

---

[55] See the replies of James Shiel and L. Minio-Paluello, to the hypothesis of P. Courcelle, references in Chapter 8.

[56] Zach. *Ammonius* or *de Mundi Opificio* PG85, 1016A.

[57] Ammonius' largest work was a commentary on Aristotle's *Int.* which survives, and from which extracts are given in Chapters 8 and 16.

[58] Asclepius *in Metaph.*; Philoponus *in An. Pr.*; *in An. Post.*; *in GC*; *in DA*. See H.D. Saffrey, 'Le chrétien Jean Philopon et le survivance de l'École d'Alexandrie au VIe siècle', *REG* 67, 1954, 405; L.G. Westerink, *Anonymous Prolegomena to Platonic Philosophy*, Amsterdam 1962, xi-xii; Étienne Evrard, 'Jean Philopon, son *Commentaire sur Nicomaque*' 1965, 596. Many findings from the first two of these articles are usefully summarised in G. Verbeke, 'Some later Neoplatonic views on divine creation and the eternity of the world', in Dominic O'Meara (ed), *Neoplatonism and Christian Thought*, State University of New York 1981, 45-53; 241-4.

[59] Asclepius *in Metaph.* 89, 4-5; 90, 27-8; 171,9-11; 186,1-2; 194,23-6; 226,12-15; Philoponus *in Phys.* 54,9-55,26; 189,10-26.

[60] L. Taran has shown that, in a discussion parallel to that of Asclepius, Philoponus has suppressed reference to Ammonius' belief ('Asclepius of Tralles' commentary to Nicomachus' *Introduction to Arithmetic*', *TAPS* n.s. 59, part 4, 1969, 11).

[61] Zach. *Ammonius* or *de Mundi Opificio*, PG85, cols 1005-1144.

[62] This has been argued by Philip Merlan, 'Ammonius Hermiae, Zacharias Scholasticus and Boethius', *GRBS* 9, 1968, 193-203.

[63] Thus they both object that certain arguments against a beginning or end of the universe would prove too much, since they would imply that individual persons do not begin or end (Zacharias 1040B-C; 1041A; 1084A-C; Philoponus *Aet.* pp. 92; 128-9). But Zacharias considers a number of possible replies.

[64] Damascius, *Life of Isidorus*, at Photius 242§292 (= Fr. 316, Zintzen).

have Christians like Philoponus and Zacharias among his pupils, but he had the Christians Elias and David among his successors. The non-Christians Ammonius and Simplicius agreed with the Christians Philoponus and Elias in ascribing to Aristotle the view that God is causally responsible for the existence of the universe,[65] even though not in the sense of giving it a beginning. Conversely, the Christians Synesius and Elias conceded to the non-Christians that the universe is beginningless and endless.[66] We need not believe Zacharias' portrayal of Ammonius as coming over to the Christian side on this last issue.[67] Nor have I accepted the common view that Boethius' *Consolation* moves over to the Neoplatonist side.[68] All the same, Boethius provides another example of a man who felt able to combine the Christian doctrines found in his earlier works with the Neoplatonist doctrines found in his *Consolation*.

It will be evident how fruitful Proclus' eighteen arguments against the eternity of the universe proved to be, both in Athens and in Alexandria. I hope this will appear more evident when I come in the next chapter to Philoponus' replies.

Islam and the Latin West

Islamic thinkers, from the ninth century onwards, were thoroughly familiar with the arguments of Philoponus and of his opponents. They repeated and built on them, with the 'theologians' favouring Philoponus' arguments for a beginning, and their opponents the 'philosophers' developing the other side of the case.[69] It was thanks to this that Bonaventure, Albert and Thomas Aquinas were able to offer similar arguments in the thirteenth century.[70] Thomas Aquinas, as we shall see, obtained some of his most ingenious ideas from what he found in the Islamic and Jewish traditions. Some leading scholars of the medieval West have created the impression that the 'infinity' arguments in favour of a beginning were invented by Bonaventure.[71] But in fact Bonaventure was simply repeating Philoponus' arguments and even using the same illustrations. Philoponus has not always got the credit he deserves, partly because of the anathema which was imposed on him in A.D. 680, just over a hundred years after his death. Many philosophers nowadays, if they know of the 'infinity' arguments at all, will know them only from

[65] Simpl. *in Phys.* 256, 16-25; 1363, 8-24; *in Cael.* 271, 13-21; Philoponus *in Phys.* 189, 10-26; *in GC* 136, 33-137, 3; cf 286, 7; Elias *in Cat.* 120, 16-17.

[66] Synesius *Ep.* 105 and Elias *in Cat.* 187, 6-7 make it endless.

[67] Zach. *Ammonius* or *de Mundi Opificio* PG85, 1113B; 1116B.

[68] *Cons.* 5.6.

[69] See again Davidson, op. cit., for the fullest references, but a number of the passages are translated by H.A. Wolfson, *The Philosophy of the Kalam*, Cambridge Mass. 1976, 410-23; 452-5.

[70] References are given by Wolfson, op. cit., 455-65, and some of the passages are translated in Vollert, Kendzierski and Byrne.

[71] This is the impression created, consciously or otherwise, by E. Gilson, *La Philosophie de Saint Bonaventure*, Paris 1924, 184-8; John Murdoch, 'William of Ockham and the logic of infinity and continuity', in Norman Kretzmann (ed.), *Infinity and Continuity in Ancient and Medieval Thought*, Ithaca New York, 1982, 166. See also G.J. Whitrow, 'On the impossibility of an infinite past', *BJPS* 29, 1978, 40, n.1.

Kant's *Critique of Pure Reason*. But Kant's version, in the first antinomy, is only a faint echo of the ancient originals.

Philo on the creation of time and the material universe

I shall finish the chapter by fulfilling my promise to consider the view of Philo. In Philo's opinion, it is generally agreed, God is the creator of the ordered *kosmos*. It ought to be agreed, I shall argue, that Philo makes him creator not in the sense of giving the *kosmos* a beginningless existence, but in the sense of giving it a beginning, although here already there is some dissent.[72] But something is admittedly less clear. For God created the *kosmos* by imposing order on formless matter, and it is hard to tell whether that *matter* existed before the *kosmos*, and whether it existed beginninglessly. Despite the opposition of Wolfson, the majority view seems to be that formless matter had no beginning.[73] It is then disputed whether, in that case, God is the producer of beginningless matter.[74]

There are numerous discussions of the creation in Philo, but they tend to concentrate on the creation of the ordered *kosmos*, so that the status of its matter can be inferred only directly. The most systematic treatment of *matter* is found in the two treatises on Providence, *de Providentia* 1 and 2, mostly available only in Armenian, from which Aucher made a Latin translation.[75] If these can be relied upon, then I think a considered view emerges, whether or not Philo always stuck to it elsewhere. It has been alleged that the main passage, *Prov.* 1. 6-7, is itself full of contradiction,[76] and it has been very variously interpreted. But I believe that, when one or two points have been properly understood, its main thrust becomes clear enough. Not only the ordered *kosmos*, but also formless matter had a beginning, and indeed began simultaneously. Furthermore, time began with them, and there was no 'before'.

The passage, I shall argue, attacks two alternative views, both of which deny a beginning to the material universe, before going on to give Philo's own view. The first view to be attacked does not differentiate between the ordered *kosmos* and its matter, but simply declares that there was no beginning to the ordered *kosmos*. There is a German translation from the Armenian by C. Hannick, which is said to be more faithful and to smooth over the difficulties in the Armenian less.[77] I shall translate from Aucher's Latin, since the divergences are not very relevant to the present discussion, although I shall draw attention to them when they could make a difference.

---

[72] David Winston, *Philo of Alexandria, The Contemplative Life, The Giants and Selections*, London 1981, 7-21.
[73] For bibliography, see Wolfson, *Philo*, vol. 1, 301; Winston, op. cit., 304, n.25; May, op. cit., 1-21, esp. notes on pp.9; 17; 18. Wolfson gives his view, op. cit., 300-12.
[74] May, op. cit., says not; Winston says that he produces it, but as one might cast a shadow, rather than as a deliberate piece of creation.
[75] Now conveniently available in the edition of Mireille Hadas-Lebel, Paris 1973.
[76] W. Bousset, *Jüdisch-Christlicher Schulbetrieb in Alexandria und Rom*, Göttingen 1915, 146.
[77] Printed in M. Baltes, *Die Weltentstehung des platonischen Timaios nach den antiken Interpreten*, vol. 1, Leiden 1976, 88-90.

1.6 It often happens, if a person wanders carelessly around in his observations, that he will believe this *kosmos* has existed and endured from eternal years without beginning, and that therefore it has had no origin (*principium; Anfang*) in creation, but has had perpetual existence, and can by no means be destroyed. But when we have adduced the clear observations to be cited below, people will not be able to bring forward that universal sophistical argument, which is drawn out into a long digression, by which they try to show that God did not at any time begin the creation of this *kosmos*, but was always attentive to the creating of this very beautiful universe. For they say it is not fitting for Divinity ever to be without activity, since that is idleness and inertia, but that God established all things without beginning. In this, they do not recognise the absurdity of such a hypothesis; for in wanting to remove a minimal charge against God, they press against him a maximal one.

What is this maximum charge? I suggest the idea is that, if God established all things without beginning, then, however far back in time you go, you will find he has already completed his work. This means that he has been idle for indefinitely much longer. There is an analogous complaint by Philo in *de Opificio Mundi* 2.7 about the different view that the universe has not been created at all: that too leaves God with nothing to do.

Philo must think he has by now given *some* answer to the first view, to the idea of a beginningless *kosmos*, because he goes on in the next paragraph, 1.7, to say that only one possibility remains for his opponents, and then to describe the *alternative* and *incompatible* view which distinguishes formless matter and says that at least that had no beginning, even though the ordered *kosmos* did.[78] The German opening is only marginally different: 'The contrary is the case, since there is no alternative: they say ...'

1.7. So the remaining possibility is for them to say that matter which lacked order, form, or shape was endowed by him with quality and form, and then took on shapes which were not formerly in it. For God, in their view, did not even begin to create it [*sc.* matter]. But if that cunning creation produced by God established the lovely form of the world just out of matter and matter acquired from that its very beautiful appearance, are they saying that God did <only> this one thing, when he began to create the *kosmos*, and when matter, which before was accustomed to wander without rule or order, acquired its lovely order and embellishment? Did matter take the place of a basic principle for God, <when he was founding the *kosmos*>?

Philo's reply to this second view may be that it gives God only *one* task instead of two (and so, presumably, makes him idler) when he creates the ordered *kosmos*, because he does not then create its matter as well. At the time the ordered *kosmos* is made, matter is already available as its basis. In Hannick's German version, I find the force of the objection harder to grasp: 'how did God begin to create the *kosmos* if matter was in disorderly and irregular motion without true sequence, but thereafter the *kosmos* received

---

[78] I cannot agree to the connexion of thought suggested by Winston, op. cit., 16. He takes Philo's complaint so far to be that matter has been made independent of God, and he suggests that that will force the opponents – but surely it would not – to take the view described in the next paragraph, 1.7, that matter must have been disorderly once.

beauty and adornment <with matter having come to be its basis(?) – Armenian unclear>?'

I shall pause to resist a certain alternative to the interpretation I have given of the two passages just quoted. David Winston in effect sees Philo's complaint as focused in the last of the quoted sentences: 'Did matter take the place of a basic principle for God?'[79] On Winston's view, the idea is not merely that matter is already present when God imposes order (which leaves him with only one task at that time), but that it has always been present as a 'basic principle', and therefore has an existence *independent* of God's. I must agree with Winston that this independence could be one facet of the opponents' views, for hitherto it has been neither asserted nor denied by them (it is not asserted when they say that God did not ever *begin* to create matter). Because it has not been asserted or denied, it would be quite legitimate for Philo to raise it as a *question* in 1.7 whether they hold to this additional objectionable idea. But I am not so ready to accept Winston's further suggestion that this objection is already intended by Philo in 1.6, and that it is his main point of difference from the opponents described there. For the objection against the opponents in 1.6 is that they substitute a maximal charge against God in place of a minimal. That maximal charge ought to arise out of something to which they are explicitly committed, and it ought to be obvious why it is maximal. The independence of matter, however, is not something to which they are explicitly committed – on the contrary, the opponents of 1.6 do not even mention matter – and the whole emphasis of the passage is on the lack of a beginning, not on independence at all. Nor yet is it obvious why their substituting the ordering of matter for the creation of matter would make the charge against God 'maximal'.

In the remainder of 1.7, I believe Philo gives his own view, and this must differ from both the views just rejected. Yet I would complain that some commentators assign Philo the first rejected view[80] and some the second.[81] Winston would not agree that he assigns Philo the first rejected view, because he takes that first view to imply that matter is independent of God; but I have just been arguing that that cannot be the intended point of difference. To supply a view different from the two rejected, I should expect the next passage to postulate a beginning of the universe, including its matter. Let us see whether it does so:

1.7 (cont.). But the Creator gave order to matter immediately by his thinking. For God did not begin to think before acting; nor was there ever <a time> when he was not acting, the species <or ideas> being with him from the beginning (*ex origine; von Anfang*). For the will of God is not later [*sc.* than his thinking], but is always with him, since the processes of nature never fail him. So it is that he always creates by his thinking, and gives a beginning of existence (*principium existendi; ein Beginn*) to perceptible things. So the two things are always found together: God's acting by divine plan and perceptible things

---

[79] I presume that this is the sentence on which he would rely (*Philo of Alexandria* 14-16), although his translation of the passage on p.109 is closer to the German and omits this sentence.
[80] Winston, loc. cit.
[81] E.g. May, loc. cit.

receiving a beginning of existence. For it is impossible that anything should be benefited with some benefit, and that the benefit should not be from the benefactor, in such a way that each shares in the good: the benefactor who gave and the one who received the benefit.

The first half of this passage seems to supply the very beginning we were looking for, and to add the further information that there is, in Philo's view, no delay between the beginning of matter and its being ordered into a *kosmos* – there was no period when matter was waiting to receive form. Of course, the notion of a beginning is sometimes given a *non-temporal* sense by Philo, but it would be madness to take it that way here, when what the argument required was a *temporal* beginning. It is by making the universe, including its matter, begin that Philo avoids the two views he has just rejected. Moreover, he also solves the problem of idleness which one of those views was trying to solve. How does he do this last?

The implied solution is that, since the Creator 'gave order to matter immediately', there was no period of idleness when matter already existed, but God had not yet imposed order. Admittedly, Philo does not here deal with the question whether there was a period of idleness before matter and the *kosmos* were created. But we are reminded a few paragraphs later of what answer he could give to that. For in other works, he says that time and the *kosmos* are linked together, so that there was no time before the *kosmos*,[82] and in *Prov.* 1.20, where he cites Plato in support of his own views, he ascribes this opinion to Plato – that time began with the heavens.

In giving this account of Philo's solution to the idleness question, I am departing from certain commentators. For they suggest that Philo excludes earlier idleness by ascribing to God the creation of a prior world of intelligible ideas, to serve as a pattern for the later world of intelligible things.[83] Certainly *Prov.* 1.7 talks of God's thinking (*intelligere*), and of the ideas being with him from the beginning, while 1.21 ascribes to Plato the view that God has always been the creator of intelligible things. But the suggestion is not that God creates intelligible ideas *before* he creates the perceptible world. On the contrary, 1.7 seems to insist on the opposite when it says, 'God did not begin to think before acting [i.e. before giving order to matter].' That leaves only two likely options, that God's creation of the intelligible world is timeless, as God himself is sometimes said to be,[84] or that it began together with his creation of matter and the *kosmos*. The surrounding context seems to assume the latter, and this is reaffirmed in *de Opificio Mundi* 7.27-9. It is not Philo, then, but Origen who tackles the idleness problem by postulating that God's creation of the intelligible world *preceded* his creation of the perceptible world. In Chapter 15, I shall try to trace the new steps which Origen had to take.

There is one important point left concerning the passage in which Philo cites Plato (incorrectly, as I shall argue in Chapter 17) in support of his own

[82] See *de Opificio Mundi* 7.26; *Leg. Alleg.* 1.2 and 2.3; *de Sacrificiis Abelis et Caini* 18.65; *Quod Deus ...* 6.31-2.

[83] Baltes, op. cit., 37; May, op. cit., 19-20.

[84] *Quod Deus ...* 6.32.

views. In *Prov.* 1.20, Philo ascribes to Plato a beginning of time, in 1.21 a beginning of the *kosmos*, and the creation by God of both *kosmos* and intelligibles. But a problem arises in 1.22, where some commentators believe he ascribes to Plato a pre-existent matter which exists *before* the *kosmos*, and perhaps even *beginninglessly* before.[85] I do not myself think that he commits himself to this; it would make Plato's view in 1.22 diverge from his own in 1.7. Moreover, the view he ascribes to Plato is that matter came forth *together with* (*cum*) embellishment. It is only *in itself* (*per se*) that it lacks embellishment, but that allows that it may always have had embellishment added. Admittedly, Philo has to say that Moses made water, darkness and chaos exist *before* (*ante*) the *kosmos*, because he thinks that is how Genesis presents things. But, in view of his classification of the Pythagoreans and Aristotle in *Aet.* 3.10-12, he is not likely to think here that the many Greeks he goes on to mention, including Pythagoras and Aristotle, all made matter exist before the *kosmos*, so there is no need to think he believes this of Plato. Even with regard to Moses, 1.7 has shown that Philo would offer a philosophical reinterpretation of the 'before'. The passage reads as follows:

> 1.22. Plato knew that these things [*sc.* the four elements and other things mentioned in the preceding paragraph] were created by God and that matter, in itself (*per se*) lacking embellishment, came forth in the *kosmos* together with (*cum*) embellishment. For these [*sc.* matter, the four elements, etc.] were the prime causes out of which the *kosmos* itself was made; since Moses too, the lawgiver of the Jews, said that water, darkness and chaos existed before (*ante*) the *kosmos*, while Plato <named> matter, the Milesian Thales water, that other Milesian Anaximander the infinite, Anaximenes air, ...

I have been arguing that, in Philo's *Prov.* 1, the *kosmos* and its matter are made by God to begin, and to begin together, and that Plato is thought to agree. The other important passage is in *Prov.* 2, where Philo at least *envisages* the same hypothesis that the *kosmos* and its matter are made by God to begin. For in 2.46, Alexander, who is attracted by Greek views, asks Philo, *inter alia*, how anything could come into being out of nothing (*quomodo ex nihilo aliquid fiet*). Philo is trying to win over Alexander with care, and so presents to him two hypotheses. On the first hypothesis (2.48), by which Alexander is attracted, the universe (*universum*) is uncreated and everlasting (*ingenitum, sempiternum*), and matter is uncreated (*ingenita materia*). On this hypothesis, God did not generate a primary matter for ever (*materiam primam non generavit sempiterne*), but took matter and used it to create other things. This paragraph already raises a difficulty for those who believe that Philo never even *envisaged* the rival hypothesis that God created the matter of the *kosmos*. In 2.50, Philo explicitly turns to his rival hypothesis, and here we have the original Greek, preserved by Eusebius, which differs little from Aucher's rendering of the Armenian, except in including the following description of the rival hypothesis: 'if in reality it did come into being' (*ei dē gegonen ontōs*). Is Philo talking of the ordered *kosmos* coming into being out of an uncreated matter, or

[85] Winston (before); May (beginninglessly).

of the universe, *including* its matter, coming into being? Eusebius classified the passage in the second way, as envisaging the creation of *matter*, and this is very plausible, seeing that 2.46 characterised the alternative hypothesis as involving no creation of matter. 2.50 goes on to say that, no less than an ordinary craftsman, God measured the quantity of matter so that it would be right.

My conclusion is that, if we can rely on *de Providentia* 1 and 2, Philo does imply that God gave the universe, including its matter, a beginning. Every conceivable doubt has been raised about the two treatises, as to whether they are by Philo, or whether they have been tampered with at one or other stage of their transmission.[86] But equally, just about every doubt has been contested, and the latest editor, Mireille Hadas-Lebel, accepts none of them. I shall therefore view the suspicions as not yet proven, and shall assume that for the time being we must take the evidence from *de Providentia* 1 and 2 seriously.

Outside the *de Providentia*, Philo does not always stick to the view that matter has a beginning. This at *Aet.* 2.5 he declares that nothing comes from non-being, and at *de Opificio Mundi* 5.21 the argument requires not only that matter exists *before* the *kosmos*, but that it is not created at all.[87] Still, by and large, it is hard to assess whether the other treatises disagree with the *de Providentia*, because the main arguments on the subject are inconclusive. They are best treated in a footnote,[88] except for the one which will play a role in what follows. This is the argument that Philo repeatedly speaks in *de Opificio Mundi* (2.7-9; 5.21-6.23), and in other works, as if the *kosmos* were formed out of *pre-existent* matter. This argument is indecisive, because we have seen that *Prov.* 1.7 makes us reinterpret the reference to pre-existent matter at *Prov.* 1.22. And something comparable happens here. For Philo insists that despite the talk of six days, and despite his own chronological presentation, the creation was really simultaneous for everything, the intelligible world included (*de Opificio Mundi* 3.13; 7.26-8; 22.67; *Legum Allegoriae* 1.2; *de Sacrificiis Abelis et Caini* 18.65). This means that for matter too the talk of pre-

---

[86] P. Wendland catalogued the doubts as to whether the treatises *de Providentia* are by Philo, one question being whether they are too Greek (*Philosophische Schrift über die Vorsehung*, Berlin 1892). H. Diels conjectured that *Prov.* 1 might have been denuded of a dialogue form before the translation into Armenian, while L. Massebieau tried the opposite hypothesis that *Prov.* 2 had its dialogue form added (Diels 4; L. Massebieau *Le Classement des oeuvres de Philon*, p.90). H. Leisegang, following P. Wendland and W. Bousset, alleged that interpolations were introduced at this stage ('Philon von Alexandrie', PW 20, 1, 1941, col.8). It has been said that there are signs of Christian tampering at the next stage in the Armenian version, although H. Lewy claims that the Armenian translations of Philo are slavishly accurate, wherever they can be checked against the Greek (see Henry Chadwick, 'St Paul and Philo of Alexandria', *Bull. John Rylands Library* 48, 1965-6, p.292, n.6; H. Lewy, *The Pseudo-Philonic De Jona*, London 1937, 17). Finally, Aucher's rendering of the Armenian is agreed to be free, but it can be checked, for *Prov.* 1.6-7, against Hannick's, and for *Prov.* 2.50 against the Greek in Eusebius.

[87] 5.21 gives as the motive for creation the bringing of beauty to unbeautiful matter, as if that unbeautiful matter would have existed on its own, but for God's intervention.

[88] Of the three main considerations remaining, the first is the set of passages which appear to be speaking for or against creation out of nothing. Reasons for thinking most of these passages inconclusive were given above. The second argument tries to show that matter is not only pre-existent, but also uncreated, on the ground that it would be too *ugly* for God to create. But this is not so, unless it would have to exist at first in its unformed state. The third argument is an argument from *silence*: the claim that the Bible is silent has been noted above; that Philo is relatively silent about the origin of matter is true of other treatises, but not of the *de Providentia*.

existence is not on its own decisive: it would allow matter to have been created and *simultaneously* ordered into a *kosmos*, as it is in Augustine.[89] As it happens, Philo does not take trouble to stick to the view of the *de Providentia*. But what I would say is that the *de Providentia* offers the most conscious and systematic treatment of the origin of matter.

On another issue, I think the evidence is decisive even outside the *de Providentia*: Philo surely did not make the orderly *kosmos* beginningless. It has been said that he must have, or else he would have no defence against the charge of making God change his will, when he decided to impose order.[90] But Philo would have a defence, namely that since there was no earlier time, there was no earlier state of God's will. Meanwhile the evidence for a beginning of order in Philo seems to me too extensive to be set aside.[91]

One question remains – the meaning of Philo's idea that time (*chronos*) began with the ordered *kosmos*. Could he mean only that *measured* time began, while none the less allowing sense to the idea of a *before*, in what *we* should regard as a temporal sense? Once again, I doubt if he was consistent. At least two arguments require that there should be no 'before'. One is Philo's reply in *Prov.* 1.7 to the 'idleness' difficulty, a reply which would fail, if there were a 'before' in which God could be idle. The other argument is the one at *Legum Allegoriae* 1.2, that the creation did not occupy six days or any kind of time (*katholou chronōi*), since all time (*sumpas ho chronos*) is a system of days and nights. The argument would be spoilt, if there were a 'before' over which the creation could be spread. On the other hand, Philo will have to allow a 'before', if he allows matter to exist before the *kosmos*. We have seen that the argument at *de Opificio Mundi* 5.21 requires him to allow that, whereas *Prov.* 1.7 forbids him. The remaining evidence strikes me as inconclusive.[92] I would say, then, that Philo has not given careful thought to any idea of a duration *before* measured time. He presupposes sometimes its existence and sometimes its non-existence.

[89] Formless matter is not *chronologically* prior to the six-day ordering of the world (Aug. *Conf.* XII.9; *de Gen. ad Lit.* I.15,29), and that six-day ordering is not itself spread out in (ordinary) time (Chapter 2 above).   [90] Winston *Philo of Alexandria* 16-17.

[91] The case for a beginning of the ordered *kosmos* is as follows. (i) There are four passages which, to every appearance, are attacking those who deny it (*de Opificio Mundi* 2.7-12; 61.171; *de Somniis* 2.43.283; *de Confusione Linguarum* 23.114). (ii) There are several assertions, as noted above, and by Winston, p.106, that time had a beginning. (iii) There are several passages, as noted in the text above, and by Winston, p.9, which deny that the creation required a stretch of time, and strongly suggest instead that it occurred at an instant. (iv) *Aet.* 4.14-16, in agreement with *Prov.* 1.20-2, goes out of its way to reject the interpretation according to which Plato did not intend a beginning of the *kosmos*. Winston notes this, and gives as his answer that Philo wants to dissociate himself from certain ideas historically associated with the ascription to Plato of a beginningless *kosmos*, and therefore (confusedly) dissociates himself from the ascription itself (*Philo of Alexandria* p.14 and n.37). (v) I am not aware of any compensating gain from the effort of reinterpreting the passages which Winston agrees speak of the creation in temporal terms (op. cit., p.17 and n.49 cite *de Opificio Mundi* 4.16 and 19 and *passim*; *de Mut. Nom.* 4.27; *Leg. Alleg* 2.1.2). Winston further (p.7) takes *Prov.* 1.22 to ascribe to Moses a matter existing *before* the *kosmos*; presumably he would eventually reinterpret this too.

[92] Thus Philo leaves it open at *Prov.* 2.57 whether there is an infinite time over and above the system of days, months and years. Infinite time was introduced into the discussion only by the Hellenising Alexander (2.53). Nor can we attach significance to loose uses of 'before', 'was' (*Leg. Alleg* 2.1.2; *de Opificio Mundi* 6.23; *de Somniis* 1.13.76), and even (once) of 'time' (*chronos, de Decalogo* 12.58), if they are easily replaceable.

CHAPTER FOURTEEN

Infinity Arguments in Favour of a Beginning

I shall devote this chapter to Philoponus' counter-attack on the pagans and, in particular, to his infinity arguments for a beginning of the universe. He launched his counter-attack by turning Aristotle's widely received views on infinity against their author and against the pagan tradition which had accepted them.

Aristotle against actual and traversed infinities

Aristotle's account of infinity in *Phys.* 3.5-7 is highly original. He allows infinity to exist only in a restricted sense:

> For in general infinity exists through one thing always being taken after another, what is taken being always finite, but ever other and other.[1]

The effect of this account is to define infinity in terms of finitude: however large a finite number you have taken, you can take more. Infinity is an *extendible finitude*. One corollary is that infinity is connected with a *process*: 'its being is always in a process of coming to be or passing away.'[2] Another corollary is that his predecessors have got things the wrong way round. They thought of infinity as something which is so all-embracing that it has nothing outside it. But the very opposite is the case: infinity is what always has something outside it.[3]

This conception of infinity would have a lot of appeal nowadays. For example, the modern idea of *approaching a limit* is very much in the spirit of Aristotle. It is said that the series $\frac{1}{2} + \frac{1}{4} + \frac{1}{8}$ etc. approaches 1 as a limit, and this way of talking enables everyone to avoid the naughty word 'infinity'. The idea is that you can get *as close as you like* to 1 by adding more fractions to a *finite* collection. And this talk of getting as close as you like by a *finite* operation is very much Aristotle's own. But Aristotle is closer still to those modern finite mathematicians who deny that there is ever more than a finite number of points in a line: the points are brought into existence only as they are marked off.[4] I say that Aristotle is closer to this, but in order to bring this

---

[1] Phys. 3.6, 206a27-9. [2] *Phys.* 3.5, 206a32-3; cf. 3.7, 207b10; b14-15.

[3] *Phys.* 3.6, 206b33-207a2; cf. 206b17-18; 207a7-8.

[4] D. Hilbert and P. Bernays, *Grundlagen der Mathematik*, Berlin 1934, 15-17.

out, I shall have to draw attention to an ambiguity in the notion of an extendible finitude.

Is Aristotle's idea that you can always go on *creating* more divisions, or that you can always go on *recognising* more of the divisions which exist already? If they exist already, their number is greater than any finite number. But I believe that Aristotle is committed to what I might call the *finitist* view, that there is never more than a finite collection of divisions in existence. Many of his phrases support the finitist interpretation. He says, for example, that infinity exists through a process of one thing *coming into being* after another (3.6, 206a21-3; 30-3; 3.7, 207b14), and he uses the future tense when he says that in something infinitely divisible there *will* be a smaller division (3.6, 206b18; b20). Whatever his point about divisions, he will hold the *same* view about numbers, for he thinks that there is infinite number only in the sense in which there is an infinite number of divisions (3.6, 206b3-12).

Certainly, Aristotle would allow only a finite number of *actually* existing divisions. An *actually* existing division is one that has been marked out in some way, either physically or mentally. But the situation is more complex when we ask how many *potentially* existing divisions there are. The very question is ambiguous, for potentially existing divisions might be understood as divisions that could be made, although they have not yet been made, or they might be understood as points at any finite number of which divisions can be made. Certainly, the *Physics* passage envisages that there are divisions which could be made over and above those which actually exist: why else should it insist that infinity always has something outside it? But here and in *GC* 1.2, 316a10-317a12, it is also implied that the number of makeable divisions is not more than finite. And the *GC* passage finishes (317a9-12) by implying that the number of points (*sēmeia*, *stigmai*) in a line is not more than finite either. Nor is it an accident that Aristotle reaches this verdict. For in the *GC* passage, he fears the threat that a more than finite number, whether of points (317a4-7) or of makeable divisions (316a25-34; 316b14-15; 316b26-7), will imply something he considers absurd, that an extended line is composed of nothing but unextended points. As for the discussion in the *Physics*, Aristotle cannot there afford to admit any collections which are more than finite, if his analysis of infinity is to surmount the problems which it is intended to surmount. One such problem is whether infinity does not have *parts* which are infinite. Aristotle complains that his predecessor Anaxagoras is committed to this.[5] Modern theory recognises that in a sense

---

[5] Aristole (*Phys.* 1.4, 188a2-5) interprets Anaxagoras as committed to holding that there is an infinity of bodies and that in each there is an infinity of portions of flesh, blood and brain. Simplicius elaborates the criticism, and puts it by saying that Anaxagoras is committed to infinity times infinity (*in Phys.* 172, 31-173,3; 460,10; 461,8), that is, to the idea that Aristotle thinks impossible: an infinity whose parts are infinite. It is true that Anaxagoras believes that each thing, and every portion in it, contains an infinity of portions (see frs 3; 4; 6; 11 DK). But he also insists (frs 3 and 6) that the number of portions is *equal* in large things and in small. This last idea was later rejected by the Platonic atomists. They argued that small components must have *fewer* parts than the whole, and so they concluded that *neither* can have infinitely many (ps.-Aristotle *Lin. Insec.* 968a2-9). Another predecessor who avoided an infinity with infinite parts was Melissus. He did so by declaring that the One, though infinite in magnitude (fr. 3), is incorporeal and so avoids having parts (fr. 9), an idea which Aristotle notices at *Phys.* 3.5,

it is possible: the whole numbers contain an infinite sub-set of odd numbers. But Aristotle declares that the same thing cannot be many infinites, and his analysis of infinity is taken to avoid this outcome.[6] If it is to do so, the finitist interpretation needs to be applied to all collections, even to collections of potentially existing divisions.

If Aristotle's view of infinity is finitist, I believe it will be perfectly adapted for some cases, but inadequate for others. If, for example, we take pairs of whole numbers, they will never be separated from each other by more than a finite number of whole numbers. The most we can do, by selecting pairs still further apart, is to obtain an extendible finitude of Aristotle's sort. On the other hand, my view would be that the totality of whole numbers is more than finite. I shall later suggest that there may be an asymmetry between the past and the future, in respect of years traversed: that the series of future years traversed, starting from now, may best be viewed as an extendible finitude in Aristotle's manner, while the series of past years, if there was no beginning, should be viewed as more than that.

I shall have to consider whether Aristotle holds consistently to his finitist view of infinity. But one thing I can say straight away is that the *Physics* analysis had probably not yet been worked out at the time he wrote *Cael.* 1.12. For there, as Sarah Waterlow has pointed out,[7] one infinite, namely, the whole of time, is described as not being less than anything, as having nothing greater than it, and as being in a way definite. This fits badly with the idea that infinity always has something outside it. At the same time Aristotle describes that which (like the future) is infinite in one direction only, and which ought to provide him with the clearest instances of potential infinity, as not being infinite at all. He further claims that, unlike the whole of time, it is not of any definite length, and this too will clash with the *Physics*, if he means what the analysis of *Phys.* 3 is meant to rule out, that an infinite future starting in 1980 is longer than an infinite future starting in 1981. The *de Caelo* treatment of infinity appears to clash again with that of the *Physics* by allowing (1.2, 269a18-23) that lines can be extended indefinitely far; the *Physics* allows infinite divisibility, but not infinite extendibility. Quite apart from the *de Caelo*, it may be wondered whether Aristotle holds consistently to his finitist analysis even *within* the *Physics*, and I shall have to raise a doubt shortly about *Phys.* 8.8, 263b6. Other passages in the *Physics*, however, seem to me less troublesome.[8]

I want now to focus on two implications of Aristotle's finitist account

---

204a20-6. Democritus was less guarded: for post-Aristotelian attacks alleging that he is committed to infinity times infinity, see Robert B. Todd, 'Some concepts in physical theory in John Philoponus' Aristotelian commentaries', *ABG* 24, 1980, 151-70. According to these attacks, Democritus allows an infinity of void spaces along with the infinity of atoms.

[6] *Phys.* 3.5, 204a20-6.

[7] *Cael.* 1.12, 281a33-b1; 283a4-10. The point is made by Sarah Waterlow, in *Passage and Possibility: a study of Aristotle's modal concepts*, Oxford 1982, 69-78.

[8] Admittedly, at *Phys.* 4.12, 221a17-18; 26-30; b3-5, Aristotle maintains that (infinite) time includes (*periechei*) all things which are in time. But this need not contradict the conception of infinity as always having something outside it. For Aristotle's idea is not that time includes the sum totality of things in time, but rather (221a27) that for any *given* thing which is in time a duration will be found which is longer than its duration. A finitist reading may also perhaps be

which are particularly relevant to Philoponus' attack. One is that infinity is merely *potential* and never *actual*;[9] the other that it can never be *traversed*, that is, gone right through.[10] Both points require a word of discussion. The new application of the word 'actual' is different from its application to divisions. It is because infinity is never more than finite, and yet is extendible, that it should be thought of as existing potentially, and never actually (*energeiāi*). An actual infinity, as I understand it, would be *more* than a finitude.

The other principle, that an infinity can never be *traversed*, is asserted frequently. But it undergoes a qualification later, when Aristotle discusses Zeno's paradox of the half-distances. According to this paradox, we cannot reach any destination, because we should first have to go half way, then half the remaining distance, and half the remaining distance, *ad infinitum*. Aristotle offers the following solution. Although we cannot traverse an infinity of *actually* existing divisions (*entelecheiāi onta*), we can traverse an infinity of potentially existing ones (*dunamei sc. onta*):[11]

> So that if someone asks whether you can traverse an infinity either in time or in length, we must say that in a way you can, and in a way you cannot. For you cannot traverse an infinity of actually existing [divisions], but you can of potentially existing ones.

This last concession is startling: how can Aristotle think himself free to allow more than a finite number of potential divisions? That they are more than finite is implied by the contrast with the finite number of actual divisions. To allow such an infinite collection would be a possible policy decision, but it would conflict with that of *Phys.* 3 and *GC* 1.2, where makeable divisions and points are not more than finite. Moreover, we have seen that that verdict was very fully motivated by the desire to avoid the Anaxagorean problem of one infinity containing others, and the problem of a line consisting of nothing but points. If Aristotle is not to reopen these problems, and if he is to remain consistent, he needs to say in *Phys.* 8.8 not merely that his divisions *exist* potentially (*dunamei onta*), but also that their *infinity* is potential as well (*dunamei apeira*), and that in the sense which I earlier defined of not being more than finite. Aristotle's unexpected concession will prove relevant to subsequent controversy.

So too will one particular *application* of the principle that infinity cannot be traversed: Aristotle uses this principle in order to prove that there cannot be infinite *number*, except in the restricted, potential sense already described. Number is countable; if, then, there were infinite number in any strong sense, one could count, and in this sense traverse, an infinity.[12]

---

given to the passage to which Matthew Neuburg has drawn my attention, in which Aristotle says that what has been changing must already have changed an infinite number of times (*Phys.* 6.6, 237a11; a16).

[9] *Phys.* 3.6, 206a14-23.

[10] *Phys.* 3.5, 204b9; 6.2, 233a22; 6.7, 238a33; 8.8, 263a6; b4; b9; 8.9, 265a20; *Cael.* 1.5, 272a3; a29; 3.2, 300b5; *Metaph.* 2.2, 994b30; *An. Post.* 1.3, 72b11; 1.22, 82b39; 83b6.

[11] *Phys.* 8.8, 263b3-6. See Chapter 21 for Zeno's paradox.

[12] *Phys.* 3.5, 204b7-10.

Philoponus turns Aristotle against himself

The two restrictions on infinity, that it is never actual or traversed, create a difficulty for Aristotle. For he argues that there is no beginning for time, motion, the universe, or the generations of man. But if there has already been an infinity of days, does this not provide an example of an infinity which exists *actually*, in a stronger sense than Aristotle wants? And will not the infinite set of past days have been *traversed*? This was the point on which Philoponus was to fasten nearly nine hundred years later, in his defence of the Christian belief in a beginning. In *de Aeternitate Mundi* (Rabe, pp. 9-11) and elsewhere,[13] he makes the following complaints. If the universe has no beginning, then an infinity of years or generations will already be both actual and traversed. Next, that infinity will shortly have been added to, which is absurd. Moreover, besides addition, we could get multiplication of infinity. This last complaint is most graphically illustrated in Philoponus' attack on Aristotle, as recorded by Simplicius (*in Phys.* 1179, 15-21). If Saturn has performed an infinity of revolutions already, then Jupiter, the moon and the fixed stars will have revolved many times that infinite number of times. Finally, Philoponus makes use of Aristotle's proof from *Phys.* 3.5, 204b7-10, that there cannot be an infinite number, because number is countable.[14] I have already indicated that these arguments were used again and again by Islamic and Jewish philosophers from the beginning of the ninth century onwards, and repeated in thirteenth-century Paris.[15] Let me begin by quoting two of the leading passages from Philoponus.

> So since past time will be actually infinite, if the *kosmos* is uncreated, the individuals which have come into being in that infinite time must also be actually infinite in number. Hence, if the *kosmos* is uncreated, the result will be that there exists and has occurred an actually infinite number. But it is in no way possible for the infinite to exist in actuality, neither by existing all at once, nor by coming into being part at a time, as we shall show more completely, God willing, in what follows. For after refuting all the puzzles designed to show

[13] The same arguments are used by Philoponus in his commentaries on Aristotle's *Physics* (428,14-430,10; 467,5-468,4) and *Meteorologica* (16,36ff.), portions of which are translated by S. Sambursky, 'Note on John Philoponus' rejection of the infinite' in S.M. Stern, Albert Hourani, Vivian Brown (eds), *Islamic Philosophy and the Classical Tradition*, Essays Presented to Richard Walzer, Oxford 1972, 351-3, and by Todd, op. cit. The arguments concerning addition and multiplication are used again in *Aet.* 619, and in Philoponus' attack on Aristotle, ap. Simpl. *in Phys.* 1179, 12-26.

[14] Philoponus *in Phys.* 428

[15] See, as above, Davidson, op. cit.; H.A. Wolfson, *The Philosophy of the Kalam*, 410-34; 452-5. Davidson cites thirteen Islamic and Jewish sources for discussions of the argument that infinity cannot be increased, thirteen for discussions of the argument that infinity cannot be traversed and of its variants, and four for use of the argument that number is finite, because countable. To these may be added the unpublished treatise of Avicenna, summarised in English by S. Pines, at the end of 'An Arabic summary of a lost work of John Philoponus', *IOS* 2, 1972, 320-52. The most accessible of the sources in English translation are: (1) Averroes *Tahāfut al-Tahāfut* (ed. Bouyges), which replies to, and incorporates, most of Ghazālī *Tahāfut al-Falāsifa*, translated by S. van den Bergh, London 1954. See pp. 16-19 (Bouyges' pagination) on increasing infinity and pp. 19-21 on traversing infinity. (2) Maimonides *Guide For The Perplexed* 1.74 (7th argument), which adds that Farabi offered replies to these arguments.

that the *kosmos* is everlasting, we then establish for our part that the *kosmos* cannot be everlasting. And I shall add to the exposition Aristotle himself establishing this particular point. I say that the infinite cannot in any way exist in actuality, and I think this is clear from the following. Since the infinite cannot exist all together and at once, for the very same reason it cannot emerge into actuality by existing part at a time. For if it were at all possible for the infinite to exist part at a time, and so to emerge in actuality, what reason would there be to prevent it from existing in actuality all at once? For saying that it is brought to birth in actuality part at a time, and counted, so to speak, unit by unit, one after another, would appear much more impossible than saying that it exists all together and at once. For if it exists all at once perhaps it will not have to be traversed unit by unit and, so to speak, enumerated. But if it comes into being part at a time, one unit always existing after another, so that eventually an actual infinity of units will have come into being, then even if it does not exist all together at once (since some units will have ceased when others exist), none the less it will have come to be traversed. And that is impossible: traversing the infinite and, so to speak, counting it off unit by unit, even if the one who does the counting is everlasting. For by nature the infinite cannot be traversed, or it would not be infinite. Hence if the infinite cannot be traversed, but the succession of the race has proceeded one individual at a time, and come down through an infinity of individuals to those who exist now, then the infinite has come to be traversed, which is impossible. So the number of earlier individuals is not infinite. If it were, the succession of the race would not have come down as far as each of us, since it is impossible to traverse the infinite.

Moreover, suppose the *kosmos* had no beginning, then the number of individuals down, say, to Socrates will have been infinite. But there will have been added to it the individuals who came into existence between Socrates and the present, so that there will be something greater than infinity, which is impossible,

Again, the number of men who have come into existence will be infinite, but the number of horses which have come into existence will also be infinite. You will double the infinity; if you add the number of dogs, you will triple it, and the number will be multiplied as each of the other species is added. This is one of the most impossible things. For it is not possible to be larger than infinity, not to say many times larger. Thus if these strange consequences must occur, and more besides, as we shall show elsewhere, if the *kosmos* is uncreated, then it cannot be uncreated or lack a beginning.[16]

Suppose (Philoponus says) the spheres do not revolve at equal speeds, but one takes thirty years, another twelve, and another in turn less, so that the sphere of the moon takes a month, and that of the fixed stars a day and a night. Suppose too that the movement of the heavens has no beginning, then the sphere of Saturn must have revolved an infinity of circuits, that of Jupiter nearly three times as many, while the circuits of the sun will be thirty times those of Saturn, those of the moon three hundred and sixty times as many, and those of the fixed stars more than ten thousand times. Is this not beyond all absurdity, if the infinite cannot be traversed even once, to entertain ten thousand times infinity, or rather infinity times infinity. Hence it is necessary (he says) that the revolution of the heavens should have had a beginning of its existence, without

[16] *Aet.* (Rabe) pp.9,14-11,17.

having existed previously, namely, when the heavens themselves (his words, not mine) began to exist.[17]

It is an interesting twist in Philoponus' attack that he declares a successive infinity harder to accommodate than a simultaneous infinity, not easier as Aristotle supposed, precisely because it immediately raises the problems of traversal, addition and multiplication.[18] We may be reminded of Augustine's claim that God can count all the innumerably many things without any succession (*alternatio*) in his thought:[19] an infinite count is evidently thought more possible, if it is not successive.

Simplicius replies to Philoponus' attack by drawing attention to an idea which crops up in Aristotle *Phys.* 3.6 (206a33-b3) and 3.8 (208a20-1).

> But in extended objects, this [taking of one thing after another] occurs in such a way that what is taken stays behind; whereas in the case of time and of men, what is taken ceases to exist, though in such a way that the series does not fail.
>
> Time and movement are infinite, and so is thinking, in such a way that what is taken does not stay.

Simplicius takes Aristotle's point to be that, because past days and past generations of men do not stay, but have perished, you do not get an infinity of them existing. Moreover, he adds, you will not find an infinity being increased, because there is no infinity there in the first place.[20]

At first sight, Simplicius' point seems arresting: what made it possible for an infinity of divisions to be a merely *potential* infinity was that the divisions not yet made did not *exist*. Has not Simplicius shown that years are like divisions, in that no more than a finite number exists? Unfortunately, we have seen that the situation is more complex, because Aristotle is willing to think of points and potential divisions as entities of a sort, capable of forming collections. And he would have to allow the same for past years – all the more so because the sense in which past years no longer exist is only the rather weak sense of no longer being present. They are still entities enough to form a collection, and Aristotle ought therefore to avoid their forming an actually infinite one, just as he does in *Phys.* 3 and *GC* 1.2 for potential divisions and points. Otherwise, he will be back with the Anaxagorean problem, which he hoped to avoid, concerning an infinity whose parts are infinite, and with the problem of a line composed of nothing but points. Admittedly, we have seen him allowing an actual infinity of potential divisions in *Phys.* 8.8, but first, that was argued to be an inconsistency, and secondly, potential divisions have less claim on existence than past years, so hardly set a precedent for them.

Things will be no better if Simplicius' point is transposed and phrased in terms of actuality instead of existence, so as to say that past years are no longer *actual*. We should then have had at least three senses of actuality:

---

[17] 6th of 6 books against Aristotle, as recorded by Simpl. *in Phys.* 1179, 15-21.
[18] *Aet.* 9-11 as above, and *in Phys.* 429,20-430,10.
[19] *City* XII.17: *innumera omnia sine cogitationis alternatione numerantem.*
[20] *in Phys.* 506, 3-18; cf. 1180, 29-31.

divisions are actual when *marked out*, years when *present* and infinities when they are *more than a finitude*. But the answer would be that the status of the years does not settle the status of their infinity. Even through the past years in a beginningless universe are not actual, their infinity must be. And for some purposes Aristotle needs to avoid any actual infinity.

My conclusion so far is that Philoponus' arguments are successful as an objection to Aristotle and the pagans. But the question remains whether *we* can answer his arguments by freeing ourselves in some way from Aristotelian ideas. To this question I shall now turn, and I shall start with Philoponus' arguments about *increasing* infinity.

Philoponus against increasing infinity

In order to answer the adding and multiplying objections, we must see what is right and what wrong about them. We can do so without entering at all into the complication of transfinite numbers. There are perfectly good analogues of adding and multiplying in relation to infinity. The only restriction is that in a certain sense these processes will not have the usual consequence of making the collection larger. In order to see why not, we can imagine an infinite series of past years terminating at the present and a corresponding infinite series of past days. Suppose we imagine the column of past years stretching away from our left eye infinitely far into the distance, and parallel to it, stretching away from our right eye, the column of past days, also receding infinitely far. The two columns should be aligned at the near end, starting at the present, and the members of the two columns should be matched against each other one to one. I can now explain the sense in which the column of past days is not larger than the column of past years: it will not *stick out beyond the far end of* the other column, since neither column has a far end.

Provided we understand this, it will not too much matter whether we talk of adding or multiplying. The context may make it very natural to say that in a year's time an extra year will have been *added* to the collection. But no objection can be raised on this basis to the hypothesis that there has been an infinity of years. For we are not committed to the only thing that is objectionable, that is, to the idea that the collection of years will soon be, or that the collection of days already is, larger in the sense just proscribed.

It was in the fourteenth century that some of these points began to be appreciated, as John Murdoch has shown.[21] An attempt then began to find a sense in which one (denumerable) infinite set might be called greater than another, and a sense in which it might not. It might be called greater in the sense of containing all the members of the other and some members besides (*preter*, elsewhere *praeter*). But it would not be right to talk of one infinite set containing members beyond (*ultra*) the other: 'besides' is legitimate, 'beyond'

[21] 'Mathesis in philosophiam scholasticam introducta: the rise and development of the application of mathematics in fourteenth-century Philosophy and Theology', *Arts Libéraux et Philosophie au Moyen Age, Actes du Quatrième Congrès de Philosophie Médiévale*, Paris 1969, 222-3; 'The "equality" of infinites in the Middle Ages', *Actes du XIe Congrès International d'Histoire des Sciences*, Warsaw-Cracow 1968, vol. 3, pp. 171-4. Philoponus, however, is not here given the credit for applying the infinity question to the hypothesis of a beginningless universe.

is not. The word 'beyond' is the one I used myself in explaining the situation. Ideas like this were developed in turn by Henry of Harclay, William of Alnwick, William of Ockham and Gregory of Rimini.

The mediaeval discussions explain nicely the sense in which the infinite set of past years can be thought of as having grown larger by next year: next year's collection will contain the same members, and one more besides (*praeter*). But a more complex account would be needed for explaining the sense in which the infinity of past days is greater than the infinity of past years, or, as Nicholas Denyer has nicely suggested to me, for explaining the sense in which a man who has spent 364 days of every past year in hell has spent more time there than the man who has spent one day of every past year in hell. It may well be that each can claim to have spent days in hell *praeter* (besides) those spent by the other. So the sense in which one infinity is greater than the other will be better brought out by saying that, however large a *finite* period we take, the ratio of days in hell remains at 364:1.

I know of no attempt before the fourteenth century which succeeds in finding a good sense for the idea that one infinity might be larger than another. But there had been gallant attempts to defend the idea of an infinite past along *different* lines. One of them exploits the point made by Aristotle[22] that ratios hold only between *finite* quantities. Thus Averroes[23] concedes that you cannot have one infinity of revolutions larger than another, but claims that there is no danger of getting this. For the revolutions of Saturn and of the sun constitute (potential) infinities and so, he claims, have neither beginning nor end. Lacking beginning or end, they do not stand in any ratio of larger to smaller; only finite sequences of their revolutions stand in such a ratio. This ingenious argument travels in the opposite direction from the preceding suggestion. For it does not consider using the difference in ratio between *finite* sequences in order to legitimise talk of one infinity exceeding another.

Aristotle's *de Caelo* contains a suggestion which would answer Philoponus' problem about addition, as Simplicius noticed.[24] For, as already remarked, Aristotle there denies that what is infinite in only one direction really is infinite (*apeiron*). The reason, which is not given, may be that it has at least one boundary. But, at any rate, the implication is that the past is not infinite, and hence that the arrival of an extra day does not make an addition to infinity. This solution, however, is not allowed for in the *Physics* where the ban on unidirectional infinities is missing.

Before I finally leave Philoponus' problem about increasing infinity, I should point out one extra way in which it is wounding to Aristotle. For in effect Philoponus is arguing that the pagan denial of a beginning gives us the very thing which Aristotle called impossible: an infinity whose *parts* are infinite. At least it will do so, if the period down to yesterday is a *part* of the period down to today.

[22] *Cael.* 1.6, 274a8.
[23] *Tahāfut al-Tahāfut* 1 (Bouyges), pp. 18-19, tr. van den Bergh pp. 9-10.
[24] *Cael.* 1.12, 283a4-10, exploited by Simpl. *in Phys.* 1180, 6-10; 1182, 12.

Philoponus against an actual and traversed infinity

I have now explained my answer to Philoponus' arguments about addition and multiplication: an infinite past will not lead to infinity being exceeded in the *objectionable* sense. But what about Philoponus' other argument, that an infinity cannot be both actual and traversed? This idea is still found compelling by many people, and it has been defended in recent years by Pamela Huby, William Lane Craig and G.J. Whitrow.[25] What I want to say is than an actual infinity of past years can perfectly well have been traversed. But if I am to break down resistance, I must examine some of the sources of temptation which may seem to make that impossible. I shall distinguish no less than eight.

(i) One supposed difficulty is that, if an infinity of days had to pass before the arrival of today, then today would never arrive. This would certainly be so, if there was a first day, and then an infinity of days to cram in before today. But of course no first day is envisaged by those who postulate a beginningless universe; so there is ample room for a preceding infinity.

(ii) The first mistaken objection is related to a second one about counting. We might try to imagine that the years have always been subjected to counting as they arrived. If the universe had no beginning, then earlier than any year we care to name, the count should already have reached infinity. But, the objection goes, it is absurd to suppose that this infinite count could be completed. What this objection overlooks is that counting differs from traversing in a crucial respect, for counting involves taking a *starting* number. This is, indeed, part of the reason why it would be so difficult to complete an infinite count. We will not be able to complete it, unless we can accelerate in our counting in Zenonian fashion, taking half as much time for each successive act of numbering. There are no such obstacles, however, to completing an infinite lapse of past years, precisely because it involves no *starting* year. Moreover, this difference, the absence of a *starting* year, has a second consequence. For it means that after all we cannot imagine that the beginningless series of past years has been subjected to counting in any straightforward way; for it has no first member to match the first number used in counting. The counting argument, then, must fail, though it has been very popular. It was frequently used by Philoponus,[26] and it has been repeated in modern times.[27]

But now I must face an objection.[28] If the only obstacle to completing an infinite count is that conventional counting takes a *starting* number, what about counting in a *backwards* direction? Ought it not to be possible for a

---

[25] Pamela Huby, 'Kant or Cantor? That the universe, if real, must be finite in both space and time', *Philosophy* 46, 1971, 121-32; and 48, 1973, 186-7. W.L. Craig, *The Kalam Cosmological Argument*, London 1979, esp. 83-7; 97-9. G.J. Whitrow, *The Natural Philosophy of Time*, 1961, 2nd ed., Oxford 1980, 27-33; 'On the impossibility of an infinite past', *BJPS* 29, 1978, 39-45; Review of Craig, *BJPS* 31, 1980, 408-11.

[26] *Aet.* 9-10, as above; *in Phys.* 428-9; ap. Simpl. *in Phys.* 1178; in Arabic summary 3rd treatise, translated by S. Pines, op. cit. (in n. 15).

[27] Huby, op. cit., 1971, p. 128.

[28] I am grateful to Norman Kretzmann for this objection.

beginningless being to count off the years, descending from the higher numbers and finishing, say, in this century with the years four, three, two and zero. A backwards-counting angel might then sigh with relief in 1982 and say, 'Thank heavens, I have reached year zero; I have just finished counting infinity.' If this is *not* possible, then how can the *traversal* of an infinity of years be possible?

My answer to this is that something like the backwards count would indeed be possible *in principle*. I am not at all sure that it ought to be called *counting*, but it is conceptually possible that God should have included a beginningless meter in his beginningless universe, to record how many years remained until some important event, say, until the incarnation of his Son. At zero B.C., the meter would register zero, but the counting would never have been begun. Rather, for every earlier year, the meter would have displayed a higher number. Whether or not this should be called *counting*, there is no logical barrier to it, I believe;[29] and therefore no logical barrier has been exhibited to the traversal of an infinity of past years.

(iii) The counting objection is akin to Kant's argument in the first antinomy, which has also been endorsed recently,[30] that the universe must have a beginning because an infinite series can never be completed by a successive synthesis. Admittedly, if the number of years is infinite, this will not be the result of completion by successive synthesis, that is, of adding to a finite collection. For that would suggest that mere addition (without any Zenonian tricks of acceleration) could take you from a finite number of years to an infinite. But this is not to rule out an infinity of years, only one method of acheiving it. In fact, the hypothesis is not that the number of years has ever passed from being finite to being infinite, but that it has always been infinite.

(iv) A further objection used by Bonaventure has reappeared in modern literature.[31] If we think backwards from the present, we will not find an infinitieth year. But this objection, like the last, represents an *ignoratio elenchi*. For those who believe that there has been an infinity of years do not mean that one of them occurred an infinite number of years ago. Infinity is a property of the collection as a whole, not of one member. This kind of situation should be familiar from other cases: a large crowd need not be a crowd of large people – it may be a crowd of midgets. If a collection could not be infinite without one of its members being the infinitieth member, we would get the absurd result that the set of whole numbers is not infinite.

Thomas Aquinas already recognised *part* of the truth, when he defended the logical possibility of a beginningless universe, by saying that the distance back to any designated day would still be finite. Unfortunately, he saw this as

---

[29] I am not persuaded that it involves the oddity alleged by Fred I. Dretske, 'Counting to infinity', *Analysis* 25 suppl., 1965, 99-101. He rightly says that, however far back you go, a backwards counter would already have counted an infinity of numbers; but he wrongly infers that the counter would have *finished*. That is to move illegitimately from 'an *infinity* of numbers' to '*all* the numbers'.

[30] Craig, op. cit., pp. 103; 109; 189.

[31] Bonaventure, *in IV Libros Sententiarum Magistri Petri Lombardi*, tomus 2, in librum 2, dist. 1, p.1, a.1, q.2, argument (3) (pp. 107-8 in Byrne); Richard Bentley, sixth of the *Boyle Sermons* 1692, in Bentley's *Works*, vol. 3, ed. Alexander Dyce, London 1838, 135; Huby, op. cit., 1971, 127; Craig, op. cit., 98; 200-1.

establishing that in a beginningless universe an infinity of days would not have been traversed.[32]

A variant on Bonaventure's argument has appeared recently.[33] It is conceded that the pagans are not committed to a *first* year which would have been the infinitieth year ago. But it is alleged that they are committed to there being many past years separated from the present by an infinite gap. I do not believe that they are so committed, for once again the infinity of the whole number series does not involve there being *any* whole numbers separated by an infinite gap from 1. But as to why this misconception should have arisen, I shall have more to say shortly.

(v) An objection discussed in the Islamic world was that the infinite, by definition, cannot come to an end, and so cannot be completed.[34] The simplest answer to this is that an infinite series can easily have one end, as, for example, the series of positive whole numbers does at zero. And if the infinity of past years is regarded as ending at the present, this will only give the series one end.

(vi) A more subtle objection has been put to me in discussion by Pamela Huby, and it is suggested by the work of G.J. Whitrow, who is in turn inspired by the Boyle sermons of Richard Bentley, delivered in 1692.[35] This objection points to the fact than an infinity of *future* years starting from now would always remain potential and never be completed. So should we not in consistency say the same about an infinity of *past* years starting from now? The same question could be raised about *any* set of years that runs forward from a given date, say, from A.D. 1700. But, for convenience, I shall consider the set of years running forwards from the *present*.

In reply, it is important to bring out why there is a disanalogy between the past and the future. Past years do not start from now. If anything, it will be our *thoughts about* past years which start from now, if we choose to think of them in reverse order, but the years themselves do not. Indeed, on the pagan view, they do not have a start at all. And this is important, because it means that when we say that they *have* been gone through, and therefore assign them a *finish*, we are not thereby assigning them *two* termini. This is what leaves us free to think of them as forming a traversed series which is more than an extendible finitude. Contrast a series of future years to which we assign a start, say, at the present. Whenever we think of such a series as *having* been gone through, we will automatically be assigning it a *second*

---

[32] *ST* 1, q.46, a.2, objection (6) and reply (pp. 65 and 67 in Vollert).

[33] Whitrow, opera citata.

[34] So the opponents envisaged by Avicenna in ch. 4 of an unpublished treatise, summarised in English by Pines, in the appendix to 'An Arabic summary of a lost work of John Philoponus' (see p. 348); also ascribed to others by Averroes *Tahāfut al-Tahāfut* (Bouyges), pp. 19-22 and 31, tr. van den Bergh, pp. 10-11 and 17.

[35] Whitrow, *The Natural Philosophy of Time*, 2nd ed., p. 29, citing Bentley, op. cit. Bentley gets very close to the truth: he sees that the years through which our souls survive, if they start from some moment of creation, could at most be a *potential* infinity; he sees that we impose *two* termini, if ever we talk about a *completed* set of years since their creation; his mistake is only to suppose (pp. 133-4) that we must also impose two termini, if we speak of a set of *past* years terminating at the present. I am grateful to Nicholas Denyer for first drawing my attention to Bentley.

terminus, a finish as well as a start. And this is what prevents the future series of traversed years from being more than finite. The asymmetry to which I am pointing depends crucially on the occurrence of the perfect tense: '*having* been gone through'. It is not future years *as such* which have a different infinity from past years, but future years which have been *traversed*, for the traversed ones will have *two* termini.

I think that failure to appreciate the nature and extent of the disanalogy between past and future series of traversed years has provided one motive for the view noted under (iv) that an infinite past would involve events infinitely far removed from the present.[36] Admittedly, a *future* set of years which started from now would become actually infinite only if, *per impossibile*, it attained to a year that was infinitely far removed. But the same ought not to be said about the *past*.

(vii) A very ingenious argument against an infinite past has been built by Craig out of a suggestion originally made by Bertrand Russell and endorsed by Whitrow.[37] It involves Tristram Shandy, who is to be imagined as keeping a diary, but as recording his life at the snail's pace of one day recorded for every year lived. If there had been an infinity of years, it is alleged, then we would get the absurd result that he could catch up, for he would have had time for an infinite number of entries. And the absurdity of this is intended to cast doubt on the conceivability of an infinite past.

This argument has several things wrong with it, I believe. First, it confuses the idea of *infinitely many* with the idea of *all*. There would have been time for an infinity of entries, but not for all. These are not the same: for example, there is an infinity of odd numbers, but these are not all the whole numbers. Secondly, Russell's original claim is not correct, when he says that no part of the biography will remain unwritten. For what if Tristram Shandy records only every 1 January, and then skips to the next 1 January, leaving permanent gaps? He must avoid such gaps, if he is to bring it about that no part of the biography remains unwritten. Moreover, it is hard to see how he *can* avoid gaps, if we add to Russell's story the idea that he has lived without beginning in the past, so that the diary has no beginning. For he then has no possibility of filling up the diary in a systematic way from its opening pages – there being no opening. There is yet a further thing wrong with the argument. For we can admit that, if the diary has a beginning, and if it is kept without gaps, and kept for ever, then no day will remain for ever unrecorded. But it does not at all follow from this that a time will eventually arrive when *all* the days *have* been recorded. Nor does it follow that they will *already* have been recorded, if Tristram Shandy has *already* lived an infinity of years.

(viii) I now come to a final kind of argument which is even more ingenious – ingenious, but desperate. It seeks to discredit the possibility of any *actual* infinity, that is, of any infinity which is more than an extendible finitude, by

[36] Whitrow, opera citata.

[37] Craig, op. cit., 1979, 98; Bertrand Russell in 'Mathematics and the metaphysicians', in *Mysticism and Logic*, London 1917 (p. 70 of the 1963 edition), revised from an article written in 1901 and published in *The International Monthly*; also in *Principles of Mathematics* 1903, 2nd edition London 1937, 358-9; Whitrow, opera citata, 1961 and 1978-80.

alleging that all actual infinities must lead to absurdity. The most spectacular example, used by two recent authors,[38] concerns the case of Hilbert's hotel. The hotel, named after the notable mathematician D. Hilbert, contains infinitely many rooms, but every single one is full. Along comes a late traveller and says, 'I know your hotel is full, but can't you fit me in somewhere?' 'Certainly,' says the manager, 'I can accommodate you in room number one.' And then in a loud voice he declares, 'Will the occupant of room number one step into room number two? Will the occupant of room number two step into room number three? And so on *ad infinitum.*' There is a temptation to think that some unfortunate resident at the far end of the hotel will drop off into space. But there *is* no far end. It is like the column of whole numbers which we considered before: the line of residents will not *stick out beyond the far end of* the line of rooms.

Once it is seen like this, the outcome should no longer seem an absurdity which can discredit the idea of an actual infinity. It should instead be seen as an explicable truth about infinity. It may be a surprising truth, even an exhilarating and delightful truth, but a truth for which we can perfectly well see the reasons, when we reflect on the idea of sticking out beyond the far end.

One author offers a cluster of puzzles in the attempt to discredit the idea of actual infinities. But they are best considered in a footnote, since they can all, I think, be shown as explicable in the light of principles which have already been explained, and none shows an incoherence in the idea of an actual infinity.[39]

In the Islamic and Jewish worlds, an objection was raised on the other side, to show that it *must* be possible after all to traverse an infinity. For any distance we traverse can be represented as divisible into an infinity of segments, e.g. a half, plus a quarter, plus an eighth, and so on. In order to avoid this, some, we are told, postulated indivisible atoms and others divisible 'leaps'. Sa'adia, however, preferred the Aristotelian reply that these segments do not constitute an *actually* existing infinity, and it is the latter which cannot be traversed.[40]

---

[38] Used by Huby, op. cit., 1971, 128, though not directly in connexion with time, and by Craig, op. cit., 84-5.

[39] Craig (op. cit., 83-7) envisages a library with infinitely many books, and is surprised that you can make various additions, subtractions and subcollections without getting any collection of a different size from the original. What remains unrecognised is the reason, namely, that none of the resulting collections would stick out beyond the far end of any other. And this also provides the reason why certain other subtractions would after all reduce the size of the collection. Craig further assimilates 'infinitely many' to 'all', when he supposes that the numbers on the spines of an infinitely numerous book collection would exhaust *all* the whole numbers. Why should not just the *odd* numbers be used? Even if all integers had been used initially, it does not follow that it would be impossible to accommodate a new book with a number of its own. One would simply re-number the original collection, perhaps using odd numbers and giving the new book an even number. Finally, more analysis is required for the claim that if you close the gaps after subtracting books, the shelves will remain full. If 'full' merely means that there will be no gaps, this is a tautology. But it is false, if it means that there will be no room for reinsertion: that would apply only to a full shelf with two ends.

[40] Sa'adia ben Joseph, *Kitab al-Amānāt*, ed. S. Landauer, Leiden 1880, 1.1, pp. 36-7 (translated by S. Rosenblatt as *Book of Beliefs and Opinions*, New Haven 1948, 45).

Let me take stock: I have considered eight sources of resistance to the idea that an actual infinity of past years could have been traversed, and I have argued that they are all wanting. To that extent, then, I have come out against Philoponus, and maintained that the universe has not been proved to have a beginning. I personally would leave it open whether the universe had a beginning or not. None the less, I would recall that I have not altogether sided against Philoponus. For I have maintained that his arguments do succeed *ad homines* against the pagans. Indeed, he has found a contradiction at the heart of paganism, a contradiction between their concept of infinity and their denial of a beginning. This contradiction had gone unnoticed for 850 years. Moreover, the materials for beginning to answer Philoponus' puzzle about increasing infinity were not even assembled until Henry of Harclay and others, some 800 years later. We can therefore see Philoponus as being at the centre of a 1600-year period. For the first time, he put Christianity on the offensive in the debate on whether the universe had a beginning. This might well be called a turning point in the history of philosophy.

E.A. Milne's time-scales

The foregoing arguments may raise a doubt: they have been couched in terms of our own time-scale of years, days and hours. But, in Chapter 6, the question was raised whether there might not be a community which, instead of hours, adopted *clours* as their unit, a clour being the time taken for a cloud to cross the sky. Suppose we now add a supposition derived by Richard Swinburne from E.A. Milne,[41] that clouds have been accelerating or decelerating since some past date at a Zenonian rate. For example, whatever their mileage in the first hour, it might have been double in the next half hour; or whatever their mileage in the most recent hour, it might have been double in the preceding half hour. Given a suitable rate of change, we might find that a finite number of past hours corresponded to an *infinite* number of past clours. Does this mean that there is room for disagreement on whether the universe began?

Like Swinburne, I think not: the moral is rather that the system of clours, which was argued in Chapter 6 to be ludicrous, would in these circumstances have to be abandoned altogether. For suppose there was a first event without predecessor: this is not something about which they would be free to disagree. They would be free to say, and we should agree, that it occurred an infinity of clours ago. But that should be proof enough to them that their clours are *not* of equal length, as they supposed, and that they will not serve as time-units.

There would be other embarrassments too. For if their clours have been accelerating at a Zenonian rate, then *all* past events will equally be an infinity of clours ago; while if the clours have been decelerating at that rate since the fall of Babylon, then at least all events prior to that fall will be an

---

[41] E.A. Milne, *Kinematic Relativity*, Oxford 1948, esp. 224f. The application of Milne's idea to arguments about a beginning is made by Richard Swinburne, *Space and Time* ch. 15, 1st ed., London 1968, 296, and I am grateful to him for drawing my attention to the issue.

infinity of clours ago. None could be represented on a calendar in which the clours were shown as equally spaced down to the present.

Avicenna allows an infinity of souls from past generations

I shall finish the chapter with two more kinds of infinity argument, one concerning infinite causal chains, and the other concerning an infinity of souls. The latter connects with two discussions which we have encountered in Greek thought already. First, it will be recalled, Simplicius defended the possibility of a beginningless universe, by urging that earlier members of the series have *perished*, so that we do not get an *actual* infinity. Versions of this argument appear in the Islamic world in Avicenna (A.D. 980-1037) and perhaps before him in Farabi (*c.* A.D. 873-950).[42] But in the case of *souls*, Avicenna is unable to appeal to the idea of perishing, because he believes that individual souls *survive* after death. Since he also believes that these souls come from an infinity of past generations, it was pointed out that there must be an *actual* infinity of souls of the dead all co-existing simultaneously.[43] This provides the second link with Greek thought. For, as noted in Chapter 12, the Neoplatonists had anticipated exactly this problem.[44] Like Avicenna, they denied a beginning to the human race, and they foresaw the threat that this would give us an actual infinity of souls – souls which would have co-existed before incarnation, if they had no beginning, and after incarnation, if they had no end. But they used this threat as a way of reaching the conclusion that there must be a *recycling* of a finite number of souls. Only through such reincarnation could one avoid Avicenna's result: an actually infinite collection.

Avicenna's response to the problem was entirely different from the Neoplatonist recycling. It was to argue than an actual infinity of souls was possible. His defence of its possibility rests on his ingenious method for discrediting certain *other* infinities.[45] Suppose we wish to discredit some proposed infinite length, which is bounded at one end. We can imagine a second length, which starts off aligned, but has a finite segment removed from the bounded end; and we can further imagine the original length and

---

[42] For Avicenna, see ch. 8 of the unpublished treatise summarised in English by Pines (p. 349), and *Najāt* (*The Cure*), Cairo 1938, 124-5. The argument is reported also without names by Ghazālī in *Tahāfut al-Falāsifa*, ap. Averroem *Tahāfut al-Tahāfut* (Bouyges), pp. 23; 273, tr. van den Bergh, pp. 12 and 162. For Farabi, see Maimonides *Guide* 1.74, 7th argument, which may be ascribing to Farabi the view that earlier members of the series exist only in imagination; and also Averroes, *Epitome of the Metaphysics*, Arabic version in *Rasā'il* (*Works*) Ibn Rushd, Hyderabad 1947, 128-9, translated into German by S. van den Bergh, Leiden 1924

[43] The objection is recorded e.g. by Ghazālī *Tahāfut al-Falāsifa*, ap. Averroem, *Tahāfut al-Tahāfut* (Bouyges), pp. 25-9; 273-4 tr. van den Bergh, pp. 13-15; 161-2; by Maimonides, *Guide* 1.74, 7th argument; and by Aquinas *ST* 1, q.46, a.2, objection (8) and reply. The response of *accepting* an actual infinity of souls is recorded by Aquinas, loc. cit., and explicitly ascribed to Avicenna by Ghazālī, loc. cit.

[44] Aug. *City* XII.21; Sallustius *de Diis et Mundo* ch. 20; Olympiodorus *in Phd.* 10.1.2-5 (Westerink).

[45] The method is given in *Najāt*, 124 (Latin translation by N. Carame, *Avicennae Metaphysicae Compendium*, Rome 1926). I am here following the clarifying account given by Davidson, op. cit., p. 380.

the curtailed one re-aligned at their bounded ends. Will they coincide at their opposite ends? Only now does Avicenna's objection go off the rails. He should deny that there are any opposite ends, but instead he protests that the opposite ends cannot coincide (since that would make the curtailed length equal to the original), and he assumes that the curtailed length will be exceeded to a finite extent by the original. From that he concludes that the curtailed length is finite (because it is exceeded), and the original length finite (because it is only a finite amount longer). The argument is misguided, although the technique of alignment marks an advance in sophistication.

But what has this all to do with an infinity of souls? The point is that souls do not have position or order as spatial entities do. One cannot therefore use Avicenna's technique of re-alignment and comparison of length in order to discredit the idea of an actual infinity of them. Avicenna thus has a double argument: past times and motions do not all exist together, and so do not exist in actuality. Souls of the dead admittedly do exist in actuality all together, but they lack an order. In neither case, therefore, can one employ the argument against actual infinities.[46]

Avicenna is effectively answered by Ghazālī (A.D. 1058-1111). For Avicenna had claimed that, since past entities are non-existent, they are neither finite nor infinite in number. Ghazālī replies to a similar claim that even non-existent entities will be odd or even in number.[47] Again, in response to the argument about souls lacking an order, Ghazālī replies that we can imagine the souls of the dead to have an order by imagining the birth of just one soul every day and every night.[48] In fact, Avicenna's argument about order is in any case vitiated by the flaw in his technique for discrediting infinity.

Aristotle against infinite causal chains

The remaining arguments have to do with infinite causal chains, and here once again Philoponus turns Aristotle's own views against him. For Aristotle holds that there cannot be an infinite chain of causes. He argues this for each of his 'four causes' in *Metaph.* 2.2, using in 994a3-19 the argument that only the first in the chain is fully entitled to be called 'cause', and an infinite chain would have no first. The best known application of his principle concerns final cause or purpose, and comes at the beginning of the *Nicomachean Ethics* (1.2, 1094a20-1), and before that in his early dialogue *Protrepticus* (fr. 12 Ross). We must desire something for its own sake, and cannot desire everything for the sake of something further, or there would be an infinite

---

[46] Avicenna, *Najāt*, 124-5. Farabi also referred to position when he said that, since time does not have position or exist in actuality, one cannot exclude its infinity in the same way as one excludes that of a straight line (ap. Averroem *Epitome of the Metaphsics, Rasā'il* Ibn Rushd, Hyderabad 1947, 128-9). It is doubtful, however, that he had anticipated this part of Avicenna's argument.

[47] *Tahāfut al-Falāsifa*, ap. Averroem, *Tahāfut al-Tahāfut* (Bouyges), p. 23, tr. van den Bergh p. 12.

[48] *Tahāfut al-Falāsifa*, ap. Averroem, *Tahāfut al-Tahāfut* (Bouyges) p. 274, tr. van den Bergh, p. 162.

regress, and desire would be empty and vain. But in the *Physics* (7.1 and 8.5), Aristotle goes much further. He supplies a whole battery of arguments against infinite chains of movers, repeating only as one argument among many that it is the first member of the chain which is most entitled to be called cause (8.5, 256a4-21; 257a27-33). Again, in *GC* 2.5, 332b30-333a15, he argues that there could not be an infinite sequence of elements, each being transformed into its successor. Aristotle had been anticipated, as regards the final cause, by Plato, who argues in *Lysis* 219C that there cannot be an infinite regress of things each loved for the sake of its successor. And Aristotle's arguments against an infinite regress of movers were copied later by the Stoics.[49]

There is a reason why Aristotle does not notice the threat posed to his belief in infinite chains of fathers and sons. For when he rules out infinite chains in the *Physics*, he is concentrating on chains of *movers*, and further on chains of movers which are *not* separated in time. His favourite example in *Phys.* 8.5 is of a hand moving a stick which moves some further object, and here cause and effect are in play *simultaneously*. There is a reason too for his choosing a simultaneous chain as his example. For his ultimate interest here is in explaining the motion which in *Phys.* 8.1 he has argued to have no beginning. If it has no beginning, its cause will not exist *before* it.

I shall select just one passage (256b3-257a14) out of *Phys.* 8.5, because it introduces the idea of an *accidental* causal chain which was to prove relevant in subsequent controversy. I shall take the passage somewhat differently from other interpreters. It is introduced as producing the same result as the preceding arguments. In other words, it is meant to show that the motion in the universe is not due to an infinite chain of movers, but ultimately to something self-moved, or (an alternative not yet stressed, but the one to be endorsed in the end) to something unmoved. The argument falls into three parts. First, could there be an *accidental* chain of movers which was infinite? No, there could not be *any* accidental chain, in the sense intended here (256b3-13). Secondly, there is interposed a merely probabilistic argument for an unmoved mover (256b13-27), based on the fact that we find the other possibilities exemplified: moved movers and moved non-movers. Finally (256b27-257a14), if there cannot be infinite (or even finite) accidental chains, can essential chains be infinite? No.

The first of these three arguments is the most relevant, and what matters for our purpose is the sense in which Aristotle is speaking of an accidental chain. I think we may envisage a hand moving a hockey stick which moves a ball, all three items being in motion. The question is then whether it is through (*dia*, 256b6) being moved that the hockey stick moves the ball, or whether it could still have moved the ball, even if it had remained motionless. In the latter case, the hockey stick does not depend for its motive power on the fact that it is moved, and in that sense the chain of moved movers is only accidental. Aristotle replies that it cannot be accidental, for then it would in principle be possible that the ball should be propelled by the stick, even if the stick and the hand and everything else in the universe remained still. But

[49] Sextus *M* 9.76.

that would be a case of motion starting up after universal rest, and that was ruled out as impossible in *Phys.* 8.1. Here is the passage (256b3-13):

> Besides what has been said, the same results will follow, if we look at things this way. Suppose that everything which is moved is moved by something which is itself moved. Either this happens to things accidentally, so that a thing is indeed moved when it causes motion, but does not in every case cause motion *through* being moved, or else this is not so, but it is moved essentially. First, then, if it is moved accidentally, there is no necessity for what causes motion (reading *kinoun* in b8) to be moved. But in that case it is clearly possible that at some time nothing in existence should be being moved, since what is accidental is not necessary, but is capable of being otherwise. If, then, we posit that what is possible actually takes place, nothing impossible should result, although something false may. But that motion should fail to exist is impossible, since it has been proved already that motion necessarily exists always.

The above interpretation is not the usual one. Some MSS read *to kinoumenon* (what is moved) in l. 8, instead of *to kinoun* (what causes motion). Aristotle's point would then appear to be that there is no necessity for what is moved (e.g. the hockey stick) to be moved: it might well stay still. But then so would the ball, and so might everything else in the universe. But in that case motion would cease everywhere, and it is that coming to a *halt*, rather than (as I claimed) a *starting up*, which is said to have been shown impossible in *Phys.* 8.1. Even without the alternative MS reading, ll. 7-13 might still be understood in this way, if they were taken in isolation. But in fact they cannot easily be so understood. For one thing, the question in l. 6 is not this new question of whether the hockey stick is obliged to undergo motion, but the question I introduced originally of whether it causes motion *through* (*dia*) being moved, in other words whether it is obliged to undergo motion *in order to cause motion*. In the second place, this interpretation is confirmed in a later portion of argument at 256b27-9, where Aristotle explains what he means by essential chains, as opposed to accidental chains:

> But if what causes motion is moved not accidentally, but of necessity, and would not cause motion if it were not moved, then ...

The phrasing reveals that by contrast an accidental chain is one in which the mover (e.g. the hockey stick) could still cause motion, even if it were not moved; that is, does not depend for its motive power on the fact that it is moved.

Philoponus against infinite causal chains

I shall now explain how Philoponus once again turns Aristotle's arguments against him. He attacks Aristotle's view in *Phys.* 8.1 that motion has gone on for an infinite length of time, and he does so by arguing that, on Aristotle's own principles, the movements which are going on now cannot be dependent on an infinite causal chain of prior movements. Philoponus exploits Aristotle's ban on infinite causal chains in Book 6 of his treatise against

Aristotle, and in Book 3 of a work which Pines believes to be his later non-polemical work. Pines has translated into English an Arabic summary of the second, but does not mention that an extremely similar summary was provided for the first by Simplicius.[50] I shall translate the relevant text of Simplicius (*in Phys.* 1178, 7-35), so that readers who care to consult Pines can observe how close the two passages are.

> As if he [Philoponus] had refuted those arguments, he dares to offer a proof himself in his own person that motion cannot be ungenerated. For the proof he assumes three premises. (1) The first is that if something which comes into being necessarily requires something preexistent for its coming into being, as a ship requires timbers, then it will not come into being, if those other things have not come into being first. (2) The second premise is that an infinite number cannot exist in actuality, nor can it be traversed in counting, nor can anything be greater than infinity, nor infinity be increased. (3) The third premise is that a thing cannot come into being, if what is needed for its doing so is an infinity of preexisting things, one arising out of another. For, says Philoponus, it was on this basis that Aristotle himself showed, in his treatise on coming into being, that the elements of bodies cannot be infinite, not even in the number of individuals, if one arises out of another. For infinity cannot be traversed; so fire, for example, would not have time to come into being if it had to do so after an infinity of other things. With this agreed in advance, then, says Philoponus, suppose that the particular movement of this fire has a beginning and end, and requires for its own coming into being that another movement should do so first. Suppose, for example, that the body of the fire had by some movement to come into being out of air which was undergoing change. And suppose again that, before the movement of the air which has changed into fire, there was first some other movement, say, of water, by means of which the water changed into air. And suppose that before that there was another movement, and so on *ad infinitum*, if the world and the transformation of things into each other had no beginning. In that case, there would have to have been an infinity of movements first, in order that this particular fire might come into being. For it would not have come into being, if an infinity of things had not existed first, because of the first premise, so he claims. So if it is impossible for an actual infinity of movements to have arisen, because of the second premise, it will then not have been possible for the movement of an individual fire to take place, both for this reason and because of the third premise, which says that a thing will not come into being, if an infinity of things have to exist before it does so. Hence, if the movement of an individual fire has taken place, then, by the necessity of contraposition, there will not have occurred before it an infinity of movements.

Up to this point the summaries of Philoponus' two treatises are very close to each other, but thereafter the Arabic summary of the later work goes on to elaborate a particularly graphic example: the chain of fathers and sons leading down to Socrates must be finite.

Accidental chains permissible after all

The irony here lies in Philoponus' trick of turning Aristotle's ban on infinite

[50] Pines, 'An Arabic summary ...'.

causal chains against him. But there is a second irony to come. It lies in the method subsequently used for defending Aristotle's belief in an infinite chain of fathers and sons. Although Islamic philosophers continued to deny that essential causal chains could be infinite, some of them allowed that there were accidental causal chains and that these could be infinite.[51] Thus Aristotle's denial of a beginning was rescued through the rejection of his other denial: the ban on accidental causal chains. Averroes[52] ascribes to 'certain philosophers' an argument which, in its initial stages, is reminiscent of the Christian Zacharias:[53] Certain philosophers had argued that the cause of a given man's existing is God; the existence of the man's father is necessary for his existence only in the way that an instrument is necessary; and the existence of more remote ancestors is not necessary at all, unless accidentally. In this way, the chain of fathers and sons can be regarded as accidental, and there is therefore no barrier to its being infinite. The sense in which the chain is accidental, however, is different from that which I associated with Aristotle. For the question is not so much whether one thing depends for its causal efficacy on another as whether it depends on another for its existance. Maimonides may have in mind yet another sense of 'accidental', when he reports the acceptance of infinite accidental chains. At any rate, he offers as the only example a chain in which one thing comes to exist when its predecessor ceases, as in the sequence of days.[54]

Thomas Aquinas and his teacher Albert the Great both endorse the acceptance of infinite accidental chains,[55] and Thomas explicitly acknowledges the Islamic sources for the idea.[56] But he reverts to a conception of accidental chains slightly closer to Aristotle's. He does not believe that there has in fact been an infinite chain of fathers and sons. But he none the less thinks that only Scripture, not philosophy, can rule it out. For such a chain would be accidental in the sense that each man begets not as the son of someone else but simply as a man. He thus does not depend on his ancestors for his act of begetting, even if he does for his existence. The infinite chain of fathers and sons, being accidental in this sense, is therefore at least logically possible.

I believe the truth of the matter is that Aquinas should have gone further. For I see no real logical objection to infinite causal chains, whether

[51] So Maimonides reports, *Guide* 1.73, 11th proposition, and see below for Averroes. Thomas Aquinas reports that Avicenna and Ghazālī had both said that an accidentally infinite multitude was possible (*ST* 1, q.7, a.4). One place where Avicenna attacks the ban on infinite causal chains is the unpublished treatise summarised in English by Pines (op. cit.). Islamic acceptance of accidental causal chains may go back to a lost work of Farabi, *The Changeable Beings*: see H.A. Wolfson, *The Philosophy of Kalam*, 430; 433-4.

[52] *Tahāfut al-Tahāfut*, (Bouyges) pp. 19-21; 268-9 (tr. van den Bergh, pp. 10-11; 158-9); cf. *Kashf* (Müller), p. 37, ll. 4-11; *in Phys.* 8, comm. 15 (both translated by Wolfson, *The Philosophy of the Kalam*, 428-30).

[53] For God as cause and father as instrument, see Zach. *Ammonius* or *de Mundi Opificio* PG85, 1041B-C.

[54] Maimonides *Guide* 1.73, 11th proposition.

[55] Aquinas, *ST* 1.46.2, reply to (7); *contra Gentiles* 2.38, reply to 5th argument; *Scriptum Super Libros Sententiarum Magistri Petri Lombardi*, in 2, dist. 1, quaest. 1, art. 5, solutio, ad quintum (2nd occurrence); Albertus Magnus, in *VIII Phys.*, tract. 1, ch. 12.

[56] *in 2. Sent.*, dist. 1, quaest. 1, art. 5, *ST* 1, q.7, a4.

accidental or not. It might be thought that there is an objection for anyone who follows Aristotle (as I do) in thinking that a cause is a factor capable of *explaining* the effect.[57] For is there not an objection to infinite chains of *explanation*? Certainly, one cannot explain Z or Y, if that explanation requires for its completion that Y be explained by X, X by W, and so on *ad infinitum*. But nothing remotely like this is the case. In an infinite chain, one cause may be perfectly capable of explaining its successor, without the explanation having to refer to any predecessor.

---

[57] Sorabji, *Necessity, Cause and Blame*, ch. 2.

CHAPTER FIFTEEN

Arguments Against a Beginning

The infinity arguments provided the best case for a beginning of the universe. I now turn to the case against. Here there are so many arguments that I can only select a few of particular influence and interest. I shall consider the 'Why not sooner?' argument, four arguments on God as creator, the argument that nothing comes from nothing, and what I shall call the 'idleness' argument.

Why not sooner?

The point of the 'Why not sooner?' argument against the universe having had a beginning is not merely that, if it began, we do not know the reason why it did not begin sooner. We can hardly expect to know everything. Rather, the point is that in the nature of the case there could not be a reason. If you say that the universe began, you must admit that it is in principle inexplicable why it did not begin sooner.

The difficulty can be made to look worse. If there was enough reason for the universe to begin at some given time, then there was surely enough reason for it to begin earlier, since the reasons available will not have been augmented, seeing that nothing was in existence, nor were any changes occurring.

It may be said that the universe could have begun without any reason. But although that is logically possible, it is natural to feel that a reason is called for, given the difference between the earlier absence and the later presence of a universe. By contrast, if the universe has existed without any beginning, we probably will not feel the same need to think that there is an answer to the only remaining analogous question, 'Why is there something, rather than nothing?'

That there could not be an answer to the question, 'Why not sooner?' is brought home by an argument of Simplicius:[1] normally we explain why something is delayed, by pointing to earlier causal sequences, in order to show that the time was not yet ripe. Socrates could not have been born that much sooner, because his parents had not yet met. When we are inquiring about the whole universe beginning, however, there is no earlier causal sequence by reference to which we can explain why things have to wait.

The 'Why not sooner?' argument is found as early as Parmenides (born *c.*

---

[1] *in Phys.* 1176 and 1177, 29; *in Cael.* 137-8.

515-510 B.C.), who says, at least on one common interpretation:

> And if it did begin from nothing, what need would have driven it to grow later rather than earlier?[2]

Parmenides is not talking about the universe exactly. The suggestion which I have followed in Chapter 8 is that he is discussing whatever can be spoken or thought of, or whatever we inquire into,[3] and is denying that it can have had a beginning.

Aristotle (384-322 B.C.) used the 'Why not sooner?' argument once at *Phys.* 8.1, 252a11-19, to attack the view that there could have been rest for an infinite period and then motion, and perhaps again (but see Chapter 17) at *Cael.* 1.12, 283a11, to deny that something could come into existence after not existing for an infinite amount of time, or perish after always having existed. He also records a spatial analogue to these arguments (*Phys.* 3.4, 203b25-8): if there is an infinity of space beyond the heavens, it must surely contain an infinite number of worlds. For why should matter be planted here rather than there if there is an infinite amount of space?

A major step towards answering the 'Why not sooner?' argument was taken seven hundred years later by Augustine (A.D. 354-430). He combines it with the related question,[4] 'What was God doing before he made heaven and earth?' Surely he cannot have been *idle*. To this question Augustine records the answer, 'He was preparing hell for people who pry into mysteries.'[5] But his own answer to both questions is that there *was* no time before the creation.[6] The ground he gives elsewhere is that God created time in creating movement, whether the mental movement of the angels,[7] or other movement.[8] For it does not make sense to suppose that time exists in the absence of movement, and God himself is unmoving. The widespread belief that time depends on movement, or on other changes, was the subject of Chapter 6. We may be tempted to imagine time ticking away and offering God a choice of dates for the creative act. But, according to the arguments discussed there, if nothing is happening at all, then we must reject this picture as illusory.

For good measure, Augustine adds an *ad homines* argument against those of his opponents who believe that there is a finite amount of matter planted in an infinite space. Surely they should have been worried by the parallel question: 'Why here and not elsewhere?' In consistency they should have postulated an infinity of worlds to fill the infinite space.[9]

---

[2] *The Way of Truth*, fr. 8, ll. 9-10 DK.

[3] These are the suggestions of G.E.L. Owen and Jonathan Barnes (references in Chapter 8).

[4] I shall return to this question, in discussing the 'idleness' arguments at the end of the chapter. We have already encountered the question in Chapters 12 and 13 in Philo and Origen. Augustine reveals that it had been used as an argument against a beginning by Manichaeans (*de Genesi contra Manichaeos* I.2.3). It appears from Irenaeus that it had earlier been used as a Gnostic argument (*adv. Haereses* 2.28.3).   [5] *Conf.* XI.12.

[6] *Conf.* XI.13; 30; *de Genesi contra Manichaeos* I.2.3; *City* XI.5-6.

[7] *de Gen. ad Lit. Liber Imperfectus* III.7-8; *de Gen. ad Lit.* V.5.12; *City* XII.16.

[8] *Conf.* XII.8-12 and 19.   [9] Aug. *City* XI.5.

Augustine's answer, that there was no time before the creation, is the best of the solutions offered by Jews and Christians. Elsewhere, however, he gives it as only one of two answers, the other being that God also created, outside of time, a world of patterns that can be grasped only by the mind.[10] This had been the solution of Origen, and I shall return to it at the end of the chapter. Another solution which was entertained alongside this one by Origen was discussed in Chapter 12: God was creating earlier *physical* worlds. Finally, Irenaeus had responded that we simply cannot know the solution.[11]

When, in his better solution, Augustine denies time before the creation, he is not the first to do so. What matters is his *applying* that denial to the present problem. Augustine's originality has been the subject of hostile comment: some have denied that he was original;[12] Edward Gibbon more cruelly said of him, 'His learning is too often borrowed and his arguments are too often his own.'[13] It is true that his *learning* is borrowed, but I believe that his *arguments* are often original and clever. The source of his learning in this instance is easy to guess. Ultimately, the idea that time began with the *kosmos* goes back to Plato's *Timaeus*, but by no means everyone read the *Timaeus* that way. The important thing was that Philo, who is a source for so much in Christian tradition,[14] did read Plato that way, and moreover endorsed the idea himself, as we saw in Chapter 13. From him, therefore, the idea passed down to Augustine's mentor Ambrose,[15] after being accepted by many Christians and Gnostics on the way.[16] But Augustine uses the idea differently from Philo. For when Philo tackles the question whether God was idle before he formed the *kosmos* out of matter, he gets only as far as saying that there was no time when unformed matter was waiting to be formed by God; God created matter and formed it simultaneously (*Prov.* 1.7 Aucher). So far Augustine would agree, but it does not answer the next question which is whether there was a period of idleness before this simultaneous activity. Philo's belief in a beginning of time would enable him to answer this suggestion, and perhaps he is silent because he is presupposing the answer. But, at any rate, he *is* silent, mentioning the beginning of time only a few paragraphs later in 1.20. What Augustine has done is to articulate what is not articulated in Philo.

---

[10] *de Genesi contra Manichaeos* I.2.3.

[11] *adv. Haereses* 2.28.3.

[12] W. Theiler, *Porphyrios und Augustin*, Schriften der Königsberger gelehrten Gesellschaft, vol. 10, Halle, 1933, 57; A.C. Lloyd, 'Nosce teipsum and conscientia', *AGP* 46, 1964, 198-200.

[13] *The Decline and Fall of the Roman Empire*, ch. 28, fn., a reference for which I thank Colin Haycraft.

[14] This is not the only time that we shall find Philo originating a line of thought by seeking to harmonise the Middle Platonism and Stoicism of his period with the Bible. H.A. Wolfson described him as the originator of a way of doing Philosophy which lasted among Jews, Christians and Moslems until Spinoza, and devoted his life's work to studying this period of thought. (For this remark, see H.A. Wolfson, *Philo*, vol. 2, ch. 14, 'What is new in Philo?').

[15] *de Fide* 1.2, sec. 14; 1.9, sec. 58.

[16] Gnostics are cited by Clement *Stromateis* 6.16; and the view is endorsed by Athanasius *contra Arianos* 1.12.13; Basil *in Hex.* 1.6; Gregory of Nyssa *contra Eunom.* 1 and 8, PG 45, cols 364A-368A; 792D. Milic Čapek cites for modern endorsements Lemaître, Renouvier, F.C.S. Schiller and De Witt Parker, but he is mistaken in thinking Augustine the only precursor (*The Philosophical Impact of Contemporary Physics*, Princeton 1961, 353-4).

If we want to find a closer anticipation of Augustine's solution, we might do better to look at Gregory of Nyssa, another person who accepted a beginning of time, and who worked only shortly before Augustine. Gregory tackles a question analogous to Augustine's: if God generated his Son at a certain time, why did he not generate him sooner? In order to answer, he refers not so much to the beginning of time as to God's timelessness. The answer is that God is prior to all time, and where there is no time, questions of earlier and later make no sense. The whole discussion does not represent Gregory's own viewpoint; for it is his opponent, Eunomius, not he himself, who thinks that the Son had a temporal beginning:

> If, as our opponent says, he generated the Son when he willed, and he always willed the good, and had the power to go along with his will, then the Son must be considered as being always together with the Father, since the Father always wills the good, and has the power to secure what he wills. Or if we should adapt also our opponent's next point to the truth, it is easy to fit that too to our own view, I mean: 'let there be no inquiry among sensible people into why not sooner.' For talk about sooner has a temporal import, being contrasted with afterwards and later. And when there is no time, the words for temporal distance are equally abolished. Now the Lord is before times and before the ages. So inquiry into sooner and later is useless among intelligent people in the case of the creator of the ages. For such words are empty of all content, if they are not applied to time. Hence, since the Lord is before times, there is no room for sooner and later with him.[17]

Does this discussion in Gregory detract from Augustine's originality? I think not: for one thing, it is doubtful that Augustine had access to Gregory.[18] At this date, Gregory had not been translated into Latin, and Augustine was not sufficiently fluent in reading Greek until later in his life.[19] Of course, Greek discussions might have been retailed to Augustine. But Gregory's argument would, in any case, need considerable adaptation.

Augustine's answer, that there *was* no sooner, is a clever one, and it has been accepted by recent philosophers.[20] But I think it needs more investigation, for it is not quite true to say that we cannot make sense of the 'Why not sooner?' question. We can do so by thinking backwards from the present. Let us suppose for convenience that the solar system is as old as the universe and that both have lasted a million years. The 'Why not sooner?' question can then be put as the question: 'Why did not the universe begin a million and *one* years ago?' To bring this about, God would only need to have tacked on an extra revolution of the earth around the sun at the beginning of the series. To imagine him doing so is not to imagine extra time without

[17] *contra Eunom.* 9.2, PG 45, col. 809B-C.

[18] So e.g. B. Altaner, *Kleine Patristische Schriften*, Berlin 1967, 285, repr. from *Revue Bénédictine* 61, 1951. In Chapter 7, I declared myself unpersuaded by J.F. Callahan's suggestion that Augustine is influenced by Gregory in his idea that time consists of three states of mind.

[19] For details see P. Courcelle, *Late Latin Writers and Their Greek Sources*, Cambridge Mass. 1969, ch. 4, translated from the French of 1948. Augustine would have had access, via Latin, to some of the other writers listed in the text.

[20] E.g. by W.H. Newton-Smith, *The Structure of Time*, London 1980, 107.

extra change; on the contrary, the extra year would have involved an extra revolution of the earth. Why did none occur? Leibniz was to recognise in the eighteenth century that the question could be interpreted in this way[21]

However, as Leibniz himself observes,[22] once we give this interpretation to the question 'Why not sooner?', it becomes possible in principle that there should after all be an answer. An extra revolution of the earth, in order to be a distinct revolution, would have to differ in some ways, possibly bad ways, from other revolutions. It would also be likely to cause some differences, again possibly bad, in subsequent revolutions. Further, it would alter the temporal distances between events, so that mankind, for example, would have had more time to degenerate between his initial creation and his confrontation with the trials of the present year. All these considerations could supply God with a reason for rejecting the extra revolution. This is not to say that we can look into God's mind, or guess what his reason might be. All that is needed is to show that there could in principle be a reason, and this is shown by the fact that the extra revolution would involve some genuine difference.

Augustine, writing in Latin, did not put a stop to discussion. On the contrary, Philoponus, writing in Greek on behalf of Christianity in A.D. 529 ignores his replies. He raises the 'Why not sooner?' question as an extrapolation from one of Proclus' arguments. He answers it by posing a counter-problem: the one discussed in the last chapter about how there could have been an infinity of days.[23] The relevance of this reply is made clearer by Richard Bentley, who elaborates it into two steps: the opponents of Christianity will not be satisfied by *finitely* sooner, since that leaves the problem unresolved; but their only alternative is to think of creation occurring *infinitely* sooner, and that, for Philoponus' reasons, is incoherent.[24] Simplicius, however, Philoponus' opponent, is not satisfied. He seeks to answer the infinity problem, and to reinstate the 'Why not sooner?' question. Indeed, he insists on the unanswerability of that question in the way already indicated, by reminding us that there is no earlier causal sequence by reference to which an answer can be given.

The 'Why not sooner?' argument against a beginning reappears among Islamic thinkers. It appears in Arabic in an early ninth-century source, pseudo-Apollonius of Tyana, and later in Avicenna. Ghazālī reports it from there, and it was known to Averroes and to his contemporary Ibn Tufayl.[25]

---

[21] Leibniz, *Fifth Letter to Clarke*, §56 in H.G. Alexander (ed.), *The Leibniz-Clarke Correspondence*, Manchester 1956.

[22] Leibniz (loc. cit.) observes that from the fact that God did not tack extra events on to the beginning of the series, we can infer that it was not agreeable to his wisdom to do so.

[23] Philoponus *Aet.* (Rabe) p. 11, l. 25.

[24] Richard Bentley, 6th of the t2Boyle Sermons, 1692, in Bentley's *Works*, ed. A. Dyce, vol. 3, London 1838, 136-7.

[25] U. Weisser, *Das Buch über das Geheimnis du Schöpfung von Pseudo-Apollonius von Tyana*, Berlin 1980 (*Ars Medica* 3.2), Book 1.3.4. Avicenna (980-1037) *Najāt* (p. 418, 10-11). The argument appears as the first of the proofs by 'the philosophers' that the universe had no beginning in Ghazālī (1058-1111) *Tahāfut al-Falāsifa*, ap. Averroem (*c.* 1126-1198) *Tahāfut al-Tahāfut* (Bouyges) pp. 4ff., tr. van den Bergh, pp. 1ff. For Ibn Tufayl, see L.E. Goodman, *Ibn Tufayl's Hayy Ibn Yaqzān* (Eng. transl) New York 1972, p. 131.

Ghazālī's discussion is particularly interesting. He reproduces, even if only in connexion with a different argument, the Augustinian solution that there *was* no sooner.[26] Against the 'Why not sooner?' argument, Ghazālī repeats Philoponus' strategy of responding with the counter-problems about infinity.[27] He also reveals that 'the philosophers' had anticipated Leibniz's method for making sense of the question 'Why not sooner?', in terms of more revolutions preceding the present.[28] They differed from Leibniz, however, in maintaining that the mere *possibility* of an earlier creation established the *actual*, not the possible, existence of an earlier time.

Thomas Aquinas (*c.* A.D. 1224-1274) returns to Augustine's handling of the problem and elaborates it. He gives rather similar treatments to the questions 'Why not sooner?' and 'For what fresh incentive was God waiting?'[29] His answer to the first question is that you cannot ask 'Why not sooner?' because there are no time distinctions in the absence of a universe; you can only ask 'Why did not God wish the universe to be eternal (in the sense of beginningless)?'[30] But the latter question has an answer: it manifests God's goodness and power more clearly if he gives the universe a beginning, because it shows more clearly that the universe depends on him.[31] Similarly, you cannot ask for what fresh incentive God was waiting, since he was not acting in time. You can only ask why he wanted time and the universe to have the duration which he assigned, and corresponding points could be made about the spatial position of matter.[32] Earlier in the century, St Bonaventure (*c.* A.D. 1217-1274) had offered a different explanation of God's not wanting the universe to be beginningless: eternity was too good for a created thing.[33]

[26] Ghazālī, ap. Averroem *Tahāfut* (Bouyges) pp. 71-2, tr. van den Bergh p. 41, in the reply to the *second* of the proofs by 'the philosophers'. For the solution that there was no sooner, H.M.E. Alousi (*The Problem of Creation in Islamic Thought*, Baghdad 1968, 247-9) cites Ka'bī and Māturīdī (both early tenth century), and for a counter-attack Avicenna *Najāt* and *Ishārāt*. Augustine's Latin would not have been available to the Arabs, and this is presumably why he is not given credit by commentators on the Islamic material for applying Philo's denial of an earlier time, to the 'Why not sooner?' problem. None the less, by that date, Augustine's influence could have filtered through Greek sources.
Less clear-cut than Ghazālī's is Maimonides' view in *Guide* 2.13. He allows us to say that God existed before the creation of the universe, but insists that this thought depends on a 'supposition or imagination of time', and not on the reality of time. H.A. Wolfson points out (*The Philosophy of Spinoza*, Cambridge Mass. 1934, vol. 1, ch. 10) that Albo (*c.* 1380-*c.* 1444) claims to be following Maimonides, when he speaks of an unmeasured duration which is conceived only in thought and which existed before the creation of the world. (For Plato's rather different unmeasured time before creation, see Chapter 17.)
[27] Ghazālī, *Tahāfut al-Falāsifa*, ap. Averroem, *Tahāfut al-Tahāfut* (Bouyges) pp. 16-17, tr. van den Bergh, pp. 8-9.
[28] Ghazālī, op. cit., ap. Averroem, op. cit., (Bouyges) pp. 83-4, tr. van den Bergh, pp. 48-9.
[29] For the first question see Aquinas *CG* 2.32, argument (5) (p. 30 in Vollert); for the second, ibid., argument (4) and *de Potentia Dei* q.3, a.17, argument (13) (pp. 28-9 and 47 in Vollert).
[30] *CG* 2.35 and *de Potentia Dei* q.3, a.17, answer (pp. 36 and 53 in Vollert).
[31] *CG* 2.38 and *ST* 1, q.46, a.1 reply to objection (6) (pp. 44 and 63 in Vollert). This kind of view is found at least as early as Philo, *de Opificio Mundi* 171.
[32] *Compendium of Theology* 1.98 and *de Potentia Dei* q.3, a.17, reply to (13) (pp. 56; 68-70 in Vollert).
[33] Bonaventure *in IV libros Sententiarum Magistri Petri Lombardi*, tomus 2, in librum 2, dist. 1, p.1, a.1, q.2, reply to (5) (p. 112 in Byrne).

When the 'Why not sooner?' argument recurs in Leibniz and in Kant in the eighteenth century, it is put to new uses. Leibniz employs the argument and its spatial analogue to attack Newton's idea of absolute time and space, as was observed in Chapter 6.[34] Kant puts the 'Why not sooner?' question to a different use again.[35] He presents it as the Antithesis to the First Antinomy. His aim is to show that the arguments for and against a beginningless universe are equally insuperable, unless one takes his own view that the world exists only in so far as it is constructed.

By an irony, the 'Why not elsewhere?' question was at one time used for the *opposite* purpose of proving a beginning of the universe. For the Ash'arite Juwaynī argues that where there are alternative possibilities, as he says there are for the location of the physical world, there must be a will which decides which possibility is realised. So far Juwaynī's claim looks like an adaptation, I would suggest, of Aristotle's assertion in *Metaph.* 9.5: that where there is a capacity to produce opposite results, the doctor's capacity, for example, to kill or to cure, there must be some other factor which decides (*einai to kurion*) which result is realised, and this factor is desire or choice.[36] The rest of Juwaynī's argument advances rather rapidly: if something exists in one of its possible states because of a *will*, then it is created in time.[37]

The theological arguments

The 'Why not sooner?' argument does not have to take a theological form. For the question 'Why not sooner?' can be raised against an atheist who gives the universe a beginning. But I shall now turn to a set of four arguments which are theological in character. They consider the possibility of God's creating the universe. It may seem that such arguments could be of interest only to believers, or to people sympathetic to religious belief. But I think it will emerge that these arguments throw a flood of light on other subjects too: on causation, on the will and on idealism.

What I should do is to recall from Chapter 13 the qualification of my claim to be considering whether the totality of things could have had a beginning. The ensuing arguments are concerned with whether the totality of things *other than God* could have begun.

Several of the arguments treat God as a changeless being. It may surprise many present-day Christians that this was once a Christian orthodoxy. But, of course, if God is outside time (and I shall come in Chapter 16 to reasons for thinking that), then he *must* be changeless. And even some of those who have regarded him as existing at *all* times have ascribed changelessness to him. This view may be not only unfamiliar, but hard to understand. The

---

[34] See the references there.

[35] Kant, *Critique of Pure Reason*, ed. A 1781, ed. B 1787.

[36] 1048a10-11.

[37] Juwaynī's argument is related by Averroes *Kashf* (Müller) pp. 37-40 (English translation by H.A. Wolfson in *The Philosophy of the Kalam*, 437-8), and by Maimonides *Guide* 1.74, 5th argument.

earliest set of arguments I know[38] for God being changeless is given by Plato in *Republic* 2, 380D-318E: something in the best condition is least likely to be altered by something else. And it will not alter itself, since any alteration would be a lapse from the best condition. We shall find a great many arguments, including Plato's, against God's changing his will in Aristotle's *de Philosophia*.[39] A further influential work is Philo's *Quod Deus Immutabilis Sit* (*That God is Unchangeable*),[40] and Thomas Aquinas offers a set of three arguments for God being changeless in the *Summa Theologiae*.[41]

Even if the idea of God being changeless is nowadays less familiar, I would not go along with the idea that earlier Christian acceptance of it can be explained as a piece of Neoplatonist contamination.[42] We need to know what made it appeal to Neoplatonists and, for that matter, to earlier Platonists, and what made that bit of Platonism appeal to Jews and Christians. And I refer to *arguments* both here and in Chapter 16. But as well as arguments, there may be an *emotional* appeal for some people. Martha Nussbaum has imaginatively described Socrates' desire, as represented in Plato's *Symposium*, for an object of love that will be *stable*, instead of letting us down, as human loves so painfully can.[43] And in Augustine there are two passages in which he treats it as self-evident, without further reason, that life, wisdom and anything else is superior, if it is unchangeable.[44]

The first argument I want to discuss is the 'changeless will' argument. If God's will is changeless, how can he have willed the universe to exist after allowing it not to exist? The second is the 'changeless cause' argument. If God is changeless, how can he have caused the universe to begin, without the aid of some change or novelty to trigger the new beginning? Closely related is the third argument from 'sufficient cause'. God's will is presumably a sufficient cause for the universe to exist. But how can a sufficient cause *delay* its effect? Fourthly, the argument from 'immaterial cause' asks how God can be the cause of matter, if he is himself immaterial.

It may seem that several of these arguments could be answered at a stroke by taking a leaf out of Augustine's book, or Philo's, and insisting that, in the absence of change, there can have been no time 'before' the creation. In that case, God will not have changed his will from some supposedly 'earlier' condition. Nor yet can he be said to have 'delayed' his effect. To suppose otherwise is to make two disputable assumptions, both that God is a temporal being, and that there was time for him to exist in before the creation.

But rather than taking such a short way with the arguments, I shall, at least for a start, grant their assumptions about time. For it will turn out that

[38] Still earlier, Xenophanes had denied both m⌐ ₁on and rest to God. Simplicius explains (*in Phys.* 23, 9-14) that this is because motion and res. involve a relation between two things.

[39] pp. 281-2. Plato's argument reappears in fr. 16 (Ross); cf. *Metaph.* 12.9, 1074b21-7; God must always think of the same thing, for any change would be for the worse.

[40] Philo *Quod Deus* ...

[41] Aquinas *ST* 1, q.9, a1.

[42] Richard Swinburne, *The Coherence of Theism*, Oxford 1977, 215.

[43] Martha Nussbaum, 'The speech of Alcibiades: a reading of Plato's *Symposium*', *P.Lit* 3, 1979, 131-72.

[44] *Conf.* VII.1; *de Doctrina Christiana* I.8.8.

there is plenty to be learnt from doing that, even if we resort eventually to denying one or other of the assumptions.

Willing a change and changing one's will

The first problem, then, is this: would not God have to change his will, if he first allowed there to be no universe, and then made it begin? And how can he change his will, if he is changeless? One reason against taking the 'short way' with this argument is that the problem of God changing his will comes up in many contexts where the 'short' solution is unavailable. It comes up, for example, in connexion with God's responses to prayer and his many other interventions in the world. For these too may seem to imply a change of will.

There is a classic reply to these problems: that willing a change is not changing one's will.[45] If I will to change my clothes after lunch, this does not mean that after lunch my will must change. The same applies to God: if he does will changes, he can will them changelessly. In the *City of God*, discussing the creation of man (XII. 15 and 18) and of the world (XI.4), Augustine argues that when God creates things, he acts with an eternal or sempiternal (*aeternus, sempiternus*) will and design (*voluntas, consilium*), not with a new or changed one. The same idea had already appeared in the *Confessions*.[46]

In an intervening work, *de Genesi ad Litteram*, and in *City*,[47] the idea of changelessly willing a change is applied to the answering of prayer. If a prayer is answered, God foreknew that the man would pray in such a way that the prayer would be granted (Augustine here treats God as knowing and willing not timelessly, but in advance). Similarly, if Ezechias was to die, so far as certain inferior causes were concerned, but was given by God an extra lease of life, then God foreknew that he would do this for Ezechias, and kept this in his will. There was no change in his will or knowledge. This time, it is clear where Augustine's view comes from. For his treatment of prayer is found in a much earlier Christian: Origen (*c.* A.D. 185-253). Origen, like Augustine, wavers on whether God's changeless will should be viewed as timeless, or as existing in advance.[48] But in his treatise *On Prayer*,[49] he plumps for the latter. His point is that God's will is changeless, and he wills changelessly that, for example, my crops should be threatened with failure, that I should then fall on my knees and pray, and that I should then receive whatever answer he thinks fit to my prayer. Origen thus supplies in

---

[45] I am indebted to Norman Kretzmann for opening up this line of thought for me by drawing my attention to Eleonore Stump's excellent article, 'Petitionary prayer', *APQ* 16, 1979, 81-91, and to work of his own, now incorporated into Eleonore Stump and Norman Kretzmann, 'Eternity', *JP* 78, 1981, 429-58, at 447-53.

[46] XI.10; XI.30; XII.15; XII.28. For denial of a new *voluntas* or *consilium* in contexts other than creation, see *City* XXII.2; *de Ordine* II.17.46. Cf. also *Conf.* XI.6. Augustine makes the point not only about the divine will, but also about the divine understanding: it is known in the eternal reason, where nothing begins or ceases when things should begin or cease: *Conf.* XI.8; similarly *City* XI.21.

[47] *de Gen. ad Lit.* VI.17.28; *City* XXII.2.

[48] See Chapter 8 for passages in which Origen treats God as timeless.

[49] *On Prayer* 5-6.

connexion with prayer the basic idea which Augustine applies to creation, that willing a change does not imply changing one's will. It is ironical, then, that Origen should have been criticised at the end of the third century A.D. by Methodius with the allegation that he made the world beginningless, in order to avoid a change in God. For Methodius' own device for avoiding a change in the creator is far less satisfactory than the device to which Origen gave birth.[50]

There is, however, one difficulty in Origen's solution, that it seems to present everything as determined in advance, including the question of whether I will pray. Origen senses the difficulty, but in the treatise *On Prayer* he is able to cite in reply nothing more satisfactory than the old Stoic arguments that what is determined may yet be up to us. A better solution would have been to introduce some conditionality into what God wills: he wills that *if* I pray, *then* my prayer will be answered. It has been suggested that this kind of solution is to be found in Descartes.[51]

Origen's idea is not the only thing which Augustine recalls in his treatment of the subject. He echoes something still earlier in his use of the expression 'new design' (*novum consilium*), for this expression is used, as we shall see, in Cicero's report of Aristotle's treatise, extant only in fragments, the *de Philosophia*.[52]

After Augustine, there is a fresh development in Philoponus. In arguing against a beginning, the Neoplatonist Proclus had protested in his fourth argument that God is unchanging, and in his sixteenth that so also is God's will. In each case, Philoponus gives the classic reply that willing a change does not imply changing one's will.[53] The novelty is that he relates this point about the will to one about knowledge which will concern us in Chapter 16. Knowledge is parallel to the will. For just as there can be a changeless will for a willed change, so also there can be definite knowledge of an indefinite thing.[54]

The problem, and the classic response to it, were known to Philoponus' Christian contemporary, Zacharias of Mitylene, and to subsequent Islamic and Jewish thinkers.[55] But the clearest formulation of the response is given by Thomas Aquinas:

It must be said that the will of God is altogether unchangeable. But concerning

---

[50] Methodius' claim is merely that *resting* from labour does not constitute an actual alteration (*alloiōsis*) in God, so neither should the *commencement* of creative labour: *ex Libro de Creatis Excerpta*, PG 18, col. 336C, translated by William R. Clark in the Anti-Nicene Library.

[51] I am grateful to John Cottingham for drawing my attention to this idea, although I am not sure whether I see it in Descartes' *Conversation with Burman*, edited by Cottingham, Oxford 1976, p. 34, to which he has kindly directed me.

[52] Cic. *Acad. Pr.* 2.38 (= fr. 20 in W.D. Ross' translation of Aristotle fragments).

[53] *Aet.* (Rabe), pp. 78; 568; 613ff.

[54] ibid., p. 568.

[55] Zach. *Ammonius* or *de Mundi Opificio* PG 85, 1116A-B. Maimonides *Guide* 2.14, 6th method, ascribes to 'the philosophers' the argument that God cannot have created the universe, since his will is changeless. Ghazālī replies that the universe could come into being because of an *eternal* will: *Tahāfut al-Falāsifa*, ap. Averroem *Tahāfut al-Tahāfut* (Bouyges) pp. 7; 13, tr. van den Bergh pp. 3; 6-7.

this we should reflect that it is one thing to change one's will, and another to will a change in things. For one can with one and the same will changelessly and steadfastly will that now one thing should come about and later its contrary. The will would change only if one began to will what he did not will before, or ceased to will what he did will before.[56]

This solution was not always retained, however. Berkeley, for one, treats the problem as insoluble, in his *Third Dialogue Between Hylas and Philonous*.

How does a changeless cause make something begin?

The argument just presented turned on a point about willing; there is another which concerns the notion of causing something to begin, and which goes back ultimately to Aristotle. On the present hypothesis, God is supposed to be changeless. Can a changeless cause make something begin, if no new factor is brought into the situation? We can perfectly well understand that something changeless might play a major role in explaining why something starts up. Perhaps, for example, there is a permanent gas leak in my cellar. One day, as I go down to investigate with a lighted candle, there is an explosion: why? The existence of the gas leak may for many purposes prove the most significant part of the causal exaplanation, even though it is a relatively changeless factor. None the less, the permanent leak would not have brought about an explosion without a triggering action, such as my intrusion with a lighted candle.

Aristotle insists on the need for a trigger. A thing can come into existence or begin to move, or grow, or change in quality, he says, only if some prior motion occurs. He enunciates this principle in three passages,[57] and in one passage, to be summarised in Chapter 17,[58] he uses it as part of his proof that there cannot have been a beginning of motion in the universe: any supposed beginning of motion would need to be triggered by a prior motion.

If a trigger really is required, then a changeless God would never be able to make the universe begin. For suppose we try to imagine him existing before the universe: if he is changeless, nothing within him can act as a trigger. And since the universe will not yet have been created, there will be nothing else available to serve as a trigger either. If, on the other hand, we postulate that he is not chronologically prior to the universe, this seems only to make things worse. For the absence of a trigger will be even clearer.

But let us examine more closely the claim that a triggering action is required. One reason for believing this has been dismissed already: it is the mistaken idea that where a change is willed, the will must change. But even though no change of will is required, another kind of change is often needed. When I will to change my clothes after lunch, my hands will need to do some work, and that work constitutes a change. Clearly, however, the case of God is different, since his will is sufficient on its own to produce a result, and he

---

[56] *ST* 1, q.19, a.7; cf. *CG* 1.83; 3.91; 98; 4.70; *Scriptum Super IV Libros Sententiarum Magistri Petri Lombardi* 1, dist. 39, q.1, a.1.

[57] *Phys.* 8.1, 251a8-b10; 8.7, 260a26-261a26; *GC* 2.10, 336a14-b17.

[58] *Phys.* 8.1, 251a8-b10.

does not need the cooperation of a physical change. The case is more like that in which I will to start thinking about something after lunch, or will, on going to bed, to wake up at six in the morning. I shall not need to sing myself a song to wake myself up at six, nor to perform any other action. But even here, it may be protested, some change has to take place. For I must dimly register events in my body or outside my room which warn me that the time has come.

A more promising example of human willing is supplied by Ghazālī, and at first sight it may seem to provide what is needed. A Muslim husband can say to his wife, 'I divorce you – with effect from tomorrow'. It may then appear that the divorce will take effect tomorrow with no further triggering events. But Ghazālī scrupulously points out, on behalf of his opponents 'the philosophers', that even here the divorce depends on a further event, namely, the arrival of tomorrow.[59] Thomas Aquinas takes up the point, and he may have known the contents of Ghazālī's *Tahāfut al-Falāsifa*. For although no Latin translation is known earlier than A.D. 1328, after Aquinas' time, when it appeared incorporated in Averroes' *Tahāfut al-Tahāfut*, none the less Wolfson has argued in a different connexion that Aquinas' teacher Albert had some knowledge of the work.[60] At any rate, Aquinas' remarks seem to echo Ghazālī's. A person who wishes to do something tomorrow, he says, has to wait at least for one change to occur: the arrival of tomorrow and the changes involved in that arrival. Boethius of Dacia repeats the argument, and seeks to ram it home, by pointing out that if the universe has not been created there can be no such thing as the arrival of the right time, since no changes would be occurring. So God would lack the necessary trigger of the right time arriving.[61]

But this suggests a new reply. It is being conceded that in sufficiently special circumstances, or for a sufficiently powerful will, the only trigger needed might be the arrival of the right time. But we are now invited to think that the creation is not preceded by time. In that case, although Boethius of Dacia will be correct that the right time *cannot* arrive, it will be equally true that we do not *need* it to arrive. Since no time is passing, no arrival has to be awaited. Why should we suppose, then, that there is any remaining need for a triggering event at all?

There is also a second reply available to those who want to deny the need for a triggering change, and that is that the arrival of the right time would in any case be a change only in a very minimal sense. The passage of time may

---

[59] Ghazālī, *Tahāfut al-Falāsifa*, ap. Averroem *Tahāfut al-Tahāfut* (Bouyges) pp. 11-12, tr. van den Bergh, pp. 5-6.

[60] Wolfson, *The Philosophy of the Kalam*, pp. 595-6.

[61] Aquinas *ST* 1, q.46, a.1, objection (6) (p. 60 in Vollert). Boethius of Dacia, *Aet.*, argument (13) and defence of (13) (pp. 40 and 42 in Sajó). Varieties of triggering argument appear in Maimonides, *Guide* 2.14, 6th method; Bonaventure *in IV libros Sententiarum Magistri Petri Lombardi*, tomus 2, in librum 2, dist. 1, p.1, a.1, q.2, arguments (1) and (2) (p. 105 in Byrne); Aquinas, *CG* 2.32, arguments (1) and (2); 2.33 argument (3); 2.34 argument (2); *de Potentia Dei* q.3, a.17, argument (26); *ST* 1, q.46, a.1, objections (5) and (6) (pp. 28; 31; 33; 49; 60 in Vollert). Boethius *Aet.*, arguments (4), (9), (10), (11), (13) with replies (pp. 37-40, 56, 58-9 in Sajó). Compare also the Aristotelian Themistius (*c*. A.D. 317-388), in his commentary on Arist. *Metaph.*, 14, 21-15, 10.

involve accompanying changes, but the sense in which it is *itself* a change is a very attenuated one.

On the other side, opponents of changeless causation may protest that Ghazālī's example is being misdescribed. For it is not a stable feature like the husband's will which brings about the divorce, but a sudden act such as his decision or his announcement, and this does not even look like a changeless cause.[62] But in reply, it should be noticed that the husband's announcement does not have to be a fleeting occurrence: he may make it in writing, and it may be the *persistence* of the husband's will and of his announcement which is causally efficacious. Admittedly, his will and his written announcement have to *come into being* at some point of time, but this, and the fact that there has to be an announcement at all, arise from the circumstances of the human condition. There is no need to suppose that these features would have to carry over to the divine case.

Can a sufficient cause delay its effect?

There is a closely related argument. We are to imagine God existing in time before the creation of the universe. Now his will is presumably a *sufficient* condition for the universe existing, and the argument is that a *sufficient* condition cannot delay its effect. Hence if God has existed without beginning, the same should be true of the universe. Arguments of this kind appear in Greek thought, in Islam and in thirteenth-century Paris.[63]

The new argument differs from the preceding one at least in this respect. It is much more obvious that it can be answered, if we once agree that the creation is not preceded by time. For if time did not pass, there will have been no 'delay'. However, we should see if there is another answer available, parallel to some of those attempted before.

In Ghazālī's example of the divorce, the will is virtually a sufficient condition of the divorce taking effect after a delay. I say only 'virtually' sufficient, because the will has to be announced, there has to be a background of legal conventions, the parties have to stay alive, and so on. But in the case of God's will, this is not needed, for his will does not depend on such things as announcements and conventions, so that it looks as if God's will could be wholly sufficient. To this it may be objected that God's will cannot be sufficient on the present hypothesis, since it is being thought of as

---

[62] I am grateful to John Watling for a somewhat similar line of objection: the announcement might be viewed as a trigger accompanying the stable cause.

[63] For a comparable set of arguments in Greek thought, see Themistius *in Metaph.* 14,21-15,10 and 15,35-16,5. For Islam, see the discussion in Ghazālī and Averroes in Averroes, *Tahāfut al-Tahāfut* (Bouyges) pp. 7; 10-12, tr. van den Bergh, pp. 3-6. Alousi (op. cit., p. 238) cites in favour of such arguments Farabi, Avicenna, Bahmanyār, Averroes, Abū al-Barakāt, and against unnamed theologians referred to by Ghazālī, Shahrastānī and al-Rāzī. In thirteenth-century Paris, the argument is considered by Bonaventure *in IV libros Sententiarum Magistri Petri Lombardi*, tomus 2, in librum 2, dist. 1, p.1, a.1, q.2, argument (5) and reply (pp. 106 and 112 in Byrne); Aquinas, *Aet.*, argument (5) (pp. 21-2 in Vollert); *CG* 2.32 argument (3) and reply in 2.35 (pp. 28-9 and 35 in Vollert); *de Potentia Dei* q.3, a.17, argument (4) and replies to (4) and (6) (pp. 45-6, 54-5 in Vollert); *ST* 1, q.46, a.1, objection (9) and reply (pp. 60-1 and 64 in Vollert); Boethius *Aet.*, argument (7) and reply (pp. 38 and 57 in Sajó).

existing in time before the creation, and in that case it depends on a further condition which is now familiar, the arrival of the right time. But one reply, to mention no others, is that the arrival of the right time could itself be subject to God's will. Thus extrapolation from Ghazālī's example of the divorce suggests that there need be no obstacle to God's will being sufficient, and yet delaying the creation.

An alternative way of defending the feasibility of the creation would be to argue that what is sufficient is the *persistence* of God's will up to the moment of creation. That would mean that after all there was no delay between the sufficient cause and its effect. It is interesting that we have already encountered the question of a sufficient cause delaying its effect in an utterly different context: that of Shoemaker's universal freezes interrupting the history of the universe (Chapter 6). The first events after a freeze might be the delayed effects of sufficient causes that had occurred before the freeze. Shoemaker himself expresses anxiety about this, and prefers the alternative idea just mentioned that there is no delay, because the sufficient cause of unfreezing is the *persistence*, for the period of the freeze, of the forces which were in existence when it started.

There is another context, as Mark Stevenson has pointed out to me, in which many philosophers happily accept that a sufficient condition delays its effect. For this is implied by most versions of causal determinism. According to one version, the total state of the universe at any one instant is a causally sufficient condition of its state at any later instant, however late that later instant may be.

Can an immaterial cause produce a material effect?

The last argument I shall mention of those which turn on the nature of God, will be more fully treated in Chapter 18. Gregory of Nyssa (*c.* A.D. 331-396) asks how God can create matter, seeing that there is nothing material in his nature. The difficulty turns, presumably, on the (mistaken) assumption that cause must be like effect. Gregory answers by giving an idealistic account of matter. To create matter, God need only assemble groups of properties, which are themselves not material things, but intelligible concepts.[64]

Nothing comes from nothing

The most important remaining argument against a beginning of the universe does not depend on theological considerations. It is provided by the principle that nothing can come from nothing. As I explained in Chapter 13, this was not intended to exclude the kind of view we find in Neoplatonism, that a *beginningless* creation could be produced by God without being produced *out of* anything. Rather, the principle is intended to exclude something's *beginning* without its being the case that there was anything previously there out of which it came. In this sense, which excludes a beginning for the material universe, the principle was accepted by almost everyone outside the Judaeo-Christian

---

[64] The three texts are quoted in full in Chapter 18.

tradition. A few possible exceptions and a sample of the numerous endorsements are best listed in a footnote.[65] In the text, I shall consider only those who made a significant contribution to the discussion, the most important of whom was Aristotle with his concepts of matter and form.

Even nowadays, although we accept that *matter* can be newly created, or (as in an atomic explosion) utterly destroyed, there is still a principle of the conservation of *energy*. Moreover, the Judaeo-Christian tradition was not unanimous in violating the principle that nothing comes from nothing. For I drew attention at the beginning of Chapter 13 to a whole series of authors in the Judaeo-Christian tradition who made God fashion the world out of a beginningless matter.

Discussion of the principle that nothing comes from nothing started very early in Greek philosophy. Thus Parmenides of Elea (born *c.* 515 B.C.) made a special contribution to the subject, because he actually offered *reasons* for the principle that nothing comes from nothing: you cannot say that what is discussable arose out of nothing, first because you cannot speak or think of nothing, and secondly because of the 'Why not sooner?' question:[66] 'If it did begin from nothing, what need would have driven it to grow later rather than earlier?' Parmenides concludes:[67]

Thus coming into being is extinguished and ceasing to be is not to be heard of.

Aristotle's contribution to the subject begins in *Phys.* 1.7-8. He summarises the position of Parmenides and other Eleatics by saying that nothing can come to be, since what exists cannot come to be, because it already is, and meanwhile nothing can come to be from what is not.[68] In his reply, he brings the notion of *matter* to the fore. When a bronze statue is brought into existence, the bronze matter exists before the change, and the statue has to come from it.

In another work,[69] he gives attention to the *form*, for example, to the statue's shape. In many cases of creation, the form as well as the matter pre-

---

[65] Endorsements: Anaximander fr. 1 (Presocratic numbering as in DK); Heraclitus frs 30; 31; 90; Xenophanes frs 29; 30; 33; and A33; Parmenides fr. 8, ll. 7-9; Melissus fr. 1, Empedocles frs 8; 12; Anaxagoras frs 10; 17; and ap. Arist. *Phys.* 1.4, 187a26; Democritus ap. Diog. Laert. 9.44; Arist. *Phys.* 1.4, 187a26-b1; 1.8, 191a24-33; b13-17; Themistius *in Metaph.* 15, 8; Epicurus, *Letter to Herodotus* 38-9; 40-1; 54; also ap. Diog. Laert. *Lives* 10.2, and ap. Sext. *M.*10. 18-19; Lucr. *de Rerum Natura* 1. 149-264; 540-50; Boethus and other Stoics, ap. Philonem. *Aet.* 16; Plut. *de Proc. An.* (*Mor.* 1014B); *Quaest. Conv.* (*Mor.* 731D); *adv. Col.* (*Mor.* 1111A; 1112A; 1113C); Plot. *Enn.* 2.8.6(5); Procl. ap. Philop. *Aet.* (Rabe) p. 338.

Possible exceptions: Aristotle, Alexander and Themistius on the creation of forms *ex nihilo* will be considered below. The poet Hesiod was attacked, first by Epicharmus (fr. 1) and later by Epicurus (ap. Diog. Laert. *Lives* 10.2; and ap. Sext. *M* 10. 18-19) for saying that first of all Chaos came into being, and for being therefore unable to say *from what* it came. The Aristotelian author of *MXG* (wrongly attributed to Aristotle) attacks Xenophanes' contention that God could not have come into being either from like or from unlike, by saying (*MXG* 4) that there is a third possibility: his coming into being from what is not. Plato's account of creation in the *Timaeus* was interpreted only by a few as implying creation *ex nihilo*, as will be seen in Chapter 17. For Porphyry and Hierocles, see Chapter 20.

[66] *The Way of Truth*, fr. 8, ll. 7-10.   [67] fr. 8, l. 21.

[68] *Phys.* 1.8, 191a27-31.   [69] *Metaph.* 7.7-9.

exists. For when a biological specimen is generated, the form pre-exists in the (male) parent, and when a bronze statue is made, the form pre-exists at least in the mind of the artist.

As regards matter, Aristotle offers an infinite regress argument against its being created.[70] His objective is limited, and he may mean only that when we make a bronze statue, we are not obliged at the same time to make its matter (the bronze), or else an infinite regress would arise. For in making the bronze, we should be obliged to make the bronze's matter, and so on. This does not exclude the possibility that this particular piece of bronze was at some time created out of the elements, earth, air, fire and water. Moreover, any particular parcel of air may have been created out of a parcel of water. The regress argument cannot rule this out. On the other hand, it might well be put to a further use (even though this would go beyond Aristotle's original purpose), in order to show that matter *as a whole* cannot have gone through a process of being created. For such a process would presuppose *other* matter.

In the *de Caelo* (3.2, 301b30-302a9), Aristotle has a further argument. Although one body can be created out of another, it cannot be created out of nothing, because that would require a pre-existing vacuum. And vacuum can be shown to be impossible.

Despite these objections to creation out of nothing, Aristotle came to be interpreted later in a different sense. I shall not dwell on Hippolytus' claim that Aristotle is committed to creation out of nothing, for that is based on a fanciful extrapolation.[71] I shall also leave until Chapter 20 the record of Arabic and later misconceptions about Aristotle and his successor Theophrastus accepting a creation of the world out of nothing. What is more important is that Aristotle makes certain concessions about form. When a substance, instead of being created, merely acquires a new quality or quantity, that new property need not have pre-existed, except potentially. The newly acquired form of whiteness, for example, need not have been in existence before.[72] Moreover, in four passages Aristotle discusses the forms of artifacts, and allows that they too can exist at one time without existing at another.[73] There is only one restriction: such forms begin to exist without going through a *process* of coming into existence.[74] For a *process* would involve a *further* matter and form combining to produce them. There are many further casual references to forms being created or destroyed, again presumably without any *process*.[75] Analogues for these ephemeral forms are already supplied by Plato's talk of the coldness in snow, the heat in fire and the tallness in Simmias, all of which

---

[70] *Phys.* 1.9, 192a27-32; *Metaph.* 7.8, 1033a24-b19; 7.9, 1034b10-19; 12.3, 1069b35-1070a4. For the *indestructibility* of some matter, see *Metaph.* 2.2, 994b6-9.

[71] Hippolytus, *Refutation of All Heresies* 7, 15-18. See Catherine Osborne, *Antiparathesis: An Approach to Presocratic Philosophy*, Ph.D. Dissertation, King's College, Cambridge, 1983.

[72] *Metaph.* 7.9, 1034b16-19; 8.5, 1044b21-4.

[73] *Metaph.* 7.15, 1039b20-7; 8.3, 1043b14-23; 11.2, 1060b23-8; 12.3, 1070a13-26.

[74] *Metaph.* 7.15, 1039b23-7; 8.3, 1043b14-18; 8.5, 1044b21-4.

[75] *Phys.* 1.9, 192b1-2; 7.3, 246a16; 246b14-16; *Metaph.* 11.2, 1060a21-3; *DA* 2.5, 417b3; *GA* 2.1, 731b31-5; *GC* 1.10, 328a27-8; 2.11, 338b14-17; *Cael.* 3.7, 306a9-11. The evidence is well discussed by Robert Heinaman, 'Aristotle's tenth aporia', *AGP* 61, 1979, 249-70, and A.C. Lloyd, *Form and Universal in Aristotle*, Liverpool 1981, 25-6.

are liable to perish.[76]

These concessions seem to imply that after all form can be created out of nothing. Aristotle makes no parallel concessions about matter. But a certain concession might seem to be required, in the form of a modification of his infinite regress argument. For if matter, like form, can be created without a *process*, then no regress need be implied by the suggestion that when we make a thing by a process, we also make its matter.

The Aristotelian Alexander of Aphrodisias (fl. *c*. A.D. 205) may have taken up Aristotle's concession. For a work ascribed to him, and entitled *On Coming-to-be*, survives in Arabic, as does a summary and a description of it, the last having been translated into English.[77] And another Islamic source ascribes to him a work called *On the Coming-to-be of Forms out of Nothing*.[78] The surviving work sets out to explain a view which it ascribes to Aristotle, that things may come to be out of nothing. Nature can create new *forms* out of nothing in a given matter. Admittedly, it is hard to accept as Alexander's the next claim, that the divine first cause can create matter itself out of nothing. But even if this shows the work to be spurious, it is still quite possible that Alexander made the first claim, that nature creates *forms* out of nothing. In another spurious Alexandrian work, there is an even more extreme statement about form. It does not exist *per se*, but can supervene on matter through the movements produced in the latter by the efficient cause.[79] Islamic sources attribute belief in the creation of forms *ex nihilo* to a later Aristotelian as well, namely, Themistius (fl. A.D. 340s-384/5).[80]

If Aristotelians did make concessions to creation *ex nihilo*, this will make the Christian position outlined by Augustine a less startling departure. God created both the angels[81] and formless matter itself,[82] not out of matter, but out of nothing. Admittedly, for God's creation of other things there had to be formless matter available. But even so, this formless matter did not have to pre-exist in time.[83]

The Aristotelian concession on *forms* is taken up in a significant way by Philoponus in his reply to Proclus. For Proclus had denied creation *ex nihilo*,[84]

---

[76] Plato *Phd.* 102A10-107B10.

[77] Text no. 16 in A. Dietrich's list in *Göttingen Nachrichten*, phil.-hist. kl., 1964, 93-100. The text is to be found in a Carullah MS., 1279, fols 64v-65r. It is referred to and described by Ibn Miskawaïh, *al-Fawz*, in a passage translated into English by J.W. Sweetman, *Islam and Christian Theology* vol. 1, London 1945, 116-17. The summary is by 'Abdallatïf, and is edited by Badawi in *Plotinus apud Arabes*, 207. Fritz Zimmermann has been kind enough to show me translations of both text and summary.

[78] Referred to by Ibn abï Usaybi'a (i.70.22, Müller).

[79] Form is *mē'on kath' hauto en huparxei*, pseudo-Alexander *in Metaph.* 721, 3. I am grateful to Robert Sharples for this reference, and to him and Fritz Zimmermann for help in sorting out the Arabic references.

[80] Averroes *Commentary on Aristotle Metaphysics Lambda* (Bouyges) pp. 1491-1505, translated into English by C.F. Genequand, *Ibn Rushd's Commentary on Aristotle's Metaphysics Book Lām*, Bodleian Library MS. D. Phil. c. 2407, pp. 178-90, citing Themistius at Bouyges p. 1498-9 = Genequand p. 184.

[81] *City* XII.16.   [82] *Conf.* XII. 8; *de Genesi contra Manichaeos* I.6.10.

[83] *Conf.* XII. 3-12; *de Gen. ad Lit.* I.15.29.

[84] ap. Philop. *Aet.* (Rabe) p. 338.

and in reply Philoponus insists that Aristotle had allowed it for forms.[85] He cites the example of a particular whiteness,[86] which he takes, presumably, from *Metaph.* 8.5, 1044b23. Philoponus' conclusion is that what is possible for form ought also to be possible for matter.

Proclus had taken up Aristotle's infinite regress argument about matter, and used it in order to show that matter cannot have had a beginning.[87] But Philoponus replies that no regress arises, for if matter needed anything in order to come into being, it would not need more *matter*. Rather, things spring out of what is *unlike*, not out of what is like themselves.[88]

Philoponus is answered in his turn by Simplicius, who defends the principle that nothing can come from nothing. His new step, which we have already met, is to support the principle by reference to the 'Why not sooner?' question, which he does in a different way from Parmenides. The 'Why not sooner?' question becomes unanswerable, says Simplicius, unless everything springs out of something else in an orderly sequence. For it is only by referring to such a sequence that we can explain the dating of any given occurrence.[89]

At least two of these arguments reappear in Islam: the argument on one side that *forms* can come into existence *ex nihilo*,[90] and the argument on the other that *matter* cannot be created, because it could only be created out of further matter.[91] Modern Islamic scholars cite Alexander as a source for the first argument, but seem less aware that Philoponus followed Alexander, or that Proclus provided the second.

Despite the appeal of the view that nothing can come from nothing, it was perhaps a good thing that Aristotle made room for doubts to grow. For in the end, I believe, creation out of nothing cannot be excluded as involving any logical impossibility. A recent work by the physicist Steven Weinberg marginally favours a beginning for the physical universe. For if the present expansion and contraction of the universe had been preceded by an infinity of earlier ones, then, at least according to present physical beliefs, the ratio of photons to nuclear particles in the universe ought to be infinite, which it is not.[92]

The idleness arguments

One final set of arguments has been scattered through Chapters 12, 13 and 15, and its threads should now be drawn together. We have seen that Philo,

---

[85] Philop. *Aet.* 340, 365,3 and ap. Simpl. *in Phys.* p. 1141. See also Philop. *in Phys.* 54, 24-5; 55, 2-3; 191, 9-33.

[86] *Aet.* 347.

[87] Proclus, 11th argument, ap. Philop. *Aet.* p. 404, ll. 21-3, translated by Thomas Taylor, *The Fragments that Remain of the Lost Writings of Proclus*, London 1825, p. 58. The argument is further recorded by Philop. in *Aet.* 446, 13-21, and (with further elaboration) 339, 2-19.

[88] *Aet.* 448-51.

[89] *in Phys.* 1176 and 1177, 29; *in Cael.* 137-8.

[90] Alousi op. cit., p. 302 cites Māturīdī, Ibn Miskawayh, Quāsim, Ibn Ḥazam, and Baghdādī.

[91] Alousi, op. cit., pp. 273-4, cites Taftāzānī and Qastālī.

[92] Steven Weinberg, *The First Three Minutes*, London 1977, 148.

Origen and Augustine were all concerned with the objection that, on their view, God would have to have been *idle* before the creation. I have ascribed at least one solution apiece to these three thinkers (matter shaped immediately, earlier material worlds, no earlier time). But I have not yet properly discussed a fourth solution: God's creation of an intelligible world.

The idea that God must not be thought of as idle was very widespread. Plato had insisted that God cannot be lazy, but must, on the contrary, be provident,[93] and this was repeated in later antiquity as an argument for divine providence.[94] The Stoic Boethus opposed his school's periodic destructions of the world, by saying that they would leave God periodically inactive, and hence in effect dead.[95] He may have been inspired by a slightly different Epicurean argument, to which I shall return in Chapter 17.[96] Galen reproduces an argument that God cannot have created the world, since that would make him idle earlier.[97] And 'idleness' arguments against creation reappeared in Islam. For the 'philosophers' and Averroes used them against the Ash'arite theologians, who accepted a beginning of the universe.[98]

Philo makes a particularly large contribution to the 'idleness' question. For one thing, he shows that 'idleness' arguments might be used in either of two opposite directions. For, on the one hand, to represent the world as uncreated is to leave God with nothing to do.[99] On the other hand, if God made the universe begin, he would have been idle beforehand.[100] I have suggested that Philo exploits the double-edged character of idleness arguments in *Prov.* 1.6. I have also said what I take his solution to be in *Prov.* 1.7, and I have resisted the alternative interpretation of Baltes, that Philo appeals for his solution to the creation of an earlier intelligible world. None the less, Philo does provide the *materials* for such a solution. For he does refer to the creation of an intelligible world at *Prov.* 1.7 and 1.21, and he does speak at *de Opificio Mundi* 4.15-5.20 *as if* the intelligible world were created first, even though he tells us to reinterpret this way of talking, and insists (*de Opificio Mundi* 7.26-8; similarly *Prov.* 1.7) that the intelligible world was created *simultaneously* with the material world. Philo provides further encouragement for the new solution by giving a helpfully full description of what the intelligible world (*noētos kosmos*) was like.[101] God made it, to serve as a pattern (*paradeigma*) for making perceptible things. There are as many perceptible kinds (*genē*) as there are intelligible. The intelligible world consists of ideas (*ideai*), and we may compare the plan in an architect's mind. In the same way, God first thought up the imprints (*tupoi*) and put together an intelligible world which

[93] *Laws* 901A-903A.

[94] Nemesius, *de Natura Hominis* ch. 44, PG40, 804A-C; Simpl. *in Epicteti Enchiridion* 102, 18ff. I owe these references to Robert Sharples.

[95] ap. Philonem *Aet* 83-4.

[96] Aëtius *Placita* 1.7.8-9 (Diels 300a18; 301a7); Lucr. *de Rerum Natura* 5.156-80; Cic. *de Natura Deorum* 1.21-2.

[97] *On Medical Experience* 19, translated from the Arabic version by R. Walzer.

[98] Maimonides *Guide* 2.14, 7th method; Averroes *Tahāfut al-Tahāfut* (Bouyges) pp. 96-7; 169, tr. van den Bergh pp. 56; 101.

[99] *de Opificio Mundi* 2.7.

[100] *Prov.* 1, translated into Latin from the Armenian by Aucher, §§6-7, pp. 4-5.

[101] *de Opificio Mundi* 4.16-5.20; 7.29.

exists in the divine reason (*logos*). Philo thought this view was to be found in Plato's *Timaeus*,[102] although in fact it is only in the *Republic* that Plato makes God create the Ideas.[103] But Philo was influenced by the intervening interpretation of Middle Platonists, already documented by his contemporary Seneca,[104] according to which the Platonic Ideas are thoughts in the mind of God, and he further equated these Ideas with the Stoic 'seminal reasons',[105] which will be discussed in Chapter 19.

In order to represent this intelligible world as an *earlier* creation, Origen has to make some new moves.[106] He cites the verse from the Psalms, 'In Wisdom hast thou made all things', understanding 'Wisdom' to refer to Christ.[107] He further says, in Rufinus' Latin rendering, that the intelligible patterns have existed in Wisdom *without beginning* (*sine initio*, *On First Principles* 1.4.5).[108] In the present passage he treats them as having existed *always*, but, in his more precise moments, he takes the Trinity to be timeless,[109] and so would perhaps make the patterns in Wisdom timeless too.

Origen is now ready to supplement his other solution of the idleness problem. Not only did God create *material* worlds earlier than the present one, but he also created an earlier (or a timeless) *intelligible* world. He offers this twice as a solution to the idleness problem,[110] and although he describes the intelligible world in terms very similar to Philo's,[111] I have argued that the solution is not in Philo, but is new with Origen.

From Philo to Augustine there was a continuous tradition of belief in a separate intelligible heaven,[112] and Augustine discusses a 'heaven of heaven',

[102] *Prov.* 1.21 Aucher.

[103] 597B-C.

[104] *Ep. Mor.* 65, 7. See John Dillon, *The Middle Platonists*, London 1977, 95; 138.

[105] Philo *Legatio ad Gaium* 55.

[106] *On First Principles* 1.4.3-5, and fr. 10 (Koetschau) from the Letter of Justinian to Menna. The fragment is incorporated in *On First Principles* 1.4.5 in Koetschau's edition, tr. G.W. Butterworth (= vol. 5 of the Prussian Academy edition of Origen's works).

[107] For the equation of Wisdom with Christ, see e.g. *On First Principles* 1.2.2; *Commentary on John* 1.19 (22), 111-15; 19.22(5), 146-50.

[108] Since the intelligible patterns have no beginning, they are evidently distinct from the intelligible intellects which God creates, since these do have a beginning (*On First Principles* 2.9.1-2; 4.4.8), although the distinction is not always made clear, especially as the intellects, like the patterns, are said to constitute a sort of intelligible heaven (e.g. *On First Principles* 2.9.1).

[109] *On First Principles* 1.3.4; 2.2.1; 4.4.1.

[110] *On First Principles* 1.4.3-5 and fr. 10 (Koetschau).

[111] He puts the point here by saying that the genera and species (*genē, eidē*), even if not the individuals, have existed always. All things have existed without beginning 'according to prefiguration or preformation' (*secundum praefigurationem et praeformationem*) in God's beginningless Wisdom. Elsewhere he talks of prefigured species, origins and reasons (*species, initia, rationes, On First Princples* 1.2.2). God's Wisdom is an intelligible world (*noētos kosmos, Comm. on John* 19.22(5), 146-50), and in it, as in the mind of an architect (*Comm. on John* 1.19(22), 111-15), are imprints, reasons and thoughts (*tupoi, logoi* or *rationes, theorēmata, On First Principles* 1.2.2; *Comm. on John*, two references above and 1.34(39), 244; 1.38(42), 283; 2.18(12), 126; *contra Celsum* 5.39; *Comm. on Epistle to the Ephesians*, fr. 6, in *JTS* 3, 1902, 241), which serve as patterns (*imagines, Comm. on Song of Songs* 3, p. 208 of Prussian Academy edition, vol. 8) for all the visible things on our earth. These invisible patterns are in the invisible heavens (ibid.).

[112] Theophilus of Antioch, *Libri ad Autolycum* 2.13, PG 6, col. 1073B; Clement *Stromateis* 5.14; Plot. 3.2.4(6-7); 5.5.8(9); 5.8.3(27-36); 6.7.12(1-19); Gregory *in Hex.* PG 44, cols. 81B-D; 84C-

somewhat like that of Origen, in *Confessions* XII.[113] Moreover, as already noticed, he refers to God's timeless creation of this intelligible world, as an additional answer to the question what kept God from idleness.[114] Although he added to this his other, more sophisticated solution, he had less impact than we could wish on the Greek discussions. Accordingly, a little over a century later, in the early sixth century, we find Zacharias, Bishop of Mitylene, relying on the old solution of the 'idleness' difficulty: he ascribes to his Platonist interlocutors,[115] and endorses in the name of Christianity,[116] the reply that God earlier created an intelligible world.

D; Basil *in Hex.* 1.5. An intelligible heaven is also mentioned in the *Selecta in Psalmos*, 18, 2 and 148, 4-5, in PG 12, 1240D and 1680A, attributed by some to Origen and by others to Evagrius Ponticus. See Jean Pépin, 'Recherches sur le sens et les origines de l'expression *caelum caeli* dans le livre XII des *Confessions* de S. Augustin', *Bull. du Cange* 23, 1953, 185-274, repr., with additions, in his *Ex Platonicorum Persona*, Amsterdam 1977.

[113] There is the same uncertainty in some passages whether Augustine is populating it with objects of thought as well as with thinkers. See Jean Pépin, op. cit., 269-74; R.P.A. Solignac, in *Oeuvres de Saint Augustin*, Bibliothèque Augustinienne, vol. 14, *Les Confessions, Livres VIII-XIII*, Paris, 1962, 592; J. de Blic, 'Platonisme et Christianisme dans la conception Augustinienne du Dieu Créateur', *Recherches de Science Religieuse*, Paris 1940, 172-90.

[114] *de Genesi contra Manichaeos* I.2.3.

[115] Zach. *Ammonius* or *de Mundi Opificio* PG 85, 1120A.

[116] Zach. op. cit. 1088A.

CHAPTER SIXTEEN

Timelessness Versus Changelessness in God

In Chapter 15, I have considered some defences of the idea of a *changeless* God. But the God of Augustine, Boethius, Anselm and Aquinas is not only changeless; he is timeless. The idea of a timeless God has recently come under heavy criticism. But I should like to see to what extent the idea of a *timeless* God is more vulnerable than the idea of a temporal, but changeless, God. The idea will turn out to be more vulnerable in some respects. But it is hard to identify what these respects are, and there is a danger of underestimating the skill of Christian thinkers at meeting the more obvious objections.

Four reasons for thinking God timeless

One question has been whether the idea of God's timelessness, as opposed to his changelessness, is of any importance. Thus Anthony Kenny says, in reference to Nelson Pike's book, *God and Timelessness:*

> I agree with the general conclusion of Pike's book which is that the doctrine of the timelessness of God is theologically unimportant and inessential to the tradition of western theism.[1]

Kenny adds that, in comparison with the doctrine of timelessness, the doctrine of God's *changelessness* 'is far more deeply entrenched in the tradition of western theism'. In fairness to Pike, it should be said that he views the issue as a very complex one, and presents his conclusions only tentatively. Others, however, have argued that the idea of timelessness or of God's timelessness is the product of a few simple confusions.[2] And there has been some tendency, already noticed in Chapter 15, to cite the influence of Neoplatonism, as if contamination from that source is enough to explain any

---

[1] Anthony Kenny, *The God of the Philosophers*, Oxford 1979, p. 40, referring to Nelson Pike, *God and Timelessness*, London 1970.

[2] J.R. Lucas cites the spatial analogy, a confusion between three different senses of the word 'present', and a confusion between instants and intervals (*A Treatise on Time and Space*, London 1973, 300-6). M. Kneale cites the failure to consider whether it is merely pointless, or actually meaningless, to pick out one particular time in the existence of something like God ('Eternity and Sempiternity', *PAS* 69, 1968-9, 227-31).

absurdity.[3] A feeling that the idea of God's timelessness is just silly may sometimes have encouraged the search for a meaning of 'eternity' other than timelessness. But I want to draw attention to four ways in which the idea of God's timelessness is after all important.

First, we should not underestimate the effect of mystical experience in which all sense of being in time is lost. I tried to explain in Chapter 11 how this experience, coupled with Plotinus' metaphysics of matter, or with Augustine's metaphysics of time, could easily lead to the view not merely that one loses a *sense* of time, but that time can actually be transcended. Moreover, in having this experience, mystics have supposed that they were approaching God.

Secondly, a very big difference is made by the doctrine, attributed in Chapter 15 to a good many thinkers from Philo to Augustine, that time had a beginning, along with the moving creatures that God created, since time cannot exist in the absence of motion. This immediately makes it impossible to say that God exists in time. For, first, this would now imply that he too, like time, had a beginning and a finite past. And, secondly, it would imply that he depended for existence on his own creatures. For he could not exist without time, nor time without motion and the moving creatures created by him. This is a very strong reason indeed for thinking God timeless.[4]

Thirdly, a reason unavailable to antiquity arises from the Theory of Relativity. There is no one answer, according to this theory, to the question what distant events are going on simultaneously with events here. There are different answers for different observers, depending on their state of motion or rest. What answer can there be, then, for a being, God, if we are to assume that he has no state of motion or rest at all? The theory allows no way of saying that one set of events is present for him rather than another, and so provides a further reason for thinking of him as timeless. This problem will not arise for those who give God a spatial location.

Fourthly, there is a powerful reason in Boethius for wanting God's knowledge to be timeless.[5] Only so, he argues, can we avoid the knowledge *in advance* which would restrict our freedom. For if we make God's knowledge of our doings to be both infallible and existent in advance, then what he foresees as happening will have been *inevitable* all along. It is then doubtful that we can be either morally responsible or free. I have argued for these conclusions elsewhere.[6] But they are not obvious, and they deserve a little explanation. They arise from the peculiar combination of infallibility and knowledge in advance.

[3] It may be unfair to single out Richard Swinburne, *The Coherence of Theism*, Oxford 1977, 215, and Lucas, op. cit., 302. I think Lucas is taking it that Augustine was not yet a Christian when he was bowled over by Neoplatonism; but if so, that view is controversial, and has been criticised in Chapter 11.

[4] It would be no good protesting (see Chapter 6) that Augustine needs to place God only outside some kind of *measured* time – God is the creator of *all* time, angelic and sidereal (see Chapter 2).

[5] *Consolation* 5.6. This reason is given also by Pike and Swinburne. For yet other reasons, see Pike, op. cit., ch. 9; Swinburne, op. cit., 218-22; W. Kneale 'Time and Eternity in Theology', *PAS* 61, 1960-1, 101.

[6] Sorabji, *Necessity, Cause and Blame*, 112-13. The point is appreciated by J.R. Lucas, *The Freedom of the Will*, Oxford 1970, ch. 14; and by Pike, op. cit., ch. 4.

Boethius: timeless knowledge avoids determinism

If God were not *infallible* in his judgment of what we would do, then we might be able so to act that his prediction turned out *wrong*. But this is not even a possibility, for to call him infallible is to say not merely that he *is* not, but that he *cannot* be wrong, and correspondingly we *cannot* make him wrong. But even so, as Boethius recognises, we could still have acted otherwise than we do, if his knowledge had not existed *in advance*. Thus, for example, Boethius says, if a being *sees* what we are doing *contemporaneously* with our doing it, it in no way follows that we could not have acted otherwise, however infallibly he may know what we are doing. The restriction on freedom arises not from God's infallibility alone, but from that coupled with the *irrevocability* of the past. If God's infallible knowledge of our doings exists *in advance*, then we are *too late* so to act that God will have had a different judgment about what we are going to do. His judgment exists *already*, and the past *cannot* be affected.

Boethius does not see quite all of this. He does not, for example, diagnose the threat as coming from the *irrevocability* of the past. But he does see that to assign God timeless knowledge rather than foreknowledge is to remove the threat, even if he does not show us how to make sense of the idea of timeless knowledge.

Boethius' solution represents the culmination of a chain of ideas going back at least as far as the Neoplatonist Iamblichus. Ammonius, who says he will offer a view in accordance with *(kata)*[7] that of Iamblichus, reports the latter as arguing that knowledge need not have the same status as the thing known.[8] Ammonius and his teacher Proclus, as well as Boethius, believe that knowledge takes its character from the knower, not from the thing known, and they offer many illustrations.[9] One is that, although the future is *indefinite* and *contingent*, divine knowledge of it can be *definite* and *necessary*. All three take this to reconcile the falsity of determinism with their belief that God or the gods have definite knowledge of the future.[10] A further illustration of the general principle, which is given by Proclus and spelt out in detail by Ammonius, is that the gods have *timeless* knowledge of things *temporal*.[11] Boethius, who had contented himself with the point about *definite* knowledge in *Consolation* 5.5, makes the point about *timeless* knowledge in 5.6, and makes it more central than it had been in his predecessors. Moreover, he enhances

---

[7] *in Int.* 135, 14

[8] *in Int.* 135, 14-19; 136, 11-12.

[9] Procl. *Elements of Theology* proposition 124 (English translation by E.R. Dodds); *in Tim.* (Diehl) 352, 5-27; *in Prm.* (Cousin) 956, 10-957, 40; *de Decem Dubitationibus circa Providentiam* q. 1-2 (Boese) 5-8; *de Providentia* (Boese) 63-4. (These last two treatises on Providence were translated into English from Latin by Thomas Taylor in *The Six Books of Proclus, The Platonic Successor* (etc.), London 1816; and *Two Treatises of Proclus, the Platonic Successor*, London 1833 and 1841. Some Greek text has now been assembled by H. Boese, Berlin 1960, and there is a French translation by Daniel Isaac in the Budé series which takes the Greek into account); Ammonius *in Int.* 135,12-137,11. Boethius *Consolation* 5.4 has the general principle; 5.5 applies it to *definite* knowledge; 5.6 to *timeless* knowledge.

[10] Procl. *Prov.* (Boese) 63-4; *de Dec. Dub.* q. 2 (Boese) 6-8; Ammonius *in Int.* 131,4-137,11; Boethius *Consolation* 5.5.

[11] Procl. *Elements* proposition 124; *in Tim.* (Diehl) 1.352, 5-27; *Prov.* (Boese) 63; *de Dec. Dub.* q. 2 (Boese) 7; *in Prm.* (Cousin) 956,10-957,40; Ammonius *in Int.* 133,16-30; 136,15-25.

it, by making the further point, which so far as I know is new, that somebody who *sees* your action, but not in advance, in no way restricts your freedom.

Interestingly enough, the idea that God has timeless knowledge of temporal things is already expressed, if less crisply, by Augustine[12] before the time of Proclus, and before that is at least hinted at by Plotinus and Philo.[13] Moreover, in a passage to be quoted below, Augustine makes the point which Boethius borrows, that we must ascribe to God knowledge (*scientia*), not foreknowledge (*praescientia*).[14] But Augustine does not apply the point to the question of human freedom, and in general his discussion of how to square our freedom with God's knowledge of the future is much less satisfactory than that of Boethius.[15] Not only is Boethius' discussion of the issue philosophically superior, but it has a unique poignancy. For he wrote his *Consolation of Philosophy*, which has consoled so many others,[16] in prison, awaiting execution on a charge which he denied, at about the age of forty-four, with his task of bringing Greek thought to the Latin world incomplete. This is the context in which he considers whether is is the subject of ill fortune or of divine providence, and, if of the latter, whether that implies determinism. The passage immediately follows the one translated in Chapter 8:

> Since, therefore, every judgment comprehends its subject in accordance with its own nature, and God's state is always eternal and present, his knowledge too stays fast, transcending every temporal motion, in the simplicity of his presentness. Embracing the infinite lengths of past and future, it considers everything as if it were going on now in a simple mode of awareness. So, if you want to weigh the presentness with which he discerns everything, you will more rightly judge it to be not a foreknowledge (*praescientia*) as of the future, but the knowledge (*scientia*) of a never failing instant. Hence it is called pro-vidence (*providentia*) rather than pre-vision (*praevidentia*), because it looks forth from a position far removed from things below as if from the highest summit of things. Why, therefore, do you insist that things illuminated by the divine light should become necessary, when not even men make things necessary when they see them? Does your gaze add any necessity to the things you see present?

Timeless versus changeless agency

My claim has been that the idea of a timeless God is by no means pointless; but is it coherent? The idea of a timeless cause has not been confined to theology. It has been a recurrent view of philosophers that time is a mere phenomenon causally produced by a more fundamental timeless reality, and

---

[12] *Conf.* XI.8; *City* XI.21.

[13] Plotinus is talking of the World Soul, when he says 4.4.12(18-32) that it knows the future, and a little later, 4.4.15-16, that its knowledge is timeless. For Philo see *Quod Deus* ... 6.29-32. These passages will be further discussed below.

[14] Aug. *ad Simplicianum* II.2.2.

[15] His treatment of this subject is surveyed by Joseph van Gerven, 'Liberté humaine et prescience divine d'après Saint Augustin', *Rev. Phil. de Louvain* 55, 1957, 317-30.

[16] There were translations by King Alfred (ed. Sedgefield 1899 and 1900), Chaucer (ed. Skeat, in *Chaucer's Complete Works* vol. 2, Oxford 1894) and Queen Elizabeth I, although that by Alfred is more of a paraphrase adapted for his own purposes.

there is a recent example of this view in contemporary physics.[17] None the less, the idea of a timeless God has come under heavy attack. Many people have objected that a timeless being could not *act*.[18] But some of the objections would prove too much: they would, if sound, exclude the idea of temporal, but changeless agency, and that can no longer be so lightly dismissed after the defences canvassed in Chapter 15. If timeless agency is to be ruled out, we may need to find an objection which turns on the special character of timelessness, as opposed to changeless temporality.

This is not the usual view. J.R. Lucas, for example, takes it that a *changeless* God could no more act than a *timeless* God.[19] Nelson Pike argues that a timeless being could not respond, because responses are located in time *after* that to which they are responses.[20] This argument appears to overlook the possibility envisaged by Origen, that a temporal, though changeless, God might *changelessly* will a prayer followed by an outcome 'in response'. Such a God would not have to take action *after* the event. I am not sure either whether to agree with W. Kneale's argument that 'to act purposefully is to act with thought of what will come about after the beginning of the action'.[21] Could a powerful enough being bring it about through his changeless will that there be a sequence of A after B, without thinking of this sequence as present, past, or future, and without performing an act which has any beginning?

Objection to timeless will

Despite these *caveats*, there is an objection to the idea of timeless agency, to which I, at least, can see no answer. For it is not clear what it can mean to speak of a *will* which does not exist at any times. Of course in ascribing a will, or anything else, to God, we should be prepared to find that the ascription will not have quite the same implications as an ascription to humans. That has already received concrete illustration from the discussion in Chapter 15 of the very different way in which a changeless will would have to operate. But there must be some similarity to human will, if there is to be any point in using the word 'will' at all in application to God. And it is hard to see how something which existed at no times could begin to have the requisite similarity. The consequence, if a timeless God cannot will, is that he cannot act either. Here is the specific objection to timeless agency which does not turn on its changelessness.

The objection generalised to other attributes

The difficulty can be generalised. For it is hard to see the sense of ascribing to

---

[17] For a survey of such views, see Milic Čapek, *The Philosophical Impact of Contemporary Physics*, Princeton 1961, 164; and for the recent example, David Bohm, *Wholeness and the Implicate Order*, London 1980, 210-12.

[18] References in Chapter 9.

[19] Lucas, op. cit., 302.

[20] Pike, op. cit., 128.

[21] Kneale, op. cit., 99.

God such other attributes as knowledge, thought, judgment, love, compassion, or any form of consciousness, if he is timeless, although I have described (Chapters 8 and 11), in connexion with Plotinus, how his metaphysics could lead him to *suppose* timeless thought possible.

If the idea of timeless consciousness is incoherent, this will be the clearest objection to a timeless God. But if we can imagine this fundamental objection waived, then some very much more subtle and interesting objections will need to be considered. They concern God's knowledge and power, and they suggest that some knowledge and some power will be unavailable to God, equally whether he is timeless, or temporal but changeless. They will not seek to deprive such a God completely of knowledge or power, but they will imply that he cannot be omniscient or omnipotent.

Does timeless knowledge exclude omniscience?

The first objection is that, if a timeless God can have knowledge at all, there will still be a class of propositions which he cannot know, namely, those which are expressed by means of the 'flowing' or 'token-reflexive' time words.[22] For the special feature of these time words is that they date things by reference to the thinker's own location in time (which is why Russell called them 'egocentric'), and on the present hypothesis God would have no location in time. Thus, to take an example, God could not judge about us, 'Their hour of need is *now*,' for to do so would be to date our hour of need by reference to his *own* location in time. This difficulty would not interfere with his judging, 'Their hour of need is on 1 December 1978,' for that judgment simply dates one circumstance (the hour of need) by reference to *another* (the birth of Christ). But God could not add the thought, '1 December 1978 is *today*.'

There may be several objections to this claim. First, someone may protest that such token-reflexive expressions as 'now', 'today', or the tenses could enter his thought within *quoted* speech, as follows: 'They judge (tenseless), "Our hour of need is now".' The reply is that God still could not know whether this human judgment occurred in the past or future, for that would be to date it by some temporal location of his *own*.

A second objection would be that, just because God has no temporal location, there is no fact to be known corresponding to his saying, 'Their hour of need is now.' That is true, but it is also true that there is a fact to be known corresponding to *our* saying, 'Our hour of need is now,' and he cannot know it.

The standard objection[23] is that the very *same* fact which we express with token-reflexive words like 'now' could be expressed by a timeless being

---

[22] In recent philosophy, this point has been made by A.N. Prior, 'The formalities of omniscience', *Philosophy* 1962 (repr, in his *Papers on Time and Tense*, Oxford 1968, ch. 3, esp. p. 29); Norman Kretzmann, 'Omniscience and immutability', *JP* 63, 1966, 409-21; Nicholas Wolterstorff, 'God everlasting' in Orlebeke and Smedes (eds), *God and the Good*, Grand Rapids, Michigan, 1975.

[23] Pike, op. cit., ch. 5, summarised p. 95; similarly Norman Kretzmann (retracting an earlier view) with Eleonore Stump in their 'Eternity', *JP* 78, 1981, 455-8.

without his using any token-reflexive expressions at all. Unlike Nelson Pike, who is one of the most prominent exponents of this objection, I treat the *tenses* as token-reflexive.[24] So I presume that the timeless being's judgment would have to take a form like that indicated above: 'Their hour of need is (tenseless) on 1 December 1978.' And I have already argued in Chapter 9 (with the example of the hand grenade) that these words would not express the *same* fact as we would express, if on 1 December 1978 we said, 'Our hour of need is now.' A symptom of the difference was that for humans one utterance would lack the action- and emotion-guiding force of the other.[25] This difference is not expunged by the fact, which will become clearer shortly, that for God the situation as regards action- and emotion-guiding force is entirely different.

A fourth objection which was raised in a subtle paper by Castaneda, makes play with the word 'then'.[26] Adapted to the present context, it would maintain that God could know that, on 1 December 1978, we know that our hour of need is *then*. And that will mean that God also knows that our hour of need is *then*, so that it sounds as if he knows exactly the same proposition as we know. This illustrates what is alleged to be a general principle, that if *A* knows that *B* knows a proposition, then *A* knows that same proposition, whether or not he can express it in the same words as *B*. The advantage of introducing the word 'then' is that, even though it is not used in this context in a token-reflexive manner, it has an especially close connexion with certain token-reflexives. For we could, on 1 December 1978, express what we know by saying, 'Our hour of need is *now*.' Admittedly, God cannot express what he knows in the same way, but the general principle proposed above asks us to acknowledge that he knows the same proposition as we can express through the token-reflexive 'now'.

My difficulty is that I think the proposed general principle false. I agree with the claim made above in connexion with 1 December 1978 that God knows that our hour of need is *then*, but only if this is taken as meaning that he knows our hour of need is on 1 December 1978. If it is so taken, then I further agree that he knows the same proposition as we know. But what about the special proposition which we know on 1 December 1978, as opposed to all other dates? I would not agree that he knows that proposition, for it is one which we can express by saying, 'Our hour of need is now.' And I have reasons (to do with action- and emotion-guiding force) for thinking that that is not the same as the proposition which God knows, that our hour of need is on 1 December 1978. Consequently, this gives me reason for thinking that Castaneda's proposed general principle is false, unless some independent argument can be found to show that it is true.

I believe, then, that if a timeless God could have knowledge at all, he could

---

[24] Nelson Pike treats them otherwise, op. cit., p. 88.

[25] Arthur Prior uses something like this point in his denial that a timeless God can be omniscient, loc. cit.

[26] H.-N. Castaneda, 'Omniscience and indexical reference', *JP* 64, 1967, 203-10, endorsed by Swinburne, op. cit., 162-7. I have adapted Castaneda's discussion, which was originally conducted not in terms of God's timeless knowledge, but in terms of his changeless knowledge and his knowledge of other persons.

not be omniscient. On the other hand, although any limitation on God's knowledge may seem startling for orthodox Christian belief, I am not sure that this limitation is one that matters.[27] At any rate, it is not one that interferes with his ability to intervene in the world. For, on the view that he is changeless, it is not supposed that he decides when to intervene by thinking, 'Now is the time.' Rather, he intervenes in the way described in Origen's treatise *On Prayer*, through having a changeless plan for a sequence of changes. In his response to prayer, for example, he changelessly knows and intends the sequence danger, prayer, response. But in intending this sequence he does not need to think of it in token-reflexive terms. Perhaps the conclusion should be, then, that strict omniscience is not *needed* by God.

The Islamic objection: does changeless knowledge exclude omniscience?

It was argued by certain Islamic thinkers – and the argument has been reinvented by Norman Kretzmann – that omniscience is impossible equally for a temporal, though changeless, God.[28] The argument is ascribed by Ghazālī to the so-called 'philosophers'. They allowed God to know various temporal relations involved in the theory of eclipses, but not to judge about an individual eclipse that it was past, present, or future. The reason given is an interesting one, namely, that if God made judgments of the latter type, he would have to *change* from thinking 'The eclipse is present,' to thinking 'It is past.' That is not all: the philosophers considered the objection,[29] closely analogous to one raised in modern discussions, that God might after all be said to have *one and the same* knowledge, first of the future event, then of the present event, and finally of the past event. They rightly reply that this is not so. For the thing known is different, and even if it were not, the knower's relation to the thing known would change, and hence so would his state of knowledge. Ghazālī vainly tries to reply that the change would not be in God, but only in the thing known, so that it would be like the case of something moving from your right to your left, without your changing at all.

The original argument that God cannot know whether an eclipse is past, present, or future is found in Avicenna, from whom van den Bergh translates a relevant extract. He adds that besides Avicenna and the 'philosophers'

---

[27] I am here fulfilling a promise to correct my brief remarks in *Necessity, Cause and Blame*, 126.

[28] Ghazālī *Tahāfut al-Falāsifa*, incorporated in Averroes, *Tahāfut al-Tahāfut* (Bouyges) pp. 456-7, tr. van den Bergh pp. 275-6. Kretzmann, 'Omniscience and immutability', 409-21, mentions the Islamic original. I would dissent only from his suggestion that the argument would have been unavailable to mediaeval Christian philosophers. For it is given by Ghazālī also in what he intended as a companion to the *Tahāfut* (or *Destruction of the Philosophers*) namely, his *Intentions of the Philosophers*. This was translated into Latin in the twelfth century by Gundissalinus as the *Metaphysics* of Ghazālī, and treated (wrongly) as representing Ghazālī's own views. The argument appears on pp. 71-3 of J.T. Muckle's edition of the Latin: *Algazel's Metaphysics*, Toronto 1933. I do not agree with Kretzmann's subsequent retraction of his argument in Stump and Kretzmann, 'Eternity', 455-8. The crucial claim (number 3b', p. 457) is that an eternal God can know the dating of a time and the time's being experienced as present. I would accept this only in a very restricted sense, and not in the sense which I think is implied by the further formulation (number 3b″, p. 457) that he can know which events are actually happening now.

[29] Averroes *Tahāfut al-Tahāfut* (Bouyges) pp. 458-9, tr. van den Bergh p. 278.

some of the Mu'tazilite theologians also denied God's knowledge of past, present and future for exactly the same reason, namely, in order to avoid his changing.[30] The controversy was still raging in the time of Thomas Aquinas.[31]

Is the argument sound? I think that it is, and hence that the omniscience of a changeless God can be defended only by allowing him to change in knowledge in the way just discussed and by refusing to count this as a relevant kind of change.

The reply to Ammonius on change in judgments of past, present and future

The Islamic argument reads like a continuation of, and a reply to, a late Neoplatonist discussion which I have already mentioned, which took place in Iamblichus, Proclus and Ammonius, followed by Boethius. Proclus, going beyond Plotinus and in opposition to the followers of Aristotle, insists that divine providence involves the gods having definite knowledge of *everything*.[32] Of the problems which this raises, one has already been discussed: how can divine foreknowledge be squared with the contingency of future events? But another, which had already been hinted at by Aristotle,[33] is this: how can the gods know things that *change* without themselves changing, at least in their own minds? Proclus repeatedly exploits Iamblichus' distinction between the status of the knowledge and that of the thing known, so as to say that the gods can have *changeless* knowledge of the *changeable*.[34] Ammonius' special contribution[35] is to stress that for such *changeless* knowers nothing can be past or future. For if it were, the knowers would have to change in their minds. None the less, Ammonius insists, the gods do know all that is past and future. Only they do so in the way appropriate to themselves, that is, changelessly, the objects being presented to them in a single eternal now.

It is at this point that the Islamic argument fits on; for the Islamic 'philosophers' agree that normal thinkers have to change in their thoughts when they think in terms of past and future. And they further agree that consequently God, who is changeless, sees nothing as past or future. But their new move is to insist, against the whole intention of Proclus and

---

[30] S. van den Bergh, vol. 2, pp. 151; 155-6, of his translation with commentary of Averroes' *Tahāfut al-Tahāfut*. He refers to Avicenna's *Shifa (Healing)* and *Najāt (Cure)*. For the Mu'tazilites, he cites Shahrastānī *Religions and Philosophical Sects*, ed. Cureton p. 60, on Jahm, and Ibn Ḥazm *On Religions and Philosophical Sects*, 2, 130, on the Mu'tazilites generally.

[31] Aquinas *ST* 1a, q. 14, a. 15, objection 3 and reply, reports the nominalists (Abelard and Peter Lomard?) as holding that the *same* thing is signified by the differently tensed sentences 'Christ is born', 'Christ will be born', and 'Christ was born', and hence that God's knowledge of the nativity does not change. Thomas denies that the same thing is signified, but none the less claims that God knows, without himself changing, that these enunciations change their truth value.

[32] Procl. *Prov.* chs 62-5; *de Dec. Dub.* q. 1, chs 2-5; *Platonic Theology* 1.21; *in Prm.* (Cousin) 958, 30-961,18.

[33] Aristotle made God confine his thoughts to one and the same thing, so as to avoid mental change, *Metaph.* 12.9 1074b21-7.

[34] Procl. *Elements* proposition 124; *in Tim.* (Diehl) 1.352, 5-27; *de Dec. Dub.* q. 2, chs 7-8.

[35] To be translated below.

Ammonius, that there is therefore something which God does not know. I have argued that they are right.

It is worth quoting the two related passages in Ammonius, partly because they are among the many in late Neoplatonism that have not, to my knowledge, been translated, and partly because, although Boethius mentions the idea that God's knowledge of the future must be *unchanging*,[36] he does not stress it so strongly as Ammonius. He is more concerned with God's knowledge being *definite*, that is, not merely knowledge that something *may or may not* happen.[37] Even in Ammonius, the discussion starts with the point about definiteness:

> The gods cannot be ignorant of our affairs, nor can their knowledge be indefinite, so that they guess how things are going to turn out. For, first, as Timaeus taught us and as Aristotle himself affirms in his theology, and before them Parmenides, not only Plato's Parmenides but Parmenides speaking in his own words:[38] with the gods nothing is either past or future, at least if each of these lacks being, the one being no longer and the other not yet, the one having changed and the other being of a nature to change; and if there is no way to connect things like that with things that really have being and that do not admit of change even in their thought. For what is altogether unchangeable must take precedence over what changes in any way, in order that the latter may even remain in a state of change. So with gods who are related to things as their source, there is no possibility of the past or future being contemplated. Everything with them is situated in the single, eternal now.[39]

The intended conclusion is that the gods know future events *definitely*, because they do not see them *as* future. This does not, however, make the events themselves definite and determined, since, as we have seen, Ammonius denies that knowledge need have the same status as the known. It is now that this last denial is applied to *changeability*: although things past or future are subject to temporal *flux*, the gods' knowledge of them does not have to change:

> Since this is so, we must say that the gods know all that has happened, or is, or will be, or is about to be, in the way that is appropriate to gods, that is, with a single, definite and unchanging knowledge. ...

> For they must know divisible things in a way which is indivisible and unextended, and pluralised things unitarily, and temporal things eternally and generated things ungeneratedly. For we shall not allow anyone to say that the knowledge of the gods runs along with the flow of things nor that anything with them is past or future. Nor shall we allow that 'was' and 'will be' are used among them, words which, as we have heard from Plato's *Timaeus*, signify some change.[40]

[36] He touches on it, glancingly, in the passage from *Consolation* 5.6 translated earlier in this chapter.

[37] Boethius illustrates such indefinite knowledge with the example of Teiresias telling Odysseus that he and his crew either will or will not kill the sacred bulls of Helios, *Consolation* 5.3.

[38] These historical attributions need not be taken as accurate: see Chapter 8.

[39] *in Int*. 133, 15-27.

[40] *in Int*. 136, 1-3 and 15-20 (the last lines were quoted in a different connexion in Chapter 8).

Ammonius anticipated by Augustine and others

I have so far spoken as if these ideas about changeability were the invention of late Neoplatonism. But this is not so; for in Augustine there are two passages that preserve God's changelessness by arguing that he knows the past and future not *as* past or future:[41]

> What is foreknowledge (*praescientia*) but knowledge (*scientia*) of the future? But what is future to God who transcends all times? For if God's knowledge possesses (*habet*) the things themselves, they are not future to him, but present, and so it can no longer be called foreknowledge, but only knowledge. On the other hand, if the future is not yet with him, just as in the order of temporal creatures it is not, and instead he anticipates the future in his knowledge, he will be aware of it twice over, once in accordance with his foreknowledge of the future, and differently in accordance with his knowledge of the present. In that case, something temporal will happen to God's knowledge (which is the most absurd and false supposition), for he cannot know after their arrival the things he foreknew before their arrival, except by noting them twice, that is, by foreknowing them before they occur and by knowing them when they have occurred. In that way there will come about something far removed from the truth, that something temporal happens to God's knowledge. For temporal things which are foreknown will also be sensed when they are present, although they were not sensed before they occurred, but only foreknown. But if nothing new is going to happen to God's knowledge, even upon the arrival of things whose arrival was foreknown, but rather that foreknowledge will remain just as it was before the things foreknown arrived, how can it still be called foreknowledge? For it is not knowledge of the future, since the future which he saw is now present and a little later will be past, and knowledge of the past or present cannot in any way be called foreknowledge. Thus we return to the result that what was foreknowledge, when these things were future, becomes knowledge, when the same things are present. And since that which was previously foreknowledge comes subsequently to be knowledge in God, he suffers mutability and is temporal, even though he is God, who truly and supremely is, mutable in no part, and temporal in no new motion. Therefore it is fitting not to speak of God's foreknowledge (*praescientia*), but only of his knowledge (*scientia*).

> It is not that God's knowledge varies in any way, so as to be affected in one way by things which are not yet, in another by things which already are, and in another by things which have been. For he does not in our manner look forward to what is future, or contemplate what is present, or look back at what is past; but works in some other way removed far above the habits of our thinking. He sees altogether unchangeably, without his thought changing from this to that, in such a way that he comprehends in a stable and sempiternal present all the things that happen in time, whether they are yet to be in the future, or already present, or already over in the past. He does not see them in one way with his eyes and in another with his mind, since he does not consist of mind and body. Nor does he see them in one way now, in another way

---

[41] They are, in chronological order, Aug. *ad Simpl.* II.2.2; and *City* XI.21. Cf. the much earlier statement of Pantainos in the second century A.D. that God knows intelligible things non-intellectually and perceptible things non-perceptually: ap Clementem Alexandrinum, fr. 48 in vol. 3 of Prussian Academy edition, ed. Stāhlin, taken from Maximus Confessor.

beforehand and in another way afterwards. For his knowledge of the three
times, present, past and future, does not change in diversity like ours, since
with him there is no change or shadow of movement.

Augustine's own statement is itself, as so often, an inspired crystallisation
of his predecessors' ideas. For they had repeatedly said that nothing is past
or future to God, that past and future are, or are as if, present to him, and
they had used Augustine's formula about God possessing (*tenere*) past and
future. Statements along these lines are ascribed to the Stoics, and are found
in the first century A.D. in Philo and Seneca. There are varying formulations
in Lucan, in Clement of Alexandria and in Nemesius, as also in Plotinus,
who would in addition have increased everyone's awareness of God's status
as eternal rather than temporal.[42] None the less, these statements are not
applied to the issue of preserving God's changelessness, except in Philo,
Seneca and Plotinus. Nor is it made quite clear that the future's being present
to God involves his not thinking of it *as* future. Philo and Plotinus come closest
to expressing this thought, but others clearly do not have it in mind, and it is
only Augustine who brings it out fully.

I have implied that the foregoing ideas are not yet fully developed in
Plotinus. Mario Mignucci has drawn my attention to a passage in which
Plotinus even denies the later view that God can have knowledge of change
without himself changing,[43] and Proclus implies that this view is not yet to be
found in Porphyry either.[44] Plotinus does take care in many passages to make
divine knowledge changeless,[45] and he does in one of these passages allow
that the World Soul knows how the world will be in the future.[46] But he has
no interest in maintaining that there is divine knowledge of *everything* that
passes from future to past.[47] The souls of the stars do not need to know the
details of their movements,[48] nor does God have knowledge of their periods.[49]
The kind of knowledge needed for creation is a beginningless knowledge of
beginningless creatures,[50] not a knowledge of all that happens. Plotinus
therefore has little incentive to consider how changeless knowledge can be
directed to the passage of time.

Changeless and timeless power

I shall finish with a problem about God's *power*, and how this can be
changeless. For there seems to be a *change* in God's power to prevent

[42] For the Stoics, see Cic. *de Div.* 1.127 and Calcidius *in Tim.* 160 (p. 193 Waszink). See also
Philo, *Quod Deus ...* 6.29-32; Seneca, *Nat. Quaest.* 2.36; Lucan, *de Bello Civili* 5, 177; Clement
Stromateis 7, 37, 5 (cf. 6, 17); Plot. 6.7.1(45-58); Nemesius *de Natura Hominis* ch.44, PG40, 801B. For
discussion of Philo, see Whittaker, *God, Time, Being,* second study, 33-66.
[43] 4.3.25(13-27, esp. 20-4).
[44] *in Tim.* (Diehl) 1.352, 13.
[45] 4.3.25; 4.4.7-13 and 15-16; 6.7.1-3.
[46] 4.4.12(18-32).
[47] So R.T. Wallis, 'Divine omniscience in Plotinus, Proclus and Aquinas', in H.J. Blumenthal
and R.A. Markus, eds, *Neoplatonism and Early Christian Thought,* London 1981, 223-35, esp. 226.
[48] 4.4.8.
[49] 4.3.25(20-4).
[50] 6.7.1(45-58); 6.7.3.

something, in fact a *loss* of power, after the thing has happened. Aristotle quotes the poet Agathon as saying:

> Of this alone even God is deprived: making what has been done not to have happened.[51]

Thomas Aquinas agrees, and cites Augustine and Jerome, as well as Aristotle, in support. In an example which he borrows from Jerome, God is unable, after someone has lost her virginity, to save her from losing it. Yet one would suppose that, before the loss occurred, he was able to save her. So is there not a change and a loss in divine power? Thomas discusses the question more than once,[52] and says that there is not:

> With regard to the denial that someone can do something, I answer that we must say that it can happen from two causes. It can happen either from a defect in power, as a man who does not have the power of vision is said not to be able to see, or on the part of the object which does not have it in its nature to be possible (*non habet rationem possibilis*), as a man with sight is said not to be able to see a sound, since it is not visible. ... Similarly, with regard to someone's not being able to do all that he could once do: we must say that this too can happen from two causes. It can happen either because he loses some power which he had, and that is not applicable to God – indeed, the Master's [Peter Lombard's] solution goes that way in the source book – or because of a change in the object, which loses the status it previously had of a thing in whose nature it was to be possible (for an active power is active in respect of some doing). Hence when something is already determined, as when it is actually present, or has become past, it no longer has it in its nature to be possible, and so it is said that God cannot bring it about, it being the same thing, but referred to by different assertibles, because the time is different.

> To the second point we must say that God can do everything, so far as concerns the perfection of divine power, but some things do not fall under his power because they fall short of having it in their nature to be possible. Thus if we attend to the changelessness of divine power, God can do whatever he could do before. But some things which once had it in their nature to be possible, while they were waiting to be done, fall short of having it in their nature to be possible, now that they are done. And in this way God is said not to be able to do them, because they are not able to be done.

Thomas is not concerned in these articles to dissociate God from *all* change, but he does want to deny that his power (*potentia*) is diminished. He thinks this denial compatible with the admission that, at least in one way,

---

[51] *NE* 6.2, 1139b10-11. What interests me is that when the past has happened in a certain way (call it *f*), God does not have the power to make it fit the description 'not-*f*'. This provides a contrast with the future; for even if, as a matter of fact, the future is going to happen in way *f*, it does not follow that God lacks the power to make it fit the description 'not-*f*', but only that he is not going to exercise that power.

[52] *Scriptum Super Libros Sententiarum*, in lib. 1, dist. 44, q. 1, art. 4, 'Whether God can do all that he could do once' (the reference to the diminution of his power is at 5 '*sed contra*'; my extract is from the *solutio*); also *ST* 1, q. 25, art. 4, 'Whether God can make the past not to have been' (the reference to the diminution of his power is at 2, and I have translated the reply to 2).

God is not able (*non posse; non potest*) to do what he once could do. The argument that his power is not diminished turns on locating all diminution in the woman rather than in God.

It can be admitted straight away that, when the woman loses her virginity, no change occurs *in* God. To take an analogy, suppose that the measles virus can kill men; it is lethal. It would still perhaps be lethal, even if the human race died out, for that would only deprive the virus of the *opportunity* of killing men. But suppose instead that the human race changes so as to become permanently immune to the virus. That change would have taken place in men, not in the virus, and yet the virus would have lost its former power to kill. So something *about* the virus would have changed, even though the change did not take place *in* the virus. Is it not the same with the loss of virginity? The change admittedly does not take place *in* God, when the woman loses her virginity, and that loss becomes past. None the less, has not something *about* God changed, and has he not lost a power of prevention, albeit though no decline on his part?

There may still be a temptation, because the change does not take place *in* God, to say that what he has lost is an opportunity, not a power. But this temptation must confront the implausibility of saying that God still has the power to save her virginity. One thing which I am inclined to say is that the *woman* at least has lost a power: the power to be saved from losing her virginity. If God has not lost the power of saving her, then we have one of those situations discussed in Chapter 7 in which corresponding active and passive powers can exist one without the other. Perhaps a vessel is capable of receiving water in the desert where water cannot reach, and a virus may remain lethal when there are no beings left to kill. But I doubt if it can remain lethal when beings have become permanently immune, and I doubt equally whether God can retain the power to save a woman who has permanently lost the power to be saved. Even if this is disputed, Thomas has in effect conceded the point which is of immediate concern, that something about God has changed. For although he does not allow us to talk of a diminution of power, he does allow us to say that God is not able to do what he could do once.

The question was discussed by other thinkers, and John of Mirecourt, for one, connected the suggestion that God's power might be *diminished* by the passage of time with the suggestion that his knowledge might be correspondingly *increased*, both of which suggestions he denied.[53] Various solutions were, or might be, canvassed, besides that of Thomas Aquinas. For example, it might be denied that God ever had the power to save the woman, but that solution would dramatically reduce the range of God's powers. At the opposite extreme, a number of authors maintained that it was logically possible for God (of his 'absolute' power, rather than of his 'ordained' power) to make the past not to have been. This view has been credited to Peter Damian, Gilbert de la Porrée, William of Auxerre, Thomas Bradwardine and Gregory of Rimini,[54] and it would provide a solution to our problem, but

---

[53] *Lectura Sententiarum* lib. 1, q. 39, quoted on p. 250 of William J. Courtenay, 'John of Mirecourt and Gregory of Rimini on whether God can undo the past', *Recherches de Théologie Ancienne et Médiévale* 39, 1972, 224-56, continued in 40, 1973, 147-74.

[54] Courtenay, op. cit.

only a highly paradoxical one. It may be wondered whether a change in God's powers could more easily be avoided, if those powers were viewed as *timeless*. But the very idea of a power being timeless is called into question, if it ceases to be possible for the power to have its effect. None the less, an approach of this type is advocated by Thomas Aquinas in connexion with one particular case. If God has predestined Peter to be saved, then it is an irrevocable fact that he has predestined him. Someone cannot be predestined at first, but not predestined later. Does this not mean that God is no longer able *not* to predestine Peter? Thomas tries to reply by drawing a now familiar distinction:

> To the fourth point we must say that the act of predestination is measured by eternity, and does not pass into the past. So it always and unchangingly has it in its nature to be possible, in so far as it proceeds from the liberality of the divine will. But on the part of the effect, it does pass into the past, and accordingly does lose the status of a thing in whose nature it is to be possible.[55]

Resumé

I have argued that there are strong reasons for wanting to think of God as timeless, and that the objections are fewer than is often supposed. Admittedly, if a timeless God had knowledge at all, he would not be omniscient, but that may be equally true of a changeless God, and in any case the restriction need not matter. Admittedly again, there are difficulties about timeless power, but then so too are there about changeless power. However, there are certain respects in which the idea of God's timelessness is more vulnerable than that of his changelessness. In particular, there are serious objections to ascribing to a timeless God such attributes as will, knowledge, thought, judgment, love and compassion.

---

[55] *Scriptum Super Libros Sententiarum*, loc. cit., *ad* 4.

CHAPTER SEVENTEEN

Plato and Aristotle on the Beginning of Things

It will be clear from the last three chapters how much influence Plato and Aristotle exercised over the subsequent debate on creation. I cannot, in a short space, give an account of the whole of that debate, but it is worth outlining the contributions of Plato and Aristotle. And that will incidentally reveal some of the other main lines of argument in the ensuing centuries. The discussion will be historical in this chapter, and readers whose interest is more on the philosophical side may prefer to pass on.

The dispute over Plato's Timaeus

Plato's meaning in the *Timaeus* has been the subject of endless dispute. But he warns us, at *Tim.* 29C-D, that his account of the physical world (as opposed to the eternal) is only a likely story, and that there may be inconsistencies in it. So it is not surprising that different interpretations of his story soon developed. I shall divide the main interpretations into three groups instead of the usual two.[1] First, time began together with the ordered *kosmos*, and there was nothing before that. Secondly, *orderly* time began together with the ordered *kosmos*, but before that there was *disorderly* matter, motion and time. Thirdly, nothing began, the talk of beginning being merely a metaphor.

There are not many who support the first interpretation quite unequivocally. But the person who is most enthusiastic about it is the Christian Philoponus in the sixth century A.D. For this interpretation enables him to represent Plato as endorsing the Christian belief in a beginning of the physical universe. He regards Aristotle as his ally in this interpretation, and he attacks Proclus' denial that Plato intended a beginning.[2]

As regards Plato's ascription of a beginning to *time*, Philoponus reports this in *de Aeternitate Mundi contra Proclum* (Rabe) pp. 117-118 and 141-2. It is here

---

[1] The first two interpretations are commonly amalgamated. The fullest account of interpretations (but excluding Christian ones) is given by M. Baltes, *Die Weltentstehung des platonischen Timaios nach den antiken Interpreten*, 2 vols, Leiden 1976.

[2] Philoponus *Aet.* (Rabe), pp. 135-242. The first and third interpretations agree that the *Timaeus* account of chaos before God organised the *kosmos* is a fiction. So when Philoponus calls it a fiction in his earlier work, the commentary on Aristotle's *Physics*, 575, 7-11, this need not be a sign that at that stage he preferred the third interpretation (Proclus') to the first.

that equivocation may appear to enter, for in the midst of reporting the beginning of time, he *seems* to assume that the creation of time is after all preceded by a *before* (*prin*, 117,22; 141,18), and that whatever comes into being, time included, is non-existent *previously* (*proteron*, 142,1). But I do not think that this is really a weakening of his interpretation, because of his views on words like 'before' and 'previously'. He believes that these temporal words can be used in non-temporal senses.

This view emerges when he tries to answer Proclus, who uses the old Aristotelian argument that time cannot have a beginning. Otherwise you would be forced into saying that there was a *time when* (*ēn pote hote*) there was no time. Philoponus' reply is that it is perfectly all right to say that time does not *always* exist, and to talk about *when* (*pote*) it does not exist.[3] You can even use the formula which Proclus condemned: 'There was a *time when* there was no time.'[4] All you need to do is to give the temporal words 'always' and 'when' non-temporal senses. Simplicius further tells us that Philoponus thinks you can say things like, 'time was non-existent before it came into being', or 'time was not always', without the 'was', 'before', or the 'always' referring to a time.[5] Philoponus confirms all these points in his commentary on the *Physics*.[6]

This should remove the uncertainty about Philoponus' intention in *de Aeternitate Mundi*, pp. 117-118 and 141-2. We need not doubt that he is ascribing to Plato an absolute beginning of time, since the unexpected temporal words would be understood by him in non-temporal senses. None the less, it must be admitted that the point which most engages him is the beginning, not of time, but of the *kosmos*. It is because they concede this that he is prepared to leave Plutarch and Atticus undiscussed,[7] even though in vital respects their interpretation of Plato differs from his.

Philoponus is justified in claiming Aristotle's support. For Aristotle takes Plato to mean that the heavens and the *kosmos* began,[8] and that time began with the heavens.[9] Moreover, he take the beginning of time quite literally, and attacks it as impossible. On the other hand, his account is not quite so clear-cut as this might suggest. For he also records the incompatible idea that there was disorderly motion *before* the creation.[10] Like Plato himself, he allows these two ideas to be recorded, without any attempt to adjust one to the other.

Aristotle's pupil Eudemus of Rhodes makes the incompatibility a matter of explicit complaint.[11] Since there was disorderly motion before the heavens were created, Plato ought to have said there was time, but in fact he denied this.

[3] *Aet.* (Rabe) 104,21-107,17 ('always') is supplied by Rabe at 105,7 and can be supported by the analogy of 106,22; 107,10 and 12).
[4] *Aet.* 116, 1-24.
[5] ap. Simpl. *in Phys.* 1158, 31-5.
[6] *in Phys.* 456,17-458,16.
[7] *Aet.* (Rabe) 211, 11; 519, 23.
[8] Arist. *Cael.* 1.10, 280a28-32; 3.2, 300b16-18; *Metaph.* 12.6, 1071b37-1072a3.
[9] *Phys.* 8.1, 251b17-19.
[10] *Cael.* 3.2, 300b16-18; *Metaph.* 12.6, 1071b37-1072a3.
[11] Eudemus ap. Simpl. *in Phys.* 702, 24 (= Eudemus fr. 82b Wehrli).

Another person who records the beginning of time in Plato is Philo
Judaeus in the first century A.D. But he shields Plato from the charge of self
contradiction, since in the same passage he suggests that matter did not pre-
exist the ordering of the world for Plato, but was brought forth together with
that order (*cum ornatu ipso*). This is the interpretation for which I argued in
Chapter 13.[12]

Others who record the beginning of time follow Aristotle in recording also
the conflicting ideas. Diogenes Laertius in the third century A.D. certainly
belongs to this group,[13] and so perhaps does Cicero in the first century B.C.[14]

The second interpretation of Plato seeks to resolve the apparent conflicts
recorded by most of these authors, by saying that there is one kind of time
which began, but that it was preceded by another kind of time. The idea of
two kinds of time is already found in the mouth of Velleius, spokesman for
the Epicureans in the *de Natura Deorum* of Cicero, written in the first century
B.C.,[15] although Velleius, like Eudemus, is *attacking* Plato, rather than
interpreting him. He admits that days, nights and years can be described in
Plato's manner as created along with the heavens. But he protests that Plato
ought to accept that something else passed before the creation. The
something else can be called ages (*saecla*). There would have been an
unmeasured eternity from infinite time, which can best be understood as a
sort of extension (*spatium*).

Something like this is taken up, not as a criticism of Plato, but as an
interpretation of what he actually meant, by the two Middle Platonists
Plutarch (*c.* A.D. 46-after 120) and Atticus (*c.* A.D. 150-200).[16] On this
construction, *orderly* time was created along with the heavens, but it was
preceded by disorderly time. Plutarch offers parallel solutions for other
discrepancies in Plato about whether soul or matter were created.[17] There
were two souls: one kept things in disorderly, the other, coming later, in
orderly motion. Equally, matter exists in two forms, an earlier disorderly,
and a later orderly one.

There is a little doubt whether the sources are accurate in their account of
Plutarch's view on time. For in *Quaestiones Platonicae* 1007C, Plutarch offers a
more nuanced interpretation. There was not *time* before the creation, but
only a disorderly movement which was, as it were, the *matter* (*hulē*) of time.[18]
He none the less uses temporal words in relation to this disorderly movement
('was', 'before').

---

[12] Philo *Prov.* 1.20 and 22 Aucher. Otherwise, Plato's view will not align with Philo's in 1.7.

[13] Diog. Laert. *Lives* 3 (Plato) 73; 76-7.

[14] Cicero translated the *Timaeus*, and the surviving fragment includes both evidence for the
beginning of time and of all perceptible things and evidence for God pre-existing the orderly
kosmos and taking over disorderly matter. In his other writings, Cicero supports the idea that
Plato gave the orderly *kosmos* a beginning (*Tusc.* 1.53; 1.70), and had it made out of matter (*Acad.*
2.118).

[15] *de Natura Deorum* 1.21.

[16] Plutarch and Atticus, ap. Procl. *in Tim.* (Diehl) 1.276,31-277,7; 3.37,7-38,12; cf. 1.283, 27-
285,7; 1.381,26-384,5; and ap. Philop. *Aet.* (Rabe) 211, 11; 519, 23; Atticus ap. Aeneam
Gazaensem *Theophrastus* (Colonna) 46, 16.

[17] *de Proc. An. in Tim.* 1014A-1017A.

[18] Cf. Arist. *Phys.* 4.14, 223a27.

There are traces of the disorderly, and subsequent orderly, time in Augustine.[19] And there are signs that Atticus' contemporary Galen may have followed this second interpretation.[20]

The third interpretation, that Plato intended no beginning either for time or for the *kosmos*, was already known to Aristotle.[21] Indeed, it may have been devised to protect Plato from Aristotle's criticisms. Some of the criticisms were redirected to show, not that Plato was wrong, but that the Aristotelian interpretation of him must be mistaken. According to the third interpretation, Plato spoke of a beginning only for paedagogic purposes, as a mathematician uses diagrams, to make things easier to understand. This is the view taken probably by Speusippus, Plato's successor as head of the Academy, and certainly by Xenocrates, the third head, and by another early Platonist, Crantor (died *c.* 290 B.C.).[22]

None the less, the third interpretation took some time to gain ground. Aristotle's successor, Theophrastus, said it might 'perhaps' be true, but thought that Plato more probably intended a beginning of the *kosmos*.[23] The interpretation was rejected, however, by all those I have so far mentioned, and by many others besides.[24] Among others to reject it I would mention one whose interpretation is too idiosyncratic to fit into the threefold classification above. Thus in the second century A.D. the Middle Platonist Severus interpreted Plato's view of time in the light, not of the *Timaeus*, but of the *Statesman*, where the *kosmos* is said to revolve in alternating directions. Severus supposed that a new *kosmos* was envisaged by Plato for each new revolution, but that in another sense Plato's *kosmos* was uncreated, since there was no beginning to the series of revolutions.[25]

Despite opposition, the metaphorical interpretation of Plato was embraced by Eudorus of Alexandria (fl. *c.* 25 B.C.).[26] Thereafter, although his Middle Platonist successors were divided, the majority[27] accepted this interpretation. And then with Plotinus (A.D. 205-269/70) and Neoplatonism, nearly all

[19] *City* XII.16.
[20] Galen *Compendium Timaei* 4, ll. 1-13, in *Plato Arabus* ed. Kraus-Walzer, has an eternal precosmic motion of matter with a disorderly soul, as in Plutarch.
[21] *Cael.* 1.10, 279b32-280a1.
[22] Xenocrates and Crantor are named by Plutarch *de Proc. An.* 1013A; Speusippus in a scholium (Cod. Paris. 1853(E), p. 489a Brandis = Speusippus fr. 54b Lang).
[23] ap. Philop. *Aet.* (Rabe) 145, 20; 188, 9 (= Diels 485, 17).
[24] Other rejecters include the Aristotelian Theodorus of Soloi (judging from Plutarch, *de Def. Or.* 427A-E); Atticus' pupil Harpocration, ap. Scholia cod. Vat. f. 34r, in Kroll (ed.), *Proclus in Platonis Rem Publicam*, vol. 2, Leipzig 1901, p. 377, ll. 15-23; Numenius ap. Calcidium *in Tim.* 295, p. 297, 7-16 (Waszink); Alexander of Aphrodisias, ap. Simpl. *in Phys.* 1121, 28 and ap. Philop. *Aet.* (Rabe) 211,26-225,12. Zacharias of Mitylene takes over Philo's interpretation of Plato, according to which, before the creation of the physical world, God created an intelligible world, *Ammonius* or *de Mundi Opificio*, PG 85, 1120A.
[25] ap. Procl. *in Tim.* (Diehl) 1.289, 7-13; 2.95, 29-96,1.
[26] ap. Plut. *de Proc. An.* 1013B.
[27] Plut. op. cit. 1013A-1014A. See Baltes for references to pseudo-Timaeus Locrus, unnamed authors in Philo, the Ammonius who taught Plutarch, Aëtius, Albinus and Apuleius. That Albinus belongs here is suggested by his *Didaskalikos* ch. 14 (in vol. 6 of C.F. Hermann's edition of Plato 169, 26 ff.) and by Procl. *in Tim.* (Diehl) 1.219, 2. But because his chs 12 and 13 use such Platonic phrases as 'before (*pro*) the coming to be of the heavens' (Hermann p. 167), Albinus' position has been a matter of controversy (e.g. R.E. Witt, *Albinus and the History of Middle*

Platonists endorsed it.[28] We have thereafter to look to Christians like Philoponus or Zacharias for other interpretations.

The Timaeus text

My own belief is that, as Plato himself warned, he has not been consistent, and that consequently none of the interpretations fits everything he says. None the less, one will appear closer than the others, when we consider the text of the *Timaeus*. At first sight, the creation of time, described in *Tim.* 37D-40C,[29] would seem to imply a creation *ex nihilo*, especially as Plato says in 37D6-E4 not only that days, nights, months and years came into being, but that so also did 'was' and 'shall be'. Moreover, the days, nights, months and years constitute *parts* of time. The immediately following pages constantly repeat that time began to exist (38B6; 38C2-6; 38E3-39A3; 39C1-D7; 39E3; 40B8-C2). Besides the passages which discuss time, there is another which maintains that the *kosmos*, and anything else which is perceptible, must have a beginning:

> As regards the whole heaven, then, or the *kosmos* (let us give it whatever name would be most acceptable to it), we must begin by considering the question which is recognised as the first to be asked about anything: did it exist always (*aei*), having no origin by which it came into existence (*archē geneseōs*); or has it come into being (*gegonen*) starting from some origin (*ap' archēs tinos arxamenos*)? It has come into being; for it is visible and tangible and has a body. All such things are perceptible, and perceptibles which are apprehended by opinion conjoined with sense perception, are evidently things which go through processes of becoming (*gignomena*) and are generated (*gennēta*).[30]

After the evidence for a creation of time, it is disconcerting to find that some of the passages speak as if there was after all time before the creation. Thus the passage just cited, 28B2-C2, implies that what is created has not existed *always* (28B6), so that there was evidently time *before* its creation. Again, 39E3 talks of what happened *up to* (*mechri*) the coming into being of time. And 38E3-5 speaks as if time had to pass before the heavenly bodies could arrive at the right time-generating movements. Indeed, the whole passage from 31B to 36E speaks as if many things had to be created and arranged by God *before* the creation of time. Although 34B-C warns that the order of exposition is not always the same as the order of the events expounded, since soul was not created *after* body, none the less this remark is not explicitly applied to time, whose creation is presented as if it occurred *after* that of the world's soul and body.

Platonism, Cambridge 1937, 120; J.H. Loenen, 'Albinus' metaphysics: an attempt at rehabilitation', *Mnemosyne* 9, 1956, 301-2; John Whittaker, 'Timaeus 27 D5 ff.', *Phoenix* 23, 1969, 182-4).

[28] Baltes gives references for Plotinus, Porphyry, Iamblichus, Calcidius (a Christian), Macrobius, Hierocles, Sallustius, Proclus and some of the Platonist ideas in Zacharias' dialogue.

[29] Part of this is translated in Chapter 8 above.

[30] 28B2-C2. Cf. 29A5: 'The *kosmos* is the best of the things which have come into being (*gegonotōn*)'.

Even if we do not press the idea that time was one of the later things to be created, we must cope with the passages which suggest that there were things in existence before God created the heavens (which are supposed to be coordinate with time), and before he started any creative activity at all. Thus at 48B3-6 and 52D2-4, Plato says:

> We must consider the actual nature of fire, water, air and earth before (*pro*) the heavens came into being (*genesis*) and the qualities which existed before (*pro*) that. For nobody up to now has explained the coming into being (*genesis*) of these [elements].

> Let this on my verdict be the account given, elaborated as regards its main points: the existence (*einai*) of being, of space and of coming into being, three distinct things, even before (*prin*) the heavens came into being.

There follows an account of how a chaotic state preceded the creation. There were materials already in existence, but subject to chaotic movements, and the creator's task was to instil order. The passage 53A2-B5, and its recapitulation at 69B2-C2, are shot through with words implying that this was an earlier *time*: then (*tote*, three times), when (*hote, hotan*), before (*prin, pro toutou*), and a galaxy of past tenses.

Quite early in the *Timaeus*, at 30A2-6, it is already made clear that God's creative work took the form of imposing order on a pre-existing disorder:

> For god wanted all things to be good, and as far as possible nothing to be poor. So he took over all that was (*ēn*: past tense) visible, which found no rest, but moved disharmoniously and without order; and he brought it (*ēgagen*: past tense) out of disorder into order, thinking order better than disorder in every way.

We may add to the list of entities which existed *before* the creation God himself. For at 34A8-B1, the divine creator is contrasted with the divine world which he created in terms which suggest that he existed *before* his creation:

> All this, then, was the plan of the god who exists always (*ontos aei*) for the god that was to exist at some time (*pote*).

Things are just as bad if we turn to works other than the *Timaeus*. There too there is conflicting evidence.[31] In the face of all this conflict, I think the

---

[31] Thus *Rep.* 499C8 talks of *infinite* past time; but *Theaet.* 175A says only that every man has countless myriads of ancestors; while *Laws* 781E-782A offers alternatives: either the human race has no beginning, or it has existed for an enormous length of time. *Phdr.* 245C5-246A2 (cf. *Laws* 10.894Eff.) argues that all motion is ultimately due to *soul*, so that presumably the disorderly motion of the *Timaeus* cannot after all have occurred *before* soul was created. The same *Phaedrus* passage further maintains that, as a first principle of motion, soul cannot have come into existence, a statement which seems to conflict with the *Timaeus* account of its creation. A further apparent discrepancy arises from *Tim.* 38C-39D, where Plato makes the heavenly motions the very standards of temporal duration and hence (presumably) of speed; yet in *Rep.* 529D he declares that, since the heavenly bodies belong to the perceptible world, their velocities are not quite regular.

interpretation of Plutarch and Atticus is the one to which Plato could shift with least disturbance to his account. That is the interpretation according to which there are two kinds of time, and correspondingly two kinds of soul and of matter.[32] Even so, I doubt if this is what Plato actually said. For, apart from other difficulties, he does not say that time already existed in another form, but simply that time came into being. For another thing, it is not only days, nights, months and years (that is, the measured amounts of time) which came into being upon the creation of the heavens, but also 'was' and 'shall be' (*Tim.* 37E3-4). And further we are told that all perceptible matter must come into existence (28B2-C2), which ought to mean that even chaotic matter did so.

In some respects, my conclusion is close to that of Gregory Vlastos, who agrees that Plato's easiest course would be to distinguish orderly and disorderly time, and yet that he does not do so explicitly.[33] But there are certain features of Vlastos' interpretation which I would not accept. First, he maintains that no one in antiquity appreciated this distinction,[34] which seems unfair to 'Velleius', Plutarch, Atticus and the others who have been discussed above. Secondly, he does not agree that the interpretation which I have been associating with Plutarch and Atticus is in any way impeded by Plato's assertion that 'was' and 'shall be' have come into being as characteristics of created time,[35] nor yet by Plato's claim that anything perceptible must have a beginning.[36] In Vlastos' view, the creation of 'was' and 'shall be' need not imply that the difference between 'was' and 'shall be' began with orderly time; it might be just a fresh *instance* of 'was' and 'shall be' that began then. However, I cannot myself see the demiurge as creating a *fresh* instance of 'was' and 'shall be', if, as would be agreed, his role is simply to impose order on a pre-existing temporal succession. As regards Plato's other claim, that anything perceptible must have a beginning, Vlastos questions whether matter in its original chaotic state was in any straightforward sense *perceptible*, and he shows that Plato's cursory reference to chaotic matter as *visible* at 30A is not on its own conclusive. But the main account of chaotic matter at 52D-53C is so sensuous that I am left in no doubt, and it includes the statement at 52E1 that space had every sort of diverse appearance to the sight (*pantodapēn idein phainesthai*).

The hardest of the three interpretations to defend, I believe, is the metaphorical one. It encountered two particularly difficult passages, and invented two devices for dealing with them: an insistence on non-temporal senses for temporal words and emendation of Plato's text. The Middle Platonist Taurus (fl. A.D. 145) distinguished no less than four senses of

---

[32] A variant of the last interpretation might suggest that Plato was not ruling out a time before creation, but only the possibility of describing it with words as *anthropocentric* (see Chapter 3) as 'was' and 'shall be'. Against this, Plato breathes no word of such a doctrine. Secondly, he connects 'was' and 'shall be' with the creation of the *heavens*, not of *humans*. Thirdly, he is willing to use tenses *retrospectively* (53A8; 69B6; 7) in relation to the period before creation.

[33] Gregory Vlastos, 'The disorderly motion in the *Timaeus*' and 'Creation in the *Timaeus*: is it a fiction?', in R.E. Allen, ed., *Studies in Plato's Metaphysics*, London 1965, 379-419, reprinted respectively from *CQ* 1939 and *PR* 1964.

[34] Vlastos, 'Creation', 413.   [35] Vlastos, 'Creation', 411-12.   [36] Vlastos, 'Creation', 404-5.

'created' (*genētos*), with corresponding senses of 'has come into being' (*gegone*).[37] A thing is created, if it is of the same genus as what is literally created, if it is composite although it has never in fact been combined, if it is for ever undergoing processes of generation (this is already mentioned by Philo),[38] or if it depends for its existence on something else (this is already in Crantor),[39] as does the moon's light upon the sun. Armed with these special interpretations, and with another word 'origin' (*archē*), Taurus reconstrues[40] one of the passages quoted earlier (*Tim.* 28B2-C2), where Plato says of the *kosmos*:

> Did it exist always (*aei*), having no origin by which it came into being (*archē geneseōs*); or has it come into being (*gegonen*), starting from some origin (*ap' archēs tinos arxamenos*)? It has come into being.

Unfortunately, Taurus' attempt to construe these words non-temporally is full of difficulties, many of which were forcefully pointed out by Alexander of Aphrodisias.[41]

The other text which particularly attracted attention was *Tim.* 27C4-5, where Plato says:

> We who are going to hold a discussion about the universe (*to pan*) in some way (*pēi*), whether/how (*ē*, or *hēi*), it has come into being (*gegonen*), or is indeed (*ē kai*) ungenerated (*agenes*), must pray ...

Literalists can follow the most natural interpretation, according to which two *contrasted* alternatives are raised for discussion: the world has come into being *or* it is ungenerated. But non-literalists need to get rid of the contrast implied by 'or'. On the whole, they are forced into emending the text, in order to do so.[42]

[37] The main non-literalist moves are recorded by Philop. *Aet.* (Rabe) pp. 135-242. Taurus' four senses are given on p. 145, 13 and are translated into English by Thomas Taylor, *Ocellus Lucanus On the Nature of the Universe; Taurus the Platonic Philosopher On the Eternity of the World; (etc.)*, London 1831; and by John Dillon, *The Middle Platonists*, London 1977, pp. 242-4.

[38] Philo *Aet.* 14.

[39] Crantor ap. Procl. *in Tim.* 1.277, 8.

[40] Taurus, ap. Philop. *Aet.* 147, 13-21. Compare the hint in his contemporary, Albinus, ap. Procl. *in Tim* 1.219, 7.

[41] ap. Philop. *Aet.* 214,10-20; 215,4-216,6; and ap. Simpl. *in Cael.* 297,26-298,16. One difficulty is that Plato *contrasts* '*gegonen*' (it has come into being) with 'it has existed *always*', whereas the non-literalists would like to make having come into being *compatible* with having existed always. Another difficulty is that the non-literalists can hardly apply to the *Timaeus* senses of 'destroyed' (*phtharton*), which will parallel those for 'created', when the creator is made to say (41A-B) that what has been put together can be taken apart, but will be preserved from such destruction by his good will.

[42] Philop. *Aet.* 191-3; Procl. *in Tim.* 1.218-19. Some understood the *ē* as an *ei*, or replaced it with an *ei*, and so substituted 'if' for 'or'. The question raised is then in what way the world came into being, *even if* it is (in a different sense) ungenerated. Other non-literalists got rid of the 'or' by reversing the word order, and reading *kai hēi* instead of *ē kai*. The question is then *in what way* the universe has come into being and *in what way* it is ungenerated, and it was urged that this would fit with the use of *pēi* (in some way). A third move was to change the *agenes* (ungenerated) to *aeigenes* (ever in process of generation), but this would not help, unless the 'or' could also be removed.

Plato's influence

Plato's influence was partly due to his positive arguments. We shall see many of these being exploited, for example, in Aristotle's *de Philosophia*, and all the arguments utilised there were put to work many times again. But his influence was also due to his very unclarity and suggestiveness, which left room for so many subsequent interpretations. It allowed Philo, as was seen in Chapter 15, to introduce the idea of a beginning of time and of an intelligible world of patterns enjoying a non-temporal priority. It also permitted the quite different Platonist *denial* of any beginning and the reinterpretation of the word 'created'. This conception of a beginningless creation introduced many novel ideas about the causal relations in which things can stand. And Plato's importance is partly due to these Platonist reinterpretations which were not necessarily faithful to his original conception. I shall return in Chapter 20 to the distinctive Platonist contributions to the topic of causality.

Aristotle

Aristotle was influential in a quite different way. There is no doubt what conclusion he was aiming at, and he produced a dense array of arguments against the idea of a beginning, thus providing a source book for later controversy. Some of his views, we saw, were turned against him, such as those concerning infinity, or creation *ex nihilo*, and these have been treated at length in two earlier chapters. His own positive arguments, however, have only been more briefly discussed. I need not repeat what I have said about them in Chapters 14 and 15 nor recapitulate his distinctions of non-temporal senses (Chapter 8), or his treatment of endless recurrence (Chapter 12). But it will give a better picture of his position and influence, and provide a background for some of the arguments already encountered, if I record five passages or groups of passages, which go against a beginning of things.

Physics 8.1

In *Phys.* 8.1, Aristotle's subject is *movement,* and he aims to show that it has no beginning or end. He starts (251a8-b10) with one of his more ingenious arguments, in which, as was seen in Chapter 15, he treats change as requiring a *trigger*: motion requires a pre-existing mobile,[43] and this in turn presupposes prior movement. For either the mobile came into being (but creation presupposes a movement), or it existed always. If it existed always, either it always moved, or it began to move. But even its beginning to move would presuppose a movement (e.g. the approach of a mover), to put an end to the state of rest.

Aristotle further argues (251b10-28) that there must be movement when there is time; and time is shown to have no beginning by several

---

[43] Philoponus objects to the claim of pre-existence: could not some fire come into existence and begin moving upwards without delay? Simplicius replies that there would have pre-existed the air, or other matter, out of which the fire came (Simpl. *in Phys.* 1133).

considerations. First, there is a hint of the argument already encountered, to which I shall return, that a beginning of time would imply an absence of time *previously*; but the talk of 'previously' implies that after all there was time. Time is shown to lack a beginning, secondly, by the testimony of everyone except Plato, and thirdly (cf. 4.13, 222b6-7) by the reflection that any instant is a boundary of some preceding and following time. (Aristotle does not explain why some instant should not be the boundary solely for some following time).

Finally, he argues against those who postulate motion after infinite or recurrent periods of universal rest: why should motion start up just when it does rather than earlier (the 'Why not sooner?' argument of Chapter 15)? Why too should motion and rest alternate, and why have the particular durations they have?

de Caelo 1

A second discussion occupies a large part of *Cael.* 1 (chs 3-4 and 10-12). It is concerned not with movement, but with creation and destruction. But much of it involves outmoded empirical theory, or the testimony of earlier thinkers,[44] and the most challenging discussion, from a philosophical point of view, does not come until 1.12, which has been the subject of some very able analyses.[45] As commonly interpreted, the chapter argues for two main theses. The first (281b3-32) is that a *kosmos* which exists at all times is actually *incapable* of non-existence. How does Aristotle make the illegitimate move from the temporal 'all times' to the modal 'incapable'? The move proved so influential that it is all the more striking to find that it depends on at least one mistake. And yet there is much disagreement about what the mistake is. Aristotle recognises that a thing can at the same time possess the capacities for existing and for not existing, but he does not recognise that it can at the same time (say, 1981) possess the capacity for existing, say, in the year 2000 and the capacity for not existing in the year 2000. Through not allowing that the capacities can *relate* to the same time (2000), as well as being *possessed* at the same time (1981), he draws a wrong conclusion about a *kosmos* which exists at *all* times. Such a *kosmos* will exist in 2000, and *a fortiori*

[44] The element of which the stars and spheres are made has no opposite (otherwise it would have an oposite to its motion, but there is no opposite to the circular). Lacking an opposite, it is not liable to creation or destruction, for these are into and out of opposites (1.3-4). Further (1.10) although all thinkers make the present orderly arrangement of heaven and earth to be created, there are reservations. For some Platonists (as we have seen) say that Plato uses the idea of creation only metaphorically. Moreover, those who postulate a succession of *kosmoi*, through the alternate combining and dissolving of materials, are in effect denying the universe a beginning. In fact, so it is argued, creation requires a prior constitution (*sustasis*, 1.10, 280a23-7) of things. 1.11 goes on to distinguish different senses of 'created' and 'destroyed' and to discuss different kinds of possibility, in preparation for the more elaborate arguments of 1.12.

[45] Jaakko Hintikka, *Time and Necessity*, Oxford 1973, ch. 5, revised from *Ajatus* 20, 1957, 65-90, reprinted in *AA* 3; C.J.F. Williams, 'Aristotle and corruptibility', *Religious Studies* 1, 1965, 95-107 and 203-13; Sorabji, *Necessity, Cause and Blame*, ch. 8; Sarah Waterlow, *Passage and Possibility: a study of Aristotle's modal concepts*, Oxford 1982, esp. ch. 4; Lindsay Judson, forthcoming in *Oxford Studies in Ancient Philosophy* 1, 1983. My comments on Waterlow and Judson are based on pre-publication lectures which they gave, rather than on the printed versions.

has a capacity for existing in 2000. It cannot then, he thinks, have a capacity for not existing which relates to the *same* time. And since it *always* exists, there is no time left in relation to which it can have a capacity for not existing. Aristotle draws a parallel conclusion for the idea of there *never* being a *kosmos*: if matter is *never* (temporal word) organised into a *kosmos*, then it is *incapable* (modal word) of being organised. The argument is that it has a capacity for not being organised which relates to every time, and hence there is no time left in relation to which it can have a capacity for being organised.

Aristotle's mistake so far, I believe, is to deny that the opposed capacities can relate to the same time. But why does he make this mistake? He fears, at 281b19 and b22, that, if we let the capacities relate to the same time, then we will be allowing as possible the contradiction of their being exercised simultaneously. It is a mistake to fear this, but why is it a mistake? The answer which I gave briefly elsewhere,[46] is that it involves an error about what should be held constant and what should be imagined as different, in envisaging counterfactual situations. This is a recurrent problem for those who do philosophy. If something exists always, but we want to imagine it as exercising a capacity for not existing in the year 2000, then our imaginations must not hold constant in the counterfactual situation the fact of its always existing, or else we will indeed get the contradiction which Aristotle seeks to avoid of its existing and not existing simultaneously. Why does Aristotle believe that this contradiction will arise? I would give an explanation in two parts. First, he has noticed, quite correctly, that the time to which the opposed capacities relate (2000, in my example) is the time of their *exercise*, and this correct insight is the *positive* incentive which leads him to fear (wrongly) that, if the time is one and the same, we shall be committed to the possibility of a simultaneous exercise. Secondly, he does not discover the right antidote to this fear, which is to get clear on the confusing issue of what differences are to be expected in counterfactual situations. This second factor is viewed by Lindsay Judson as the main source of error throughout Aristotle's chapter.

On the usual interpretation, the chapter has a second theme, starting perhaps at 282a25. It is one which Aristotle had already announced in 1.10, 280a28-34 and at the opening of 1.12, 281a28-b3. He wants to attack the view of Plato's *Timaeus* that the orderly *kosmos* had a beginning, but (through the grace of God) will not end. Aristotle has a battery of arguments for the conclusion, which many of his predecessors had accepted and which continued to be influential thereafter, that what begins must end, and that what ends must begin. The argument already discussed from the earlier part of his chapter remains relevant, on the present interpretation, because many of the new arguments are parallel to it, or rely on it. Judson's interpretation has the advantage of making the two parts of the chapter closer than that, and he may be right, but I do not wish to anticipate his publication.

Of the new arguments, one (283a11) presents an initial appearance of being independent, for it looks like a fresh occurrence of the 'Why just then?' question, parallel to that at *Phys.* 8.1, 252a11-19. But it turns out that the

---

[46] Sorabji, *Necessity, Cause and Blame*, 129.

'Why just then?' question has here been adapted to fit in with concerns from earlier in the chapter. I believe Aristotle is asking why Plato's matter should lose some incapacity for organisation. If there is no reason for it to do so at óne moment rather than another, then it must instead for an infinite time have had the capacity to be organised, even before it came to be organised into a *kosmos*. Aristotle's objection to this recalls the earlier part of his chapter: what if the capacity had been actualised? Would we not have got the contradiction of simultaneous organisation and disorganisation?

For an argument which is genuinely independent, we should look to the one in 283a4-10, which Sarah Waterlow had discussed illuminatingly.[47] Aristotle demands, rather misguidedly, that capacities should be defined by reference to definite amounts. The capacities to lift weights, which were offered as illustrations in *Cael.* 1.11, were capacities to lift definite amounts, and similarly here the capacity of matter to be organised should be a capacity to be organised for a definite (*hōrismenon*) time. The whole of time can, for this purpose, be viewed as definite in a way (*hōristai pōs*). But a period which is infinite in only one direction is not definite in any way at all. Presumably, Aristotle is thinking that, for example, an infinite future is of no *single* length, since one that begins in 1980 is longer than one that begins in 1981. If so, we have seen in Chapter 14 that this view is too simple, and that it is any case corrected in the *Physics*. But the conclusion here drawn from it is that matter does not have the capacity, in the manner of Plato's *Timaeus*, to be first disorganised for a period which is infinite in one direction and then organised for a period which is infinite in the other.

A time before time

A third passage bearing on creation is *Metaph.* 12.6, 1071b8, where Aristotle hints again, as in *Phys.* 8.1, 251b10-13, at an argument which we have encountered also in Proclus. You cannot say time began, for that would imply that *previously* (*proteron*) there was no time, and the word 'previously' implies that there was time after all. This argument got repeated again and again with variations in antiquity and in Islamic thought, and it has been repeated again today.[48]

Augustine came very close to giving the right answer in his emphatic

[47] loc. cit.

[48] For versions of the argument see the following. Velleius, spokesman for the Epicureans in Cic. *de Natura Deorum* 1.21. Peripatetics reported by Philo *Aet.* 10.53. Albinus *Didaskalikos*, in the Teubner edition of Plato, vol. 6 (Hermann), p. 169,26. Sextus *M* 10.189. Alexander of Aphrodisias, *de Temp.* (Latin and Arabic editions cited in Chapter 2: in the Latin, see p. 97, ll. 1-4. The two versions are rather different, but both are clearly concerned with this argument). Themistius *in Metaph.* 12, p. 14, 1-20. Aug. *City* XII.16; *de Gen. ad Lit.* V.5.12. Procl. 5th argument in *Aet.* ap. Philop. *Aet.* (Rabe) p. 103, tr. by Thomas Taylor. Simpl. *in Phys.* 1156-7. One of Ammonius' Platonist pupils in Zacharias, *Ammonius* or *de Mundi Opificio* PG 85, 1081A. Arguments of the same type are said to have reappeared frequently in Islam: Alousi (op. cit., p. 17, n. 41) and van den Bergh (op. cit., vol. 2, pp. 29; 100) ascribe them to al-Rāzī, Ibn Ḥazm, Avicenna and Averroes. Ghazālī ascribes such an argument to 'the philosophers', ap. Averroem, *Tahāfut al-Tahāfut* (Bouyges) pp. 64-5, tr. van den Bergh p. 37. For a modern version, see Richard Swinburne, *Space and Time*, London 1968.

assertions that there is no time prior to the creation; and in this he was followed by certain Islamic philosophers.[49] It only needs to be spelled out a little more explicitly that talk of a beginning of time or of motion does not imply some earlier *time* at which they were absent. At most it implies the *absence* of any earlier time at which they were present. Augustine actually discusses the absurd suggestion that there was a time when there was no time. But he points out that this absurdity does not follow from the suggestion that time began, but only from the different, and absurd, suggestion that time has not existed for all time.[50] It is a pity that Philoponus, writing in Greek, did not pay attention to this good discussion. His alternative line of answer has already been criticised as giving too much away. He thought that it would be all right to talk about *when* time did not exist, and about the situation *before* its creation, because he thought that it would be possible to give the temporal worlds a non-temporal sense.

Richard Swinburne, who is the modern proponent of Aristotle's argument, has (in conversation) expressed to me dissatisfaction with the suggestion that there is a hygienic way of talking about time beginning. For does the supposedly hygienic formula, 'It is not the case that there was an earlier time,' differ more than verbally from the absurd formula, 'Earlier there was no time'? I think that this new question goes beyond Aristotle's attempt to trap us in an absurd form of words – that cannot be done. Instead, it raises the more serious issue of whether we can attach sense to the idea of time beginning. That will depend on different considerations, for example, on whether we are convinced by the arguments of Chapter 6 that time must be accompanied by *change*, and on whether we are convinced that it makes sense to think of there having been a *first* change. What can be said is that if we are so convinced, we shall have succeeded in attaching sense to the idea of time beginning.

Potentiality

A fourth group of passages was used, not by Aristotle himself, but by his successors, in order to disprove a beginning. The passages have to do with potentiality. Suppose God changes to creating the universe after producing nothing. Now all change, according to Aristotle, involves something passing from a potential to an actual state.[51] And this can only happen through the agency of something which is itself in an actual state.[52] The conclusion drawn from these Aristotelian passages is that, if God changes to creating the universe, he will be dependent on some other agency to actualise his potentiality for creation.[53] A further conclusion is that God will be imperfect, if he is a creator only *potentially*, instead of always creating.[54]

[49] Aug. *City* XI.5-6; XII.16; *Conf.* XI.13 and 30. The same answer is ascribed to Ka'bī and Maturīdī by Alousi (op. cit. p. 259, note 136, referring to p. 248, note 86), and is ascribed to the theologians by Ghazālī, ap. Averroem *Tahāfut al-Tahāfut* (Bouyges) pp. 65-6, tr. van den Bergh pp. 37-8.
[50] *City* XII.16.   [51] *Phys.* 3.1.   [52] e.g. *Metaph.* 9.8, 1049b24-7.
[53] Themistius, *in Metaph.* 15, 35-16,5; Procl. 3rd argument in *Aet.* ap. Philop. *Aet.* (Rabe3 pp.

de Philosophia

Finally, Aristotle has a series of arguments, mostly against God's changing his will, in his early treatise *de Philosphia*.[55] The arguments start in fr. 16 (Ross),[56] but the fragment most closely connected with those encountered above is 19C. God would not destroy the world to save himself from world-making, for it befits him rather to turn disorder into order (cf. Plato *Tim.* 29E1; 30A5; 41A-B), and further he would not repent of what he had earlier done, since repentance is a disease of the soul. (This last theme became very common, and we have encountered it in Chapter 12 in connexion with Augustine's treatment of endless recurrence.) On the other hand, God would not destroy the world for the purpose of making another one. For if the new world were worse, then God would become worse, but that would show poor sense, and his works are in any case the product of perfect knowledge. If the new world were similar, then God would be labouring in vain, like children building sandcastles. If the new world were better, God would have become better, and must formerly have been lacking in skill and wisdom.

These arguments were repeated in late antiquity by the Neoplatonists.[57] But they were applied not only to the destruction, but also to the creation of the world. This application could be made in two ways. One was to prove indestructibility first, and then appeal to the principle, which was accepted by Presocratics, by Plato (with qualifications) and by Aristotle, that what is indestructible is not created. Another way was to compare, as better, worse, or similar, not two successive worlds, but the earlier non-existence and subsequent existence of a single world.

The last authenticated fragment of the *de Philosophia* which is relevant, fr. 20, denies that the world could have come into being by a new design (*novum consilium* – the expression picked up by Augustine, when he denied that the creation involved a change in God's will). But B. Effe claims to have

---

42-3 (Proclus threatens a regress of actualising agencies, which can only be stopped by postulating one which has always been at work without ever beginning); Maimonides *Guide* 2.14 (5th method), with reply in 2.18.

[54] Procl. 4th argument in *Aet.* ap. Philop. *Aet.* (Rabe), pp. 82 and 225, 7. Philoponus replies that this would prove too much, for (*Aet.* p. 92) a given individual is not for ever being created by God, nor (p. 97) is the *future* state of the universe. The same reply is made to some different Neoplatonist arguments by Philoponus' Christian contemporary Zacharias, who offers a fuller discussion (*Ammonius* or *de Mundi Opificio* 1040B-1041A; 1084A-1085C).

[55] For the treatment of creation in the *de Phil.* further references can be obtained from Baltes (op. cit.); B. Effe (reference below); J. Mansfeld, 'Providence and the destruction of the universe in early Stoic thought', in M.J. Vermaseren (ed.), *Studies in Hellenistic Religions*, Leiden 1979; the introduction and commentary by A.D. Nock in his translation, Cambridge 1926, of Sallustius, *de Diis et Mundo*; J. Pépin (reference below).

[56] Fr. 16 (Ross) is adapted from Plato's argument, in *Rep.*2.380D-381E, for the changelessness of God. Fr. 18 (preserved, like 19, in Philo's *Aet.*) declares that it is ungodly to say that the *kosmos* is generated or perishable and hence no better than the work of man's hands. Fr. 19a follows an influential Platonic pattern, and argues against the world's destruction by ruling out separately destruction from without and destruction from within (cf. *Rep.* 10.608D on the soul and *Tim.* 32C-33B on the *kosmos*). Fr. 19b analyses destruction as the return of a thing's parts to their natural positions, but protests that the parts of the *kosmos* are already in their natural positions.

[57] Hierocles ap. Phot. *Bibl.* cod. 251, p. 461a8ff.; Sallustius *de Diis et Mundo* §7; Ammonius, if we can trust Zach. *Ammonius* or *de Mundi Opificio*, PG 85, 1032A-1033A.

identified a new fragment of the *de Philosophia*. He draws attention to a set of Epicurean arguments which appear in three sources, and he argues that they derive from Aristotle's *de Philosophia*, and that one of the three versions (that of Aëtius) actually represents a fragment of the original.[58] The arguments address themselves in effect to the question of fr. 20, why there should be a new design (although they do not use the phrase) leading God to begin creation. Was God sleeping? But eternal sleep is death.[59] Or was he anxious to avoid fatigue? But he is not subject to fatigue. Or was he discontent with his surroundings, and did he live in darkness? But then he should have acted earlier. Or did he create for the sake of men? But it is no detriment not to be born.

If the Epicureans took over some of these ideas from Aristotle, they will have drawn a different moral. For, unlike Aristotle, they did believe that the present world order began and would end, and their point was only that the gods would have no hand in it.

Aristotle's influence

The passages cited should have made clear the influence of Aristotle on the 'why not sooner?', 'trigger', and 'changeless will' arguments of Chapter 15, as well as on a host of other arguments. But the irony remains that Aristotle exercised just as much influence by providing the materials for the 'infinity' and 'creation *ex nihilo*' arguments *against* his view.

Temporal and causal aspects of creation

I have so far concentrated on the temporal, rather than the causal aspect of creation. Aristotle denies any *beginning* to the universe, and so denies its creation in any sense which involves a beginning. But it has been asked whether he does not allow its creation in a weaker sense which implies no beginning. Here Platonism affects the interpretation of Aristotle, for we have seen that the later Platonists gave sense to the idea of a beginningless creation. Thomas Aquinas read Aristotle in this way, as allowing a beginningless creation of the universe by God,[60] and so earlier did the Alexandrian Ammonius and his followers, Philoponus, Simplicius and Elias.[61] Among recent authors, R. Jolivet has surveyed the evidence and has argued that a creation, though never explicit, is compatible with all that Aristotle says, and is even implied, not admittedly by his conclusions, but by

[58] B. Effe, *Studien zur Kosmologie und Theologie der aristotelischen Schrift Über die Philosophie*, Munich 1970, 23-31, referring to Aëtius *Placita* 1.7.8-9 (Diels 300a18-301a7); Lucr. *de Rerum Natura* 5.156-80; Velleius in Cic. *de Natura Deorum* 1.21-2.

[59] This should not be classed as an 'idleness' argument, because Epicurus believed that the gods *were* idle in relation to men. However, the Epicurean argument may have inspired some of the 'idleness' arguments recorded in Chapter 15.

[60] The most emphatic text is Aquinas *in Phys.* 8, lectio 2, n.974-5 (translated R.J. Blackwell, R.J. Spath, W.E. Thirlkel, London 1963). See also *in Cael. 1*, lectio 21, n.10; *in Metaph. 2*, lectio 2, n.295; *in Metaph. 12*, lectio 12, n.2614; *de Potentia Dei* q.3, a.5, in c.

[61] Ammonius ap. Simpl. *in Phys.* 1363, 8-24; Philop. *in Phys.* 189, 10-26; Simpl. *in Phys.* 256, 16-25; *in Cael.* 271, 13-21; Elias *in Cat.* 120, 16-17.

some of his principles.[62] Jolivet has to struggle with formidable obstacles: the apparent denial that God can deliberate or engage in action (*praxis*), and the suggestion that the only worthy activity for him is thinking of what is best. But his survey of the evidence is valuable, even though it leaves me convinced of the conventional view that Aristotle's God merely inspires celestial motion by acting like an object of love. A more promising suggestion is that of J. Pépin, that Aristotle accepted a creation at the time of his early *de Philosophia*.[63] We have just seen Aristotle, in fr. 19C (Ross), speaking of God as if he were a creator. The same is true of fr. 13 and Pépin cites 12a. Moreover, the hypothesis would explain the account of Aristotle given by Ambrose.[64] None the less, it is hard to tell from the fragments what originally went on in the *de Philosophia*. If, as Pépin suggests, Plato was an interlocutor in the dialogue,[65] then it may be Plato's views which are under discussion in 19C. Meanwhile, frs 12 and 13 only explain how *other* men come to believe in God or the gods, while 12a in any case presents God as a cause of *movement* and *order*, rather than of existence. So I do not think it can be shown that Aristotle thought of his God as causing the existence of the world even in his early *de Philosophia*.

Creation always implies *causing*, whether with a beginning or not, and in the next three chapters I want to consider its *causal*, rather than its temporal, aspect. This will complete discussions already begun in Chapters 13, 15 and 17. In Chapter 20 I shall collect some of the principles of causation that have been hinted at in the preceding discussion. Before that, I shall try to trace the origins of three theories which are bound up with creation and causality: idealism, occasionalism and the view that God is the sole creator.

---

[62] R. Jolivet, 'Aristote et Saint Thomas' or 'La notion de création', in *Essai sur les rapports entre la pensée Grecque et la pensée Chrétienne*, Paris 1955.

[63] J. Pépin, *Théologie cosmique et théologie chrétienne (Ambroise Exam. 1.1, 1-4)*, Paris 1964, 475-8.

[64] Ambrose *Hex.* 1.1.1-2, p. 3, 10-13 (Schenkl).

[65] Pépin, op. cit., 467.

PART IV

CREATION AND CAUSE

CHAPTER EIGHTEEN

Gregory of Nyssa: The Origins of Idealism

Gregory of Nyssa's theory of creation is hard to interpret. But I am going to suggest an interpretation according to which it modifies the contrast between creation out of nothing and creation from God, and does so by introducing an idealistic theory of matter. The subject of idealism in antiquity has been put on the map by a seminal lecture of Myles Burnyeat. Gregory's theory, I shall suggest, moves in the opposite direction from the best known version of idealism, that of George Berkeley. For Berkeley moves from idealism to conclusions about cause and creation; whereas Gregory moves from a view about cause to a conclusion about creation which involves idealism.

By idealism I mean, very roughly, the view that things are ideas in the mind. In Berkeley's version, physical objects like tables and chairs are simply collections of ideas, and depend for their existence on being perceived. The other things in the universe are spirits or minds which do the perceiving. One consequence for Berkeley is that the Biblical creation must have taken the form of God's making the ideas, which were already present in *his* mind, perceptible to *other* spirits.[1] As regards causal power, one set of ideas cannot be the cause of another, not the sun of heat, for example. Rather God produces these ideas in our minds, although he produces them in predictable sequences.[2]

Is there idealism in antiquity?

Burnyeat has argued not only that idealism is absent from ancient philosophy, but also that it *could* not have arisen, and he defines idealism in two alternative ways:[3]

> Idealism, whether we mean by that Berkeley's own doctrine that *esse est percipi* or a more vaguely conceived thesis to the effect that everything is in some substantial sense mental or spiritual, is one of the very few major philosophical positions which did *not* receive its first formulation in antiquity.

[1] Berkeley, *Third Dialogue Between Hylas and Philonous*. I am grateful to David Owen for reminding me of this discussion.

[2] Berkeley, *Principles*, §§25-33; 60-5; 103-4.

[3] In 'Idealism in Greek Philosophy: what Descartes saw and Berkeley missed', delivered as a lecture to the Royal Institute of Philosophy in 1978, and intended for publication in G. Vesey (ed.), *Idealism: Past and Present*, Royal Institute of Philosophy Lectures, vol. 13, publication postponed. It has now appeared in *PR* 91, 1982, 3-40; quotations from pp. 3 and 40.

The stronger thesis, that idealism *could* not have arisen, is applied only to Berkeley's version, according to which *esse est percipi*: to exist is to be perceived:

> Descartes had achieved a decisive shift of perspective without which no one, not even Berkeley, could have entertained the thought that *esse est percipi*.

I am glad that Burnyeat ventured the bolder thesis, because I think it enables us to see many things in a new light. According to Burnyeat's suggestion, both idealism and a proper refutation of scepticism had to wait for a new move from Descartes. What Descartes did was to formulate a more extreme kind of scepticism than any found in antiquity, treating everything except the mental, even his *own body*, as part of the external world, and seeking to doubt whether the external world existed at all.[4] It was only by pushing doubt so much further that Descartes was led to discover something which could not be doubted, namely, the *Cogito*: 'I think', 'I exist'. Again, it was only because the existence of everything except the mental had been doubted, that Berkeley had a motive to suggest that tables and chairs might simply be bundles of ideas in the mind. In that way, they would be things we could know about. And that solution constituted the birth of idealism.

The thesis that idealism *could* not have arisen until then depends on the assumption that it could only be motivated by the need to answer such scepticism. But idealism could arise from many motives. I shall be arguing that a certain version actually did arise in Gregory of Nyssa from a quite different one. This version may have been non-Berkeleian, but none the less it is important to notice the enormous *variety* of motivations that led people at least part of the way towards Berkeley. We have found idealism about a particular thing, namely, time, arising in Augustine on the basis of the paradoxes discussed in Part I above. Moreover, Augustine has difficulty in preventing that idealism spreading to other things, for example, to the uttered syllables whose time he estimates. He has to argue in *Conf.* XI.27 (cf. X. 8-9) that the sounds themselves do not remain in his mind waiting to be measured, but only their duration, and he would have to say that although the time taken by a movement is a mental entity, movement itself is not. Again elsewhere in Augustine an idealistic theory of matter emerges, even if only to be rejected, out of a theory of truth.[5] In Plotinus we find entirely different considerations pushing him several steps in Berkeley's direction. Even though I agree with Burnyeat that he did not get there, I hesitate to think that no motive could have pushed an ancient philosopher all the way.

---

[4] Descartes, *First Meditation*.

[5] Aug. *Soliloquies* II.6-7. I am grateful to John Procope for the reference: a stone must be a true stone, if it is to be a stone at all. But in order to be true, it must be what it appears to be, at least on one definition of truth. But then stones exist only when being perceived, because otherwise they present no appearance at all. And their interiors do not exist, because these are not perceived.

Augustine's Cogito

Before we reach Gregory, Augustine deserves a hearing on the other issue, that is, not on account of his brushes with idealism but because of his claim to have anticipated Descartes' refutation of scepticism: 'I think', 'I am'. Burnyeat discounts this claim, on the ground that Augustine puts knowledge of subjective states merely on a level with other replies to the sceptic. The special indubitability of such propositions as 'I think', 'I am', could not become clear until Descartes tried to extend doubt further. But on Augustine's behalf, it may be urged in reply that he puts forward his own versions of the *Cogito* in at least seven places,[6] and in some of these, for example, in *de Trinitate* XV. 12.21, he appears to see it as absolutely central to the refutation of scepticism. He seems to have realised its importance without the advantage of taking Descartes' intermediate step of doubting the existence of the entire external world. Descartes himself offered a different reply, when it was suggested to him that Augustine had anticipated the *Cogito*. He maintained that Augustine had not put it to the same use. But this reply too is only partly justified. For Augustine bases on one version of the *Cogito* a proof that the substance of the mind is not anything material,[7] and he uses another version as a step in his argument for God's existence.[8] One thing that must be stressed is the variety of versions which he offers, since some are closer to Descartes than others.

A.C. Lloyd has sought anticipations of Augustine's *Cogito* in Greek philosophy, since, like some others,[9] he doubts Augustine's philosophical originality. Personally, I should regard Augustine as frequently original precisely in putting old materials to new uses; so while removing the slur, I should not be surprised to find earlier sources. Lloyd's interesting reconstruction rests on a passage in an obscure writer, Oenomaus of Gadara, the Cynic, and author of a book which Lloyd renders *Swindlers Unmasked.*[10] Others have made other suggestions,[11] but I should like to offer one new one. For Sextus records an argument that you cannot deny that the soul exists, without making use of your soul, in order to make the denial.[12]

---

[6] *City* XI.26; *de Trin.* XV.12.21; X.10.13-16; *On Free Will* II.3.7; *de Beata Vita* II.7; *Sol.* II.1.1; *de Ver. Rel.* XXXIX.73.

[7] *de Trin.* X.10.16.

[8] *On Free Will* II.3.7.

[9] For details see Chapter 15 and my comments there.

[10] A.C. Lloyd, 'Nosce teipsum and conscientia', *AGP* 46, 1964, 188-200, esp. 198-200. Lloyd conjectures that Stoics may have used some analogue of the *Cogito* to show that there is an indubitable kind of perception, but had the argument turned against them by Cynics and Sceptics.

[11] Plotinus is sometimes cited, e.g. by Julius R. Weinberg, for whom see yet other suggestions in 'The sources and nature of Descartes' *Cogito*', in *Ockham, Descartes and Hume*, Madison Wisconsin 1977. But although, as was seen in Chapter 10, Plotinus makes the Intellect think itself, he does not introduce this idea in the context of scepticism. Moreover, his ground is that the Intellect is identical with its object, and this, as Lloyd points out, does not correspond to Augustine's viewpoint. As for self-awareness of a more ordinary kind, Plotinus goes out of his way to stress that we often act without it, that is unconsciously. (So Lloyd, op. cit., and Rose Marie Mossé Bastide, *Bergson et Plotin*, Paris 1959, ch. 2).

[12] Sext. *M* 9.198.

Gregory's idealism

I now come to Gregory of Nyssa's theory, which may be recalled from
Chapter 15. Gregory, a Christian father (*c.* A.D. 331-396), was influenced by
Platonism,[13] although I hesitated to accept the suggestion that he influenced
Augustine in his turn.[14] The motivation for his theory has nothing to do with
scepticism. He is worried about how a purely immaterial creator could have
produced matter. The idea which causes his worry is a thesis about *causation*,
namely, that a cause needs to be somehow like its effect. Gregory's solution is
that material objects consist of nothing but qualities like colour and
extension, and that these qualities are not themselves material things, but
mere thoughts and concepts. So to create matter, God has only to make these
immaterial qualities come together, and to dissolve it, he would evidently
have only to separate them. Gregory expounds his theory in three passages:

> This being so, no one should still be cornered by the question of matter, and
> how and whence it arose. You can hear people saying things like this: if God is
> matterless, where does matter come from? How can quantity come from non-
> quantity, the visible from the invisible, some thing with limited bulk and size
> from what lacks magnitude and limits? And so also for the other characteristics
> seen in matter: how or whence were they produced by one who had nothing of
> the kind in his own nature? ... By his wise and powerful will, being capable of
> everything, he established for the creation of things all the things through
> which matter is constituted (*di' hōn sunistatai*): light, heavy, dense, rare, soft,
> resistant, fluid, dry, cold, hot, colour, shape, outline, extension. All of these are
> in themselves thoughts (*ennoiai*) and bare concepts (*noēmata*); none is matter on
> its own. But when they combine (*sundramein*), they turn into matter (*hulē
> ginetai*).[15]

> The corporeal creation is thought of in terms of properties which have nothing
> in common with the divine. And in particular it produces this great difficulty
> for Reason (*logos*). For one cannot see how the visible comes from the invisible,
> the solid and resistant from the intangible, the limited from the unlimited, or
> what is in every way circumscribed by quantitatively conceived proportions
> from what lacks quantity and magnitude, and so on for everything which we
> connect with corporeal nature. But we can say this much on the subject: none
> of the things we connect with body is on its own a body – not shape, not colour,
> not weight, not extension, not size, nor any other of the things classed as
> qualities (*poiotēs*). Each of these is an idea (*logos*), but their combination
> (*sundromē*) and union with each other turns into a body (*sōma ginetai*). So, since
> the qualities which fill out (*sumplērōmatikos*) the body are grasped by the mind
> and not by sense perception, and the divine is intelligent, what trouble is it for

---

[13] M. Aubineau, introduction to Gregoire de Nysse, *Traité de la virginité*, Sources Chrétiennes,
Paris 1966, 97-118; P. Courcelle, 'Gregoire de Nysse, lecteur de Porphyre', *REG* 80, 1967, 402-6,
with cautions introduced by John M. Rist, 'Basil's "Neoplatonism": its background and
nature', in Paul J. Fedwick, ed., *Basil of Caesarea: Christian, Humanist, Ascetic*, Toronto 1981, 137-
220, esp. 216-20.
[14] In Chapter 7, I did not think the texts cited by J.F. Callahan were really parallel. In
Chapter 15, I suggested a much closer parallel, but was deterred from drawing conclusions by
Augustine's slowness in acquiring Greek, the language in which Gregory wrote.
[15] *in Hex*. PG 44, col. 69B-C.

the intelligible (*noētos*) being to create the concepts (*noēmata*) whose combination (*sundromē*) with each other produces corporeal nature for us?[16]

There is an opinion about matter which seems not irrelevant to what we are investigating. It is that matter arises from (*hupostēnai ek*) the intelligible (*noētos*) and immaterial. For we shall find all matter to be composed of (*sunestanai ek*) qualities (*poiotēs*) and if it were stripped bare of these on its own, it could in no way be grasped in idea (*logos*). Yet each type of quality is separated in idea (*logos*) from the substratum (*hupokeimenon*), and an idea (*logos*) is an intelligible not a corporeal way of looking (*theōria*) at things. Thus, let an animal or a log be presented for us to consider, or anything else which has a corporeal constitution. By a process of mental division we recognise many things connected with the substratum and the idea (*logos*) of each of them is not mixed up with the other things we are considering at the same time. For the ideas (*logoi*) of colour and of weight are different, and so again are those of quantity and of tactile quality. Thus softness and two-cubit length and the other things predicated are not conflated with each other, nor with the body, in our idea of them (*kata ton logon*). For the explanatory formula (*hermēneutikos horos*) envisaged for each of these is quite individual according to what it is, and has nothing in common with any of the other qualities which we connect with the substratum. If, then, colour is intelligible and so is resistance and quantity and the other such properties, and if upon each of these being removed from the substratum, the whole idea (*logos*) of body would be removed: what follows? If we find the absence of these things causes the dissolution of body, we must suppose their combination (*sundromē*) is what generates material nature. For a thing is not a body if it lacks colour, shape, resistance, extension, weight and the other properties, and each of these properties is not body, but is found to be something else, when taken separately. Conversely, then, when these properties combine (*sundramein*) they produce material reality. Now if the conception (*katanoēsis*) of these properties is intelligible (*noētos*), and the divine is intelligible in its nature, it is not strange that these intellectual (*noeros*) origins for the creation of bodies should arise from an incorporeal nature, with the intelligible nature establishing the intelligible properties, whose combination (*sundromē*) brings material nature to birth.[17]

I am not sure that *logos* should be translated 'idea', and I am not resting the case for idealism in Gregory on that, but on the talk of thoughts and concepts (*ennoiai, noēmata*). In another context, the Greek words might perhaps mean no more than things thought of, but here they must mean thoughts, concepts, or else Gregory will not have solved the causal problem he started with. The first two extracts suggest that body and matter consist of nothing but qualities which are thoughts or concepts. And Gregory imagines their being separated again in such a way that body would be dissolved.

The analogy with Berkeley's theory is already remarkable, and it is made closer by what Berkeley says about the advantages of his idealism. It saves us from a view which he considers absurd, that a corporeal substance existing outside the minds of spirits should be produced out of nothing by the mere will of a spirit. There is, by contrast, no puzzle about how a spirit can at will

---

[16] *de Anima et Resurrectione*, PG 46, cols 124B-D.
[17] *de Hominis Opificio* ch. 24, PG 44, cols 212-13.

produce *ideas*.[18] Gregory sees comparable advantages in his own theory.

There is, however, one objection to the comparison with Berkeley that needs investigation. Gregory's third passage talks three times of a *substratum*,[19] something which would be foreign to the idealism of Berkeley's earlier writings. But I doubt if Gregory's reference to a substratum can amount to very much. For one thing, he must make sure that, whatever this substratum is, it is immaterial and incorporeal. Otherwise, he will have defeated his whole purpose, which is to show how matter and body can be created by an immaterial and incorporeal God. For another thing, Gregory says in his third passage that, if matter were stripped bare of the qualities, it could in no way be grasped in idea. Evidently, then, the substratum is not a separately conceivable thing. In a comparable passage, Gregory's elder brother, Basil of Caesarea, says that there would be *no* substratum left, if you mentally stripped away the qualities.[20] In fact, the residual reference to a substratum in Gregory has clearly been taken over from Platonist theory, and it reveals much about the origin of Gregory's own views.

In *Ennead* 6.3.8(12-19), Plotinus raises the question what is left over in a perceptible substance, when in imagination we strip away all its properties. He answers that there must be some seat of attachment for the properties, that this is, as it were, a *part* of the sensible substance, and that it is *matter*. None the less, he very much downgrades this 'part', by describing matter in ll. 34-7 as a mere shadow upon a shadow, a picture, an appearance. This passage, as A.C. Lloyd has shown, led the subsequent Neoplatonists Porphyry, Proclus and Simplicius, as well as the Christian Philoponus, to treat the individual as a bundle of qualities, unique in that the same combination would not be found in another individual.[21] Often the same words are used as we find in Gregory: *sundromē* for the bundle or combination, *poiotēs* and *idiotēs* for the qualities or properties. What Gregory has done is to idealise these qualities by turning them into thoughts or concepts.

But what meanwhile has happened to Plotinus' seat of attachment for the qualities? Porphyry retains the same minimal conception of matter. For a portion of his work on matter is preserved by Simplicius.[22] In this, Porphyry approves an account according to which Plato made the matter in sensible things 'a shadow of non-being in the category of quantity'. It is this shadowy conception of matter which reappears in Gregory's brother Basil, and in Gregory himself when he talks of a substratum. But it can be seen how little work it is doing. The striking aspect of the Neoplatonist theory is that the individual is a bundle of qualities. The striking aspect of Gregory's is that a

[18] Berkeley, *Three Dialogues*, 3rd dialogue.

[19] Cf. also Gregory *in Hex.* PG 44, col. 77D-80C, where he brings in the idea of a void which receives, and is filled with, qualities.

[20] Basil, *in Hex.* 1.8, 21A-B.

[21] The Neoplatonist theory is documented by A.C. Lloyd on pp. 158-9 of 'Neoplatonic logic and Aristotelian logic', *Phronesis* 1, 1955-6, 58-72 and 146-60. He has been kind enough to supply me with extra references: Porph. *Isagoge* 7, 21-3; *in Cat.* 129, 9-10; Procl. ap. Olympiodorum *in Alcibiadem* 204 (Creuzer); Philop. *in An. Post.* 437; Simpl. *in Cat.* 229, 17. See now his *Form and Universal in Aristotle*, Liverpool 1981, 67-8.

[22] *in Phys.* 230,34-231,7.

body, or a piece of matter,[23] is a bundle of thoughts or concepts.

The connexion between Gregory and Porphyry is still closer, because Porphyry is troubled by the same question: 'How can the immaterial produce matter?'[24] Only he gives a different answer, referring to the action of the 'seminal reasons' in a tiny drop of sperm, whereas Gregory responds by idealising material bodies.

Gregory's 'substratum' evidently does not commit him to very much. But it can be added that it also does not differentiate him very sharply from Berkeley, because of a development in Berkeley's own views, to which Burnyeat draws attention. In his late work *Siris*, Berkeley distinguishes between his contemporaries' conception of corporeal substance, which he rejects, and ancient Greek conceptions of matter, including Neoplatonist ones. For the Greeks, matter is not a positive, actual being, and it is only the actual being of matter that he is out to attack.[25] The Neoplatonists' conception of matter is so negative, that Berkeley can accept it, just as Gregory accepts their concept of substratum. I would add that he can accept it without really departing from the spirit of his earlier writings.

I have been arguing for the closeness of Gregory's theory to Berkeley's, and I shall shortly suggest yet further points of convergence. But, of course, in theories differently arrived at, there are bound to be differences, differences of emphasis being as important as differences of doctrine. Berkeley claims that his theory sets forth 'the necessary and immediate dependence of all things' on God.[26] This it does the more graphically, because the point he emphasises about ideas is that they depend for their existence on being in a mind. It could have suited Gregory to bring out the world's dependence in this way too. But, starting as he did from the question of cause being like effect, he was under less pressure to attend to the status of the ideas once they were assembled. This inattention gives the two versions of idealism a very different feel. There would be more than a difference of feel, if Gregory had added that the ideas cease to be ideas during the period when they are assembled to form bodies. If that were so, his theory would not be properly idealistic at all. But rather than any such statement, all we find is the inattention of which I have been speaking. A reasonable interpretation is that the ideas have a *dual* status when they are assembled, becoming perceptible to us, while remaining still, or timelessly, intelligible to God.

Gregory's inattention creates a major difference between himself and Berkeley. But in many other respects, their theories overlap in interesting ways, with Gregory sometimes providing the additional complications, when it comes to matters concerning the creation. Berkeley holds that the ideas are eternally in God's mind. So the world's creation consists not in the creation of the ideas, but in their being made perceptible to us. Presumably, if the world were to be destroyed, this would not be through the ideas ceasing to be in God's mind, but through their losing this perceptibility for us. Gregory's view starts off similarly: for him too the creation or destruction of the

[23] For Gregory, the word 'matter' refers to *body*, for the Neoplatonists to the *substratum*.
[24] Porph. ap. Procl. *in Tim.* (Diehl) 1.395-6.
[25] Berkeley *Siris* 306, 317-20.
[26] Berkeley, *Three Dialogues*, 3rd dialogue.

material world would not consist in the creation or destruction of ideas. But he differs from Berkeley in making it consist in their assembly and disassembly. This means that he would not agree with Berkeley's formula, that all things have existed eternally in God's mind. At most, it would be the component *ideas* that existed there in a (timeless) eternity.

I do not know in fact whether Gregory intends, like Berkeley, and like many of his own predecessors, to locate the ideas within God, that is, in his mind or in his Wisdom.[27] But if he does, then one final conclusion can be drawn. Wolfson has directed attention to a striking innovation in Gregory's theory. This is his interchanging the claims that the sensible world is created *out of nothing* and that it is created *from God*.[28] Other Christians had separated these two notions, reserving the expression 'from God' for the generation of Christ from God the Father. For Father and Son, by analogy with human fathers and sons, are of one substance. Gregory is convinced, however, that Scripture requires us to believe that the sensible world comes from God,[29] and, on one construction, his theory makes this description considerably more appropriate. For the very same ideas whose concurrence forms a physical object are in God's mind, so that the physical object, in a new sense, comes *from God*. On the other hand, it is still true that God does not need any independently existing matter for his creation, and for that reason the description 'out of nothing' is still justifiable. In suggesting this interpretation, I am departing from Wolfson, who has construed Gregory's 'out of nothing' quite differently.[30] Berkeley's theory would, of course, make the description 'from God' even more appropriate. It is not just the component ideas which have existed in God's mind, but, as he says, 'all things'.

What conclusion can we draw from this discussion? I think it shows that philosophers can arrive at strikingly similar theories by different routes. Burnyeat has shown beautifully how crucial for Berkeley's arrival at idealism were Descartes' peculiar and insufficiently noticed adaptations to scepticism. I should be distorting, if I did not say what a brilliant and illuminating interpretation Burnyeat gives. Yet if we take Burnyeat's broader description of idealism, and if the suggestions I have made about Gregory are right, then Gregory arrived at idealism by an entirely different route. What if we take Burnyeat's narrower description of idealism, as the doctrine that to exist is to be perceived? This was not Gregory's version. Yet even here I am not inclined to say that he *could* not have reached it, because it would in fact have suited him to bring out the world's dependence in that way. It is understandable, but was not inevitable, that his attention should not have been drawn in that direction.

---

[27] He certainly follows his predecessors in postulating an intellegible creation, *in Hex.* PG 44, cols 81B-D; 84C-D.

[28] Gregory, *de Hom. Op.* 23-4, PG 44, 209C-213C. H.A. Wolfson, 'The identification of *ex nihilo* with emanation in Gregory of Nyssa', *HTR* 63, 1970, 53-60.

[29] Gregory *de Hom. Op.* 23, PG 44, 212B.

[30] 'Out of nothing', on his suggestion, means: from intelligible matter in God's mind which is nothing, in the sense that it is ineffable. But I see no sign of this in the text. Wolfson is influenced by the later equation in Eriugena of 'out of nothing' with 'from God, the ineffable'.

Is there idealism in Islamic thought?

So much for idealism in antiquity, but it is also worth considering the claim that there is idealism in mediaeval Islamic thought. The question will bring us back, albeit briefly, to Gregory. Alousi maintains[31] that an early Islamic thinker, Ḍirār b. 'Amr (died A.D. 815 at the latest), believed, as Alousi puts it, 'that the body is a collection of sensations which have no essential connexion outside the mind'. I doubt if the evidence is good enough, since the Islamic reports talk merely of accidents (that is, accidental properties) rather than sensations. But I would draw attention to an unnoticed piece of evidence which makes the connection between Ḍirār and idealism closer in one way than Alousi realised. The evidence is that Ḍirār's view seems from the reports to be quite closely modelled at some points on that of Gregory of Nyssa. Fritz Zimmermann has confirmed for me that the relevant text of Gregory could have been known in Islam in Ḍirār's time.[32] And he has kindly supplied me with the following abstracts concerning Ḍirār.

> Ḍirār b. 'Amr: a body [consists of] accidents so composed and assembled as to make a stable body susceptible of ... being changed from one state (or: moment) to another. These accidents are such that a body cannot but have either the feature in question or its contrary, like life and death (one of which will always be present in a body), colours and tastes (for a body will not be without a feature of either type), and similarly weight (heaviness and lightness), roughness and smoothness, hotness and coldness, moistness and dryness, and similarly compactness(?)[33]

We are further told, a few pages later, that certain other Islamic thinkers agreed with this view. However, despite the similarities with Gregory, the quoted passage goes on to say, not that the accidents are *mental* entities, as Gregory would maintain, when they are separated from each other, but rather that they are then *non-existent*. In view of this, Ḍirār's idea sounds more like that of Gregory *minus* the idealism. This is how the passage continues:

> Ḍirār rules out that these accidents should assemble to form bodies at a time when they already exist themselves. He thinks it absurd that this should happen (lit.: be done) to them at any time except at the beginning [of their existence]: they cannot come into existence except as an ensemble. Thus according to him they can be both united and existent, but not both separate and existent. For if they were both separate and existent, a colour could exist without belonging to something coloured, and life could exist without belonging to something alive.

We have already encountered a possible source for Ḍirār's idea, namely,

---

[31] H.M.E. Alousi, *The Problem of Creation in Islamic Thought*, Baghdad 1968, 291.

[32] There is a substantial tradition of works on the Hexaēmeron in Syriac literature, beginning with Jacob of Edessa (died 708). The crucial passage from Gregory is quoted verbatim by Moses bar Kepha (died 903). See F. Zimmermann, 'The Bretheren of Purity and living tradition', forthcoming in *Göttingische Gelehrte Anzeigen*.

[33] Ash'arī, *Maqālāt*, pp. 305f.

the Neoplatonist view that sensible individuals are bundles of properties. These properties are not idealised by the Neoplatonists. They are referred to by Proclus and Philoponus as 'accidents' (*sumbebēkota*). It looks, then, as if Gregory and Ḍirār have a common source in this tradition, but that only Gregory added the idealism.

In the end, Alousi rests his case for idealism in Ḍirār not on the passage quoted, but on a different report, stemming from an opponent of Ḍirār's about his treatment of the theory of 'latency'. A less hostile souce[34] tells us that Ḍirār allowed there was oil latent in olives and sesame seeds, and juice latent in the grape, although they did not interpenetrate each other, as Naẓẓām supposed. On the other hand, fire could not be latent in stone and the like, without burning it. So far the controversy seems exactly to reproduce the Epicurean reply to Anaxagoras, as presented by Lucretius, *de Rerum Natura* 1.830-920. Anaxagoras had held that there was something of everything latent (*latitare*, 877) in everything. Wheat can nourish us, because it contains portions of blood, but those portions in turn contain portions of wheat, and with infinite divisibility (844), there is no end to the division of portions. This is matched by Naẓẓām's belief in latency, infinite divisibility and interpenetration. On the other side, Lucretius replies that there cannot be fire hidden in wood, or the forests would all be burnt up (897-906). His rival theory is that there is not fire in the wood, but only atoms which, when suitably rearranged, can constitute fire. His objection to there being fire in the wood is exactly that which Ḍirār was later to give.

The report to which Alousi refers ascribes to Ḍirār a much more sweeping denial of latency. According to this source, Ḍirār denied that there was blood hidden within a man, and insisted instead that blood is created, presumably by God, only on being seen:

> Ḍirār used to contend that, in order to uphold God's oneness, one had to reject the theory of latency. For to believe in latency was to presuppose the existence in man of blood, when in fact this was something which would only be created as and when it was seen.[35]

Even if we can believe this report, however, it does not commit Ḍirār to the idealist view that blood is imply an idea in the mind which owes its existence to being seen. The suggestion seems rather to be that the blood is created *at the same time as* we see it, not *by means of* our seeing it. I do not think, therefore, that the case for idealism in Ḍirār has been made out.

What has emerged is not any idealism in Ḍirār, but his extremely close connexion with Greek thought. This will be relevant in Chapter 25, when I come to the question whether Islamic thinkers as early as Ḍirār had yet been much influenced by Greek thought. We have seen Ḍirār closely following a Neoplatonist view of sensible individuals as bundles of properties and an Epicurean attack on the Anaxagorean theory of latency.

[34] Ash'arī, *Maqālāt*, 328, 6-11.

[35] Jāḥiẓ, *Kitāb al-Ḥayawān*, ed. Ḥārun, vol. 5, Cairo 1966, p. 10, ll. 2ff. Compare the German translations by J. van Ess, in, 'Ḍirār b. 'Amr und die "Cahmīya", Biographie einer vergessenen Schule', *Der Islam* 43, 1967, 241-79.

CHAPTER NINETEEN

The Origins of Occasionalism

Islamic thought

Even if we do not find idealism in Islamic thought, we certainly do find occasionalism, and here Greek scepticism did have an influence on the way in which the theory developed. Occasionalism, in one of its versions, turns all causation into forms of creation, and gives a sense, not merely, like Berkeley's theory, of the dependence of the world on God, but of its precariousness. The theory can be illustrated by saying that the dye does not cause the cloth to become black. Rather, its presence merely provides God with the occasion for making the cloth black, so that, at least on the strictest version of the theory, all causal power is reserved for God. The example of the dye is given by Maimonides, and he provides a second example, a variant of which appears earlier in Ghazālī.[1] The hand does not cause the pen to move. Instead, God in turn creates the man's will to move the pen, the necessary power, the motion of the hand and the motion of the pen. The talk of power should not be misconstrued. None of these four items is the *cause* of another; none can be said to *act*. They merely co-exist in time. God is the only cause and agent. The reason why, as Aquinas pointed out,[2] all causation here becomes a form of creation is that it involves God creating a new accident.

In the case of the Ash'arite theologians, followers of Ash'arī (died 935), this occasionalism is presented as a corollary of a further doctrine. For the Ash'arites held that every time-atom God creates an entirely new set of accidental properties, although they may be accidents of the same kind as before. If he omits to create new accidents, the substance which bore them will cease to exist. This shows why the blackness which we think is introduced into the cloth by the dye must in fact be created and re-created every time-atom by God.

Maimonides reports further corollaries of this view: things do not have essential natures of their own, since it is God who creates all their properties. Again, things have no tendency of their own to persist – it is this which most of all would encourage a sense of precariousness. The doctrine that our continuation requires that God re-create us from moment to moment is repeated in Descartes,[3] and from there it influenced the seventeenth-century

---

[1] Maimonides *Guide* 1.73, 6th proposition, and Ghazālī *Iḥyā'* 4, Cairo, p. 214. See Majid Fakhry, *Islamic Occasionalism*, London 1958, 73.

[2] This feature is pointed out by Aquinas *CG* 3.69, and by Averroes in bk 12 of his *Great Commentary on the Metaphysics of Aristotle*, in a section translated into French by Renan, *Averroès et l'Averroïsme*, pp. 109ff.

[3] Descartes *Principles of Philosophy* 1.21; and third of the *Meditations*.

occasionalists. The best known of them, Malebranche, agreed that all causation involves creation, although there is controversy over whether he allowed the non-creative occasions of God's action to have causal efficacy in a sense.[4]

Maimonides warns us that not all the ideas I have mentioned are so characteristic of the older sect of Mu'tazilite (literally: schismatic) theologians. Thus many Mu'tazilites allow accidents to endure, once created.[5] Moreover, some theologians assume casuality, and some Mu'tazilites allow that man acts by virtue of the power created in him.[6] Some deny that God creates all accidents.[7] Some allow that things have natures of their own.[8] And it is not until Ash'arī that the idea of *habit* is given a central role, as a rival to the idea of nature.[9] None the less, there are Mu'tazilites who take an even more extreme position than Ash'arī's. Not only do they ascribe all effects in the external world to God, and deny any necessary connexion between effects, but they also allow God to produce combinations of effects which we should regard as logically incompatible.[10] Fritz Zimmermann tells me that some of the views described appear very early. For there is a treatise of about A.D. 700, which applies to the field of ethics the idea that God can combine properties as he chooses, annexing the property of goodness, for example, to murder.[11]

It may seem that occasionalist theory is far removed from any Greek influence. And indeed a major source of inspiration for it is the Koran. Thus Ash'arī cites the authority of the Koran for his claim that God creates not only man but also the deeds of man.[12] And Wolfson quotes a large number of verses from the Koran which seem to reserve almost all power for God.[13]

---

[4] Did Malebranche ever, and did he consistently, allow causal efficacy to occasional causes? Although I am not quite persuaded by Beatrice K. Rome's affirmative answer to both questions, she provides a useful collection of evidence and opinions, in *The Philosophy of Malebranche*, Chicago 1963, pp. 209-42. See *Entretiens sur la métaphysique*, 7th Dialogue, for Malebranche's most polished exposition of occasionalism.

[5] Maimonides loc. cit. A catalogue is given by H.A. Wolfson, *The Philosophy of the Kalam*, Cambridge Mass. 1976, 518-44. Ka'bī is the only Mu'tazilite cited as agreeing fully with the Ash'arites that accidents do not endure; but Abū l-Hudhayl, whom Maimonides probably has in mind, holds that some endure, others not. To these I would add Dirār, who maintains that bodies are collections of accidents. Those accidents which are *not* 'parts' of bodies cannot last for two moments (Ash'arī, *Maqālāt* 359,16-360,6), and a body will not persist if half or more of its accidents are replaced by their contraries (Ash'arī, *Maqālāt* 305f).

[6] Maimonides loc. cit. Details are given by Fakhry, op. cit., 44: Bishr acknowledges deeds which are of *our* doing, and Abū l-Hudhayl allows men to be the authors of *some* acts.

[7] They allow, however, that he creates all *bodies*. Details in Wolfson, op. cit., 535: according to Mu'ammar, God creates no accidents; according to Bishr, only some. See also Ash'arī *Maqālāt*, section translated into German by S. Pines, *Beiträge zur islamischen Atomenlehre*, Berlin 1936, 28-9.

[8] Wolfson, op. cit., 559-78 cites Mu'ammar and Naẓẓām.

[9] Details in Wolfson, op. cit., 546.

[10] See Fakhry, op. cit., 46-7 on Ṣāliḥ and Ṣāliḥī. He points out that the possibility of such combinations is allowed also by some of these who acknowledge that *we* are often the authors of actions.

[11] It is translated into German by J. van Ess, in *Anfänge muslimischer Theologie*, Beyrouth and Wiesbaden 1977.

[12] Ash'arī's *Creed* (*Ibāna*), with reference to the Koran 37, 94, translated by Fakhry, op. cit., 56-7.

[13] Wolfson, op. cit., 518-19.

None the less, there is a highly interesting adaptation of occasionalist theory in Ghazālī (*c.* A.D. 1058-1111) – interesting, among other things, because here the influence of Greek philosophy becomes particularly clear.

Ghazālī's theory

For English readers, Ghazālī's theory is most accessible in his *Destruction of the Philosophers* (*Tahāfut al-Falāsifa*), as incorporated in Averroes' reply *Destruction of the Destruction* (*Tahāfut al-Tahāfut*), since the latter is translated, with an extremely informative commentary, by S. van den Bergh. In the seventeenth discussion,[14] Ghazālī argues that there is no necessary connexion and no logical implication between the contact of a piece of cotton with fire and its burning. God, either with or without the intermediacy of angels, does everything. And he can create contact without burning, or burning without contact, although he does not in fact do so. Ghazālī's opponents rely on observation, but observation shows only that there is burning *when* there is contact, not *because* there is contact. (In other translations, this is put by saying that the burning occurs *at this time*, not *on this account*; that burning occurs *with* contact, not *through* it; with it, but not by it; or again that there is a *simultaneity*, not a *causation*.)[15] Thanks to this, Ghazālī says, miracles are possible. And the only limitation on God's power is that he cannot do various things which we should classify as conceptual impossibilities.

Ghazālī anticipates a number of objections,[16] for example, that everything must be *unpredictable*. A man might have to say:

> 'I do not know what there is at present in my house; I only know that I left a book in my house, but perhaps by now it is a horse which has soiled the library with its urine and excrement, and I left in my house a piece of bread which has perhaps changed into an apple tree.'

Again, on seeing a stranger, one might have to acknowledge:

> 'It may be that he was one of the fruits in the market which has been changed into a man, and that this is that man.'

Ghazālī offers two alternative answers. The first is that these bizarre sequences are only being described as *possible* for God. In fact, however, God creates things in a *habitual* sequence. Moreover, God creates, along with the habitual sequence, a confidence in us that the habit will continue. This reference to habit is taken over by Ghazālī from his fellow-Ash'arites. The second reply departs from the spirit of occasionalism, and has led one recent

---

[14] Ghazālī ap. Averroem *Tahāfut al-Tahāfut* (Bouyges 517-42, tr. van den Bergh, vol. 1, 316-33). Ghazālī's text is distinguished from Averroes' reply by the smaller print. There is a translation just of Ghazālī's *Tahāfut* by S.A. Kamali, *al-Ghazali, Tahāfut al-Falāsifah*, Pakistan Philosophical Congress, Lahore 1958.

[15] These are the translations, respectively, of Wolfson, op. cit., 544; L.E. Goodman, 'Did al-Ghazālī deny causality?', *Studia Islamica* 47, 1978, 91; Fakhry, op. cit., 61; Kamali, op. cit., 85; van den Bergh, op. cit., 317 (= 519, Bouyges).

[16] Ghazālī ap. Aver. op. cit. (Bouyges pp. 534-7, tr. van den Bergh pp. 327-30).

commentator to deny that Ghazālī was an occasionalist.[17] For here Ghazālī seems to concede his opponents' view that flame has a certain *nature (khalqa)* of its own, and to accept the Aristotelian idea of an *enduring* matter being endowed with successive forms. He protests only that God can add to the flame or to the combustible an extra property which does not remove its essential characteristics, but which does interfere with combustion. I doubt if Ghazālī means by this to abandon the occasionalist position. His point is only that miracles, and the possibility of a flame's not burning cotton, do not lead to unpredictability. They do not do so, whether we take the occasionalist view, or the rival view that there are natures, forms and essences.[18]

A further complexity is introduced into Ghazālī's view by his earlier claim in the third discussion[19] that only a *voluntary* agent can be said to act in any but a metaphorical sense. This means that fire cannot be an agent, but on the other hand it sounds as if it gives some efficacy to human wills, and throughout the seventeenth discussion Ghazālī is prepared to consider that angels may be intermediary agents.

To the extent that Ghazālī's stance is occasionalist, it is known to Thomas Aquinas, who attacks it in more than one place.[20]

The influence of Greek scepticism

I earlier denied that scepticism was the only route leading to idealism. But I now want to say that it *is* a source of influence on occasionalism. This has been said by others, but it is most clearly brought out by van den Bergh.[21] As he shows, Ghazālī seems to be following the ancient sceptics in passage after passage.

Ghazālī's first point is that contact with fire does not logically *imply* burning, and this kind of view is ascribed by Galen to the ancient empirical school of medicine, which was closely associated with scepticism.

> The consequent is discovered from experience, and not as something implied (*emphainomenon*) in the antecedent. And for this reason none of the empirical doctors say that this is implied in that. They say rather that this follows after that, this precedes that, and this accompanies that.[22]

There is another highly relevant passage in which Diogenes Laertius describes the view of Pyrrhonian sceptics (followers of Pyrrho):

> We perceive that fire burns, but we suspend judgment on whether it has a burning *nature (phusis kaustikē)*.[23]

[17] Goodman, op. cit., 83-120.
[18] William J. Courtenay, 'The critique on natural causality in the Mutakallimun and nominalism', *HTR* 66, 1973, 77-94.
[19] Ghazālī ap. Aver. op. cit. (Bouyges pp. 156-61, tr. van den Bergh pp. 92-6).
[20] Aquinas *in 2 Sent.* d.1, q.1, a.4; *CG* 3.69 (cf. 3.65); *de Potentia Dei* q.3, a.7; *ST* 1.105, 5.
[21] van den Bergh, on the 17th discussion, in his commentary on Averroes' *Tahāfut al-Tahāfut*.
[22] Galen *de Methodo Medendi* 2.7.
[23] Diog. Laert. *Lives* 9.104 (on Pyrrho).

The doubt as to whether it has a *nature* which makes it burn is to be compared with the Islamic attack upon *natures*. And the example of burning comes up throughout the Islamic discussions of this subject. It is usually employed by those who want to deny natures, but it is also used in an opposite way by the Mu'ammar, who insists that fire does produce burning by nature.[24] The example appears in Greek thought not only in Diogenes Laertius, but also twice in Sextus, who is himself a Pyrrhonian sceptic. He questions whether fire has a burning nature (*phusis kaustikē*, the same phrase), since it does not burn everything alike, but seems to depend on the materials with which it comes into contact. This in turn is made into part of the attack on the idea that fire is the cause (*aition*) of burning.[25] And indeed the sceptical attack on *natures* is part of a wider attack on *causation*.

The connexion with Ghazālī is closer still; for in a pair of neighbouring passages, Sextus complains that the sun dries mud, but melts wax, whitens clothes, but blackens our complexion. Hence, once again, you cannot call it a cause (*aition*), for it does not affect all things alike. If it is complained that variations in the effect are due to variations in the materials affected, the sceptic will ask why, in that case, the properties of the materials are not treated as part of the cause.[26] Precisely these examples are used by Ghazālī in the passage already discussed. His opponents say that the sun softens some things and hardens others, it whitens the fuller's clothes, but darkens his face, but they refuse to ascribe these variations to God's will. They assign them instead to variations in the materials affected, which are thereby given a greater causal role than Ghazālī is willing to acknowledge.[27]

Another move by Ghazālī's opponents was recorded above, namely, that Ghazālī would make everything unpredictable. This move too is cited by Sextus in another neighbouring passage. If there is no cause, the sceptics' opponents allege, then a horse could be formed from a man, or a plant from a horse.[28]

Greek scepticism about causes came in different versions. The empirical school of medicine, to which Sextus was close, expressed a comparatively moderate doubt: a suspension of judgment rather than an outright denial, and about hidden causes and natures, rather than about superficially obvious causes. But Galen speaks of other doctors who denied causes outright. For the details, I would refer to a masterly article by Jonathan Barnes,[29] but what I want to say now is that either version could have

---

[24] The example is ascribed to Ṣāliḥ, Jubbā'ī, Abū l-Hudhayl and 'the majority of the people of the Kalam' (theologians), as well as to Mu'ammar. References in Wolfson, op. cit., 544 (citing Ash'arī *Maqālāt* p. 312, 11-13) and 560 (on Mu'ammar); Fakhry, op. cit., 46-8 (who includes Ṣāliḥ).

[25] Sext. *M* 9.241-5; 8.197-9.

[26] Sext. *M* 9.246-51; 8.192-4.

[27] Ghazālī ap. Aver. op. cit. (Bouyges pp. 525-6, tr. van den Bergh p. 321).

[28] Sext. *M* 9.202-3; *PH* 3.18. The passage is related to the Epicurean attack on creation *ex nihilo*, in which it is argued that this would allow anything to come from anything (Lucr. *de Rerum Natura* 1.159-214).

[29] See Celsus *de Medicina*, Prooemium 27 and 54, translated in the Loeb series by W.G. Spencer, and Galen *On Antecedent Causes* 13.162 in *CMG*, suppl. 2, pp. 41-2, translated in

inspired Ghazālī's occasionalism.

They may be yet other ways in which ancient Greek scepticism influenced Islamic occasionalism. I have not seen the early Arabic treatise which allows God to annex goodness to murder, and which Zimmermann tells me has the form of a Greek treatise. But I should not be surprised if Greek scepticism helped to pave the way. For Sextus has a long argument that nothing is good or bad *by nature* in which he cites numerous shocking and rival opinions about what is good or evil.[30] Sextus' discussion has been cited elsewhere[31] as a source for the later Ash'arite view that man cannot know what is good or bad except through divine revelation.

Other Greek influences

Scepticism, however, can have influenced only certain aspects of Islamic occasionalism. It provides no motive for giving causal power to *God*, an idea which on the contrary flows out of the religious convictions of Islam. As regards God's power, at most only certain points of detail betray the possibility of Greek influence, for example Ghazālī's idea, already mentioned, that only voluntary agents can act; the view ascribed by Averroes to the so-called 'philosophers' that God is the cause of a man's existing, the father merely a necessary *instrument*;[32] or finally the view that God is still creating the world. Greek antecedents for these ideas will emerge shortly.

Alternatives to occasionalism: Augustine

Occasionalism is *one* theory which reserves causal and creative power for God, but there are *alternatives*. Augustine is acknowledged by Malebranche as a source of inspiration for his own occasionalism,[33] but Augustine is *not* an occasionalist. Instead, he restricts causal and creative power in alternative ways, of which I shall distinguish three.

First, only voluntary agents can serve as efficient causes for Augustine, rather as later for Ghazālī only voluntary agents can act. Bodies are not efficient causes.[34] I shall suggest below that Augustine is inspired by a Platonist tradition in saying this.

Secondly, God is the only creator. Augustine repeatedly says that mothers, fathers, doctors and farmers do not create, but only make use of the forces supplied by God to bring forth what he has already created. And the same is true of the angels.[35] Augustine exploits a passage from St Paul:

Jonathan Barnes, 'Ancient scepticism and causation', forthcoming in Myles Burnyeat (ed.), *The Skeptical Tradition*, University of California.

[30] *M* 11.190-6; *PH* 3.198-234; 245-9. *PH* 3.210-11 includes, but only alongside others, some shocking opinions of the gods on these matters.

[31] S. Horovitz, 'Der Einfluss der griechischen Skepsis auf die Entwicklung der Philosophie bei den Arabern', *Jahresbericht des jüdisch-theologischen Seminars zu Breslau*, Breslau 1915, 5-49, esp. 25-8.

[32] Aver. op. cit., (Bouyges pp. 19-21; 268-9; 518, tr. van den Bergh pp. 10-11; 158-9; 317).

[33] I am grateful to Anne Whelan for drawing my attention to this influence.   [34] *City* V.9.

[35] *de Trin.* III.7.12-III.9.18; *City* XII.25-6; XXII.24; *de Gen. ad Lit.* IX.15.26-7; IX.16.29; *Quaestiones in Heptateuchum* II.21.

I planted, Apollos watered, but God gave the growth. So neither he who plants, nor he who waters is anything, but only God who gives the growth.[36]

Augustine uses this passage to reserve all creative power for God. He also supplies a mechanism to explain how God can be the sole creator even of present-day individuals. Thus he takes over the idea of seminal reasons which has already been encountered in Chapter 15, and which is found in different forms in the Stoics and in Plotinus.[37] All natural kinds were created in seminal form at the very beginning of the world's history. Thereafter things unfold on their own from the seminal reasons which are already there.[38] Augustine offers seminal reasons as the mechanism by which God remains the sole creator,[39] and he also argues that they enable us to reconcile the claim of Genesis that God completed his creation on the sixth day with the statement in John 5:17, 'My Father is working up to the present (*usque nunc operatur*)'. The latter must not be taken to mean that God is now instituting (*nunc instituere*) new creatures whose kind was not created at the beginning.[40] Even the characteristic *behaviour* of the various species is due to seminal reasons.[41] It has been convincingly argued[42] that Augustine uses seminal reasons to account only for the creation of new *bodies*, not of new *souls*. None the less, Augustine is clear that for souls too God is the creator, and that God is also responsible for the conjoining of souls with bodies.[43] The doctrine of seminal reasons helps to illustrate how different is Augustine's method of reserving creative power for God from that of the Ash'arite occasionalists.[44] For the Ash'arites would presumably reject seminal reasons as being too like essential natures and as making God's influence too indirect.

The third way in which Augustine concentrates causal power is by making everything depend for its continuation on God:

If the Creator's virtue were at any time to be missing from the created things which are to be governed, at once their species would go missing, and the whole of nature would collapse. For it is not like the case of a builder of houses who

[36] I Corinthians 3: 6-7.

[37] But there had been adaptations. The Stoics had connected seminal reasons closely with God, with the beginning of the *kosmos*, and with the idea of determining causes which govern everything, while Plotinus had sought to undo these links. For the Stoics, see e.g. Diog. Laert. *Lives* 7.135-6 (= SVF 1.102; 2.580); Aëtius *Placita* 1.7.33 (= SVF 2.1027; Diels 306a3 and b3); Arius Didymus fr. 36 in Diels 468a25 and b25 (= SVF 1.107; 512; 2.596); and fr. 38 in Diels 470, 10 (= 1.497); Philo *Aet.* 9.94-103; Origen *Against Celsus* 1.37 and 4.48 (= SVF 2.739 and 1074); Seneca, *Nat. Quaest.* 3.29.3; Plot. *Enn.* 3.1.7. For Plotinus' own view, see e.g. 2.3.14-18; 3.2.2 (18-31); 3.3.1-5; 4.3.10-11 and 15; 4.4.11-12; 36; 39; 4.8.6 (7-10); 5.9.6 (10-24).

[38] *de Gen. ad Lit.* V.7.20; V.12.28; V.23.45; VI.5.7; VI.6.10; VI.10.17 (the fullest account); VI.11.18-19; VII.24.35; *Quaest. in Hept.* II.21. See the comments ad loc. of P. Agaësse and A. Solignac in their edition of the *de Gen. ad Lit.*, vol. 48 in *Oeuvres de Saint Augustin*, Bibliothèque Augustinienne, Paris 1972.

[39] *de Trin.* III.9.16; *City* XXII.24.

[40] *de Gen. ad Lit.* V.20.40-V.23.46.

[41] *de Gen. ad Lit.* IV.33.51; IX.17.32.

[42] Gerard J.P. O'Daly, 'Augustine on the origin of souls', Festschrift H. Dörrie, forthcoming.

[43] *City* XXII.24.

[44] I am grateful to Fritz Zimmermann for this observation.

goes away once he has built, but whose work stands, even though he is missing and goes away. The world could not last like this for the duration of an eye blink, if God were to withdraw his governance from it.[45]

The difference from occasionalism is again instructive; for Malebranche takes over the contrast with an architect's house,[46] but what he adds, and what is missing, so far as I know, from Augustine, is the idea that continuation depends on continuous re-creation. Augustine's suggestion may be a step in the direction of occasionalism, but he will have conceived it instead as a step beyond Platonism. For, on the one hand, Plato makes the Creator say at *Timaeus* 41A-B (available to Augustine in Cicero's Latin) that because the *kosmos* has been created, it can be destroyed, but that the Creator's goodness will prevent its dissolution. This dependence on the Creator's goodness is the point of similarity. Augustine even refers to Plato's statement, and uses an expression which suggests that he may have had access to the comments of the Middle Platonist Atticus.[47] But Augustine's difference from Platonism comes in his refusal in the passage quoted to assimilate God to the builder of a house. For in the context of Plato's discussion, Atticus, Galen, and the Christian Hippolytus and (at a later date) Philoponus, take Plato's creator to be replacing worn parts in the universe, exactly like the maker of a house, ship, wagon or other artifact.[48] Augustine evidently wants the universe to be more dependent on God than that.

The Platonists are made to take a further step in this direction in a later treatise by Zacharias, the sixth-century Christian who describes Ammonius' Neoplatonist circle in Alexandria. Not only does Zacharias record the view of Plato in *Timaeus* 41A-B, but at first he also makes his Platonist interlocutor describe God as still creating (*dēmiourgein*) the world. Thereafter, however, he makes him retreat, complaining that this is inconsistent with Platonist principles. But he none the less claims to have heard a Platonist say that, if God were so much as to divert his attention, the world would be utterly destroyed.[49]

Augustine's relation to the Platonists is one of initial acceptance and of subsequent borrowing, adapting, distancing himself and urging them to follow. I have just suggested that he borrows for his idea that only voluntary agents are efficient causes, and that he adapts and alters Plato's ideas on how the world continues in existence. In *City* XII.25, he criticises Plato for entrusting part of the work of creation to the lesser gods. But in XII.27 he

[45] *de Gen. ad Lit.* 4.12.22.

[46] Malebranche, *Entretiens sur la métaphysique*, dialogue 7, section 8, translated by M. Ginsberg, pp. 186-7.

[47] *City* XIII.17: 'God, whose will, as Plato says, no force can overcome' recalls a portion of Atticus preserved by the Christian Eusebius, *Praeparatio Evangelica* 15.6.16. In Atticus and several other authors, Plato's discussion is built up into an argument for divine providence: since the world cannot survive on its own, we must agree that there is providence, in order to account for its continuation (Seneca *Ep. Mor.* 58, 28; Atticus, op. cit. 15.6.2; Galen, *On Medical Experience* 19 (translated from the Arabic version by R. Walzer); similarly Philo *de Op. Mund.* 2.9-11).

[48] Atticus, op. cit. 15.6.13; Galen ap. Philop. *Aet.* (Rabe) 600; [Galen] *Historia Philosopha* 17 in Diels 609, 15; Hippolytus, *Philosophumena* (= *Ref.*) 1.19.5, Diels 567, 22-6; Philop. *Aet.* 228, 5-10.

[49] Zach. *Ammonius* or *de Mundi Opificio* PG 85, 1072A; 1128C.

argues that the Platonists implicitly endorse his own view that God is the creator of all things, since they say that God's mind contains the ideal Forms of all creatures. It might seem that Neoplatonism could not possibly concentrate power in a single being, since it recognises not one, but three, levels of reality, the One, the World Intellect and the World Soul. But even here Augustine says to Porphyry at one point (*City* X.29) that, in calling these three levels gods, he is really referring to the Trinity. As a matter of fact, the Platonists did restrict the spread of causal and creative power even if not quite so severely as Augustine. I shall distinguish three kinds of way in which they did so.

Alternatives to occasionalism: the Platonists

First, Plato says that only *souls*, not *bodies*, can originate motion; bodies merely transmit motion by banging into each other.[50] The Neoplatonists Plotinus, Proclus and Simplicius accept this restriction on bodies, although they elaborate Plato by recognising as active not only souls, but also the higher realities, the Intellect and the One.[51] It is this tradition which presumably influenced Augustine, and perhaps helped to encourage Ghazālī, in their remarks on voluntary agents.

Secondly, there is a threefold distinction among causes of which we find traces in Porphyry, Iamblichus and Proclus. To start with, there are causes in the proper sense (*kuriōs aitia*).[52] Then there are cooperative causes (*sunaitia*).[53] And finally one kind of cooperative cause is the merely instrumental (*organikon*).[54] The idea of cooperative and instrumental causes is suggested by remarks in Plato.[55] Proclus uses the scheme in order to downgrade matter and enmattered forms, in other words causes in the sensible world, as being only *cooperative* causes, while the only causes properly so called are the One, Intellect and Soul.[56] Proclus claims to find this view about sensible causes in Plato's *Phaedo*, presumably in 95E-99D. I do not know whether his predecessor Iamblichus is expressing agreement when he insists, against Porphyry, that the things employed in sacrifices (for example the heart of a cock) are not the *causes* of the magical results, but only the cooperative causes (*sunaitia*) or a *sine qua non*.[57] At any rate, *Proclus'* use of the concepts has the effect of removing causality in the proper sense from the sensible world. Moreover, his ideas on the instrumental (*organikon*) cause

---

[50] *Phdr.* 245C-246A; *Laws* 896A-897B.

[51] Plotinus 3.6.12 (48-57); 3.6.15: Soul, Intellect and One are active, bodies are passive, but (a complication, 3.6.7 and 12) the matter of bodies is neither. Proc. *Elements of Theology* 80: bodies are passive, only incorporeal entities are active. Simpl. *Commentary on Epictetus' Enchiridion* (Dübner) pp. 97-8: bodies are not, while souls are, among the things that can act as causes.

[52] Porph. ap. Simpl. *in Phys.* 10,25-11,3; cf. Procl. *Elements* 75; *in Tim.* (Diehl) 1.1,24-4,5; *in Prm.* (Cousin) 1059, 3-19. I am grateful to Robert Sharples for several of the references in this chapter.

[53] Same references and Iamblichus *de Mysteriis* 5.8-10.

[54] Porph. loc. cit.; Procl. *Elements* 75 and *in Tim.* 1.261, 15.

[55] *Phd.* 99A-C; *Tim.* 46D; *Statesman* 281D-E.

[56] Procl. *in Prm.* 1059, 3-19; *in Tim.* (Diehl) 1.1,24-4,5.

[57] Iamb. *de Myst.* 5.7-10.

may be part of what lies behind the view which we find in Zacharias, and which we found earlier in the chapter among the so-called 'philosophers' of Islam, that God is the cause of existence, while fathers are only the instruments (*organa*).[58]

A third Platonist view has to do with one particular type of causality, namely, creation. This time, Proclus clearly goes beyond Iamblichus, since Iamblichus allows that his gods (who, on one interpretation, are all at the level of the One)[59] create non-eternal beings only indirectly: they are not causes (*aitioi*) but only pre-causes (*proaitioi*) of these.[60] Proclus does much more to concentrate creative power in the One, by using a variety of principles such as the principle that the cause is greater than the effect. Thus at *Elements of Theology* 56-7, he fills this principle out in the following way: the Good, that is, the One, operates before the Intellect, in that it produces the Intellect. But it also produces all the same things as the Intellect produces and produces them in greater measure (*meizonōs*), in that it is the cause of the Intellect's ability to produce. Finally, it continues to produce things at a lower level than the Intellect can reach, for it produces matter, which Intellect does not. This enables Proclus to say, in contrast to Iamblichus, that all things are created by the gods *immediately* (*amesōs*), and not at several degrees of remove. This is true even of matter, he says, for the One (even if not the gods of other levels) is everywhere.[61] There are signs that Proclus was not the first Neoplatonist to concentrate creative power in a single being. Origen[62] (the Neoplatonist, not the Christian) and Syrianus[63] may have done so before him, and Simplicius[64] continued to do so afterwards. If Augustine had known of these developments, he might have thought that they were some improvement.

---

[58] Zach. *Ammonius* or *De Mundi Opificio* PG 85, col. 1041B-C.

[59] See John Dillon, *Iamblichi Chalcidensis in Platonis Dialogos Commentariorum Fragmenta*, Leiden 1973, 412-16, but see the reply by H.D. Saffrey and L.G. Westerink in their edition of Proclus' *Platonic Theology*, vol. 3, Paris 1978, ix-lii. According to Dillon, Iamblichus places all the other gods at the same level as the One. Proclus calls these gods henads, but does recognise gods at lower levels. However, they are gods only through participation in henads, which are the gods proper. (For Proclus, see e.g. *Elements* 128-9; 162-5; for Iamblichus e.g. Procl. *in Tim.* 1.308, 19-23).

[60] Iamblichus as elaborated by Ammonius, *in Int.* 134, 3-6; 136, 4-8.

[61] *in Tim.* 1.209, 13-26.

[62] Origen is credited with a work called *The King is the Only Creator* (Porph. *Life of Plotinus* §3). There is disagreement whether he is referring to the One, or eliminating it and recognising only the Intellect. If he is eliminating the One, this will not be, as Praechter argued, because of any Christian influence on Alexandrian Neoplatonists, but simply because of a Middle Platonist background. So John M. Rist, 'Basil's "Neoplatonism": its background and nature', in Paul J. Fedwick (ed.), *Basil of Caesarea: Christian, Humanist, Ascetic*, Toronto 1981, 137-220, esp. 156; 165-6; 169. However, a reference to the One is argued by A.C. Lloyd in the *Cambridge History of Later Greek and Early Medieval Philosophy*, ed. A.H. Armstrong, Cambridge 1967, 316; and Ilsetraut Hadot, *Le Problème du néoplatonisme Alexandrin: Hiéroclès et Simplicius*, Paris 1978, 48-9; 115-16.

[63] Syrianus *in Metaph.* 59, 17 applies similar causal principles to the two supreme entities, the One and Being.

[64] Simpl., *Commentary on Epictetus' Enchiridion* (Dübner) pp. 97-8: it is the god of gods who creates everything.

CHAPTER TWENTY

Principles of Causation among Platonists and Christians

In the last two chapters, I have surveyed three types of theory on cause and creation. I shall now put together some of the principles of causation which have emerged from this and from the earlier discussion. The Platonists and their Christian contemporaries were highly original in their ideas about causation. In some cases, but not in all, I think their innovations constituted an improvement.

Causation without changes

One question which I am now in a position to discuss is whether causation involves *changes*. The Greeks progressively examined the different points at which change may be present in causal situations, and revealed how it may also be absent. This investigation culminated in the findings of Middle Platonism and Neoplatonism. But I must start with a retraction. In an earlier book, I attributed to Aristotle the idea that the efficient cause is 'that whence comes the origin of a change'.[1] I stopped too soon: what Aristotle actually says is that the efficient cause is 'that whence comes the origin of change *or stability* (*ēremēsis, stasis*)'.[2] An example in the same chapter illustrates the point: the presence of the navigator (which no doubt involves a lot of activity and change) is the cause of the ship's safety (that is, of its *not* suffering a certain change). Conversely, the absence of the navigator (which involves no particular activity or change – he may be asleep) is the cause of a certain change, namely shipwreck.[3] Even though shipwreck is a new state of affairs, safety is not. And I would add that there are other parts of Aristotle's theory in which efficient causation does not seem to involve a new state of affairs.[4]

---

[1] Sorabji, *Necessity, Cause and Blame*, ch. 2, esp. 40-4.

[2] *Phys.* 2.3, 194b30; 195a23; Although Aristotle might be thinking only of an origin of *coming* to rest, comparison with other passages suggests that he would include also an origin of *remaining* at rest, or being stable. Thus see the account of nature in *Phys.* 2.1 as an inner origin of change *or resting* (*ēremein, stasis*, 192b14; b22).

[3] *Phys.* 2.3, 195a11-14.

[4] For example, the transparent spheres which carry the stars have souls. And these souls, inspired by the divine intelligences, are efficient causes of the spheres moving. Since this motion has no beginning, it is not a new state of affairs. However, the example is less clear than the subsequent Platonist ones, since motion is itself a form of change, and involves the occupation of new places.

Avicenna was later to pride himself on shifting away even from this reduced reference to change, by defining the Aristotelian efficient cause as the origin of *being*, not of change.[5] It has been suggested[6] that some of Aristotle's followers had already begun that shift.

Aristotle's example provides an answer to some of the questions about the role of change in causation. A first question is whether the *effect* has to be a change or a new state of affairs. I shall concentrate throughout on particular causes and particular effects (his smoking caused his cancer, rather than smoking causes cancer). The example of the ship's safety demonstrates that a particular effect does not need to be a change or a new state of affairs. There is an admirable treatment by Jonathan Barnes of the ancient sceptical discussions of cause, which brings out how the point was reinforced by Galen. For Galen draws attention to the existence of a whole class of such causes, by giving them a name: preservative (*phulaktika*) causes are distinguished from active or productive (*poiētika*) causes, precisely by the fact that they *prevent* change, rather than producing it. For the Stoics too the so-called cohesive causes (*sunektika*) are causes of a thing's holding together in existence, not just causes of change, and they further give the example of the stones so wedged as to keep an arch in being. Similarly the sceptic Sextus cites, although he does not endorse, the example of a supporting pillar which causes a lintel to remain immobile. In none of these cases is the effect a change or a new state of affairs.[7]

On a second question there is less agreement: does a cause itself have to engage in change, or even do anything at all? It is Michael Frede who has led the way here: in his classic discussion of the Stoics on cause, he suggests that it is their innovation to insist that a cause has to engage in activity (*energeia*), and to *do* something (*dran, facere*).[8] It was an idea that proved influential, and Sextus presents it as a common view,[9] even though he, and perhaps the Stoics too, recognise that other things are loosely called causes. Aristotle's voice is more ambiguous, but he could be thought of as anticipating the Stoics, in so far as he complains against Plato that Platonic Forms cannot cause anything, since they do not engage in activity (*energein*).[10] But is it a sound suggestion, either as a report on, or as a recommendation for, usage? Jonathan Barnes has endorsed it for the kind of cases discussed by Sextus.[11] But it is only in

[5] Avicenna *Metaphysics* 6.1, ed. S. van Riet in the series Avicenna Latinus, p. 292, 20-2 of Avicenna, *Liber de Philosophia Prima*.

[6] Michael Frede, 'The original notion of cause', in Malcolm Schofield, Myles Burnyeat and Jonathan Barnes, eds, *Doubt and Dogmatism*, Oxford 1980, 232-3, but only as a tentative interpretation of the puzzling passage at Clement of Alexandria *Stromateis* 8.9 (26, 4). For a different interpretation, see Jonathan Barnes, 'Ancient skepticism and causation', in Myles Burnyeat, ed., *The Skeptical Tradition*, University of California Press, forthcoming.

[7] Jonathan Barnes, 'Ancient skepticism and causation'. For Galen, Barnes refers to *Art of Medicine* vol. 1, pp. 365-6 (Kühn); *Synopsis Librorum de Pulsibus* 9.458 (Kühn); *Causae Continentes* 1.3; 9.2-3 (*CMG*, suppl. orient. 2, pp. 133, 15-19; 140, 6-17). For Sextus he cites *M* 9.229. For the Stoics, see M. Frede, op. cit., 243, and for their example of the arch, Clement *Stromateis* 8.9.

[8] See Sext. *PH* 3.14; Clement *Strom.* 1.17 and 8.9; Seneca *Ep.* 65, 4 and 11; M. Frede, op. cit.

[9] Sext. *PH* 3.14.

[10] *Metaph.* 12.6, 1071b14-17, discussed by Julia Annas, 'Aristotle on inefficient causes', *PQ* 32, 1982.

[11] Barnes, op. cit.

the thinnest sense that the pillar mentioned by Sextus has to *do* anything in order to keep the lintel immobile, or that Aristotle's absent navigator has to *do* anything in order to be the cause of the shipwreck. The one has to stay immobile, and the other has to be absent, but neither of these is much of a doing, and certainly neither is an engaging in change. Similarly a fracture might cause a loss of fluid not so much by actively doing anything as simply by *permitting* a flow. We could collect a whole group of such cases under the heading of *permissive* causes. In relation to Aristotle's navigator, the Stoics would deny that a cause can be *absent*,[12] and in regard to the fracture, they would insist that a cause must be a *body* – a quality can be a body, for the Stoics, but a fracture could not. None the less, there are other examples to ask them about; for it would seem that a soldier's button could cause a bullet to ricochet simply by being *present*, and that the stones in an arch could cause the arch to stay up simply by being *wedged*. What is the *doing* here? Even in relation to the Stoics' favourite examples, it is hard at first to understand the demand for *activity*. For Chrysippus claims that the *perfect* and *principal* cause of a cylinder's rolling, when pushed, and of its continuing to roll thereafter, is its own nature and volubility of form (*volubilitas formae*).[13] The pushing is not a *principal*, but only an *antecedent*, cause of the rolling. But what would Chrysippus say that volubility of form has to *do* when it causes the rolling? He needs to avoid the trivialising answer that it simply causes the rolling. Perhaps his reply would be to present his view as a theory in physics: the cylinder's nature is the pervasive *pneuma* or gas within it in a certain state of tension. What it *does*, when it causes the rolling, is to alter its state of tension, and a corresponding story might be told about the soldier's button and the pillar or arch. If that is the story, one consequence will be that the *pneuma* in the motionless pillar or arch *maintains*, rather than alters, its state of tension; so the doing will *still* not be an *event* or *change*.[14] Moreover, in each example, the doing will be required only by physical theory; it will not be implied by the concept of cause itself.

Meanwhile, if we step right outside Stoic theory, we shall have little reason to think that every cause must *do* something. Aristotle allows efficient causes to include such things as a raid or thirst. These may *be* doings, or *involve* doings, but it would be inappropriate to say that they must *engage* in doings in order to be causes. The atomists make void, as much as atoms, cause movement.[15] And if we include among effects a state like the sponginess of a particular object, they might say that it is caused by an atomic structure, without thinking that that structure has to *do* anything.

A third question about the role of change has been discussed in Chapter 15. It concerned Aristotle's idea that any new state of affairs has to be triggered off by the occurrence of a change, even if no change occurs within

---

[12] Clement *Strom.* 8.9.

[13] Aulus Gellius *Noctes Atticae* 7.2.11; Cic. *de Fato* 42-3.

[14] Even if Samburksy is right, *Physics of the Stoics*, London 1959, 31-2, that Stoic *pneuma* is always in motion, at least the *pneuma* in the pillar will not have to *change* its pattern of motion.

[15] Epicurus *Letter to Herodotus*, preserved in Diog. Laert. *Lives*, 10.44; Aristotle *Phys.* 4.7, 214a24-5; although the reference may be to void's facilitating movement in general, rather than to its causing any particular motion.

the causal agent. I tried to show how Christians might resist the need for a triggering change, in the case of their changeless God creating the universe. A fourth question discussed in that same chapter was whether such a God would not at least have to undergo a change of will, and I argued that he would not.

So far the need for change has been denied at one point after another in the causal story, but it has not yet been denied that there need be any change at all. Even the Christian account of creation allows a quasi-change, for it conceives of the created universe as having a beginning, and that is a sort of change. It is therefore a further and a final step when the Platonists postulate causation without any change at all. On their view, God or the One is indeed causally responsible for the existence of the physical universe and of the other levels of reality below himself, the World Intellect and the World Soul. But all these lower levels are themselves beginningless, so that there is not even the quasi-change of a beginning. To show that their view makes sense, they continually offer homely examples to show that there need be no change or beginning in a causal relationship. Surprisingly enough, Christians use the same examples, for in another context they too want to postulate this kind of causation: Christ the Son is beginninglessly generated from God the Father. The illustrations include the sun's illumination of the moon, sun and light, shining bodies and light, body and shadow, fire and light and a foot which has for ever and without beginning caused an imprint in the dust.[16]

It may be clearer that there is a causal relationship, that the foot causes the dent to persist, if I substitute a springy cushion for the dust. For then the removal of the foot would undo the effect. In the story as now told, change is still relevant to there being a causal relation, but it is a merely *imagined* change: we think that the cushion *would* spring up, *if* the foot were removed. Perhaps we should also suppose changes to be going on elsewhere in the universe, in order, for one thing, that we may imagine time to be passing. But the point will be that none of these changes need come within a hundred miles of the foot and the cushion. Or at least the concept of cause does not require that they should. Admittedly, modern science may reveal that some of the Platonist examples do involve changes. For we believe nowadays that light is cast through the successive emission of photons. But I do not know whether there would be parallel theories about the imprint in the cushion.

---

[16] Platonist views: Taurus ap. Philop. *Aet.* (Rabe) 14,24 and 147,8; Sallustius *de Diis et Mundo* 7: Aug. *City* 10.31; Aeneas of Gaza *Theophrastus* (Colonna p.45,21-46,23); Zacharias of Mitylene *Ammonius* or *de Mundi Opificio* PG 85, cols 1077B-1080C; 1096B; Philop. *Aet.* (Rabe) 14,20-8. Jewish texts: *Wisdom* 7.26; Philo *de Migratione Abrahami* 8.40. Christian texts: Hebrews 1.3; Origen *On First Principles* 1.2.4; and a fragment cited ad loc. in Koetschau's edition, tr. Butterworth, namely, fr.1 of *Homily on the Letter to the Hebrews*, in Lommatzsch's edition, vol. 5, p.297 (= Pamphilus' *Apology for Origen* 3, in Lommatzsch 24.328); Dionysius of Alexandria Athanasius *de Sententia Dionysii* PG 25, cols 501C-504B, 508A-B; 512C; 516C; Theognostus ap. Athanasius *de Decretis Nicaenae Synodi* PG 25, col.460C; Alexander of Alexandria *Epistolae de Ariana Haeresi* PG 18, cols 557C; 565C; 576A-B; Athanasius *contra Arianos* 1,8; 1.12; 2.34-6; 4.2 (PG 26, cols 25C-28C; 37A; 220A-225A; 469C; 480C-D); Basil of Caesarea *in Hex.* 1.7 (PG 29, col 17B-C); cf. *de Spiritu Sancto* PG 32, col.186B-C; Gregory of Nyssa *de Fide* (PG 45, col.140A-B); *Oratio de Deitate Filii et Spiritus Sancti* (PG 46, col.560C-D); Ambrose *Hex.* 1.5; the analogy is not fully approved by Gregory of Nazianzus *Oratio* 32 (PG 36, col.169B).

And at any rate, we do not have to believe that there are such changes, in order to be assured that the foot causes the imprint to persist, or that the sun causes something to be illuminated.

It may be objected that I am misrepresenting the Platonists' intention. For Michael Frede has pointed out that the Neoplatonists were influenced by the Stoic idea that a cause must be *active* (*poioun, drastērion*).[17] And Plotinus does, indeed, say that the One has an *activity* (*energeia*) in virtue of which it produces the lower levels of reality.[18] But this activity does not involve change. Plotinus is reluctant to describe it in any way, even, as was seen in Chapter 10, as a kind of thinking. Admittedly, when lower levels of reality produce things in their turn, this is said to be by means of thinking.[19] But, even then, the thinking involved, in the case of the Intellect and in the case of Nature (the lowest phase of Soul), is described as a static and changeless thinking.[20] Nor should the wrong conclusion be drawn from Plotinus' talk of procession and reversion, and from the commentators' talk of emanation, in describing the dependence of one level of reality on another. For these terms are to be understood only as metaphors describing a static relationship.

It is not always appreciated how early the examples of changeless causation appear. As a source, commentators have considered an unknown post-Iamblichan commentator, or Porphyry, or scattered hints in Plotinus,[21] and the particular example of the foot has been connected with Porphyry.[22] But I have not seen it remarked that the illustration of the footprint may be earlier than Porphyry, and that the example of the moon's light goes back as far as the Middle Platonist Taurus.[23] It could have been suggested by reflection on Plato's rather different comparison in *Republic* 509B between the sun and the Form of the Good. In the Judaeo-Christian tradition, the examples are also found very early. Already in *The Wisdom of Solomon* there is a description of God's wisdom as an effulgence (*apaugasma*) from eternal light (*phōs*). In Christian thought, God's Wisdom tends to be equated with Christ, and we find in Paul's Letter to the Hebrews that it is God the Son who is called the effulgence (*apaugasma*) of God. Origen cites both scriptural texts, *Wisdom* and *Hebrews*, and thereafter the idea of Christ as an effulgence (*apaugasma*) is

---

[17] Frede, op. cit., 219, cites Iamblichus ap. Simpl. *in Cat.* 327, 6ff.; Damascius *in Philebum* 114, 6W.

[18] E.g. Plot. 5.4.2 (27-39).

[19] Plot. 3.8.3-5; Porph. ap. Procl. *in Tim.* (Diehl) 1.396, 5; Iamblichus *de Myst.* 3.28; Procl. *Elements* 174; *in Prm.* (Cousin) 791, 14-28; *in Tim.* (Diehl) 1.352, 8; 362, 2; *de Dec. Dub.* 8, 32-5. Compare the view of Philo, documented in Chapter 13, which reflects Middle Platonism: God creates by thinking (*intelligendo facit, de Prov.* 1.7 Aucher).

[20] For Plotinus' account of Nature producing the sensible world by contemplation, see 3.8.3-5.

[21] References in J. Pépin, *Théologie cosmique et théologie chrétienne*, Paris 1964, footnote on pp. 279-80. Cf. M. Baltes, *Die Weltentstehung des Platonischen Timaios nach den antiken Interpreten*, vol. 1, Leiden, 1976, 163-6.

[22] P. Courcelle, *Late Latin Writers and Their Greek Sources*, Cambridge Mass., 1969, p. 187, translated from the French of 1948.

[23] Taurus ap. Philop. *Aet.* (Rabe) 147,8; as for footprints, Courcelle cites Porphyry's talk of footprints (*ichnē*) of the forms with reference to the creation of the world (ap. Procl. *in Tim.* (Diehl) 1.383, 19; and ap. Philop. *Aet.* (Rabe) 547, 8). But see the passage translated below from Plot. 6.7.17, in which life is described as a footprint (*ichnos*) of its bestower, although the reference is admittedly not to the life of the *physical* world.

used repeatedly against the heresy of Arius. The idea is that Christ does not begin to exist, but is co-eternal with the Father, in the way than an effulgence coexists with the light source.[24]

The Platonist view may be relevant to present-day controversy. For Donald Davidson's is now perhaps the most influential account of causation, and he has argued that, in a true singular causal statement, such as 'the flood caused the famine', the causal relation reported holds between *events*. Moreover, the causal relation never holds between *facts*.[25] Admittedly we may talk of *things* and say that a *brick* broke a window. But this is permissible only because we can expand the account of the cause to embrace an *event*, the *movement* of the brick.[26] On the most natural understanding, an event is something which involves *change*, and again on the most natural understanding, the Platonists' causal statements are singular ones, in that they have a singular subject, the sun or foot, and that the object acted on is an individual, the moon or the cushion. I do not know whether Davidson intended to cover *all* singular causal statements, and I do not know whether he would after all count the beginningless dent in the cushion as an *event*. My aim is not to be polemical. Rather, the idea that every particular causal relation integrally involves change is tempting in itself, and might readily be suggested to the reader by Davidson's impressive and influential analysis. What I believe the Platonist examples show is that this tempting idea is false.

Causation, creation and time

The second area in which Platonists and Christians made innovations concerns the relation of creation and cause to *time*. I talked in Chapter 17 about the Platonist recognition, as early as Crantor – two generations after Plato – that we can make sense of the idea of creation without a temporal beginning. What I did not do was to bring out fully the enormous influence of this idea. It appears most startlingly in Thomas Aquinas, who says that a beginning of the universe does not follow from the fact of creation; it does not even follow from the fact of creation out of nothing.[27] It is simply an article of Christian faith. In the interim, the Platonist view was recorded by Augustine, who comments that the Platonists speak of the world as made (*factum*) and as having an origin (*initium*), without meaning to attribute to it any temporal beginning.[28] In Islam, we find Avicenna, in a passage referred to more than once, saying that motion had an origin (*initium*), in the sense of proceeding from the Creator (*creatoris*), but not a temporal beginning.[29] And Averroes

[24] References in note 16. Evidence for the early material is assembled by W. Böhm, *Johannes Philoponus: Ausgewählte Schriften*, Munich, Paderborn, Vienna 1967, 44-50; cf. E.R. Goodenough, *By Light, Light*, New Haven Connecticut, 1935.

[25] Donald Davidson, 'Causal relations', *JP* 64, 1967, 691-703.

[26] Donald Davidson, 'Agency', in R. Binkley, R. Bronaugh, A. Marras (eds), *Agent, Action and Reason*, Oxford 1971, 10 and 15.

[27] Aquinas *Aet.* §§6, 7 and 8 (tr. by Cyril Vollert *et al.*, op. cit.); *Scriptum Super Libros Sententiarum Magistri Petri Lombardi*, in 2, dist. 1, q. 1, a.2; solutio; *ST* 1, q.46, a.1 and 2.

[28] *City* X.31; XI.4.

[29] *Metaph.* 9.1.

too allows that the world can be created by God, without this implying a beginning.[30] Again, there is for Christians an analogy with what they want to say about the Son of God. Although he may not be described, thanks to the Council of Nicaea (A.D. 325), as 'created', he is to be characterised as 'begotten' or 'generated' (*gennētos, genitus*), and as having an 'origin' (*archē, initium*), in a non-temporal sense. Yet his being generated from the Father as origin in no way implies that he had a beginning.[31]

The other point at which Platonists and Christians introduced something new has to do with the temporal relation of cause to effect. I have pointed out elsewhere that Aristotle is aware, despite some occasional wavering, that a cause need not precede its effect, but may be simultaneous, as, in his view, is eclipse and the alignment of heavenly bodies.[32] Barnes has pointed out that this was fairly widely appreciated, only Plato and Sextus constituting exceptions,[33] and he has also drawn attention to a very interesting discussion in Sextus of how a cause must be simultaneous, not indeed with its effect, but with *the causing of* its effect.[34] This is already suggested by Aristotle's discussion of how the builder building is simultaneous with the house being built, although the builder is not simultaneous with the house.[35] But the Platonists break entirely new ground. For, in effect, they imply that there need be no temporal relation at all. Thus in some sense Intellect, although a timeless entity, creates soul. There is a parallel with the Christian belief that a timeless God can create a temporal world. I do not think that this innovation is something for which the Platonists and Christians can take credit, since it needs to be made intelligible *how* something timeless can stand in causal relations, and this proved in Chapter 16 a harder task than showing how something *changeless* can.

Creation not out of matter

There is a third aspect of causality in regard to which the Platonists had more to offer than has been recognised. In Chapters 13 and 15, I commented that creation not out of matter is, in a sense, accepted by Neoplatonism. The Neoplatonist Hierocles said that God needs no matter for his creation.[36] This has been used as evidence for a distinct school of Neoplatonism in Alexandria under Christian influence.[37] But, in fact, the statement that God needs no

---

[30] Averroes in *Tahāfut al-Tahāfut* (Bouyges) pp. 124; 162, tr. van den Bergh pp. 73 and 97.

[31] Origen *On First Principles* 1.2.4; 1.2.9; *Homily on Jeremiah* 9.4 (cited by Koetschau, ad *On First Principles* 1.2.4); Basil of Caesarea *adv. Eunomium* 2.17-18; Gregory of Nyssa *contra Eunomium* 1.33; 1.42; 8.5 (PG 45, cols 393-7; 461-4; 784-800; Ambrose *de Fide* 1, ch. 9, sec. 60.

[32] Sorabji, *Necessity, Cause and Blame*, 43.

[33] Barnes, op. cit., citing Plato *Philebus* 27A; Sext. *PH* 3.25-8; *M* 9.232-6.

[34] *PH* 3.25-8; *M* 9.232-6.

[35] *Phys.* 2.3, 195b16-21.

[36] Hierocles *de Providentia*, ap. Phot. *Bibliotheca*, codex 214, 172a22ff. Bekker (= vol. 3, p. 126 Henry) codex 251, p. 460b22ff. Bekker (= vol. 7, p. 189 Henry); 461b7ff. Bekker (= vol. 7, p. 192 Henry).

[37] This was the hypothesis of K. Praechter, in a series of publications from 1910 to 1913; it is answered by I. Hadot, *Le Problème du néoplatonisme Alexandrin: Hiéroclès et Simplicius*, Paris 1978, who deals with this piece of evidence on pp. 86-9.

matter for his creation is accepted also by Plotinus' disciple Porphyry.[38] Indeed, it is the most orthodox Neoplatonism and follows from Plotinus' own theory. Matter is itself created, and no independent stuff is needed for its creation. So far from existing independently, it scarcely exists at all, and so far from being required for creation, it is itself created only at the ultimate point where creation peters out. There is no need, then, to invoke Christianity to explain God's dispensing with matter, and, indeed, it would be unwise to do so. For Hierocles incorporates into his account an extra element not accepted by orthodox Christianity, namely, that the creation has no beginning in time.[39]

This Neoplatonist conception of creation without matter might be described as a kind of creation *out of nothing*, albeit one with no beginning. But Porphyry prefers to put it by saying that God generates things 'from himself' (*aph' heautou*),[40] rather than out of nothing, and Plotinus also talks of coming *from* the One.[41] There was a distinction in Christianity also between 'out of nothing' and 'from God',[42] the former being used for the material universe, the latter being reserved normally for Christ the Son, on the grounds that he is of one substance with God the Father. But we saw in Chapter 18 how and why Gregory of Nyssa deviated from this way of talking, and allowed the material universe to come *from* God.

Gilson appears to overlook the Platonists when he argues that the Greeks do not have the conception of a creator who relies on nothing independently existent.[43] He supports his statement by pointing out that Plato's demiurge does not create matter, but only rearranges matter which already exists, while Aristotle's God is only a mover and not a creator at all. Instead, he finds the idea of a creator who genuinely accounts for existence, without relying on something existent, in Avicenna. From there it was taken by Albert the Great and by Peter of Auvergne, respectively the teacher of Thomas Aquinas and the Rector of his University, and it also appears in Thomas himself.[44] But in justice to the Platonists, it should be pointed out that Avicenna was inspired by the Arabic epitome of Plotinus' *Enneads* which went under the name of *The Theology of Aristotle*.[45] Moreover, Avicenna's view

---

[38] Porph. ap. Procl. *in Tim.* (Diehl) 1.300, 1; 396, 6 and 23; 2.102, 7.

[39] Hier. *de Prov.* ap. Phot. *Bibl.*, codex 251, p. 460b22ff. Bekker (= vol. 7, p. 189ff. Henry); 463b30ff. Bekker (= vol. 7, p. 198 Henry).

[40] Porph. ap. Procl. *in Tim.* (Diehl) 1.300, 2.

[41] 5.8.12.

[42] For the history of the distinction in Christian and non-Christian thought, see John M. Rist, 'Basil's "Neoplatonism": its background and nature', in Paul Fedwick (ed.), *Basil of Caesarea: Christian, Humanist, Ascetic*, Toronto 1981, 167; H.A. Wolfson, 'The meaning of *ex nihilo* in the Church Fathers, Arabic and Hebrew Philosophy and St Thomas', in *Medieval Studies in Honor of Jeremiah D. Ford*, Cambridge Mass. 1948, 355-70; and 'The identification of *ex nihilo* with emanation in Gregory of Nyssa', *HTR* 63, 1970, 53-60.

[43] Étienne Gilson, *The Spirit of Mediaeval Philosophy*, London 1936, ch. 4.

[44] Étienne Gilson, 'Notes pour l'histoire de la cause efficiente', *Archives d'Histoire Doctrinale et Littéraire du Moyen Age* 1962 (published 1963), 7-31.

[45] In P. Henry and H.-R. Schwyzer, edition of Plotinus *Enneads*, vol. 2, Paris, Brussels 1959, translated from the Arabic by G. Lewis.

was closer to the Neoplatonist than to the Christian position, since he held that the world was produced by the Creator without any temporal beginning.[46] Furthermore, *The Theology of Aristotle* had exerted an influence well before Avicenna (A.D. 980-1037). It was known, for example, to Farabi (*c.* A.D. 873-950), and, what is more, it inspired Farabi to claim that Aristotle made the universe to be created out of nothing in the sense of having a beginning.[47] Although this view cannot in fact be found in Aristotle or in Plotinus, Farabi found it, or thought he found it, in *The Theology*, and was thereby enabled to align that text with the common view of orthodox Muslim and Christian theology, according to which the universe began. Avicenna was, then, a long way from being the first to make common cause with Christianity in accepting creation not out of anything.

Cause unlike effect

One final pair of causal principles deserves discussion: the venerable, but misguided, principles that a cause must be *like* effect, and that it must be *greater* than its effect. The first principle, that of *likeness*, has always exercised appeal. We considered one version of it as the source of Gregory of Nyssa's theory of creation (Chapter 18). Even the atomists tried to find some analogy between macro-qualities and the micro-qualities which caused them.[48] The second principle, that of *greaterness*, was encountered in Chapter 19, and I would trace it back at least as far as Plato's *Lysis*.[49] Subsequent stages in its history have been charted in an invaluable article by A.C. Lloyd,[50] up to its use by Descartes for proving the existence of God. Plotinus accepts the principle that a cause must be *greater* than its effect, and we might therefore expect that he would equally accept the principle that it must be *like* its effect; for although the cause will possess the same characteristics in greater degree, it will at least possess the same characteristics. This is, indeed, how Lloyd

[46] Thomas Aquinas ascribes to Avicenna the view that God makes the (beginningless) world not out of anything (*ST* 1, q.46, a.2, ad 2). The relevant work of Avicenna is the *Metaphysica*, which was one of the portions translated into mediaeval Latin from his *Shifā* (*Healing*). The Latin is available as Avicenna *Liber de Philosophia Prima*, ed. S. van Riet, Leiden 1980, in the series Avicenna Latinus. There is a French translation from the Arabic by M.M. Anawati, Quebec 1952, and a German one by M. Horten, Halle and New York 1907. Gilson quotes references to God as creator from book 9, chapters 1 and 4; the main discussion of God as creator comes in book 6.

[47] Farabi, *Harmony of Plato and Aristotle*, ch. 11, ed. Nader, Beirut 1968, German translation by F. Dieterici in *Alfārābī's Philosophische Abhandlungen*, Leiden 1892, p.39. Robert Sharples has pointed out to me that belief in a creation of the world out of nothing is (incredibly) ascribed also to Aristotle's successor Theophrastus by a modern edition of Lactantius, as recorded in PL 6, col.124D, and the same belief features in a work in Arabic translation, ascribed to Alexander, as was noted in Chapter 15.

[48] See Theophrastus *de Sensibus* in George M. Stratton, *Theophrastus and the Greek Physiological Psychology Before Aristotle*, London 1917; and compare Plato *Tim.* 55D-56C; 61C-68D.

[49] In *Lysis* 219Eff. Plato introduces the principle that what makes a thing dear to us is itself more dear.

[50] A.C. Lloyd, 'The principle that the cause is greater than the effect', *Phronesis* 21, 1976, 146-56.

understands Plotinus and the Neoplatonists.[51] But so far as I can see, Plotinus, to his credit, does not interpret his principle that way at all at 6.7.17. After all, the One is not *very* like the Intellect. And the point he makes here is that a thing need not possess what it gives. The One lacks life and form, which it gives to Intellect, precisely because it is greater than life. This denial that the life-giving One must possess life is all the more striking, because Plato had supposed in the *Phaedo* that the soul which gives life to the body must be living.[52] The relevant lines deserve to be quoted:

> It is not necessary that a thing should possess what it gives. Rather, we should think in such cases that the giver is greater, and the given less than the giver. For that is what coming into being is like for things which have being. First, there must be something in a state of actuality, and the subsequent things must be in potentiality what comes before them. The first is beyond the secondary things, the giver beyond the given, because it is superior. If, then, there is something prior to actuality, it is beyond actuality, and so beyond life. If, therefore, there is life in the Intellect, the giver gave life, but is itself more beautiful and estimable than life. The Intellect, therefore, possesses life, without need of any complexity in the giver, and life is a footprint (*ichnos*) left by the giver, but [there is] not life in the giver. ... The Intellect therefore is itself a footprint of the One. But since it is a form and is developed and multiplied, the One is without shape or form; for that is how it produces form. If it were itself a form, the Intellect would have been its *logos*.[53]

There are traces here of a resemblance theory. For Plotinus is repeating one of Aristotle's concessions to resemblance theory, namely, that the thing affected is potentially what the agent is actually, and so becomes like it.[54] And the idea of a footprint may suggest, although it does not require, similarity. But the main point is that, in certain cases, the need for the cause to be greater actually *removes* any straightfoward resemblance. This denial of the need for resemblance is, I believe, a positive merit, even if it springs out of a view which I would not accept.

Platonists and Christians compared

I shall finish with a comparison between Platonists and Christians, for it will have emerged that their views on causation are often very close. Of course, there are differences, and Christian writers may reasonably want to stress these. But we have seen with other topics how close Platonism and Christianity sometimes came to be in Origen, Augustine, Boethius and the circle of Ammonius at Alexandria. On the present topic too the differences are not altogether easy to state.

We have seen that Platonists share with Christians the idea of a creator

[51] On the basis of independent evidence. He notes, however, that the resemblance theory can be modified, when it is combined with the 'greater' theory, and he gives an example of this from ps.-Dionysius.

[52] *Phd.* 104D-105E.

[53] 6.7.17(3-14; 39-43); cf. 6.9.6(30-3; 54-5).

[54] Arist. *DA* 2.5, 417a18-20.

whose work requires nothing independently existent. It is a major difference, of course, that the Platonists accept creation without a beginning. But this is not a difference in the *concept* of creation, according to the passages cited above from Thomas Aquinas, for his *concept* implies nothing about whether there is a beginning or not. It is rather a difference in views about creation, views which Christians derive from Revelation. Moreover, we have seen that this difference vanishes when Christians talk not about creation, but about the different relationship of generation which holds between the Father and the Son.

Many Christians distinguish their view of creation from the Platonist one by reference to the role of *will*.[55] Thus the Platonist analogy with body and shadow, or sun and light, is taken to exclude God's will and choice, by Basil, Ambrose, Aeneas of Gaza and Zacharias.[56] Thomas Aquinas distinguishes the Christian idea of creation as involving will, not mere necessity,[57] and the early fathers said the same.[58] But this way of drawing the distinction is too simple; for will does play a role, at least at the level of the Platonist Intellect, whether or not at the level of the One. The status of the One is a little uncertain. For Plotinus refers often enough to its will (*boulēsis, thelēsis, ephesis*),[59] but dismisses many of these references as inexact,[60] and in one passage denies that secondary beings exist through its will.[61] On the other hand, not all these references to will are dismissed,[62] and John Rist has recently taken them very seriously.[63] Moreover, whatever may be the case about the One, Plotinus does not feel able to be dismissive in connexion with lower levels of reality, presumably because Plato had made the demiurge exercise will (*boulesthai*) in passages of the *Timaeus* already encountered in Chapter 17.[64] Accordingly, Plotinus allows the lower levels of reality to create by will (*boulēsis*). He does so for Plato's God (2.1.1 (2 and 8)), who is perhaps here equated with Intellect, for the World Soul (3.7.11 (15-33)) and for individual souls (4.7.13 (11); 5.1.1 (4-5)). That the demiurge (who is sometimes identified with Intellect and sometimes with Soul) creates by will is repeated by later Neoplatonists.[65]

[55] Particularly useful on this subject are: I. Hadot, *Le Problème du néoplatonisme Alexandrin: Hiéroclès et Simplicius*, Paris 1978, 90-2; E.R. Dodds, edition of Proclus *Elements of Theology*, note to proposition 174; A.H. Armstrong, 'Elements in the thought of Plotinus at variance with classical intellectualism', *JHS* 93, 1973, 13-22; J. Pépin, *Théologie cosmique et théologie chrétienne*, Paris 1964, 502-6; C. Tresmontant, *La Métaphysique du christianisme et la naissance de la philosophie chrétienne*, Paris 1961, 190-4; 319-26; 364; J.M. Rist, *Plotinus: The Road to Reality*, Cambridge 1967, ch. 6; and ch. 8 of *Human Value*, Leiden 1982.

[56] In the passages cited above for the body-shadow, sun-light analogy.

[57] Aquinas *in Cael. I, lectio* 29, n. 12; *Scriptum Super Libros Sententiarum* ... , in 3, dist. 25, q.1, a.2, ad 9; *ST* 1, q.32, a.1, ad 3.

[58] Irenaeus, *adv. Haereses* 2.1.1; 2.30.9; 3.8.3; Aug. *City* XI.24; *Enarratio in Psalmum* 134, sermo 10.

[59] Plot. 6.8.9(44-8); 6.8.13; 6.8.15(1-10); 6.8.18(35-52); 6.8.21(8-19).

[60] 6.8.13(1-5 and 47-50); 6.8.18(52-3).

[61] 5.1.6(26).

[62] 6.8.21(8-19) appears to stand; see John Rist, *Plotinus*, ch. 6.

[63] Rist, *Human Value*, ch. 8.

[64] Plato *Tim.* 29E-30A; 41B; see Pépin, loc. cit.

[65] Iamblichus *de Myst.* 3.28; Hier. *de Prov.* ap. Phot. *Bibl.*, codex 214, p. 172a22ff.; codex 251, 461b7ff.; Procl. *in Tim.* 1.362, 2; 1.371, 4.

The difference from Christian accounts of creation lies not in the fact, but in the *manner*, of the will's being exercised. Thus Plotinus does not follow Plato's *Timaeus*, which repeatedly represents the demiurge as deliberating, that is, as thinking out how to achieve his aims.[66] Plotinus and his successors deny that there is any deliberation.[67] Further, the thought and will of the creative beings is turned in upon itself; it is not directed towards producing a creature and is not for the sake of the creature. The creator does not move in that direction at all, but stays in itself. It does not *seek* to create, but does so by being what it is.[68] Because of this, creation is represented as exemplifying a law which is said to spread far beyond things which possess choice (*proaeresis*), or even life. Anything which is already perfect will generate something of itself, and this is illustrated by the case of fire, snow, drugs and odorous objects. The effect is like the light which surrounds the sun.[69] It is in this last context that Plotinus denies the roll of will:

> If there is any second thing after [the One], it must exist without [the One] having been moved, or having inclined towards it, or having willed (*boulēthēnai*), or moved in any way.[70]

I think the point is not so much to deny that the One creates by willing, as to deny that what it wills is the existence of a creature.

Despite giving a role to the will, Plotinus does not hesitate to represent the creation of inferior levels of reality as necessary.[71] Some commentators speak as if the presence of necessity must imply the absence of will, but in fact the relations between will, free will and necessity are subtler than these discussions suppose. I shall not go into them now, since I have written about them elsewhere.[72]

I have found the Platonists and Christians especially interesting on four aspects of causality. Every period of ancient thought contributed to the study of causation, but the philosophical interest of some of the earlier contributions has already been brought out,[73] whereas that of the Platonists has not. I have now finished with the topic of creation, and in Part V I shall return to a subject that has not been handled since Chapters 2 and 5: the problems of the continuum and of atomism.

---

[66] Plato *Tim.* 29E-38E. See, for example, the reference to design, thinking and devising (*logos, dianoia*, 38C3; *mēchanasthai*, 37E3) in the long quotation in Chapter 6, and to wanting, thinking, and planning (*boulesthai, hēgeisthai*, 30A2-6; *logismos* 34A8) in the quotations in Chapter 17.

[67] Plot. 2.9.8(21-7); 3.2.1-3; 5.8.7; 6.7.1(28-43); Procl. *in Tim.* (Diehl) 1.321, 10-14; Ammon. *in Int.* 134, 10-18.

[68] Plot. 3.2.1-2; 5.1.6; Hier. *de Prov.* ap. Phot. *Bibl.*, codex 251, p. 463b30ff. Bekker(= vol. 7, p. 198 Henry); Procl. *in Prm.* (Cousin) 791, 14-28; Ammon. *in Int.* 134, 10-18.

[69] Plot. 5.1.6; 5.1.7(2 and 37); 5.4.1.

[70] 5.1.6(25-7).

[71] 2.9.3(1-18); 2.9.8(21-7); 3.2.2; 3.2.3(3-5); 4.8.6(12-13); 5.1.6; 5.1.7(37); 5.3.16(1).

[72] Sorabji, *Necessity, Cause and Blame.*

[73] On Plato, see e.g. Gregory Vlastos, 'Reasons and causes in the *Phaedo*', *PR* 78, 1969, 291-325 (repr. in his *Platonic Studies*, Princeton 1973); Sorabji, *Necessity, Cause and Blame*, 59; 206-8; Julia Annas, 'Aristotle on inefficient causes', *PQ* 32, 1982; On Aristotle: Sorabji, *Necessity, Cause and Blame*, chs 1-3. On the Stoics: Michael Frede in *Doubt and Dogmatism*; Sorabji in the same volume is a variant on *Necessity, Cause and Blame*, chs 3 & 4. On the Sceptics: Barnes, op. cit.

PART V

ATOMS, TIME-ATOMS AND THE CONTINUUM

CHAPTER TWENTY-ONE

Zeno's Paradoxes of Motion

The paradoxes of Zeno have at least two kinds of interest. First, they have a philosophical interest: I do not believe that they have all yet received a satisfactory solution, certainly not an agreed solution, even though there may be an impression that the work of modern philosophers or mathematicians has disposed of them. Further, I do not believe that all the necessary lessons have yet been absorbed. For often we can see that Zeno's outrageous conclusion does not follow, but in looking to discover exactly where he has gone wrong, we learn the most surprising things about space and time – things that we should not learn, if we dismissed the paradoxes as something already disposed of.

The second kind of interest is historical. For, as we shall see in Chapters 22 and 24, Zeno's paradoxes and other paradoxes in the tradition which he founded, helped to inspire the idea of atomism. Among other things, even time came to be thought of eventually as atomic.

Zeno may have been writing around 460 B.C. According to Plato, he wrote a book to defend Parmenides' view that there is only one thing in existence, by showing that those who believe that there are many things can be ensnared in paradox.[1] Later sources say that there were forty arguments in this book.[2] Aristotle preserves a set of four arguments designed to show motion impossible.[3] Since their avowed subject is not the number of things in existence, it may be easier to suppose that they come from another book: more than one book is ascribed to Zeno.[4] On the other hand, the four arguments all presuppose the existence of many things, so their moral could be taken as follows: either there is not a plurality of things, or, if you say there is, then at least you must admit Parmenides' other conclusion, that there is no motion.

The dichotomy, or the half-distances

The most challenging of the paradoxes of motion is the dichotomy, or, as I shall call it, the half-distances. It has the startling consequence that you cannot leave the room, or move from where you are. For (to reformulate Aristotle's wording) in order to reach any destination, you would first have to

[1] Plato *Prm.* 127A-128E.
[2] Proclus and Elias, quoted in DK under 'Zenon' A15.
[3] *Phys.* 6.2, 233a21-31; 6.9; 8.8, 263a4-b9.
[4] Suidas, in DK A2.

go halfway, then half the remaining distance, and half the remaining distance *ad infinitum*. In other words, you would have to traverse an infinity of sub-distances, in order to get anywhere at all. And that is impossible.

Aristotle's replies

I do not think that there is a single answer to the paradox, for the answer depends on what you consider the impossibility to be. Aristotle takes Zeno to have supposed that an infinity of sub-distances would require an infinite time. To that he gives the right answer, that we must distinguish between infinite divisibility and infinite length. The distance is infinitely divisible, not infinitely long, and therefore the time available is adequate, because it is infinite in the appropriate way, that is, infinitely divisible.

There is some independent reason for thinking that Zeno may have made the mistake which Aristotle attributes to him. For Simplicius records the actual wording of another of Zeno's paradoxes, which I shall give below (fr 1 and 2 DK). And it appears from this, although the matter is controversial, that Zeno probably made the same error of assuming that an infinity of parts will invariably make something infinitely large.

Aristotle does not rest content with his first answer to Zeno. And this is interesting because he has been accused, in a classic set of works by Harold Cherniss, of being very unfair in reporting his predecessors, particularly Plato.[5] Here we seem to have the opposite situation. When Zeno features in Plato's *Parmenides*, we learn nothing of the brilliance of his arguments – reasonably enough, since that would be irrelevant to the purposes of the dialogue. Aristotle, by contrast, not only reports and answers Zeno's four arguments on motion, but points out how Zeno's case could be strengthened.[6] The first reply to the half-distances is adequate *ad hominem*, but does not do justice to the facts. For somebody could raise Zeno's question not as one about the time available for traversing a distance, but as one simply about time. How can we live through another minute? For to do so we should first have to live through the first half-minute, then through half the remaining time, and so on *ad infinitum*. Equally, the question could be raised as one simply about space and motion, without reference to the time available: how can the infinity of sub-distances in any journey be traversed? The problem stems from the idea which Aristotle advances in *Phys.* 3.5, 204b9, and in many other places (see Chapter 14), that it is impossible to traverse an infinite *number* of anything. And to this it is no answer that the *length* is not infinite.

Aristotle's reply is that there is not an infinity of cross sections actually present in a period or a line. In order to insert one, you have to stop and do something either mentally or physically.

> So that if someone asks whether you can traverse an infinity either in time or in length, we must say that in a way you can, and in a way you cannot. For you

[5] Harold Cherniss, *Aristotle's Criticism of Presocratic Philosophy*, Baltimore 1935; *Aristotle's Criticism of Plato and the Academy*, Baltimore 1944.
[6] *Phys.* 8.8, 263a4-b9.

cannot traverse an infinity of actually existing <divisions>, but you can of potentially existing ones.[7]

An actually existing division is one that is actually marked out in some way, and Aristotle's denial that you can traverse an infinity of them fits perfectly with his account of infinity in *Phys.* 3.5-7. According to this account, infinity exists only in a restricted sense: the members of a collection are never more than finite, although it may be possible to *increase* their number without limit. If we apply this idea to the present context, Aristotle will not deny that the number of actually marked out divisions can be so increased. But he will insist that the number to be traversed will always remain finite, so there is no need to traverse more. This part of his answer is reminiscent of the view of modern finitist mathematics, that the points in a line are not an infinite collection, but are brought into existence as they are marked off.[8]

But if that were all Aristotle had to say, it would provoke the protest on Zeno's behalf that, as well as the divisions actually marked off on a line, there must be an infinity of positions at any one of which a further division could be actually marked off. And, in moving along a line, someone would have to traverse this infinity of positions. Since the number of such positions is more than finite, Aristotle surely cannot claim that all we need to traverse is a finitude.

Aristotle realises that something more needs to be conceded, and he puts his concession in terms of potentially existing divisions. The number of these is implied to be *more* than finite, since it is contrasted with the number of actually existing divisions. Whatever exactly Aristotle means by potentially existing divisions, I argued in Chapter 14 that he cannot afford to make this concession about them. For in *GC* 1.2, 316a10-317a12, he implies that the number of makeable divisions in a line is never more than finite, and in 317a9-12, he implies the same about the number of points in a line. Moreover, he is strongly motivated to say this, for he fears the threat that a more than finite number, whether of points (317a4-7), or of makeable divisions (316a25-34; 316b14-15; 316b26-7), will imply something he considers absurd, that an extended line is composed of nothing but unextended points. In *Phys.* 3, he is even more strongly motivated: he needs to rule out all collections which are more than finite. For he fears (*Phys.* 3.5, 204a20-6) that all such collections will contain *sub-collections* which are infinite in the same sense, and he believes this to be impossible. Indeed, his whole analysis of infinity as merely an extendible finitude is intended to avoid it. If he is to avoid it, he needs to apply his analysis to potential divisions too: even these ought to exist only in finite collections, or they will contain sub-collections which are more than finite.

If I am right, Aristotle cannot afford, on his own principles, to make the concession which he does make, that we traverse more than a finite number of potential divisions. Yet, I have claimed, some such concession is needed, because motion does involve traversing more than a finite number of *positions*.

---

[7] *Phys.* 8.8., 263b3-6.
[8] D. Hilbert and P. Bernays, *Grundlagen der Mathematik*, Berlin 1934, 15-17.

And now, going further, I would add that we ought to concede even more, for I see no conceptual impossibility in the idea of there being more than a finite number of *actual* divisions. Max Black and David Bostock have described a ball bouncing in such a way as to produce this result.[9] Although it could not happen in nature, we can conceive a ball taking half a minute to bounce up and down half a foot, then a quarter of a minute for the next bounce of a quarter of a foot, until at the end of a minute, it had completed an infinity of bounces. Aristotle would agree that reversals of direction constitute an *actual* division in a ball's motion. But he has an argument (albeit an unsatisfactory one) whose effect would be to rule out the proposed infinity of bounces. The argument is that a reversing object must spend a whole *period* of time at the point of reversal. For there will be a first instant of having reached the point of reversal, and (Aristotle assumes) a first instant of having left it. Since these instants cannot be next to each other (since instants never are), nor identical, they must be separated from each other by a period of rest.[10] The mistake, as I shall argue in Chapter 26, is to suppose that there need be a *first* instant of having left the point of reversal. Why should it not rather be that at any instant, however soon, after the instant of being *at* the point of reversal, our imaginary ball will already have left the point of reversal, there being no first such instant?

Before leaving Aristotle's account of the half-distances, I should mention an alternative version of the paradox which he records in *Phys.* 8.8, which we shall come across later as stemming from Plato's Academy. We are to imagine ourselves *counting* each time a person traverses one of the infinitely many sub-distances in his journey. If he can complete his journey, then we can complete an infinite count, but that is agreed by everyone to be impossible.[11] This version of the paradox is, however, unsatisfactory. For *counting* the sub-distances is unlike *traversing* them in a crucial respect. To count is to recite numerals, aloud or to oneself, while correlating them with something else. But the time needed for the recitation of a numeral does not diminish as we progress, in the same way as does the time needed for traversing diminishing sub-distances. There should be no surprise, therefore, that the task of *counting* cannot be completed; it does not follow that the task of *traversing* cannot be.

Other attempts at solution

We have now seen Aristotle's answers to the paradox, but there have been many other attempts at solution. The most challenging form of the paradox is that suggested by Aristotle's reformulation in *Phys.* 8.8: how can we traverse any distance, seeing that this involves traversing an infinity of sub-distances? One obvious point is that the paradox, as stated, allows us at least to go half way, so we need only double that feat in order to complete our

---

[9] Max Black, 'Is Achilles still running?' in his *Problems of Analysis*, London, 1954, 115; David Bostock, 'Aristotle, Zeno and the potential infinite', *PAS* 73, 1972-3, 37-53.

[10] *Phys.* 8.8, 262a31-b3; b21-263a3.

[11] *Phys.* 8.8, 263a7-11.

journey. But there is more than one way for Zeno to reply to that point. For one thing, the paradox could be reformulated, so as to say that *before* we get half way we must go a quarter of the way, and *before* that an eighth. The appropriate conclusion would then be that we could not even *start*. But this can be made clear even without reformulation. For it was only for the sake of argument that Zeno allowed us to go half way. In fact, that concession must subsequently be withdrawn. For if the paradox is effective at all, it shows that we cannot traverse any distance at all, however short. *Inter alia*, we cannot traverse the initial half-distance which, as a temporary concession, Zeno imagined us traversing.

There may be a temptation to object that, if only we imagine the journey differently divided, say, into three equal portions, it will present no appearance of impossibility. But to this Zeno could answer that, in traversing three equal portions, we should have to traverse his infinity of sub-distances. And the one will be possible only if the other is.

Another attempt at solution complains that we do not have to take an infinite number of strides, in order to traverse a distance. But, put that simply, such a reply is an *ignoratio elenchi*. Zeno is not maintaining that we must take an infinite number of strides. The paradox applies to someone who tries to proceed smoothly on roller skates. He too must go first half way, then half the remaining distance, and so on.

Some notable thinkers believe that mathematics has solved the problem. What Zeno failed to realise, on one view, is that an infinite series can have a finite sum, so the distance to be traversed, or the time needed, can be finite. Max Black cites for this kind of solution Descartes, Peirce, Whitehead, and we might add Quine.[12] But in effect it does not differ much from the point contained in Aristotle's first reply to Zeno, that infinite divisibility does not imply infinite length. As an *ad hominem* reply to the historical Zeno, it may be fair. But Aristotle points out, as indeed does Whitehead, that more can be made of Zeno's paradox than that.

A different mathematical solution is advocated by R. Courant and H. Robbins, and must be taken the more seriously for having been endorsed by Gregory Vlastos.[13] Courant and Robbins say that Cauchy's static definition of continuous approach 'disposes of Zeno's paradoxes as far as mathematical science is concerned'. In layman's language, the point could be put like this: say how close you would like to get to your destination, and we can show that by traversing a *finite* number of the sub-distances you can get closer than that. There is no need, then, to traverse an *infinite* number, because the gap separating you from your destination can, by the traversal of a *finite* number, be made smaller than any you care to name. I hope this does not distort the

[12] Max Black, 'Achilles and the tortoise', *Analysis* 11, 1950-1, 91-101 (repr. in his *Problems of Analysis*, and in Wesley C. Salmon (ed.), *Zeno's Paradoxes*, Indianapolis and New York 1970); Descartes, Letter to Clerselier, in the edition of Adam and Tannery, vol. 4, pp. 445-7; C.S. Peirce, *Collected Papers*, ed. Charles Hartshorne and Paul Weiss, Cambridge, Mass. 1931, 6.177-6.182; A.N. Whitehead, *Process and Reality*, New York, 1929, p. 107; W.V. Quine, *Word and Object*, Cambridge, Mass. 1960, p. 172.

[13] R. Courant and H. Robbins, *What is Mathematics?* New York 1941, p. 306; Gregory Vlastos, 'Zeno's race course', *JHP* 4, 1966, 95-108.

326 V. Atoms, Time-Atoms and the Continuum

argument, even though, when it is put as boldly as that, it invites an obvious complaint. For someone may reasonably protest that he does not want to get *close* to his home when he leaves work for the day; he wants to get *all the way*. And nothing short of an *infinite* number of sub-distances traversed is going to secure him that.

Max Black suggests as a solution that there is not an infinity of spatio-temporal entities in the path across which someone moves.[14] If we speak of an infinite sequence of distances, what we are calling 'distances' will not be spatio-temporal things, but merely pairs of numbers. J.O. Wisdom agrees,[15] distinguishing between a mathematical distance with its mathematical points, and a physical distance with its physical points. There is not room for an infinity of physical points within a physical distance, for physical points always have some size.

I cannot go so far as to agree that what we call 'distances' are only pairs of numbers. What I would accept is the more modest point that some idealisation is involved in Zeno's paradox. We can imagine someone trying to reach the door, but the door and his foot are each a jiggling mass of atoms with imprecise boundaries. So what exactly is the terminal point of his journey, and what the half-way point? We do talk of geometrical points, for example, of centres of mass, within physical bodies. But we are idealising when we do so, and the paradox cannot be formulated, if we refuse to idealise. None the less, although this should be admitted, I am not happy to let the solution turn on it. For I doubt if it is right to concede to Zeno that motion really does become inconceivable, once we try to consider it in an idealised way, as he does.

Among the solutions offered in antiquity, two are of especial interest. First, as we shall see in Chapter 22, atomists denied that there is an infinity of sub-distances to be traversed. For certain distances are so small as to be atomic, or indivisible, and we do not get smaller distances as the paradox supposes. An alternative and later view has been discussed in Chapter 5. It concedes that distances are infinitely divisible, but insists that we move by disappearing from one place and reappearing further on in 'leaps'. These leaps are not atomic, since they can be indefinitely small. But they share with atomic jumps the feature that only a finite number would be needed to bring us to our destination.

In modern times, the idea of atoms of space or time has again been put forward by Hilbert and Bernays and by Whitrow as a means of solving the paradox;[16] while Bergson suggested a more complex kind of quantisation. He maintained that only space was infinitely divisible; motion and time came in indivisible units, these units differing from atoms, in that they are to be viewed as qualities rather than quantities, and as dependent on

---

[14] Black's two papers appear in his *Problems of Analysis*, the first, 'Achilles and the tortoise', being reprinted from *Analysis*. This paper is reprinted also in Salmon, op. cit., along with J.O. Wisdom 'Achilles on a physical racecourse', *Analysis* 12, 1951-2, 67-72; James Thomson, 'Tasks and super-tasks', *Analysis* 15, 1954-5, 1-13; Paul Benacerraf, 'Tasks, super-tasks and the modern Eleatics', *JP* 59, 1962, 765-84; with a further reply by Thomson.

[15] Wisdom, op. cit.

[16] Hilbert and Bernays, loc. cit.; Whitrow, *The Natural Philosophy of Time*, 2nd ed., 199-200.

consciousness. Bergson was followed, with variations, by William James, A.N. Whitehead and Paul Weiss.[17] I do not believe, however, that such an elaborate solution is needed.

The proposed solution

The solution I would favour simply denies Aristotle's claim that it is impossible to traverse an infinity. In denying this, I must also depart from Aristotle's conception of infinity as something which is always incomplete, and which is to be understood in terms of the possibility of adding to a *finite* collection of divisions. Speaking in an idealised way, we can view ourselves as traversing a *complete* infinity of sub-distances every time we move. And why shouldn't we? I cannot foresee all the reasons why it might be thought impossible, and I shall consider only two.

One simple, but tempting, thought would be that the task of traversing an infinite series is never-ending. But this thought is ambiguous. Zeno's series of half-distances is never-ending in the sense that it has no last member, but all the same it is over at the point of destination.

A more interesting way of trying to show that it is impossible to traverse an infinity of sub-distances is to represent the task as parallel to certain other ones whose impossibility would be *agreed*. This was in effect part of Xenocrates' strategy, if it was he who pointed to the impossibility of *counting* the half-distances.[18] The most interesting argument along these lines was introduced by Max Black,[19] although his conclusion was the opposite one, namely that the impossible tasks were not really parallel after all to the task of making an ordinary journey. One example which Black used involved a marble being tossed an infinite number of times from a left hand tray to a right hand tray and back again. Another example, introduced into the ensuing literature,[20] involved a light appearing and disappearing an infinite number of times. The transitions were to be speeded up in Zeno's manner, say, half a minute for the first, a quarter for the seond, and so on, so that the whole operation took a minute. One impossibility, that the marble or the lamp switch would have to travel an infinite distance in a minute, seems an accidental feature of the examples. It could be removed by having the trays approach each other, or the light induced by a different mechanism. The alleged impossibility on which discussion focused had to do with the state of the marble or the light when a minute was up. The process seemed to require one of two states for the marble or light at the end of a minute, yet neither state would be possible, since every production of one state had been

---

[17]Henri Bergson, *Essai sur les données immédiates de la conscience*, Paris 1889, translated London 1910, as *Time and Free Will*, pp. 98-115; cf. *Matière et mémoire*, Paris 1896, 196-7; *L'Évolution créatrice*, Paris 1907, translated New York 1911 as *Creative Evolution*, 334-40. William James, *Some Problems of Philosophy*, New York 1911, 154-5; 181-7; A.N. Whitehead, *Science and the Modern World*, New York 1925, 186; *Process and Reality*, New York 1929, 53; 105-7; Paul Weiss, *Reality*, Princeton 1938, 233.

[18] See Chapters 14, 22 and 23.

[19] Black, 'Achilles and the tortoise'.

[20] Cited above.

superseded and cancelled by a production of the other.

In fact, however, there is no logical impossibility here, as was subsequently shown by Benacerraf.[21] No terminal state is implied, nor is any excluded, by the description so far given. For the description tells us only the state of the marble or light at every instant *within* the minute, not the state at the terminal instant. Admittedly, as Black pointed out, there would be some *discontinuity* of position if the marble appeared, say, in the left hand tray at the terminal instant; for we could not say that it had been getting closer to the left hand tray as the minute approached its end. But, first, it is not clear that such discontinuity matters. Secondly, it does not affect all the examples offered in the literature, not, for example, the light. And thirdly, we can imagine, if we prefer, that the marble simply disappears at the terminal instant, so that there is no discontinuity of position. Benacerraf pointed out that, even in the case of a person traversing Zeno's half-distances, it is not a conceptual requirement that he should be in any particular position after he has traversed all the segments. If we allow our imaginations free play, we can suppose that he shrinks to half his size, and half of that again, each time he covers a half-distance, so that at the terminal instant, when all the half-distances have been traversed, he finally disappears and is nowhere.

I conclude that one cannot in this manner cast doubt on the possibility of traversing an infinity of sub-distances. For the parallel tasks, which were supposed to be impossible, are not impossible after all. I return, therefore, to the solution which asserts that we do, without difficulty, traverse an infinity of sub-distances every time we move. But if I think the answer so simple, am I not siding with those who believe that Zeno can be dismissed? No, for I do not think this answer is simple after all. If we are to stick to it, as I believe we should, then we shall be committed to the most unexpected consequences. We shall not have finished answering Zeno, until we appreciate what the consequences are, and until we admit to him that we are obliged to swallow them. I shall draw attention to two.

Two unexpected consequences

The first emerges when we ask: how can the traversal of all the half-distances be *sufficient* to get us to our destination? For no one of the half-distances reaches the destination, or it would be the *last* member of the series, and this infinite series can have no *last* member.

An initial answer would be to distinguish between the *members* of the series and the series *as a whole*. Admittedly, no member of the series reaches the destination, but the series as a whole reaches it. At least, it 'reaches' it, not in the sense of *including* the terminal point, but in the sense of not being separated by any *gap* from the terminal point. This can easily be shown; for if anyone thinks that the series is separated by a gap from the terminal point, let him name the size of the gap. We can soon show that the series gets closer than that, and that it does so in a finite number of steps. Alternatively, the

21 Benacerraf, op. cit. For other alleged difficulties, see the papers of Black and Thomson, cited above.

fact can be shown in another way (though this may be harder to follow). If a person is still separated by a gap from his destination as he approaches it, he cannot yet have traversed all the infinitely many half-distances in Zeno's series. It follows, by the logical manoeuvre called 'contraposition', that if he *has* traversed all those half-distances, he will not remain separated by a gap from his destination.[22]

But we now have a very surprising situation. The series as a whole reaches the destination, in the sense of 'reaches' explained, yet no member of it does so. How can this be? It is, of course, quite familiar that a whole collection need not have the same attributes as its members. A crowd can be large, although none of its members is large. But *reaching* looks like a special case. For if the crowd *reaches* the steps of the White House, this would normally have to be because at least one of its members reaches those steps. What we have to learn (and this is the first lesson) is that, in the case of certain infinite series, this seemingly undeniable relationship between a collection and its members does not hold.

Recent work by John Murdoch and Edith Sylla shows that, in the fourteenth century, Walter Burley and William of Alnwick were aware of something similar. God sees that the end point of a line is separated by an interval from any other given point in the line, but not from the *totality* of points in it.[23]

For the second lesson to which I want to draw attention, I am entirely indebted to Norman Kretzmann, who has recently revived interest in an acute fourteenth-century logician, Richard Kilvington.[24] After some riots in Oxford, Kilvington was attracted, along with Thomas Bradwardine, to Durham. Bradwardine returned to become Archbishop of Canterbury, and Kilvington to become Dean of St Paul's. Kilvington wrote a book of *Sophismata*, and his sixteenth *sophisma* could be rephrased to fit our earlier discussion as follows. Imagine that the half-distances in Zeno's series are numbered, so that odd and even segments alternate. Imagine someone who has completed his journey, standing at the terminal point, and looking back over the half-distances traversed. He will not be separated by any gap from odd-numbered segments, for whatever gap you name, there will be odd-numbered segments nearer than that. Equally he will not be separated by a gap from even-numbered segments, for a corresponding reason. But if he is not separated by a gap from either, it would be natural to conclude that the odd and even segments coincide. This, however, we know they do not do. Our surprise at results like this may eventually go away as we reflect on

[22] I do not insist that he will have actually *occupied* the terminal point, because the special case of Benacerraf's shrinking traveller shows that this result, though normal, is not logically inevitable.

[23] John Murdoch, 'Naissance et développement de l'atomisme au bas moyen-âge latin', in *La Science de la nature: théories et practiques*, Cahiers d'Études Médiévales 2, Paris 1974, 11-32; Edith Sylla, 'Infinite indivisibles and continuity in fourteenth-century theories of alteration'. Sylla cites Burley *Phys.* 176ra; Murdoch cites Henry of Harclay, *Questio de Infinito et Continuo* (MSS Tortosa 88, f. 88r; Firenze BN II.II.281, f. 98r); William of Alnwick, *Determinatio 2* (MSS Vat. Pal. lat. 1805, ff. 14 r-v).

[24] Norman Kretzmann, 'Continuity, contrariety, contradiction and change', in *Infinity and Continuity in Antiquity and the Middle Ages*.

them, and there may be some who are so clear-headed that they would never have been surprised at all. But, as I indicated on pp. 148-9, I cannot help thinking that they would have missed something of value.

It will be clear how much can be gained from taking the paradox of half-distances seriously, rather than dismissing it at once as fallacious. The last two lessons are among the things to be learnt, and one can learn again, not only from embracing a favoured solution, but also from disagreeing with rival ones. I shall be returning in particular to the solution of atomic distances, and seeing what it implies for time and motion. Other paradoxes too, to which I shall be coming, can be rewarding to study. The paradoxes of stopping and starting, for example (Chapter 26), can help us to see how we need to define motion at an instant. And the paradox of division at every point (Chapter 22) provides elementary lessons in how points can be arranged on a line.

Of all Zeno's paradoxes, that of the half-distances was perhaps the most fruitful historically. We have found it inspiring not only the idea of atoms and of infinitely divisible leaps, but also (Chapter 14) a modification of the ban on actual infinities and arguments in favour of an infinite past.

Achilles and the tortoise

Zeno's second paradox of motion, as Aristotle says, is based on the same principles, and requires the same solution, as the first.[25] The fastest runner cannot catch up with the slowest. Later Greek commentators say that the examples were Achilles and a tortoise. Suppose Achilles gives the tortoise a hundred yard start, and runs ten times as fast. Then, when he has covered those hundred yards, the tortoise will be ten yeards ahead; when he has covered those ten, the tortoise will be one yard ahead; when he has covered that one, the tortoise will be a tenth of a yard ahead, and so on. The series is never-ending. But once again the idea of never-ending is ambiguous. The series has no last member. But all the infinitely many members will have been run through quite soon, and elementary mathematics can tell us where: at $111\frac{1}{9}$yards.

The moving rows

The paradox of the moving rows (or stadium) could be put in modern terms in the following way.[26] We are to imagine two trains rushing past a station in opposite directions and at equal speeds. We can assume that the trains and the station are of the same length, say four carriages long. The paradoxical conclusion which Zeno tries to derive is that half the time is equal to double.[27] The leading carriage of train C will pass all four carriages of train B in four seconds, if it spends one second opposite each. Over the same period,

[25] *Phys.* 6.9, 239b14-29.

[26] Arist. *Phys.* 6.9, 239b33-240a18. There are considerable textual difficulties, but they are well handled by W.D. Ross in his commentary on the *Physics*, Oxford 1936.

[27] *Phys.* 6.9, 240a1.

the leading carriage of train *B* will pass only *two* carriage-length portions of the station, *A*. Suppose now that it too spends one second opposite each, for Aristotle tells us three times that this is the fallacious assumption: that, given the same speed, a thing of the same length will spend the same time opposite a body of the same length, regardless of whether that body is still or moving.[28] Given this fallacious assumption, we get Zeno's paradoxical result: while train *C* is taking *four* seconds to pass train *B*, *B* is taking *two* seconds to pass two carriage-length portions of the station. Four seconds is equal to two seconds.

Paul Tannery argued in 1877 that the paradox so understood, was too feeble to have been constructed by Zeno, whereas it would have more bite, if construed as an attack on Pythagorean atomism.[29] On both counts I think he estimated wrongly. But his ingenious interpretation won widespread acceptance, even though he did not provide independent evidence for his hypothesis that there were Pythagorean atomists at the time Zeno was writing.

Tannery's innovation, expressed in terms of the example above, was to treat each carriage as atomic, that is indivisible, in length. Suppose that an atomic carriage of train *B* passes an atom-length section of the station, *A*, in one time-atom. The question is: how long does that carriage take to pass an atomic carriage of train *C*? Answer: half a time-atom; so evidently the time is not atomic after all.

Not only is Tannery's new problem not in the text, but I think he exaggerates the difficulty of answering it. The atomist answer should be that, when a carriage of *B* passes an atom-length section of station, it passes *two* atomic carriages of train *C*. But it passes these two all at one go, by disappearing from where it was and reappearing further on. There is no stage at which it has passed only *one* carriage of *C*. This involves the surprising idea that motion occurs in a jerk: the moving body is now here, now there, without ever being in between. But an atomist is committed to that in any case: even in passing a *single* atom, a moving body must jerk in this discontinuous way.

Tannery also underestimates, I believe, the force that the paradox would have had without the importation of atomism. As construed by Aristotle, the paradox turns on the relativity of motion: it is ignored whether the body being passed is still or moving. Now in Zeno's time, fallacies about the relativity of motion would not have been easy to detect. We have Plato's explicit testimony that Zeno was confused about relativity.[30] He had supposed it would be paradoxical if something could be both like and unlike, both one and many, both in motion and at rest. Plato makes Socrates point out to him that there is no paradox here, provided we specify the different respects and relations in which a thing is like and unlike, one and many, moving and resting: it is, for example, one *person* and many *limbs*. Confusions about relativity are particularly hard to clear up. One that had been generated by Zeno's predecessor Heraclitus was still worrying Aristotle in

[28] *Phys.* 6.9, 240a1-4; a12; a15-16.
[29] Paul Tannery, *Pour l'Histoire de la Science Hellène*, Paris 1877, 247-61.
[30] *Prm.* 128E-130A.

the following century: the road from Athens to Thebes may be uphill, while the road from Thebes to Athens is downhill. Aristotle thinks this is a case of the roads being identical with each other, and yet not having the same predicates.[31] He does not see that they do have the same predicates, as would be clear if only the relations were stated properly. Each road is uphill in the direction of Thebes, and downhill in the direction of Athens. Viewed in historical context, then, there is nothing surprising in the mistake about relativity which Aristotle ascribes to Zeno.

It has even been suggested to me[32] that Zeno has not make a mistake about relativity, but has rather discovered the truth that motion is relative, and thinks of its relativity as making it less than real. There would be a parallel with Democritus' argument from the relativity of colour and flavour to their unreality. Zeno's point will be that, if you treat motion as non-relative, you will get the self-contradictory result that half the time is equal to double. To avoid this, you should treat it as relative and hence not fully real.

Be that as it may, the evidence for Tannery's reconstruction is not strong; and meanwhile his reconstruction has the disadvantage that it has no textual support, but must on the contrary treat as mistaken our main source for Zeno's argument, namely, Aristotle. None the less, Tannery's reconstructed argument gains interest from the fact that it is not too unlike an argument against atomism devised by Aristotle over a century later.[33] Moreover, it is very close, as we shall see in Chapter 25, to an argument devised by the Islamic thinker Naẓẓām against the view of his atomist contemporaries that an atom has a maximum speed of one atomic space per time-atom. Naẓẓām refutes this assumption, by pointing to the *relative* speed of atoms which are moving at this pace, but in *opposite* directions.

The flying arrow

Zeno's last paradox of motion, the arrow, might be expressed in modern terms as follows. Imagine a knife stuck in the jam: so long as it does not make a slit wider than itself, it does not move. But at any instant an arrow does not make a slit in the air wider than itself. So at any instant an arrow is not moving. But what is true of an arrow at any instant must be true of it over any period; hence an arrow never moves.[34] This, at any rate, is one interpretation of the paradox.

Aristotle's reply is that time is not composed of nows, that is, I take it, of instants. And the conclusion he presumably intends is that you cannot therefore infer from what is true at an instant to what is true over a period. I doubt, however, if it is relevant whether time is composed of instants. For even with genuine components, it is not safe to assume that what is true of the components is true of the whole. A crowd composed of small people need not be a small crowd. On the other hand, a crowd composed of people making a noise will be a noisy crowd. Similarly with the instants contained in

[31] *Phys.* 3.3, 202a19-20; b13-16.
[32] By David Sedley.
[33] Arist. *Phys.* 6.2, 233b19-33.
[34] Arist. *Phys.* 6.9, 239b5-9; b30-3.

an hour, if every one precedes noon, then the hour precedes noon, but if every one lacks length, it does not follow that the hour lacks length. We have to look at the particular attribute in question (small, noisy, preceding noon, lacking length), in order to decide whether the whole and its contents will have the same attribute. The attribute which concerns us now is: *moving*. So long as we are talking of motion in the most straightforward sense, an instant fails to provide one of the necessary conditions for motion, viz. duration, and so inevitably it is not attended by motion. As soon as we turn our attention to a period, however, that necessary condition of motion is restored. So there is not the same necessity for motion still to be missing, and that supplies one answer to Zeno.

As G.E.L. Owen has insisted, however,[35] it has proved extremely useful for the purposes of mechanics to introduce a *derivative* sense in which there can be motion at any instant. For there to be motion at any instant is simply for there to be motion over a period which includes that instant. (Whether we should add, 'or is bounded by that instant' is something to be decided in Chapter 26.) In this derivative sense, we can speak of an arrow as moving at an instant (nor can Zeno forbid us at this stage of the argument, for he has not yet made out his conclusion). And then it will be not merely legitimate, but actually mandatory, to allow that the arrow moves over a period, for that is implied by the claim that it is moving at an instant. That supplies a second answer to Zeno.

I have been assuming that when Aristotle reports Zeno as speaking of 'nows', he means *instants*. But there are two alternatives, one of which is a very interesting suggestion by Jonathan Lear.[36] On this interpretation, Zeno's primary concern, when he talks of 'nows', is with the *present*, but he assumes that, strictly speaking, the present is only an instant. The present tense 'is moving' cannot then be applied to a whole *period* of time, since a period will never be present. But in that case, 'is moving' cannot be applied derivatively, in the manner described by Owen, to an instant, in *virtue* of its application to a period. 'Is moving' cannot then be applied at all.

Lear's interpretation seems to me possible, and, in that case, some different answers to Zeno will be needed. One answer, as Lear says, will be to argue, as I did in Chapter 1, that a whole period can be regarded as present with no lack of strictness.[37] Another will be to say that 'is moving' can be applied to a present instant, regardless of whether it can be applied to a period, simply in virtue of different positions being occupied at arbitrarily close instants.

There is another, less satisfactory, interpretation of the reference to 'nows', according to which they should be construed not as sizeless instants, but as extended time-atoms. Tannery first suggested that Zeno was concerned with time-atoms in the paradox of the moving rows, but his suggestion soon came to be extended to the paradox of the arrow.[38] As explained in Chapter 2, an

[35] G.E.L. Owen, 'Zeno and the mathematicians', *PAS* 58, 1957-8, 199-222.

[36] Jonathan Lear, 'A note on Zeno's arrow', *Phronesis* 26, 1981, 91-104.

[37] Aristotle, as I pointed out in Chapter 1, argues on the opposite side, *pace* Lear.

[38] M.G. Noel, 'Le mouvement et les arguments de Zénon d'Elée', *Revue de Métaphysique et de Morale* 1, 1893, 107-25.

instant is not a very short period, but the beginning or end of a period. An example of an instant would be 2 p.m., and, being a mere beginning or end, it has no size. Time-atoms are much more puzzling and controversial entities than instants. For they are supposed, like instants, to be indivisible, but, surprisingly, unlike instants, to have a positive size. I do not see any *philosophical* gain in the suggestion that Zeno might have been entertaining this more elaborate concept.

One piece of textual evidence might be urged in favour of the view that Zeno is after all envisaging time-atoms.[39] For Aristotle expresses the paradox by saying that the arrow stands still (*hestēken*, 239b30). Now it might be thought that, on Aristotle's own view, our arrow could stand still during a time-atom, if there were such a thing, but at an instant should be described *neither* as moving *nor* as resting. In that case, will not the Zenonian argument which Aristotle is reporting be envisaging time-atoms rather than instants? I think that this argument fails: for one thing, whatever Aristotle thought, *Zeno* at least may have allowed talk of rest at an instant. Secondly, Aristotle's own position is complicated. Some of his arguments that there can be neither motion nor rest at what he calls a 'now' would apply to time-atoms just as much as to instants.[40] Admittedly, two polemical passages may be ascribing to atomists belief in rest at a time-atom.[41] But that is not a sign that Aristotle thinks such an idea to be in order. On the contrary, the second passage probably takes rest at a time-atom to imply something which we *know* Aristotle rejects, namely, rest at its terminal instant.

For an anti-atomist argument like that which has been read into the arrow, we must again wait a century, until we come to one devised along somewhat similar lines by Aristotle himself.[42]

Large and small

These are the four paradoxes of motion recorded by Aristotle. Some of the paradoxes about plurality are preserved with Zeno's own wording in a passage of Simplicius.[43] I shall cite only the paradox of the large and small, because of its influence on subsequent philosophy. Zeno is attacking the view that there is more than one thing in existence.

> If there are many things, they must be both small and large, so small as to have no size, so large as to be infinite.[44]

It looks as if Zeno first argued that if there are many things, each must be sizeless, in order that it may be 'the same as itself and one'.[45] Presumably the

---

[39] See Gregory Vlastos, 'A note on Zeno's Arrow', *Phronesis* 11, 1966, esp. p. 9, n. 20a.

[40] For the arguments, see *Phys.* 6.3, 234a24-b9; 6.8, 239a10-b4; for reiteration of the conclusion, 6.6, 237a14.

[41] *Phys.* 6.1, 232a12-17; 6.5, 236a18.

[42] *Phys.* 6.1, 231b18-232a17.

[43] *in Phys.* 138, 29-141, 8, incorporating Zeno, frs 1-3 DK.

[44] *in Phys.* 141, 6-8, and almost identically 139, 8-9.

[45] *in Phys.* 139, 19.

postulation of many things is taken as implying that there are ultimately some unitary individuals to make up the collection. And if these individuals had a size, they would not be unitary, but would be each a multiplicity of parts. There are echoes of this argument in Leibniz as part of his case for monads, and in Hume and Kant.[46]

But then Zeno turns round and objects that sizeless things would be non-existent, as shown by the fact that attaching them to something or detaching them would not alter the size of the other thing. So after all, if there are many things, they must have a size. And he now argues that the size will be infinite:

> Each of the many things has a size, and they are infinite because through infinite division there is always something in front of what you are taking.[47]

> If they exist, each must have some size and thickness and one part of it must stand clear of another. And the same argument applies to the part that stands out. For it too will have a size, and part of it will stand out. It is, then, the same thing to say this once and to say it always. For no such part of it will be the last, nor will there fail to be one part bordering another.[48]

My guess is that Zeno sees the parts in this infinite division as ever diminishing. If they diminish at a suitable rate, and there is no smallest member, his mistake will be the understandable (and influential) one of supposing that an infinite collection of this kind must build up to an infinitely large total. Some commentators, however, think that Zeno is envisaging segments of *equal* size. In that case, he would be right to say than an infinity of them would yield an infinitely large total, and his error would be to suppose that things contain an infinity of *equal* parts. Others again suppose that Zeno is assuming, wrongly, that there would be a smallest segment in the series. The argument, construed in the first way, was still being used in the seventeenth century by the young Newton and in the eighteenth by Berkeley.[49] Ted McGuire has kindly shown me the relevant part of the forthcoming edition of Newton's treatise and has pointed out to me that Newton would have read the argument in his copy of Epicurus' *Letter to Herodotus*, which is in the library of Trinity College, Cambridge.

---

[46] Leibniz, in the *Journal des Savants* 1695, and in Letter to Arnauld, 30 April 1687; Kant, Second Antinomy in the *Critique of Pure Reason*; Hume, *A Treatise of Human Nature* 1.2.2. I owe the last reference to Steven Makin.

[47] Simpl. *in Phys.* 139, 16-18.

[48] *in Phys.* 141, 2-6.

[49] Isaac Newton, *Questiones Quaedem Philosophicae*, ed. Ted McGuire and Martin Tamny, forthcoming Cambridge. Berkeley, *Principles* §47, and *Notebook* B. Aristotle's pupil Eudemus, with Simplicius' approval, agrees than an infinity of entities with size and in contact with each other would make an infinitely large whole (Simpl. *in Phys.* 459,22; 462,3-5). But their conclusion would presumably be the Aristotelian one that a thing has only a finite number of parts in *actual* existence, the infinity of divisions being merely *potential*.

CHAPTER TWENTY-TWO

Arguments for Atomism

Division everywhere

Paradoxes such as Zeno's were used as arguments in favour of atomism. One paradox which was used by the atomist Democritus is explained at length by Aristotle in *GC* 1.2.[1] What looks like a briefer version appears as one half of a dilemma in a later source.[2] The dilemma is improbably ascribed by Porphyry to Parmenides. Simplicius, however, citing the authority of Alexander, ascribes it with greater probability to Parmenides' defender, Zeno.[3] It does indeed bear some resemblance to Zeno's paradox of the large and small. The fuller argument in Aristotle also bears some resemblance to one half of that paradox. It is therefore likely either to be Zeno's, or to derive from Zeno. I shall quote it in full.

316a10. You could see from this too what a difference there is between physical and dialectical inquiry. For on the existence of atomic magnitudes, the one group says that [but for this] the Triangle would be multiple, whereas Democritus would seem to have been persuaded by appropriate arguments of a physical kind. Our meaning will be clear from what follows.

316a14. It creates a puzzle, if you suppose that there is a body or magnitude divisible everywhere, and that this is possible. For what will there be which escapes the division? If it is divisible everywhere, and this is possible, then it might be all at once in a divided state, even if the dividing did not occur simultaneously. And there would be nothing impossible about this having happened. Therefore equally if it is divided by bisections or in any other way, if it is by nature divisible everywhere, nothing impossible will have happened. There would be nothing impossible, even if it were divided into ten thousand parts, each divided ten thousand times, although perhaps nobody would so divide it.

316a23. Since therefore the body is like this everywhere, let it have been divided: what will be left ? (1) A magnitude? No, that is not possible. For there would be something that had not been divided, whereas it was divisible everywhere.

316a25. But if there will be no body or magnitude left, yet the division does occur, either it will consist (2a) of points, and its constituents will be non-magnitudes, or the constituents will be (2b) nothing at all, so that it might also come into being out of nothing and exist as a composite of nothing, in which case the whole would be nothing but an appearance. And similarly if it consists

[1] 316a10-317a12.
[2] Simplicius *in Phys.* 139, 26-8; 31-2; 140, 1-6.
[3] *in Phys.* 140, 21-6.

of points, it will not be a quantity. For when the points were in contact so that there was a single magnitude, and they were together, they did not make the whole any larger. For when it was divided into two or more parts, the whole was neither smaller nor larger than it was before. Hence even if all the points are placed together they will not create a magnitude.

316a34. But suppose that, as the body is being divided, (3) something like a piece of sawdust is produced, and in this way a body escapes from the magnitude. Even so, the same will apply. For in what way is that piece of sawdust divisible?

316b2. But if what escaped was not a body, but (4) some separable form or quality, and the magnitude is points of contact endowed with this quality, it is strange that a magnitude should be composed of non-magnitudes. Moreoever, where will the points be, and will they be motionless or moving? Again, a point of contact is always a single point of contact between two things, which implies that there is something else besides the point of contact or division.

316b8. If, then, you postulate that any body, whatever its size, is divisible everywhere, those are the consequences. Moreover, if, after dividing, I put the wood or whatever back together, it will again be equal to what it was before and single. So this is the case, clearly, at whatever point I cut the wood. Potentially, then, it has been divided everywhere. What is there, then, besides the divisions? For even if there is some quality, how is it dissolved into divisions and quality, and how created out of these? And how could they exist separately?

316b14. Hence if magnitudes cannot consist of points of contact, there must be (5) indivisible bodies and magnitudes. Yet for those who adopt this postulate there are consequences no less impossible. They have been investigated elsewhere.

316b18. We must try to solve these problems, which is why the puzzle must be stated again from the beginning. On the one hand, it is not strange that every perceptible body should be divided at whatever point you take and undivided. For it will be divided potentially, and undivided actually. On the other hand, it would be thought impossible that it should be (6a) potentially divided everywhere simultaneously. For if that were possible, it might happen, not in such a way that it was simultaneously both actually undivided and divided, but in such a way that it had been divided at whatever point you took. There would then be nothing left, and the body would be destroyed into non-body, and would be created again either out of points, or out of absolutely nothing. And how is that possible?

316b28. None the less, it is apparent that it is divided into separate and ever-diminishing magnitudes which stand clear and are separated. The process of breaking up cannot then be infinite for someone who is dividing part by part. Nor can it have been simultaneously divided at every point, for that is not possible, but only so far. There must then be in it (5) invisible atomic magnitudes, especially if creation and destruction are to occur by association and dissociation. This then is the argument which is thought to make it necessary that there are (5) atomic magnitudes.

317a1. Let us explain that it conceals a fallacy, and where the fallacy lies. Since point is not next to point, magnitudes are in one way divisible everywhere, but in one way not. It is thought, when they postulate divisibility everywhere, that there is a point both (6b) anywhere and everywhere, so that the magnitude has to have been divided into nothing. For there is a point everywhere, so that it consists of points or points of contact. But in fact there is a way in which there is a point everywhere, namely, that (6b) anywhere there is a single point, and what is true of each point is true of all. But there is not more than one point anywhere <in succession>,[4] for points are not successive, so that there are not points everywhere. For [otherwise], what is divisible at the centre would also be divisible (7) at a next point. But it is not, for position (*sēmeion*) is not next to position, nor point (*stigmē*) to point, that is, division to division or contact to contact.

Democritus was writing around 430 B.C., and Aristotle implies that he used the paradox of division everywhere as an argument for the existence of atomic magnitudes, that is, of magnitudes which, although having a positive size, are indivisible. Aristotle envisages that the attempt to divide everywhere might proceed by dividing in half, then dividing each half in half, and so on (316a20; 317a10). This time, we divide *each* half in half, so the bisection is unlike that in the half-distances paradox, where at each stage we divide only *one* half. One small detail is that although the bisection of each half would involve division at an *infinity* of points, it would not, as Aristotle supposes, involve division *everywhere*. The point exactly one third of a way along, for example, would never be cut by this series, although the cuts in the series would not be separated by any gap from the third-way point.

The statement of the paradox ignores this difference between infinite divisibility and divisibility everywhere. It argues against the latter by saying that if a thing were divisible everywhere, it might all at once be in a state of such division. And that is impossible, for what would be left? Four possibilities, which I have labelled (1)-(4), are considered and rejected. (1) *magnitudes* could not be left, or the entity would not have been divided *everywhere*. But equally, (2), we could not be left with sizeless points or nothings, for neither of these could make up the original whole. Two desperate alternatives are then set aside: it is no good suggesting (3) that something like a piece of sawdust might be left over, for the original problem would arise whether that piece of sawdust was divisible everywhere. Finally, you cannot say (4) that a quality might be left over, and that the body consists of points endowed with this quality, since, apart from anything else, a magnitude cannot be made up in this way of non-magnitudes.

It is at this stage (5) that the atomist Democritus comes in, and concludes that things cannot after all be divisible everywhere, and that, to avoid their being so, we must postulate atomic magnitudes which cannot be further divided. Aristotle agrees that division everywhere is impossible, and he gives the new ground (7) that the divisions would be next to each other. But he cannot agree with the atomist solution, and he refers to the battery of objections he produced against atomism, chiefly in *Phys.* 6 and *Cael.* 3. He therefore prefers a different solution, (6), which distinguishes between being divisible *anywhere* (*hopēioun*) and being divisible *everywhere* (*pantēi*, 317a8-9), or

---

[4] It is David Sedley's clarifying suggestion that the MSS. in 317a9 may originally have read *ephexēs* (in succession) twice: *pleious de miās ouk eisin ephexēs, ephexēs gar ouk eisin.*

everywhere simultaneously (316b22; cf. 316b30-1). Magnitudes are not divisible everywhere simultaneously; they are only divisible *anywhere*. This is still not to agree with the atomists, because any segment can be re-divided into smaller ones.

Aristotle's solution parallels his treatment of infinite divisibility in *Phys.* 3.5-7. There too he denies that there can ever be a complete infinity of divisions. All we have is the possibility of increasing any finite number of divisions. *GC* 1.2 agrees and implies as well (317a9-12) that there is not a complete infinity of points. The main discordant note was encountered in *Phys.* 8.8, 263b6, where Aristotle does after all allow a complete infinity of potential divisions.

To decide whether Aristotle is right that division everywhere (or division at his infinity of points) is impossible, we need to consider what it is to divide. In one kind of division, segments are mentally or physically pulled apart from each other, and quite often Aristotle's discussion envisages this kind of division. It would be involved, for example, if something like a piece of sawdust dropped out (316a34). And the argument for atomism explicitly envisages that segments will stand clear of each other (*apechein*, 316b29, the same word as is used by Zeno in fr. 1, Simplicius *in Phys.* 141, 3). To divide a line everywhere in this way would be impossible for a simple reason: such division presupposes flanking lines to flank the gap created – no flanks, no gaps. If *per impossibile* such division had taken place everywhere, there would be no lines left to do the flanking.

But what if division took the form of a mind thinking 'there is something to the right of this point and something to the left'? No human mind could think this of *every* point in a line, but is there a conceptual impossibility, which would prevent a sufficiently powerful mind from thinking it? Certainly, the fourteenth-century philosophers mentioned in the last chapter, Walter Burley and William of Alnwick, believed that God, unlike us, was aware of every point in the line.[5] And I do not see why it should be conceptually impossible.

But let us consider Aristotle's reason, (7), for alleging an impossibility. It is an unsatisfactory one. He argues that, if a thing were divisible everywhere, the points of division would be successive (*ephexēs*, 317a9) and next to each other (*echomenon*, 317a10), and as he rightly says point cannot be *next* to point. This was explained in Chapter 1 above, by saying that, when we start from a given point, there is no way of specifying one next to it. For if the one next is alleged to be a millionth of an inch away, we can always refer to a closer point which is a *two* millionth of an inch away. On the other hand, if the next point is alleged, like the next house in a terraced row, to be *no* distance away, it will turn out not to be distinct from the original point; for, unlike the house next door, it does not even have any *parts* which are separated by a distance.

But why does Aristotle allege that divisibility everywhere would violate this principle? Why should not his opponents say that, when they talk of division at *every* point, they accept the Aristotelian principle that no two points will ever be *next* to each other? Aristotle seems to be arguing not only about actual

[5] op. cit. above, Chapter 21, n.23. The same goes for William's opponent, Henry of Harclay.

divisions, but also about points at which divisions could be made, and of these too he alleges that they would *per impossibile* be next to each other, if he allowed his opponents to locate them everywhere. Yet this allegation is surely untrue, so why does Aristotle make it? It has been plausibly suggested to me[6] that Aristotle is arguing *ad homines* against the atomists. Whether they are dividing the line into atoms, or whether they are threatening us with the absurdity of division into sizeless points (as being the only alternative to atoms), in either case they are likely to think of the resulting units as *next* to each other. And Aristotle is simply following their conception, without allowing them the benefit of a more sophisticated one. He would probably think that he could in any case rely on the other argument against divisibility everywhere which was given earlier in the passage (2). This argument, which was not merely *ad homines*, complained that divisibility everywhere would leave us with points or nothings.

The two arguments (2) and (7) share a common defect; they ignore the fact that in the series two halves, four quarters, eight eighths, ... (*ad infinitum*), *all* the segments have a positive size. That is why we never get dividing lines *next* to each other, and it is also why we never reach points or nothings.

But if it is possible to escape objections (2) and (7) in this way, it may seem that we are bound to fall into one of the opposite positions: either into (1), that there are some segments still awaiting division, or into (5), Democritus' position, that there are indivisible atoms. But this is not true. For although I have just insisted that every segment in the infinite series two halves, four quarters, eight eighths, ..., has a positive size, it is equally true that every segment in the series is punctuated by further points in the series. So none awaits division in the manner of (1), or is in principle indivisible in the manner of (5).

To see how we can in this way slip between positions (2) and (7) on the one hand and positions (1) and (5) on the other is to learn another of those lessons which we should miss, if we refused to take the Zenonian paradoxes seriously: between any two dividing points in the series there is a distance with a positive size (*pace* (2) and (7)); yet any such distance is punctuated by further dividing points in the series (*pace* (1) and (5)). If anyone feels initial surprise, as I do, that both these things are possible, it may come to seem less surprising, when he thinks of the situation in numerical terms. We can think of the fractions contained in the number one: the halves, the quarters, the eighths, the sixteenths, and so on. Clearly no fraction in this series will be sizeless, and equally clearly any fraction in the series is halved by its successor.

Appeal to numbers may also mitigate some of the surprise felt about the paradox of half-distances. Thus, if we take the series $\frac{1}{2} + \frac{1}{4} + \frac{1}{8} + \ldots$ and think of adding the fractions, it will be clear that no fraction is the one which gets us all the way to one. But it is equally clear that no gap can be specified by which the whole infinite series falls short of one.

To return to Democritus, I have admitted that he is right that division everywhere is impossible, if it takes the form which he envisages of mentally

[6] By Jonathan Lear.

or physically pulling apart. But even so, his atomism does not follow; for although you cannot in this fashion divide everywhere, there is no logical barrier to your producing in this way segments as small as you please. The argument does not show any theoretical lower limit such as Democritus wants. And to this extent Aristotle is right to insist against him that you can still divide *anywhere*.

There is one piece of unfinished business. I did not agree with suggestion (2), that division everywhere would leave you with sizeless points. But it is still worth considering the paradox's objection to this outcome, namely, that sizeless entities such as points can never compose a line, even in infinite quantity. Is this true? The answer is that an infinity of sizeless points cannot compose a line by simply being placed in a row. The situation is more complicated in at least two ways. First, there are not only points in a line corresponding to the half-way distance, the third-way distance and indeed to every *rational* number; there are also points corresponding to the *irrational* numbers. So the infinity is super-denumerable, that is, greater than the set of countable rational numbers. Secondly, the relationship between the points is more complex than that of being set in a row. For a start, the points must be arranged so that between any two there is an infinity of others. That yields the property of 'denseness'. For full-scale continuity, as that is understood in modern mathematics, an even more complex arrangement is required. Hence an ordinary infinity of points or an ordinary mode of arrangement will not yield a line, as it is nowadays conceived, but an extraordinary infinity and an extraordinary mode of arrangement will.[7]

Democritus' cone

So much for Democritus' first argument for atomism. According to Plutarch, he also produced an ingenious dilemma about a cone:

> Moreover, look at the way in which Chrysippus met Democritus, when the latter raised the following lively puzzle in physics. If a cone should be cut in a plane parallel to the base, what ought we to think about the surfaces of the segments produced? Are they equal or unequal? For if they are unequal, they will make the cone irregular, and it will take on many step-like incisions and roughnesses. But if the surfaces are equal, the segments will be equal, and the cone will evidently have acquired the properties of a cylinder, being composed of equal, not of unequal, circles, which is as absurd as could be.[8]

Unfortunately, it is not recorded how Democritus solved the problem. There are many conjectures,[9] and it is only one suggestion that Democritus intended this argument, like the last one, as an argument for atomism: there must be atomic steps, if a cone is to differ from a cylinder. The correct

[7] These matters are well explained by Adolf Grünbaum, *Modern Science and Zeno's Paradoxes*, London 1968.

[8] Plut. *CN* 1079E.

[9] Summarised in the note ad. loc. in Harold Cherniss' extremely useful edition of Plutarch's *Moralia*, Loeb, vol. 13, pt 2.

solution would be that (if no sawdust is lost) the two surfaces will be equal, although slices of any *thickness*, however small, will be unequal.[10]

Half-distances

Aristotle tells us of an argument for atomism:

> Some gave in to both arguments: to the argument that all things are one, if being signifies one thing, they conceded that being belongs to that which is not; while in response to the argument from bisection they invented atomic magnitudes.[11]

What argument from bisection was used, and by whom, in support of atomic magnitudes? Five ancient commentators name Xenocrates, the third head of Plato's Academy. And they cite as the bisection argument something like Zeno's paradox of the large and small, or the paradox of division everywhere.[12] But one of the five, Philoponus, gives in addition the paradox of half-distances.[13] Of these three paradoxes, the half-distances would qualify well, since Aristotle actually calls it the 'bisection' (*dichotomia*).[14] But so too would the paradox of division everywhere, since Aristotle envisages the division as proceeding by bisection.[15] It is only the third suggestion which I find unlikely, for the discovery of parts between other parts, which is involved in Zeno's paradox of the large, is not very naturally thought of as a *bisection*.

If the argument from bisection is the paradox of half-distances, then the 'response' of inventing atomic magnitudes will have taken the form of saying that atoms can solve the paradox; for, in moving, we do not have to traverse an infinity of diminishing half-distances, just so long as distances cannot diminish below a certain atomic size. There is independent reason to think that Xenocrates may indeed have supported atomic magnitudes in this way, by showing that they can solve the half-distance paradox. For we shall see in a moment that some Platonist author did so, judging from the *de Lineis Insecabilibus* (*On Indivisible Lines*). If, on the other hand, the argument from bisection is the paradox of division everywhere, then it will not have been Xenocrates who first responded by inventing atomic magnitudes. For we have seen from Aristotle's testimony that *Democritus* was the first so to respond. This has the slight disadvantage that we would have to take the ancient commentators as wrong in naming Xenocrates.

On the other hand, there is a consideration which marginally favours Democritus. For we need to know why Aristotle refers not only to a bisection argument, but also to *another* argument, and says that certain people gave in to *both*. It so happens that Democritus' associate, the atomist Leucippus, is

---

[10] A cut in modern parlance, is a cross section, and so is only *one* surface, but Democritus is talking about exposing *two* surfaces.

[11] *Phys.* 1.3, 187a1-3.

[12] See ad loc. the commentaries of Philoponus, Themistius, Simplicius and Simplicius' reports on Alexander and Porphyry.

[13] *in Phys.* 81, 7-16.

[14] *Phys.* 6.9, 239b22; cf. b19.

[15] *GC* 1.2, 316a10; 317a10.

connected in a single passage with two arguments of the right type.[16] To find two appropriate arguments to connect with Xenocrates and his fellow-Platonists, we should have to select as the second argument one quite unconnected with the first.[17]

Whether or not the bisection argument mentioned by Aristotle is the paradox of half-distances, we shall see that that paradox was used as a ground for accepting atomism by the atomists referred to in the *de Lineis Insecabilibus*, which I shall take next.

De Lineis Insecabilibus

This treatise, wrongly ascribed to Aristotle, but composed by some member of his school, starts by outlining a set of five arguments for atomic magnitudes, which it then attacks. As we shall see in the next chapter, the arguments are Platonist ones, and some or all may stem from Xenocrates. Two of the five will already be familiar.

968a1. Are there atomic lines, and in general in all quantities is there something which lacks parts, as some people say?

968a2.(1) For if much and large and their opposites, few and small, exist alike, but what has almost infinite divisions is not few but much, clearly the few and the little will have a finite number of divisions. But if the divisions are finite, there must be a magnitude which lacks parts, so that there will be something without parts in everything, since everything contains few and little.

968a9.(2) Again, if there is an Ideal Form of the Line, and if an Ideal Form is first among the things which share its name, and if parts are prior in nature to the whole, then the Line Itself [i.e. the Ideal Form of the Line] will be indivisible. And the same applies to the Square and the Triangle, and the other Figures, and in general to Surface Itself and to Body. For [otherwise] there will be things [i.e. parts] which are prior to these [Ideal Forms].

968a14.(3) Again, if there are elements of bodies, and if nothing is prior to elements, but parts are prior to the whole, then fire [one of the four elements] and in general each of the elements of bodies will be indivisible. Thus not only

---

[16] *GC* 1.8, 324b35-325a32, discusses some arguments of the Eleatics (Parmenides, Zeno, Melissus). The two relevant ones are designed to deny that there is more than one thing in existence (cf. 'all things are one', *Phys.* 1.3, 187a1-3). One argument does so by exploiting the paradox of division everywhere, in order to prevent there being any units to make up the plurality. The other does so by saying that a plurality of things would need empty space to separate them, but empty space has no being (cf. 'being signifies one thing' – the full, but not also the empty, perhaps). We are told that Leucippus conceded that empty space was needed at least for motion, but responded that being could after all be ascribed to empty space (cf. 'they conceded that being belongs to that which is not'), and furthermore postulated atoms.

[17] Arist. *Metaph.* 14.2, 1088b28-1089a21, says that Platonists postulated two principles, called the One and the Indefinite Dyad, in response to the Eleatic claim that, unless we ascribe being to that which is not, there will be only one thing in existence (cf. 'all things are one, if being signifies one thing', *Phys.* 1.3, 187a1-3). The Indefinite Dyad is supposed to be describable as a thing which is not, but which none the less has being (cf. 'being belongs to that which is not'). We might compare it with Aristotelian matter, which, like the Dyad, is the source of plurality.

among objects of thought, but also among objects of sense perception, there is something which lacks parts.

968a18.(4) Again, by Zeno's argument, there must be a magnitude which lacks parts, if it is impossible to touch an infinity one by one in a finite time, and if anything that moves must first reach half way, and if there is always a half of that which does not lack parts. And if anything that travels along a line does indeed touch an infinity in a finite time, and if a faster moving thing completes a longer journey in a given time, and if the movement of thought is the fastest, then thought too will touch an infinity one by one in a finite time. Thus if thought's touching things one by one constitutes counting, it is possible to count an infinity in a finite time. If that is impossible, there must be such a thing as an atomic line.

968b4.(5) Again, there must be such a thing as an atomic line, so they say, because of what is said by the mathematicians themselves, if those lines are commensurable which are measured by the same measure, and if all lines which are measured are commensurable. For there must be some length which will measure them all, and this must be indivisible. For if it is divisible, then its parts will have something that measures them, since the parts are commensurable with the whole. Thus a half portion of some measure [reading: *metrou*] will be a double portion. But since this is impossible, there must be an indivisible measure.

The argument is hereafter extended from lines to plane figures, with some technicalities, but I do not think its basic pattern is altered.

The idea behind the first of these five atomist arguments, that much and large would not be distinct from few and small, if both alike had infinitely many parts, reappears in Epicurus' Roman expositor Lucretius, then as a jibe against the Stoics in Plutarch, and again, as we shall see (Chapter 25) as an argument for atomism in the Indian Kaṇāde and among the Islamic theologians. In the seventeenth century, it recurs in Henry More and is answered by Newton in a letter to Richard Bentley.[18] It is easy to see how the argument could arise out of reflection on Zeno's paradox of the large and the small. Zeno argued, it will be remembered, that what has *infinitely* many parts will be *infinitely* large. The idea here is that what has *almost* (*sic*) an infinity of parts will still be *very* large, and the small must have some *smaller* number of parts.

The second argument, that Platonic Ideal Forms, such as the triangle, must lack parts if they are to have the priority which Plato assigned to them, has already been encountered. Aristotle complained in *GC* 1.2 that this argument was dialectical, and inferior to a physical argument like that based on division everywhere.[19]

---

[18] Lucretius *de Rerum Natura* 1.615-27; Plut. *CN* 38, 1079A-B. For references to Indian and Islamic thought, see Chapter 25. For Isaac Newton, see Letter 2 for Mr Bentley, 17 Jan. 1692-3, in *The Works of Richard Bentley* edited by Alexander Dyce, vol. 3, London 1838, pp. 208-9. Henry More is quoted, by H.A.J. Munro in his commentary on Lucretius, 4th edition, revised, London 1886, vol. 2, p. 82, ad 1.622. There is a sense in which Aristotelians would agree with the atomists' premises, for Aristotle denies that part and whole can equally have an actual infinity of parts (*Phys.* 3.5, 204a22), while his pupil Eudemus, with Simplicius' approval, implies that if there could be such a thing as an infinity of actual parts in contact with each other, they would add up to an infinitely large whole (Simpl. *in Phys.* 459, 22; 462, 3-5).

[19] *GC* 1.2, 316a10-14.

Skipping the third argument, I pass on to the fourth, which exploits the half-distances, to show the need for atomic magnitudes. But there is an elaboration, for traversing an infinity is correlated with something admittedly impossible, namely, *counting* an infinity. The general type of strategy was encountered in Max Black's argument in Chapter 21. We have already noticed a report in Aristotle,[20] that some people exploited the possibility of *counting* the half-distances as they were traversed. Now we can guess who those people are: the Platonists, perhaps Xenocrates and his followers, who are referred to in the *de Lineis Insecabilibus*. Later on, we shall find Epicurus arguing for indivisible parts on the basis of the impossibility of an infinite count. A more straightforward version of the half-distance paradox may possibly have been used by Epicurus' follower, Demetrius of Laconia (late second century B.C.), as an argument for indivisible parts. For there is a fragment in which he discusses cutting a half, and cutting half of the resulting half, *ad infinitum*.[21] But the fragment is too sketchy for us to be sure what he intended. What is certain, as will be seen in Chapter 25, is that Islamic theologians again argued for atomism, by alleging the impossibility of traversing an infinity of sub-distances.

The fifth argument might seem to be in dangerous territory. For the discovery of incommensurability looks like a threat to atomism. The facts about incommensurables are often omitted from school mathematics courses. So it may not be all that widely appreciated that no fraction and no set of decimals exactly expresses the ratio of the diagonal and side of a square, or of the diameter and circumference of a circle. The number pi, which is used in specifying the last ratio, is not equal to 22/7 or $3\frac{1}{7}$, as school teachers often say. No fraction expresses it exactly. To put the point another way, diameter and circumference have no common measure; the seventh, for example, is not really a common measure. The two are incommensurable. There cannot then be some standard *atomic* size of which both diameter and circumference are multiples. None the less, the propounder of the fifth argument believes that even the mathematicians themselves will have to concede something to atomism, provided attention is confined to lines which are actually subject to measurement.[22]

Diodorus Cronus

Diodorus Cronus devised a new argument for atomic spaces, and, as I interpret it, it is a very ingenious one. It is that atomic spaces enable us to solve a kind of paradox first introduced by Aristotle in the *de Sensu*.[23] Aristotle's version, to which I shall be returning in Chapter 26, concerns an approaching

---

[20] *Phys.* 8.8, 263a6-11.

[21] V. de Falco, *L'epicureo Demetrio Lacone*, Naples 1923, p. 97 (Pap. 1061), further reconstructed by S. Luria, *Die Infinitesimaltheorie*, p. 131.

[22] The argument, a bad one, appears to be that for *measured* lines there must be some length (A) which will measure them all, and this must be indivisible. For if it is divisible, then its parts, (B), (C), will have something which measures them (D), since parts share a common measure with the whole. Thus a half portion (B) of some measure (A) will be a double portion of (D). And this is supposed impossible.

[23] 7, 449a20-31.

object and the furthest distance at which it is visible as opposed to the nearest distance at which it is invisible. Do these distances coincide, or are they separated by a gap? Any decision seems to make the approaching object simultaneously visible and invisible, or alternatively neither visible nor invisible, when it reaches the single distance or the gap between the two distances. I shall quote the passage, although I am not concerned with Aristole's solution, which is that the two distances would coincide at an indivisible point, and that indivisible points are never perceptible.

> It is clear that every perceptible has extension and there is no indivisible perceptible. For the distance at which such a thing would not be seen is unlimited, while the distance at which it is seen is limited. Similarly also an object of smell and of hearing and all objects that people perceive when not in contact with them. There is then a last boundary of the distance at which it is not seen and a first boundary of the distance at which it is seen, and this boundary on the far side of which the object cannot be perceived, and on the near side of which it must be perceived, is necessarily indivisible. If, then, there is an indivisible perceptible, when it is placed at the boundary at which it is for the last time imperceptible and for the first time perceptible, it will turn out simultaneously visible and invisible, and this is impossible.

Diodorus' paradox must, I think, have been similar except that it will have concerned sizes, not distances, as being more immediately relevant to his atomism. His question will have been about the smallest visible size and the largest invisible. Are these sizes identical or separated by a gap? Any decision threatens to yield a size or sizes which are both visible and invisible, or alternatively neither visible nor invisible. Diodorus will have said that the problem can be solved by postulating atomic differences of size, and saying that the smallest visible and largest invisible size differ from each other by precisely one atomic unit.

The evidence that Diodorus argued for atoms in this way comes from Alexander of Aphrodisias. Commenting on Aristotle's argument about the distances, Alexander says:

> It looks for these reasons as if Aristotle was the first to raise and exploit the problem about partless entities, which was raised by Diodorus or someone. But whereas Aristotle invented the problem and made sound use of it, the others took it from him, prided themselves on it, and used it illegitimately.[24]

Earlier in the same commentary he had criticised Diodorus by saying:

> But if nothing is by its nature the smallest perceptible, nor the largest imperceptible, it could not be shown, as Diodorus thinks he shows, that anything is by its nature the smallest magnitude.[25]

Others have interpreted Alexander's remarks quite differently.[26] But the

---

[24] *in Sens.* 172, 28-173, 1.   [25] *in Sens.* 122, 21-3.

[26] David Sedley, 'Diodorus Cronus and Hellenistic philosophy', *PCPS* n.s. 23, 1977, 74-120; Jürgen Mau, 'Über die Zuweisung zweier Epikur-Fragmente', *Philologus* 99, 1955, 93-111; Nicholas Denyer, 'The atomism of Diodorus Cronus', *Prudentia* (Auckland) 13, 1981, 33-45.

advantages of this interpretation are first that it takes account of Diodorus' referring to a largest imperceptible size, as well as to a smallest perceptible one, and secondly that it takes account of the context of the first quotation from Alexander, namely, a discussion of Aristotle's analogous paradox about distances.

I must acknowledge that Alexander or Diodorus is guilty of a small oversight in the argument as I construe it. For he has no need to mention *nature* in his premises as well as in his conclusion. His conclusion is indeed that *nature* dictates some minimum size. But in the premises he could afford to admit that the smallest *perceptible* size is not laid down by nature, but varies with different observers and viewing conditions. None the less, if he has not admitted this, or if Alexander thinks he has not, it is so small and understandable an error that it hardly casts doubt on the interpretation offered.

Diodorus may have introduced several other new arguments for atomism. Thus Jürgen Mau reminds us[27] of a different context in which Diodorus argues that what is possible will at some time be actual. In that case, he might have said that, since nothing will ever be actually divided infinitely, nothing is infinitely divisible. But Mau's suggestion is at best a conjecture, and it has been quite tellingly disputed by Nicholas Denyer.[28]

There is, however, a further argument which concerns not atomic spaces, but atomic times. For, in a passage considered in Chapter 2,[29] Diodorus maintains that the present must be indivisible. He does so by taking over Aristotle's argument that otherwise the present would overlap with past and future.[30] Admittedly, he does not show why the present should be a time-atom (that is, an indivisible unit with positive length), rather than an instant (that is, a *sizeless* indivisible), as Aristotle believes. But, it was argued, that is what he *means* us to conclude.

Finally, Diodorus has an argument, already encountered in Chapter 2, to show that movement, as well as time and space, is in a certain sense atomic. A thing cannot move in the place where it is, for its place is understood as too tight-fitting to allow room for movement. Nor, obviously, can it move in the place where it is not. So at best it can only *have* moved with a discontinuous jerk.[31] A version of part of the argument is ascribed by some sources to Zeno.[32] But the *reason* for denying that a thing can move where it is, is not explicitly ascribed to Zeno, and is only analogous to what we find in Zeno's paradox of the arrow.[33] So Diodorus may well be innovating, with the aim, as I suggested in Chapter 2, of discomfiting Aristotle, who sometimes construes a thing's place as its *exact* surroundings. Moreover, Diodorus does not intend, like Zeno, to deny motion altogether. Rather, he goes on to allow that a thing

[27] Jürgen Mau, *Zum Problem des Infinitesimalen bei den antiken Atomisten*, Deutsche Akademie der Wissenschaften zu Berlin 1954, 28-31.

[28] Denyer, op. cit.

[29] Sext. *M* 10.119.

[30] *Phys.* 6.3, 234a9-19.

[31] Sext. *M* 10.85-90; 143.

[32] Diog. Laert. *Lives* 9, 72; Epiphanius, *adv. Haereses* 3, 11 (Diels, p. 590).

[33] Zeno said in the paradox of the arrow that, in order to move, an arrow needs to slit a region larger than itself, and that at an instant it cannot do so.

may *have* moved, by disappearing from one place and reappearing in another.

Strictly speaking, the movement which this argument introduces is atomic only in the *weak* sense that there is no time at which the transition from one place to the next is part way through. The argument does not prove that the motion is atomic in the *further* sense of being across an atomic distance, although this is what Diodorus,[34] for independent reasons, believed. In Chapter 5, we saw that some people postulated motion which was atomic in the first sense (never part way through), but not in the second, since it could be across indefinitely small distances. To mark this latter fact, they called their 'leaps' not 'atomic', but 'infinitely divisible'.

Epicurus

Lastly, I would mention the arguments for atomism employed by Epicurus. I must anticipate a later chapter, in order to explain that Epicurus introduced a two-tier theory of atoms. Not only were there indivisible atoms, but also within them there were minimal parts indivisible in a different sense – conceptually, as I shall say in the next chapter, rather than physically. Correspondingly, Epicurus had quite different kinds of argument for establishing the indivisibility of the atoms and of their parts. And this is part of the evidence,[35] although only *one* part of it, that two different kinds of indivisibility were involved.

The *Letter to Herodotus*, in which Epicurus summarises his views on physics, is preserved in Diogenes Laertius' Life of Epicurus. In sections 40-1 and 56 Epicurus considers whole atoms, and offers only physical considerations in favour of their indivisibility. Atoms must be indivisible, or they would by now have been pulverised and would have disintegrated into nothing. Similarly physical considerations are offered by Lucretius in his exposition. There must be bodies free of void interstices, and without void, bodies cannot be divided.[36] There must be solid and simple bodies, or everything would have been ground into powder and would need reconstruction *ex nihilo*.[37] There must be indivisible bodies, or particles would be too fine for things to grow to their full size on schedule.[38]

When we turn to the arguments for the indivisibility of *minimal parts*, we find them much more like those we have been considering, since they turn on *conceptual* considerations. Epicurus gives two in section 57 of the *Letter to Herodotus*. The first is modelled on Zeno's paradox of the large and the small. We must reject infinite divisibility, because infinitely many components, all having a size, would add up to an infinitely large body. Secondly, if there were an infinity of parts, then, in conceiving the whole, our mind could, *per impossibile*, arrive at infinity. This second argument will remind us of the ban

---

[34] Sext. *M*. 10.85-6; 143.

[35] For the evidence, see e.g. David Furley, *Two Studies in the Greek Atomists*, 21; 42.

[36] Lucretius *de Rerum Natura* 1.503-39.

[37] *de Rerum Natura* 1.540-50.

[38] *de Rerum Natura* 1.551-64.

on an infinite count, which was associated with the paradox of half-distances. Lucretius adds the familiar argument that large will not differ from small, if both alike have an infinity of parts.[39]

The character of the arguments

The arguments for Epicurus' minimal parts, and the arguments for atoms which we have considered in his predecessors, are of a wholly *conceptual* character. Often the argument turns on a paradox which càn be considered by a philosopher sitting in an armchair. Nowadays, we think that it is a question for *scientists* whether there are atoms of matter, space, or time. I do not for a moment mean that that makes conceptual considerations irrelevant. Indeed, it would take a great deal of conceptual reflection for scientists to decide, for example, whether certain modern physicists are right to talk of atoms of space and time.[40] But there is the difference that some of the modern data are observational, and could not be discovered from an armchair.

[39] *de Rerum Natura* 1.615-27.
[40] For the references to their theories, see Chapter 24.

CHAPTER TWENTY-THREE

Types of Atomism: Early Developments

The central idea of atomism is that there are indivisible bits of matter. *Atomos* means uncuttable in Greek, hence indivisible. I have already mentioned that in Greek thought, though not, as we shall see, in all Islamic or fourteenth-century Western thought, an atom is supposed to have a positive size in spite of its indivisibility. But indivisibility, as we have noticed, can mean more than one thing. And indivisible bits have been postulated not only in matter, but also in space, motion and time. I shall be especially interested in this last development. We have seen that some modern commentators think time-atoms were postulated already by the opponents of Zeno of Elea. In this chapter, we shall find time-atoms ascribed by others to Plato and to his successor Xenocrates. I shall resist these attributions, and will instead in the next chapter treat time-atoms as something introduced in response to Aristotle's objections to atomism. It was certainly easier for atomists to decide that they wanted time-atoms, when they had become clearer that they believed in *space*-atoms. And that in turn involved them committing themselves to what I shall call *conceptual* indivisibility. In this chapter, I shall trace the gradual development of these ideas. But before I focus exclusively on the idea of indivisibility, it needs to be said that atomism has often involved much more than this idea.

The connexion of atomism with scientific explanation

A persistent view has been that the indivisible atoms possess only a small range of qualities, sometimes called 'primary' qualities, and that the many 'secondary' qualities which we perceive in larger bodies are produced by the few primary qualities of the atoms. Some would add that the secondary qualities have a lower degree of reality, and are in some way dependent on our minds. At all events, it is the primary qualities which will provide all scientific explanation.

Some people still look to science today for such a theory. There is a good presentation and defence of the view in J.J.C. Smart's *Philosophy and Scientific Realism*. The view flourished in the seventeenth century with Galileo, Boyle and Locke, who provided the account best known to English philosophers in his *Essay Concerning Human Understanding*. The main ancient proponents of such a view are the pair Leucippus (fl. *c.* 440 B.C.) and Democritus (fl. *c.* 430 B.C.), followed by Epicurus (342/1-271/0 B.C.), whose ideas were put into Latin verse by Lucretius (*c.* 94-55 B.C.).

Little information remains about Leucippus' contribution, but for Democritus Aristotle informs us of the small range of qualities assigned to the atoms.[1] The fullest account of how these few qualities are supposed by Democritus to explain the many qualities of larger bodies is given by Aristotle's successor Theophrastus in his *de Sensibus*.[2] That sweet, bitter, hot, cold and colour exist only by convention (*nomōi*) not in reality (*eteēi*), we are told in Democritus' own words.[3] Epicurus' view is a matter of controversy.[4] But John Locke reaffirmed the subjectivity of secondary qualities,[5] and this reaffirmation has recently been defended by Jonathan Bennett.[6]

Atomism came to be disfavoured by the Christian church.[7] It is a symptom of this that neither of the two oldest surviving manuscripts of Lucretius carries the title at the beginning. In one, the name of Lucretius is just visible in erasure, but a different title has been substituted, no doubt to protect the manuscript from destruction. But it was not the idea of indivisibility which made the Christian church hostile to atomism. The reason lay rather in Epicurus' other views. He wished to explain the physical universe by reference to the mechanics of atoms and their primary qualities, not by reference to the purposes of the gods. Matter, moreover, was never created, but has existed without beginning. If the gods exist at all, they will not be concerned with us. We should not therefore be perturbed by stories of punishment after death. At death the atoms which compose our souls will be scattered.

Alien as this is to Christian views, Epicurus' aim was to induce tranquillity of mind by removing superstitious fear. And (in contrast to modern philosophy) the achievement of tranquillity, in various differing forms, was a major concern of nearly all the main Greek schools after Aristotle and of Christians too.[8] Christian disapproval was directed not to the aim, but to the atomist *route* to tranquillity.

That atomism need not in itself be opposed to religious belief is made clear by the case of Islam. The views of the *Kalām*, that is, of Islamic theology, are summarised in twelve propositions by the Jewish philosopher Maimonides

---

[1] *GC* 1.8, 326a1-b7; *Metaph.* 1.4, 985b4.

[2] Translated by G.M. Stratton in *Theophrastus and the Greek Physiological Psychology Before Aristotle*, London 1917.

[3] Democritus fr. 9 DK.

[4] According to Plut. *adv. Colotem* 1109A-1110D, Epicurus denied that colours belong to the nature (*sumphuē*) of bodies; rather, they are generated by arrangements and positions relative to sight, things are not coloured in the dark. In view of this, we probably should not make too much of Epicurus' *rejection* of the claim that shapes, *colours*, magnitudes and weights do not exist (*ouk eisin*), and of his assertion that these are properties either of all bodies, or of visible ones (*Letter to Herodotus* 68 = Diog. Laert. *Lives* 10.68). For this leaves him still free to take a subjective view of colour.

[5] John Locke *Essay Concerning Human Understanding* 2.8.8-26.

[6] Jonathan Bennett, 'Substance, reality and primary qualities', *APQ* 2, 1965, 1-17; and *Locke, Berkeley, Hume: central themes*, Oxford 1971, ch. 4.

[7] See K. Lasswitz's account of the attacks by Dionysius, Lactantius and Augustine in his *Geschichte der Atomistik*, Hamburg and Leipzig 1890, vol. 1, 11-30.

[8] The literature is large, but for a specimen discussion of some Christian treatments of tranquillity, with further references, see Robert C. Gregg, *Consolation Philosophy*, Cambridge, Mass. 1975.

(A.D. 1135-1204).[9] Evidently the *Kalām* actually harnessed atomism and put it to religious use. This was possible, because the theologians took over only the central idea of indivisibility, and not the idea of explaining everything by reference to the primary qualities of atoms. On the contrary, all causal power came to be attributed to God, and eventually, as we saw in Chapter 19, there developed a kind of 'occasionalism', according to which physical arrangements do no more than supply God with an *occasion* for exercising his power.[10] The sense of our total dependence on God and our precariousness was still further heightened by the suggestion that atoms have no tendency of their own to persist. Instead God simply re-creates them, by re-creating their accidental properties, every time-atom for so long as he chooses.

There had already been Greek thinkers who, without showing any religious interest, had none the less anticipated the Islamic approach of confining their atomism to a thesis about indivisibility, without going on to distinguish primary and secondary qualities. The most notable example is Diodorus Cronus, who will be further discussed in the next chapter.

Indivisibility: ambiguities

The thesis of indivisibility is still of concern to science. For although the splitting of the 'atom' has put an end to the idea that material atoms really are atomic, that is indivisible, there is still a theory that space and time come in indivisible chunks. The first preliminary, however, is to say more about the meaning of 'indivisible'.

A distinction is often drawn between conceptual and physical indivisibility, but it is subject to a dangerous ambiguity. Sometimes the words 'conceptual' and 'physical' are taken as characterising the method of division (division in the mind as against division by physical means). But I shall follow the commoner usage, according to which the words characterise the kind of impossibility involved: there may be conceptual barriers to division or there may be physical barriers. On the other conception, what is conceptually indivisible would *a fortiori* be physically indivisible. But this is not so with the conception which I shall use, for the barriers to division may all be conceptual rather than physical. In other cases, it will be the other way round: the barriers to division may all be physical, and then I shall often describe something as '*merely* physically indivisible'.

The distinction is not always a sharp one, and when I come in the next chapter to the theories of modern physics, we shall see that it is too blunt to be useful, but it will serve for discussing the ancient theories.

Who were the first atomists?

A further preliminary question is who were the first Greek atomists. I shall take the conventional view, which favours Leucippus and Democritus. But scholars have supported many earlier candidates with ingenious arguments.

---

[9] *Guide* 1.73.
[10] Maimonides *Guide* proposition 6; Ghazālī, *Iḥiyā* ;4, p. 214.

I shall simply review the list of candidates, while referring to other critics for detailed rebuttals.

Tannery suggested in 1877 that Zeno of Elea was attacking some Pythagorean atomists with his paradox of the moving rows. But we have seen that his evidence was only his belief, criticised in Chapter 21, that such a target would give Zeno's paradox more bite. Since Tannery, other evidence has been suggested for early Pythagorean atomism, but I can refer to the objections of Vlastos, Owen, Furley and others.[11]

Luria suggested a different target for Zeno's paradoxes, viz. certain mathematicians who thought that bodies were infinitely divisible into dimensionless bodily point-atoms.[12] But these mathematicians remain unidentified.

Other mathematicians who have been claimed as atomists are those who sought to square the circle, in the sense of constructing with ruler and compasses a square equal in area to a given circle, and proving that they had done so. Antiphon's method was to inscribe a polygon in the circle and then to build up its sides with triangles. He thought that in the end the sides of the enlarged polygon would be so small as to coincide with the circumference of the circle. It would then be easy to construct a square equal in area to the polygon.[13] Commentators have thought that this implied a conception of the circle as made up of atomic straight lines. But it is not clear that Antiphon drew this inference, nor, for that matter, that his work was earlier than that of Democritus.

Anaxagoras might seem a most unlikely candidate. For the first and third fragments in the Diels-Kranz collection seem to insist on infinite divisibility:

All things were together, infinite both in number and in smallness, for the small was infinite. ...

Nor is there a least of what is small, but there is always a smaller, for it is not possible that what is should fail to be.

Two reasons have, none the less, been given for finding atomism in Anaxagoras. For one thing, it is recorded that while in prison he was engaged in squaring the circle.[14] But as Luria points out, no more is implied than that he was trying, and certainly his method is not indicated. The other reason is that he talks of there being seeds in things, but recent commentators have agreed that these are not to be construed as atoms.[15]

[11] Gregory Vlastos, review of J.E. Raven's *Pythagoreans and Eleatics*, in *Gnomon* 25, 1953, 29-35; review of Kirk and Raven, *The Presocratic Philosophers*, in *PR* 69, 1959, 531-5; G.E.L. Owen, 'Zeno and the Mathematicians', *PAS* 58, 1957-8, 199-222; D. Furley, *Two Studies in the Greek Atomists*, Princeton 1967, ch. 3.

[12] S.Y. Luria, 'Die Infinitesimaltheorie der antiken Atomisten', in *Quellen und Studien zur Geschichte der Mathematik, Astronomie und Physik*, Abt. B., Studien Band 2, Heft 2, 1932-3, pp. 106-85, see pp. 106-16.

[13] Simpl. *in Phys.* 55, 6-11; Themistius *in Phys.* 4, 2.

[14] Plut. *de Exilio* 607F.

[15] Jonathan Barnes, *The Presocratic Philosophers*, London 1979, vol. 2, p. 21; Malcolm Schofield, *An Essay on Anaxagoras*, Cambridge 1980, 130-1.

Empedocles may have had a theory of particles which would never in fact be divided. But it seems that they were not atomic, but in principle divisible.[16]

Finally, there is a sentence in the twenty-eighth book of Epicurus' fragmentary treatise *On Nature*, which R. Philippson took as referring to Democritus and to atoms, and as saying that Democritus wrote about the people who first recognised them, that is, about atomists earlier than himself. But after Sedley's re-reading of the rolls, the sentence comes out looking quite different, and the reference is to Epicurus' own work on the meanings given to words by the earliest users.[17] After this brisk survey of earlier candidates, I shall now turn to Leucippus and Democritus, and chiefly to Democritus, since most of our information is about him.

Democritus

Commentators are seriously divided on whether Democritus' atoms were conceptually, or merely physically, indivisible.[18] I would suggest, however, that Leucippus and Democritus were not aware of the need to distinguish between physical and conceptual indivisibility. The later atomist Epicurus in the fourth and third centuries B.C. probably made a fairly sharp distinction, and, if so, commentators will have asked themselves, in the light of this, what was the intention of the earlier atomists Leucippus and Democritus. If those earlier atomists never made up their minds, it is not surprising that later commentators should have come up with conflicting verdicts. I have three kinds of reason for this hypothesis.

First, it fits with what I have argued elsewhere about Aristotle. Writing nearly a century later, he did not distinguish between physical and conceptual impossibilities, so I believe, and hence not between the physical and conceptual impossibility of division.[19] If this is true of Aristotle, with his great love of distinctions, it is not at all implausible that it should have been true of the earlier thinkers, Leucippus and Democritus.

A second reason for the hypothesis is that Democritus and Leucippus appear to have combined physical and conceptual reasons for indivisibility without distinction. Simplicius reports:

> Except that Leucippus and Democritus think that imperviousness (*apatheia*) is not the only reason for the first bodies not being divided, but also smallness and lack of parts (*to smikron kai ameres*); but Epicurus later does not think them partless, but says they are atoms through imperviousness.[20]

Imperviousness suggests *physical* indivisibility, while partlessness is a

[16] For the evidence on Empedocles, see Arist. *Cael.* 3.6, 305a1; *GC* 1.8, 325a6; b8; b19; 1.9, 327a10; Aëtius in Diels, pp. 312; 315.

[17] *Epicurus, On Nature, Book XXVIII*, ed. David Sedley, in *CE* 3, 1973, p. 41 (fr. 8, col. 4, ll. 4-9).

[18] For the divisibility being merely physical we can cite: P. Tannery, H. Diels, W. Kranz, J. Burnet, T. Heath, S. Pines, G.S. Kirk, A. Mourelatos, J. Barnes; for its being conceptual: J. Mau, D. Furley, H.-J. Krämer, W.K.C. Guthrie, D. Konstan.

[19] Sorabji, 'Aristotle and Oxford Philosophy', *APQ* 6, 1969, 127-35, esp. appendix.

[20] *in Phys.* 925, 13-17.

characteristic of geometrical points (as Simplicius says elsewhere),[21] and most readily suggests *conceptual* indivisibility. On the other hand again, Simplicius' report takes partlessness closely with smallness, which does not on its own suggest conceptual indivisibility at all. For what is small has a positive size, and it takes special argument to show how something with a postive size can be conceptually indivisible. Indeed, Philoponus actually contrasts smallness with partlessness.[22] There is no sharp division, then, in Simplicius' report, between physical and conceptual considerations.

This second reason could have been presented as an illustration of a third and more general reason for my hypothesis. The more general reason is that there are several considerations which support one side in the traditional controversy and several which support the other. Each party, therefore, needs to discount one half of this evidence. The advantage of the present hypothesis is that it does not have to favour one set of considerations rather than the other, but takes them all equally into account. Let us see what these rival considerations are.

Among the considerations which favour *physical* indivisibility there are three main categories. First, atoms are required to have shapes, and indeed a great variety of shapes and sizes, in order that they may account for the diversity of qualities perceived at the macro-level. It is impossible to see, in that case, how they can be conceptually indivisible, since when one is superimposed on another of a different size or shape, one atom is bound to have portions which extend beyond the other.

A second consideration suggesting physical indivisibility is that many passages give as Democritus' reasons for the indivisibility of the atoms their solidity (*nastotēs, sterrotēs, soliditas*) and the absence of void within them,[23] and these sound like physical reasons.

These two considerations in favour of physical indivisibility are good ones, but there is a third on which I would not rely. It consists of a collection of passages assembled by S. Luria, and taken by him to show that Democritus' atoms contain parts. This evidence has been well criticised by David Furley.[24]

---

[21] *in Phys.* 82, 1.

[22] *in GC.* 175, 7.

[23] Simpl. *in Phys.* 82, 1; and *in Cael.* 242, 16; Dionysius ap. Eusebium *Praep. Evang.* 14.23, 23; Cic. *de Fin.* 1.6,17; Alexander of Aphrodisias *in Metaph.* 35, 26. The first passage from Simplicius has been referred to already; it reads:

Indivisible things (*to adiaireton*) are so-called in different senses. One kind is that which has not yet been divided, but is capable of being divided like any continuum. Another is that which is not even of a nature to be divided at all because it does not have parts to be divided into, as e.g. a point or unit. Another is called indivisible through having parts and a size, but being impervious (*apathes*) through denseness and solidity (*sterrotēs, nastotēs*), like one of Democritus' atoms ... The atom is continuous and infinitely divisible and for this reason potentially many.

[24] Salomon Y. Luria, 'Die Infinitesimaltheorie'; also in his *Democritea*, Leningrad 1970, he comments in Russian on fr. 123. Furley's reply to the first publication is given in his *Two Studies*, 97-9 and 103. Luria cites as evidence for internal parts in the atom two passages from Simplicius, the one quoted in the preceding note, which will be considered below, and another from *in Cael.* 649, 2, which merely speculates what would happen *if* the atoms were divisible. He

Of the considerations on the other side favouring *conceptual* indivisibility, one concerns Democritus' arguments in favour of atomism. If atoms are to answer the paradox of division at every point or the paradox of the half distances, they will need to be conceptually indivisible. For those paradoxes exploit conceptual divisibility, and I believe that a recent argument to the contrary is mistaken.[25]

A second consideration favouring conceptual indivisibility is Democritus' description of his atoms as partless. We have seen this reported by Simplicius and it is in Stobaeus,[26] though Galen, who is probably following the same source as Simplicius, cites only smallness, not partlessness.[27]

A third consideration often cited in favour of conceptual indivisibility is less reliable. It is that Aristotle, without drawing a distinction, none the less treats Democritus as if he were committed to conceptually indivisible atoms. An instructive passage is *Cael.* 3.4, where Aristotle first says that for Leucippus and Democritus the primary magnitudes are indivisible in magnitude (*megethei adiaireta*, 303a4). He then complains that Leucippus and Democritus are obliged to conflict with mathematics (303a20), and further that they fall foul of the arguments in his own work on time and motion (303a23). This must be a reference mainly to *Phys.* 6, where the objections to atomism are appropriate only against a theory of *conceptual* indivisibility.[28]

Aristotle's treatment of Democritus was influential, and the charge of conflicting with mathematics was often repeated. But what was Aristotle's evidence? He was, of course, combating not only Democritus, but also a contemporary atomist like Xenocrates, who had committed himself to conceptual indivisibility. And he would have found in Democritus a desire to solve Zenonian paradoxes and probably references to partlessness. However, if I am right that Aristotle did not himself distinguish between conceptual

---

further cites two highly polemical passages in which Aristotle tries to pin on Democritus the view that every sphere is divisible into eight pyramids (*Cael.* 3.4, 303a30), and says that for Democritus a spherical atom owes its cutting power to constituting a sort of angle (*Cael.* 3.8 307a2; a17). A further passage in a late Latin translation of Themistius, which has descended through two intermediate languages, is corrupt (*in Cael.* tr. Landauer, 186, 26). Luria's best passage is Alexander of Aphrodisias *in Metaph.* 36, 25-7, to which I shall return below.

[25] Jonathan Barnes argues that the Zenonian paradoxes envisage, or might be taken as envisaging, a physical *division*. But, as explained (p. 352), that is irrelevant to whether they envisage a merely physical *possibility* of division. They can hardly be taken as concerned only with that. For one thing, they would then be robbed of most of their excitement. For another, why should anyone suppose there was a *physical* possibility of division at every point, or that what it would leave would be mere points or nothings? See Barnes, op. cit., vol. 2, 57; vol. 1, 245.

[26] Simpl. *in Phys.* 925, 13-17; Stob. *Ecl.* 1.14.1 p. 315 (Diels).

[27] *de Elementis Secundum Hippocratem* 1.2. There is also a tradition that it was not until Diodorus that 'partless things' (*ta amerē*) came to be used as a name for the atoms (evidence assembled by Luria in *Democritea* fr. 124). The evidence for partlessness in Democritus might be discounted on the hypothesis that it was Epicurus who first introduced this terminology (misleadingly) to describe Democritus' view.

[28] Aristotle does not really retract this treatment of Democritus' indivisibility as conceptual, when a few lines later he tries to pin on him the idea that every solid (atoms included) can be divided into pyramids, so that there will not be an infinite variety of elementary bodies, but only pyramids, or perhaps a few other shapes. This is at most something which Democritus might be forced into saying, not something he did say (303a30).

and physical indivisibility, his testimony on the kind of indivisibility involved, and the testimony of those who depend wholly on him, ought not to be given independent weight.

Despite this, there are still strong considerations on both sides of the controversy. I think this fits best with the hypothesis advocated, that Democritus had not distinguished conceptual and physical indivisibility, as those terms are explained above.

There is, so far as I know, only one other interpretation which, like this one, tries to do justice to both sets of rival considerations, rather than discounting one set. That is the ingenious interpretation of Luria presented in the pioneering paper referred to above. He briefly mentions the interpretation I have advocated, only to dismiss it,[29] and prefers to suggest that Democritus had a two-tier theory of atoms like that which others have with more plausibility found later on in Epicurus. According to this theory, atoms are themselves merely physically indivisible, but they contain within themselves minimal parts which are conceptually indivisible. Later sources, not recognising that there were two distinct tiers, will naturally have given conflicting reports on the Democritean atoms. Luria's evidence comes from passages which have already been referred to above. First, he trades on the same kind of conflict of evidence as I have exploited, but whose existence he explains as due to people confusing the two different tiers in Democritus' theory. Secondly, he offers some more direct evidence; for, among the passages which he takes to show that Democritean atoms have parts, there are several which, in his view, treat these parts as themselves partless. I have already expressed doubt about some of this evidence, and referred to Furley's reply. Luria's best piece of evidence comes from Alexander of Aphrodisias, who reports that Leucippus and Democritus have partless parts in their atoms, graspable by the mind not the senses, and that these parts are weightless.[30] But, like Furley, I find it easier to suppose that Alexander has confused the early atomist theory with the later theory of Epicurus than to suppose that all other authorities are ignorant of a central feature of Democritus' theory which only Alexander knows. Luria's two citations from Simplicius, quoted in text and footnote above, seem to me to count, if anything, against his interpretation and in favour of mine. In one, Simplicius contrasts Epicurus' theory with earlier atomism, which is the opposite of what Luria requires, and further treats smallness as akin to partlessness, thereby manifesting just the kind of confusion on which I have been trading.[31] In the other citation, Simplicius treats the atoms not merely as divisible, but as *infinitely* divisible,[32] which Luria acknowledges he does not want. If, as Luria suggests, the *infinite* divisibility is an inference drawn by Simplicius, not a doctrine acknowledged by Democritus, the same may be true of the *divisibility*, especially as it is the opposite of the partlessness, which Simplicius ascribes to Democritus' atoms in the other citation.

---

[29] Luria, 'Die Infinitesimaltheorie', p. 172.
[30] *in Metaph.* 36, 25-7.
[31] Simpl. *in Phys.* 925, 13-17.
[32] *in Phys.* 82, 1.

Plato's indivisibles

I shall consider only two other philosophers in this chapter, Plato and Xenocrates. Aristotle tells us that Plato believed in the existence of indivisible lines, although the passage has caused difficulty, because there is other testimony that he gave to points the role here given to indivisible lines.[33]

> Again from what will the presence of points in the line be derived? Plato used even to fight against this class of things [sc. points] as being something that geometers believed in, whereas he called indivisible [*atomoi*] lines the origin of the line and this he often postulated.

There is controversy whether Plato's *Timaeus* also expresses a belief in indivisible magnitudes. Plato there describes how the physical world is built out of elementary triangles. One piece of evidence that some of the triangles are indivisible is that he implies that they come in a finite number of sizes.[34] In that case, there is presumably a smallest size of triangle.

Plato distinguishes between perceptible, mathematical and ideal entities. Several considerations suggest that it is only in the world of perceptible things that indivisible magnitudes (themselves too small to perceive) are postulated. This is related to, though distinct from, the question whether their indivisibility is conceptual or merely physical. For if entities within the perceptible world are the only indivisible ones, while mathematical entities are never indivisible, then, in the absence of special argument,[35] we may expect the indivisibility to be merely physical. Conversely, if Plato ever treats conceptual indivisibility as impossible, we can infer that his indivisible entities belong to the perceptible world, rather than to the ideal or mathematical.

Without going into detail, I shall list briefly my four reasons for thinking that Plato's indivisible entities are confined to the perceptible world. (i) The *Timaeus* is discussing the triangles which compose perceptible bodies, and not some ideal triangles. (ii) In the *Parmenides*, Plato argues that a thing lacking parts, i.e. a conceptually indivisible thing, could have neither shape nor location, and could not either move or rest.[36] Of course, Plato is here propounding puzzles, and we cannot be sure what he would finally endorse. But the arguments here are particularly good ones, and several were to be taken up in Aristotle's *Physics*, so he might well be inclined to reject conceptual indivisibility. (iii) Plato recognises that some lines are incommensurable with each other,[37] and these presumably cannot consist of one or more indivisible lengths. The simplest hypothesis is that these are ideal and mathematical lines, while physical lines are the ones which contain

---

[33] Arist. *Metaph.* 1.9, 992a19-22; but see Sext. *M* 10.278; Alexander *in Met.* 55, 20; Simpl. *in Phys.* 454, 22.

[34] *Tim.* 57C8-D6.

[35] For special argument, consider the theories of modern science, described in Chapter 24.

[36] *Prm.* 137D-139B.

[37] *Laws* 7.820A-B.

indivisibles. (iv) If Plato is the author of the *Seventh Letter*, he contrasts circles drawn physically (*en tais praxesi:* by our actions), or turned on a lathe, with the ideal circle, on the grounds that the former have straight edges.[38] There is a similar reference elsewhere to the imperfection of the human, as opposed to the divine, circle.[39] We can infer that, if Plato's indivisible lines are straight, then the ideal circle cannot be made up of them. As to why human circles are so made, Plato's idea may be that we draw a circle or turn it on a lathe by depositing particles of charcoal or removing particles of wood, and these particles are not curved and do not have curved edges. The *Timaeus* shows us why not: elementary particles are all rectilinear, and, we may add, do not fall below a certain minimum size. In that case, the straight-edged circles, which we know to be physical rather than ideal, will also have edges that do not fall below that size.

My conclusion is that Plato envisages the merely physical indivisibility of items in the perceptible world. There is no suggestion in the *Timaeus* that we cannot *conceive* of parts in the smallest triangles, or that one triangle could not overlap with another. And this applies whether the triangles are conceived as thin sheets, or as something more abstract. Difficulty has been located in Plato's acceptance of infinite divisibility. But what he recognises as infinitely divisible is not physical body, but rather that element in ideal forms and physical bodies which he calls 'the great and the small'.[40] It may even be significant that he seems to have explained this infinite divisibility by using the example not of a physical body, but of a mathematical length, the cubit.[41]

If I am right, Plato's idea so far implies nothing about indivisible spaces or times. The shadow on a sundial could glide continuously part by part across the indivisible triangles which constitute the face of the dial. A water clock might emit its water one particle at a time, yet the triangles bounding the particle could slide continuously part by part out of the aperture. If, then, we are to look for evidence of time-atoms in Plato, we must turn in a different direction and go to his *Parmenides*.

Are there time-atoms in Plato?

Plato does not say explicitly in the *Parmenides* that he is talking about time-atoms, but the question whether he has them in mind has been argued out by Colin Strang and Keith Mills.[42] Strang shows that some of what Plato says would call for time-atoms, while Mills shows that other things would go against them. So I think we can at least say that Plato does not consciously embrace the idea. Interestingly enough, the Strang-Mills discussion was partly anticipated in the sixth century A.D. by Damascius. He too sees that some of the evidence fits a time-atom, while some does not, and his solution is

---

[38] *Letters* 7.343A5-9. I am grateful to Myles Burnyeat for this point, and for my treatment below of *Lin. Insec.* 968b7-8.

[39] *Philebus* 62A7-B9.

[40] See Arist. *Phys.* 203a9-10; 206b27. For further explanation of its presence in physical bodies and in Forms, see *Metaph.* 988a8-14.

[41] Porphyry ap. Simpl. *in Phys.* 453, 36.

[42] Colin Strang and K.W. Mills, 'Plato and the instant', *PAS* supp. vol. 48, 1974, 63-96.

to say that Plato is talking about that infinitely divisible leap of time in which, as we saw in Chapter 5, Damascius himself believes.

I shall glance at the evidence, but first we must decide which of Plato's entities were are talking about. For Sambursky looks for evidence of time-atoms in an entity which Plato calls 'the sudden'.[43] The sudden is introduced at *Prm.* 156C-157A and is said to divide a period of motion from a period of rest. But, as Strang and Mills agree, it sounds more like a durationless instant than a time-atom. For Plato says that when something switches from motion to rest, or from rest to motion, it is not then in any time, but is in 'the sudden', which itself is not in any time. If, as appears from this, 'the sudden' is not itself a time, nor in (that is, perhaps, coincident with)[44] any time, it does not seem to have the duration proper to a time-atom.

If, therefore, we want to find a time-atom in Plato, it will be better to look at another entity, 'the now', which is introduced a little earlier at 152B2-D2. And, indeed, it was 'the now', rather than 'the sudden', which occupied the attention of Damascius in the passages on which Sambursky was commenting.[45] Strang's evidence that this is a time-atom comes in two parts. First, 'the now' is likely to have duration. The best evidence[46] is that Plato is willing to call it a *time* (152B3), a description which he withholds from 'the sudden'.[47] But, secondly, 'the now' none the less appears to be indivisible, according to Strang. For Plato says of something which is becoming older that, when it is at the now-time, it already *is* older, and that it is not then *becoming* older. Why not, if the now-time is divisible into parts, so that the ageing object is older in the later parts of the now than in the earlier? Plato's presupposition must be that the now-time is indivisible. Here is the passage in question (152B2-D2):

> It *is* older, when, in the course of becoming older, it is at the now-time which is between was and will be. For presumably it will not skip the now when it travels from the once to the after. So when it meets the now, it then stops *becoming* older, and does not *become* but actually *is* then older. For if it were going onwards, it would never be caught by the now, since what is going

[43] Sambursky seems to be comparing 'the sudden' with the 'leaps' referred to by Damascius, which he (Sambursky) construes as atomic ('The concept of time in Late Neoplatonism', *PIASH* 2, 1968, 153-67; reprinted as part of the introduction to SP: see p. 19).

[44] The interpretation of 'in' as 'coinciding with' has the advantage that the same sense of 'in' will be applicable, when Plato says that the switching thing is '*in* the sudden'.

[45] Thus Damascius assimilates his 'leap' to Plato's 'now' in his commentary on Plato's *Parmenides = Dub. et Sol.* (Ruelle), 2.237, 15-22 (SP, p. 90, ll. 6-14, give the text, with translation on the facing page).

[46] Strang has an additional argument, which I think less satisfactory, for the now having a duration. Plato implies that a thing does its growing older not at the now, but when it is *between* the now and the after, and when it is touching both (152C3-6). I take it that the 'between' must be a divisible period, in order to leave room for growing older. Strang prefers to construe it as an instant, and uses this as a further argument to show that the now cannot be an instant, since two instants cannot be next to each other. His reason for taking the 'between' to be an instant is that otherwise it would be part of the after. But this is not so, if the after is a selected time period somewhat subsequent to the now.

[47] Mills protests that a durationless instant could easily be called a time, on the grounds that it is a position in time. But Plato denies both these designations ('time' and 'in time') to the instantaneous 'sudden'. So we should not expect him to grant them to an instantaneous 'now'.

onwards is so placed as to touch both the now and the after, letting go of the now and grasping the after, becoming between the two, the after and the now. But if of necessity anything that is becoming does not bypass the now, then when it is at the now, it always stops becoming, and then *is* whatever it may have been becoming.

So much for Strang's evidence; but Mills provides evidence just about as strong that 'nows' cannot conveniently be treated as time-atoms. He does so by raising the question when a body will make the transition from motion to rest. If there are time-atoms, then a body will not move during a time-atom, for that, being indivisible, leaves no room for motion. Rather, it will move by finding itself at different places at successive time-atoms. Correspondingly, when it comes to a halt, there will be a time-atom before which it was at different places, but after which it is in the same place. And it will be at this time-atom, if anywhere, that it makes the transition from motion to rest. But Plato wants the transition from motion to rest to be made not at this or at any time-atom, but at an instantaneous 'sudden'. Nor can it be argued that he is thinking of one of the 'suddens' which *bounds* the transitional time-atom; for which one would he choose? The earlier boundary seems too early, and the later boundary too late. If, then, Plato thinks, as he does, of the transition as being made at an *instant*, it is probable that he is thinking of time in the more usual way as continuous rather than as atomic.[48]

The upshot is that for some purposes Plato needs to have time-atoms and for other purposes to avoid them. I think this is a sign that he has not consciously intended them, and this conclusion fits with the fact that the purpose of this part of the *Parmenides* is to interest us in dialectical problems, rather than to offer definitive theories.

Damascius' interpretation: leaps

I have said that part of Strang's discussion is anticipated by Damascius, who, however, draws the different conclusion that Plato's 'now' must correspond to his infinitely divisible leaps. These leaps were the subject of Chapter 5, and it will be recalled that they differ from time-atoms in that they are supposed somehow to combine indivisibility with divisibility. In favour of Plato's 'now' having duration, Damascius anticipates Strang's argument that Plato calls it a 'time'.[49] On the other hand, in favour of its being divisible, he argues that Plato allows a thing, in its progress through time, to be simultaneously in contact with, and letting go of, the 'now'. It must, he infers, be in contact with *one* part of the now and letting go of *another*.[50] Leaps, according to Damascius, meet both of the requirements so far laid down, namely, that the 'now' should have duration, while being divisible. But they

---

[48] This is not to deny that, on the view of time as continuous, there would still be a problem about the instant of transition from motion to rest, namely, whether the body would be moving or resting at that transitional instant. But to this problem Plato believes he has an answer: it would be doing *neither* (*Prm.* 156C-157A). See Chapter 26.

[49] Damascius *in Prm.* = *Dub. et Sol.* (Ruelle), 2.237, 15-17 (SP 90, 6-8).

[50] ibid., 17-22 (SP 90, 9-14).

also meet a third requirement, that in a certain sense the 'now' should be indivisible as well as divisible. In this connexion, Damascius makes the very same point as Strang. Just as Plato says that, when a thing is at the now, it stops *coming* to be, and actually *is* whatever it was coming to be, so also we should say that a leap actually *is*. It is never part way through a process of coming into being.[51]

Indivisibles in Xenocrates and the Academy

Xenocrates was the third head of Plato's Academy, from 339 B.C. (four years before Aristotle set up his own school in the Lyceum) to 314. Numerous sources testify that he believed in indivisible lines.[52] Moreover, we have in the last chapter noticed the five arguments for atomism recorded at the opening of the pseudo-Aristotelian treatise *de Lineis Insecabilibus*.[53] Of these the second and third have ideas and terminology close to Xenocrates' own,[54] while the fourth is a version of Zeno's paradox of the half-distances, and there is, as we saw, some evidence in Philoponus that Xenocrates exploited this.[55] Although I shall leave open the exact extent of Xenocrates' influence on the five arguments, what is clear is that they come from Plato's Academy. The second, indeed, concerns an ideal Platonic Form.

A number of commentators, ancient and modern, have sought to play down Xenocrates' belief in indivisibility as meaning something unremarkable.[56] But the five arguments for indivisibles in *de Lineis Insecabilibus* between them go well beyond Plato's original idea, for they would establish indivisible magnitudes in the ideal, the mathematical and the perceptible worlds. Or if Xenocrates identified the ideal and mathematical realms, recognising only two worlds not three, then I should say that the arguments, if they are his, would establish indivisible magnitudes in both. Thus the second argument concerns the indivisibility of Ideal Forms such as the Triangle. The third explicitly deals with the indivisibility of the constituents of body in the perceptible world, as opposed to the intelligible world of Forms. The fifth argument belongs with the third, since it is concerned with lines which are

---

[51] Plato *Prm.* 152B2-E2, echoed at l. 21 by Damascius, op. cit. 242, 15-21 (SP 92, 16-23, esp. 23).

[52] See frs 43-9 in *Xenokrates*, ed. R. Heinze, Leipzig 1892.

[53] 968a1-b21.

[54] Compare Heinze frs 50-1 for the terminology of the third argument (968a14-18: *stoicheion, ameres*: element, partless). The priority of part to whole used in the second and third arguments (968a9-18) is exploited by Xenocrates in a fragment discussed by S. Pines, 'A new fragment of Xenocrates and its implications', *TAPS* 51, pt 2, 1961, 3-34. The idea of the second argument (968a9-14) that the ideal form of the line is indivisible is acribed to Xenocrates in Heinze fr. 46.

[55] Five ancient commentators assume that Xenocrates is meant when Aristotle says, at *Phys.* 1.3, 187a1-3, that some gave in to the argument from dichotomy and postulated indivisible magnitudes (see ad. loc. Philoponus, Themistius, Simplicius and Simplicius' reports on Alexander and Porphyry = Heinze frs 44-5). But whereas the others give as the dichotomy argument something like the paradox of large and small, or the paradox of division everywhere, Philoponus gives in addition (*in Phys.* 81, 7-16) the paradox of half-distances.

[56] Among those collected in Heinze frs 45-7; cf. Pines, op. cit., p. 19.

actually measured (968b7-8), i.e. presumably with lines in the perceptible world. The remaining two arguments, the first and fourth (which concerns traversing half-distances), ought to be applicable to the mathematical world (if this is distinct from the ideal). Indeed, if the fourth is to answer Zeno's paradox of the half-distances, it will have to involve indivisibility in both the mathematical and the physical worlds. In three of the arguments (the first, second and fourth) the indivisibility required, unlike Plato's, is a conceptual one.

Are there time-atoms in Xenocrates and the Academy?

Despite this extension beyond Plato's views, I am not convinced that Xenocrates and Plato's Academy went so far as to postulate time-atoms. But in a recent book which seeks to emphasise the role of Plato's Academy in the development of atomism, H.-J. Krämer offers five arguments to show that they did.[57] None of these arguments seems to me persuasive. First *Lin. Insec.* opens by asking whether there are indivisible lines, and whether in general there is something partless in *all* quantities, as some people say. The five atomist arguments follow immediately. But the reference to *all* quantities need not be intended to cover times, as Krämer suggests. The context suggests two other possibilities much more readily: first, and most obviously, the author is likely to be referring to planes and solids as well as lines. Failing that, he is likely to be referring to ideal, mathematical, and perceptible quantities.

Krämer's second point, that the first of the five arguments is general enough to apply to times, does not show that it was thought of as applying to times.

His third point is that Aristotle treats time-atoms as parallel to atoms of space and matter and attacks them all. But the question is whether in attacking the Academy's atoms of space and matter, Aristotle is not rather complaining that they have *unwittingly* committed themselves to time-atoms. Is there any reason to suppose that they accepted time-atoms *voluntarily*?

Fourthly, it is urged, the author of *Lin. Insec.* attacks time-atoms at 970b9; 971a16-20; b4. But in none of these references is it implied that the atomist opponent had thought for himself of postulating time-atoms. In the first, the author suggests (borrowing a phrase from Aristotle's *Phys.* 6.1, 231b18) that the same argument which composes a line out of atomic lines should compose a time out of atomic times. In the second, his suggestion is that 'perhaps' the same argument which composes a line out of points should compose a time out of 'nows'. Here the talk is no longer of atoms but of points, and the 'perhaps' makes it clearer still that the atomist opponent had not himself suggested this. Only five lines (971a18-21; b4) are devoted to attacking the suggestions about time, and they are all devoted to attacking the second suggestion, which is not about *atoms* at all.

[57] H.-J. Krämer, *Platonismus und hellenistische Philosophie*, Berlin 1971, 310; 350-1. Krämer sees the five arguments of *Lin. Insec.* as a compilation stemming indirectly from Xenocrates, pp. 337-8

Finally, Krämer sees Academic influence in the time-atoms referred to at a later date not only by the Platonist Plutarch, but even by the Aristotelian Strato. But I shall argue in the next chaper that Strato did not accept time-atoms, while Plutarch in his best attested reference to time-atoms (*CN* 41, 1081c), says only that the Stoics had not left themselves free to escape a certain paradox by postulating time-atoms. If he is thinking of anyone who did leave himself free in this way, I shall suggest below that it is not so likely to be an Academic as Diodorus Cronus followed by Epicurus.

There is a doctrine conjecturally ascribed to Xenocrates that sounds are heard as continuous, though produced by a discontinuous series of instantaneous blows falling on the ear.[58] Even here, however, Xenocrates is not credited with any discussion of a least audible time interval between blows.

---

[58] See Heinze, pp. 5-10, on Porphyry's commentary on Ptolemy's *Harmonics* (Wallis) p. 213.

CHAPTER TWENTY-FOUR

Atoms and Time-Atoms after Aristotle

We have not found time-atoms consciously postulated by Zeno's opponents, by Plato, or by Xenocrates. But atomism had to change drastically after it was attacked by Aristotle (384-322 B.C.). And, as Furley has argued, the acceptance of time-atoms was one of the refinements with which atomists responded to him.

Aristotelian objections to atomism

The most influential part of Aristotle's onslaught comes in *Phys.* 6.1-4 and 10. And for time atomism the crucial passage is 6.1, 231b18, where Aristotle connects atoms of magnitude, time and motion:

> The same argument applies to magnitude, time and motion: either all are composed of indivisibles and divided into indivisibles, or none are.

In much of what follows he seeks to substantiate this. He then attacks the atomist theory, and takes it that, in attacking any one kind of atom, he will be attacking the other two. Subsequent atomists accepted that, if they were to postulate atoms at all, they would have to postulate time-atoms, and the link which Aristotle forged between atoms of magnitude, time and motion was not questioned at least until Strato.

Aristotle's attacks are relevant only against a theory of *conceptually* indivisible atoms, and Xenocrates and the Academy had just provided him with such a theory. However, he himself, so I have suggested, did not notice the difference between physical and conceptual indivisibility. Furthermore, as references in the *de Caelo* show, he thinks his arguments in *Phys.* 6 valid also against the atomism of Democritus and Plato.[1]

Many other points in Aristotle's attack proved influential, but I shall select for mention just two, which will be particularly relevant to what follows. The first concerns the atomist commitment to jerky motion, which is alleged to be impossible. Moving is not possible either for an indivisible body,[2] or across indivisible spaces.[3] For if a body is engaged in moving into an adjacent space, there has to be a stage when part of the body has entered part of the adjacent space, while part of it occupies part of the original space. Clearly, this would

[1] *Cael.* 3.1, 299a9; 3.4, 303a20.
[2] *Phys.* 6.4, 234b10-20; 6.10, 240b8-241a6.
[3] *Phys.* 6.1, 231b18-232a17.

be impossible, if the body or the spaces lacked parts. In that event, the body could never be moving, but at best *have* moved with a jerk, that is, by disappearing from one space and reappearing further on. But such jerky motion would be thoroughly paradoxical. For one thing, a body which was moving by jerks across three indivisible spaces would be not moving but resting in every one of them.[4] (This last point brings us close to a certain reconstruction of Zeno's arrow which was rejected in Chapter 21).

In attacking indivisible magnitudes here, Aristotle is attacking both indivisible moving bodies and indivisible distances traversed. The former of these two attacks is already formulated in Plato's *Parmenides*, from which Aristotle took it.[5] The response of the later atomists, Diodorus Cronus and Epicurus, was simply to agree that motion must take place in this funny way by jerks.[6]

In the remaining argument, Aristotle denies that indivisibles can be arranged so as to build up a larger whole, or at any rate to compose a continuous whole.[7] In an earlier chapter, *Phys.* 5.3, he had defined three possible modes of arrangement: continuity, contact and succession. But atoms cannot, first, be continuous (*suneches*) with each other, nor, secondly, in contact (*haptomenon*), for both these modes of arrangement are defined in terms of *edges*: the edges must be either one or at least together. But an entity which lacks parts cannot have an edge as distinct from an interior, so Aristotle claims. Contact is an impossible arrangement also, because, for partless entities, it cannot be contact of part with part, nor of part with whole, while contact of whole with whole will not give us spatially separate parts, but mere amalgamation. Thirdly, if the indivisible entities are points or instants, they cannot be in succession (*ephexēs*) to each other, that is, next to each other in an order of first, second and third, because any two will be separated by further entities of the same kind; and that conflicts with the definition of successiveness given in 5.3.

Two oversights in Aristotle's objections

There are two oversights in Aristotle's objections, which are not remarked by modern commentators, but which did, I believe, have an effect on subsequent history. The first oversight arises in connexion with time. Sometimes Aristotle maintains that if motion, distance traversed, or speed are divisible, so is the time taken;[8] while sometimes he maintains the converse, that, if the time taken is divisible, then, at constant speed, so is the motion and distance traversed.[9] It is the second which introduces indivisible times; for it implies that if there is an indivisible distance, then, at a given speed, your movement to the next indivisible distance cannot take less than a certain time. We

[4] *Phys.* 6.1, 232a12-17.
[5] *Prm.* 138D2-E7.
[6] For Epicurus, see Themistius *in Phys.* 184, 9; Simpl. *in Phys.* 934, 24. For Diorodus, Sext. *M* 10.48; 85-90; 97-102; 143 (cf. 120); *PH.* 2.245; 3.71.
[7] *Phys.* 6.1, 231a21-b18.
[8] *Phys.* 6.1, 232a18-22; 6.2, 232b20-233a12; 233b19-33; 6.4, 235a22-4.
[9] *Phys.* 6.1, 232a18-22; 6.2, 233a13-21; 6.4, 235a18-22.

should think of the movement as a discontinuous jerk, and the point is that the delay between jerks will not fall below a certain minimum. But the words, 'at a given speed' constitute a most important qualification. For suppose it is possible to increase indefinitely the number of indivisible spaces traversed per minute, then the time between jerks can surely be indefinitely small, and there will not be indivisible times after all.

In any case, the argument does not seem to concern time itself, but only the time taken to traverse a given space. In order to reach any conclusion about time itself, more elaborate considerations would be needed. It might be argued, for example, that the stars which constitute our heavenly clock are taken as having a constant speed, and that if they had to traverse indivisible spaces, the time spent in each space would be a minimal time. Perhaps this is what Aristotle had in mind, although he does not mention it. But if so, it is not enough for his purposes. For it leaves open the possibility of adopting additional clocks with faster-moving indicators.

Help might be sought from a claim which Aristotle makes in another place (*Cael.* 2.6, 288b30-289a4). Not only is there a minimum time and a maximum speed for the revolution of the heavens; but also for every action, for playing the lyre or walking, there is a minimum time and maximum speed for actions of that kind. Perhaps, then, clock indicators cannot move indefinitely fast. But this will not settle the matter, for it should be possible to use two or more clocks whose jerks are out of phase with each other by amounts as small as we please. The 'ticks' of one clock can be punctuated by the 'tocks' of another and the 'tings' of a third, to give us indefinitely short intervals. I shall return to the possibility of using out of phase clocks at the end of the chapter. Perhaps Aristotle would not consider relying on clocks other than the heavenly one, since he certainly argues for the superiority of the heavenly motion to all others. But then that argument presupposes that the heavenly motion will not occur by jerks.

There is a related oversight in Aristotle's argument at 6.10, 240b31-241a1. For here he supposes that an indivisible body could perform its jerks, only (*monachōs*) if time were made up of indivisible nows. Why *only*? He has proved very satisfactorily that an indivisible body could move only by jerking. But why should not the time between jerks be indefinitely small, and why should not time itself still be infinitely divisible?

I shall argue below that Strato was the first person to challenge Aristotle's claim that atoms of one kind imply atoms of another. What he seems to have raised is the converse of what I have just raised. For he probably asked why there should not be atoms of time and motion without atoms of matter and space. I have been asking why there should not be atoms of matter, space and motion without atoms of time.

The other important loophole is to be found in Aristotle's argument that indivisibles cannot be arranged so as to form a larger whole, or at any rate not a continuous whole. For the entire argument applies more easily to sizeless points than to extended atoms. And when Aristotle reaches the relation of *succession* or *nextness* (231b6-10; b12-15), his argument is applicable *only* to points and *not* to atoms. Moreover, he takes his illustration from the case of points and instants (*not* parts, as the Oxford translation says at

231b13). The argument is that indivisibles cannot be arranged next to each other (*ephexēs*), in an order of first and second, because between any two indivisibles, there will be others, which violates the definition of *ephexēs*. This argument holds true of points, but *not* of atoms, for two atoms need not have further atoms between them. Elsewhere Aristotle seems even to acknowledge the loophole which he has left. For although he does not accept atomic times, he does imply that, if there were any, they would be successive (*ephexēs*): 'And atomic times are successive.'[10] The Aristotelian author of *Lin. Insec.* seems to have felt uncomfortable about the loophole left here.[11] For he first considers the un-Aristotelian alternative that after all succession does involve contact.[12] But failing that, we are to fall back on one of Aristotle's other arguments.[13]

There is a further curious feature of Aristotle's arguments, which suggests that, even when considering the first two modes of arrangement, continuity and contact, he has points more fully in mind than atoms. For in introducing his arguments (231a24), he qualifies his conclusion by saying that, *if anything is a genuine continuum*, it will not be composed of indivisibles, and it is only this qualified conclusion which is established by arguments such as those at 231b10-12; b15-18. This is not a conclusion that should worry believers in atoms, because they would presumably be ready, even if believers in points were not, to grant that a compound body did not constitute a genuine *continuum*. None the less, despite the modesty of Aristotle's conclusion, he tries to make his arguments about continuity and contact applicable to atoms as well as to points; only the application to atoms is less easy.

The first argument, we saw, was that continuity and contact are both defined in terms of *edges* (231a21-9): continuous things have their edges the same, while things in contact have them in the same place, but indivisibles have no edges. The absence of edges seems evident for the case of points, but in order to make it apply to atoms, Aristole has to offer a mistaken argument; he treats the edge as if it were a *part*, and concludes that no partless thing can have edges.

The second argument ruled out contact between indivisibles by saying (231b2-6) that contact must be of part with part, or of part with whole, or of whole with whole, but that the first two options are unavailable for things without parts, while the last gives us amalgamation, not contact. This argument works smoothly for points, but is harder to apply to atoms. For why should not atoms fit the earlier definition of contact by having their edges in the same place? And why should contact of this kind by edges involve contact between parts or wholes? Aristotle would presumably rely on his earlier mistaken argument that the edges are *parts*. Admittedly, this time there is a more satisfactory answer that could have been given: two atoms will not contact each other along the whole of their peripheries, but only along part, and if the peripheries have parts, presumably the atoms do as well, and so they are not atomic after all.

---

[10] *Phys.* 8.8, 264a4.

[11] *Lin. Insec.* 971b26-972a1.

[12] Aristotle allows this not for succession (*ephexēs*), but only for the different relation of neighbouring (*echomenon, Phys.* 5.3, 227a6-7).

[13] 231b15-18 is the original for 971b28-972a1.

I have concentrated on two loopholes in Aristotle's discussion, because I want to argue that they were exploited by his successors. But there are other loopholes as well. Thus only the three relations of continuity, contact and succession are considered. We may wonder why Aristotle was not able to imagine any fourth mode of arrangement. He himself makes use of an idea which could be exploited in order to introduce a fourth relationship, namely, the idea that between any two points there are others. Why should not such an arrangement of points be specified as the one which composes a continuum out of indivisibles?

I shall now turn to Aristotle's successors, and we shall soon see them exploiting the loophole which allows atoms to be arranged in succession.

Diodorus Cronus: four contributions to atomism

The scintillating Diodorus Cronus was introduced in Chapter 2. He probably overlapped with Aristotle, although his date of death has recently been brought down from *c.* 307 B.C. to *c.* 284 B.C., thirty-eight years after Aristotle's.[14] It is comparatively well known that he made one major contribution to atomism. I want to suggest that he may have made four.

The best known contribution is his acceptance of atomic motions. He had clearly taken to heart Aristotle's complaint that if you believe in atomic space or bodies, you will be committed to something having moved in a jerk without ever being in process of moving. But Diodorus replied with an argument examined in Chapters 2 and 22 that Aristotle too was committed to jerky movement. Moreover, as we saw in Chapter 2, he maintained that there was nothing unusual about our being able to use the *perfect* tense and to say 'it *has* moved', without ever being able to use the *present* tense and to say 'it *is* moving'. There are many everyday examples, he alleged, in which the perfect tense of a verb is thus divorced from the present.

It is fairly widely agreed that Diodorus made this contribution to atomist theory, namely, accepting atomic jerks. What I now want to see is what other contributions he made, if any. I have already argued in Chapter 2 that he accepted atomic times, judging from the report in Sextus *M* 10.119-20. I would, therefore, put forward time atomism as Diodorus' second contribution, although, it will be remembered, this is not the usual view.

The third possible contribution is contained in the same passage, and for that reason I shall quote it again:

> If a thing is moving, it is moving now. If it is moving now, it is moving in the present (*enestōti*). If it is moving in the present, it must be moving in a partless (*amerēs*) time. For if the present has parts, it will inevitably be divided into a past and a future part, and so will no longer be present.

> If a thing is moving in a partless time, it traverses partless places. But if it traverses partless places, it is not moving. For when it is in the first partless place, it is not moving, since it is still in the first partless place. And when it is in the second partless place, again it is not moving but rather *has* moved. Therefore a thing never *is* moving.

[14] David Sedley, 'Diodorus Cronus and Hellenistic Philosophy', *PCPS* n.s. 23, 1977, 74-120.

The point which interests me is that Diodorus here arranges his atomic spaces in an order of first and second. This exploits the loophole that we saw Aristotle leaving. When he denied that indivisibles could be so arranged as to form bodies, he omitted, in the case of atoms, to exclude the possibility that they might be arranged successively in an order of first and second, and this is the possibility which *M* 10.120 takes up.

Fourthly, Diodorus introduced some new *arguments* for atomism. As these were described in Chapter 22, I need only offer reminders. One argument, which was designed to introduce atomic *times*, is contained in the passage just quoted, and by an irony it is very much the same argument as Aristotle had used in the *Physics*[15] for the different purpose of showing that the present is a sizeless instant. Another argument, which is intended to establish atomic *spaces*, is modelled on Aristotle's puzzle in the *de Sensu*[16] about the nearest distance of invisibility and the furthest of visibility. A third argument introduces jerky *movements* by exploiting Aristotle's idea in the *Physics*[17] that the place of a thing is its *immediate* surroundings. Finally, but more dubiously, Mau attributed to Diodorus an argument for atoms based on his conception of the possible as being at some time actual.

One thing that will have emerged in reviewing these four possible contributions to atomism is the extent of Diodorus' debt to Aristotle. Thus Diodorus borrows Aristotle's argument for the indivisibility of the present. He accepts from Aristotle the idea (later attacked by Strato) that motion at indivisible times entails motion through indivisible places, and that motion through indivisible places entails having moved without being in process of moving. He exploits Aristotle's definition of a thing's place, in order to show that Aristotle is committed to a similar view of motion. And he tries to make the view palatable by taking over from Aristotle the idea that you can sometimes use the perfect tense without commitment to an earlier use of the present, and by applying this idea to the context of motion, where Aristotle would not apply it. He takes up the concession that atoms, unlike points, can be arranged in succession, and he adapts the *de Sensu* argument about distances into an argument for atomism. He would further have learnt from Aristotle the view that atomists are committed to time-atoms. Finally, in arguing for his conception of the possible as being at some time actual he uses a famous argument called the Master Argument, which looks like a riposte to Aristotle's treatment of determinism in *Int.* 9.[18]

There is one difficulty over the suggestion that Diodorus made all four of the contributions to atomism just enumerated. For certain features of these contributions are common to Diodorus and Epicurus, so that a question of priority arises. We are told, for example, that Epicurus accepted atomic motion, and that his motive for accepting it was the same as I have ascribed to Diodorus, namely, to accommodate Aristotle's arguments.[19] We shall see also that Epicurus probably accepted time-atoms and the successive ordering of minima. Diodorus was the older man, but on Sedley's dating Epicurus will

---

[15] 6.3, 234a9-19.   [16] 7, 449a20-31.   [17] 4.4.

[18] For the argument, see Epictetus *Dissertationes* 2.19, 1. I have summarised the relation to Arist. *Int.* 9 in *Necessity, Cause and Blame*, ch. 6.

[19] Themistius *in Phys.* 184, 28; Simpl. *in Phys.* 934, 18-30.

have joined him in Athens when, aged 34, he set up his school there in 306 B.C., and he will have remained with Diodorus in Athens until Diodorus moved to Alexandria in the late 290s. Who, then, can claim to be the innovator? Sextus makes clear that it was *Diodorus* who pioneered the acceptance of atomic motion, and devised the arguments to make that palatable, while Alexander reveals that it was Diodorus who adapted the *de Sensu*, to produce an argument for atomism. We also find in Diodorus, not in Epicurus, an argument that the present must be a time-atom. Most of the innovation so far discussed seems therefore to belong to Diodorus. But I shall none the less argue that, at least in what remains extant, Epicurus does more to make intelligible the idea that particles can be successive without touching. I would claim no more than that Diodorus must be credited with a good proportion of the innovations.

Epicurus and his followers

The two-tier theory of Epicurus (341-279 B.C.), according to which there are minimal parts within an atom, is expounded in §§57-9 of the *Letter to Herodotus*, the letter in which Epicurus summarises his physics. The full thirty-seven books *On Nature* survive only in fragments, some of them found in rolls which were badly charred in the volcanic eruption of A.D. 79 which buried Herculaneum. The majority, though not unanimous, view is that the minimal parts are conceptually indivisible, the atoms only physically so, and I agree with this.[20]

Epicurus' two arguments (§57) for conceptually indivisible minimal parts were cited in Chapter 22. The first argument excluded infinite divisibility on the ground that infinitely many components, all having a size, would add up to an infinitely large body. The second objected that, if there were an infinity of parts, then, in conceiving the whole, our mind could, *per impossibile,* arrive at infinity.

In §58 Epicurus introduces some important innovations. He is the only ancient writer I know of who tries to explain how conceptually indivisible parts are even possible. We are to understand the idea of a smallest *conceivable* part on the analogy of a smallest *perceptible* part. Sedley has suggested that this analogy is also used by Diodorus, but I have above interpreted the relevant passage differently.[21] Epicurus wants us to think, for example, of the smallest speck we can see on a blackboard. We can see that it has a positive size, but it is too small for our sight to discriminate a smaller size within it. By analogy, he asks, why should there not be a smallest *conceivable* size, just as there is a smallest *perceptible*?

I believe that Epicurus' discussion has a striking implication. For the smallest visible speck is probably too small for us to *perceive* any edge to it, and hence any shape. By analogy we should be unable to *conceive* an edge or shape in the smallest *conceivable* part. There has been much discussion of the

---

[20] See the arguments of David Furley, *Two Studies*, pp. 21; 42, cited above.

[21] This is how Sedley (op. cit.) interprets Diodorus' use of the argument about a largest imperceptible, and smallest perceptible, size.

shape to be ascribed to the minimal parts in Epicurus' atom. My suggestion is that they have no shape at all. If they have no shape, Epicurus will be able to escape certain geometrical objections, for example, the objection that the diagonal of a square minimal part would have to exceed the side by less than a minimal length.

I am glad to find that Sedley (followed by Konstan) has anticipated me in two of the above suggestions. For he too has suggested that Epicurus' minimal parts may be edgeless and shapeless. And he has further construed Epicurus' analogy with sense perception as I do, that is, as intended to show how conceptually indivisible parts are *possible*, rather than to prove that they actually exist.[22] This makes Epicurus' argument differ at this point from those of Berkeley and Hume. Both of these thinkers argued from a minimum extension in our ideas to the conclusion that there must be, not merely that there can be, a minimum conceivable extension in reality.[23] In this they were perhaps truer to Lucretius than to Epicurus, since Lucretius comes closer to arguing for 'must', as Sedley has pointed out to me, if by *conicere ut possis* he means 'so that you can conclude', rather than 'so that you can conjecture', at 1.749-52.

If Epicurus did argue for the possible edgelessness of his minimal parts by reference to analogies with sense perception, then there is a striking parallel in modern philosophy of physics. For Čapek has argued that atoms of time, or of space-time, should be viewed as edgeless, and he too has appealed, among other things, to analogies with sense perception.[24]

Succession without contact between Epicurus' minimal parts

I should like to take the suggestion of edgelessness one stage further. For if the minimal parts in the atom have no edges, this could be used to show how it is possible for them to be arranged successively next to each other without being in contact. We have already seen that Aristotle left a loophole when he failed to exclude this mode of arrangement, and that Diodorus made use of the loophole when he described his atomic spaces as coming in an order of first and second. I want to suggest that Epicurus used the same loophole, and further that by relieving his miminal parts of edges, he showed how it could be legitimate to do so. For a lack of edges will prevent the minimal parts from satisfying Aristotle's definition of contact (they cannot have *edges* together), but not from satisfying his definition of successiveness (two can be arranged so as to have nothing of the same kind in between). This is how they can be successive without touching.

If we turn for a moment to atoms, instead of minimal parts, these come

---

[22] David Sedley, 'Epicurus and the mathematicians of Cyzicus', *CE* 6, 1976, 23-54, n. 2; David Konstan, 'Problems in Epicurean Physics', *Isis* 70, 1979, 394-418, p. 405.

[23] Berkeley, *Principles* §§47; 85; 123-32; *Third Dialogue Between Hylas and Philonous* (cf. *An Essay Towards a New Theory of Vision* §§80-6: the same *minimum visibile* for all animals); Hume, *A Treatise of Human Nature*, bk 1, pt 2, §§1-4. The latter is discussed by Furley, op. cit., ch. 10.

[24] Milic Čapek, 'The fiction of instants', in J.T. Fraser, F.C. Haber, G.H. Müller (eds), *The Study of Time*, Berlin 1972, 332-44; cf. his *The Philosophical Impact of Contemporary Physics*, Princeton 1961, 231; 234-8.

closer to satisfying Aristotle's requirement for contact, since they do at least have edges, namely, their minimal parts, which are called edges (*perata*). But even atoms cannot be in contact in Aristotle's sense of having their edges '*together*', that is, in the same place. At most, they could be said to contact each other in the weaker sense that they have edges and that there is nothing between the edges.

That Epicurus used the loophole of postulating a merely successive relationship between conceptual minima cannot be made certain. But it is suggested by several passages. Firstly, the word *hexēs* (in succession), which is close to Aristotle's word *ephexēs*, is used along with the word 'first' (*prōton*) in §58 for the analogous minima in perception:

We see these minima in succession (*hexēs*) starting from the first (*prōton*), ...

Admittedly, it is not yet clear whether it is only our viewings which are successive (we see them successively), or whether we may understand Epicurus to mean: we see them as successive. But Lucretius avoids any such ambiguity, when he talks of the conceptual minima within Epicurus' atom as being arranged 'in order':

Then more and more similar parts in order (*ex ordine*) fill out the nature of the body in a solid array.[25]

Finally, there is a piece of evidence in another Epicurean, Demetrius of Laconia, who uses the word *hexēs* of successive times in a fragment from a Herculanean roll. Demetrius is probably talking about Epicurus, and according to two editors, V. de Falco and Enzo Puglia, his description of Epicurus' theory runs more or less as follows:

He posits something like this. The thing happens whenever, from the place where this one emerged, at the next (*hexēs*) time, which is a minimal time, the neighbouring one will follow at once.[26]

The present point of interest is that the times under discussion are described as being *next* (*hexēs*) to each other, without being described as in contact.[27] I have to admit, however, that, in connexion with times, the preference for this description is very natural and may carry no special significance. Moreover, although there is no dispute that the relationship mentioned is one of nextness or succession, it will emerge below that there is a dispute whether the times so related in this passage are really *minimal* times. So, for my

---

[25] *de Rerum Natura* 1.605-6.

[26] Demetrius of Laconia, Herculanean papyrus 1012, col. 31, 4-8, printed in V. de Falco, *L'epicureo Demetrio Lacone*, Naples 1923, p. 40; and in Enzo Puglia, 'Nuove letture nei P. Herc. 1012 e 1786 (Demetrii Laconis opera incerta),' *CE* 10, 1980, 25-53, see p. 42.

[27] We need not worry about the word translated 'neighbouring' (*echomenon*). For although Aristotle defines the *echomenon* as being in contact (*Phys.* 5.3, 227a6-7), Plato does not (*Prm.* 148E9). So the word does not necessarily carry an implication that the minimal parts are in contact with each other.

suggestion about minima, I must rely most heavily on Lucretius' use of the expression *ex ordine*.

To summarise, my evidence has been that Aristotle allowed the atomists the relationship of succession without contact, that we find them using the words for succession (*hexēs, prōton – deuteron, ex ordine*), that we never, so far as I know, find them describing their smallest conceivable units as in contact, and this even though they had every opportunity of doing so in two of the passages under discussion (*Letter to Herodotus* §58; Lucretius 1.605-6). My further suggestion has been that the perceptual analogy shows how succession without contact is possible, by robbing the conceptual minima of edges.

There are, however, many rival interpretations. Mau thinks that the relationship of minimal parts is left unclear, Furley that it is edge-to-edge contact. Krämer sees Epicurus' answer in the phrase 'measuring in their distinct way', the distinctness (*idiotēs*) being opposed to coalescence. Konstan suggests that the smallest parts of the atom are inconceivable except as parts of a whole, and that it follows that no question can arise of how one part attaches to the whole. The relevant passage of the *Letter to Herodotus* runs from §58 to §59:

> We must grasp that the minimum in perception is neither just like that which admits of traversal, nor altogether unlike. It has something in common with traversibles, but it does not have any distinction of parts. Whenever we think, because of the resemblance in common qualities, that we are going to distinguish a bit of it, one part on this side, another on that, it must be the same magnitude that confronts us. We see these minima in succession (*hexēs*) starting from the first (*prōton*), neither in the same place, nor touching part to part, nor doing anything but in their own individuality (*idiotēs*) providing the measure of magnitudes, more for a larger magnitude, fewer for a smaller.
>
> We must think that the minimum in the atom follows this analogy. For although it clearly differs in minuteness from the minimum seen in perception, it follows the same analogy. For even the idea that the atom has a size was predicated on the analogy of the things before our eyes, by our simply projecting something small onto a large scale. We must also think of the minimal uncompounded edges as providing originally out of themselves the measure of lengths for both larger and smaller things, using our reason to apply to invisibles. For what they have in common with changeable things is sufficient to accomplish this much. There cannot, however, be a meeting (*sumphorēsis*) between minima on the move.[28]

When Epicurus says that we see the perceptible minima neither in the same place nor touching part to part, he is replying, as Mau points out, to one of Aristotle's two arguments against indivisibles being arranged in contact with each other so as to build up a larger whole. That argument (*Phys.* 6.1, 231b2-6) was that contact of whole to whole is mere amalgamation, which would leave the indivisibles in the same place; while contact of part to part or

---

[28] A reason for this ban on meeting is suggested by Lucretius 1.607-8: the minimal parts cannot even *exist* on their own.

of part to whole is impossible for things which have no parts. In effect my suggestion is that Epicurus is also replying to Aristotle's other argument against contact. That is the argument (*Phys.* 6.1, 231a27-9) that since indivisibles lack edges, they cannot have edges together: the answer is that this enables them to be in succession without being in contact.

Does Epicurus accept time-atoms?

There is a further question about Epicurus' atomism: does he believe in time-atoms? When I talk of time-atoms, I do not mean to suggest a two-tier distinction between atoms and minimal parts of time. On the contrary, I am using the name 'time-atoms', only because that is the expression I have been using in connexion with other thinkers who do not have a two-tier theory. If Epicurus believes in time-atoms, he will think of them as partless minima, and will contrast them not with any shorter periods, but only with certain *longer* ones, namely, the least *perceptible* times.

That Epicurus believes in time-atoms is attested by Sextus and Simplicius.[29] Admittedly, they ascribe a 'least' and 'partless' (*elachiston*, *ameres*) time only to the *followers* of Epicurus (those *peri* or *kata* him). But phrases of this kind, as Sedley has pointed out, were commonly used to mean Epicurus and his followers. This is the clearest testimony for atomic times, but given this testimony, it is reasonable to see a reference to atomic times in certain other passages as well. Lucretius says that:

> in a single time during which we have perception, there are hidden many times whose existence is discovered by reason.[30]

This Latin exposition corresponds to the phraseology used by Epicurus himself in the *Letter to Herodotus* §§47 and 62, when he speaks of times that are contemplated by reason (as opposed to sense perception).

One commentator, while accepting the other testimony, rejects that from the *Letter*, on the grounds that the plural 'times' implies that Epicurus is talking of periods of different lengths, all below the threshold of perception, but not atomic, because not equal in length.[31] However the plural is much more easily explained by the fact that in each context Epicurus is contrasting the times which are contemplated by reason with a longer period, in one case with a perceptible time, in another with what he calls the least *continuous* time. The times contemplated by reason are mentioned in the plural, because it takes a plurality of them to match one of the longer periods mentioned.

There is one more piece of evidence for time-atoms in Epicurus. For we have seen that they are ascribed to him by Demetrius of Laconia, at least on certain readings of the fragmentary papyrus. Unfortunately, the papyrus rolls, rescued from the volcanic ash, are very difficult to read, and Sedley has

---

[29] Sext. *M* 10.142-54; Simpl. *in Phys.* 934, 26.

[30] *de Rerum Natura* 4.794-6.

[31] P. Boyancé, review of J.M. Rist's *Epicurus* in *Gnomon* 46, 1974, 754-5.

kindly reinspected the relevant portion for me and allowed me to make use of his new readings. The two editors mentioned above took the text to say:

> The thing happens whenever, from the place where this one emerged, at the next time (*kata ton hexēs chronon*), which is a minimal time (*touto* [*n d'e*]*lachiston*, de Falco; *to* [*n e*]*lachiston*, Puglia), the neighbouring one will follow at once.

But Sedley reads:

> The thing happens whenever, from the place where this one emerged, at the next time, supposedly (*kata ton hexēs chronon pou*), the neighbouring minimum (*to* [*el*]*achiston to echom* [cancelled letters]*enon*) will follow at once.[32]

What is minimal, on Sedley's reading, is not the time, as would be suggested by the letters *t o n*, but merely the neighbouring *particle*, as is conveyed by *t o*. Sedley offers this reading only as possible, not as mandatory, although one thing in its favour is that the scribe has not left himself much room for the extra letter in *t o n*.

With or without the extra testimony in Demetrius, I think the evidence of Sextus and Simplicius is good enough to warrant the ascription of time-atoms to Epicurus.[33] I would add that, once he had talked to Diodorus, and reflected on the arguments of Aristotle, he would have needed to produce some vigorous argument on the other side, if he wanted to *dissociate* himself from belief in time-atoms. None the less, I would point out that the testimony of Themistius appears to count against, rather than for, the attribution. For Themistius brings it as an objection aginst Epicurus (an invalid objection, as it happens) that his jerky movement would imply time-atoms.[34] This suggests he was unaware that time-atomism was actually endorsed by Epicurus.

The most far-reaching case against time-atoms in Epicurus is made in a recent paper by Françoise Caujolle-Zaslawsky.[35] Her main objection is that Epicurus gives to time a low ontological status, making it an attribute of an attribute, or even a mere appearance. From this it follows that time is not a substance composed of atoms and void. But it surely should not be taken to follow that time is not quantised. Nor does the author take account of the positive evidence for time-atoms supplied by Sextus and (perhaps) Demetrius of Laconia.

Her denial that there are time-atoms in Epicurus is bound up with the still bolder denial that Epicurus composes movements or sizes out of indivisible units.[36] It is conceded,[37] rather unexpectedly, that Epicurus postulates

---

[32] In Sedley, the whole sentence runs: *Touto gar geinetai hotan, hothen todę exechōrę̄* † *achth* [*en* †, *kata ton hexēs chronon pou̧ to* [*el*]*achiston to echom* [cancelled letters]*enon euthus akolouthēsēi.*

[33] Postscript. Sedley has now provided me with confirmation: an Epicurean refers to a minimal time (*chronos elachistos*) in P. Herc. 698, Fr. 23 *N*, published in W. Scott, *Fragmenta Herculanensia*, p. 290.

[34] Themistius *in Phys.* 184, 28-185, 3.

[35] Françoise Caujolle-Zaslawsky, 'Le temps Épicurien est-il atomique?', *EP* 1980, 285-306.

[36] ibid., 300.   [37] ibid., 304.

indivisibles in space. But it is not recognised that this would lead to indivisible sizes and movements. Instead, the suggestion is that it was Diodorus who postulated indivisible movements, and that even he did so only in the weak sense of saying that we cannot *recognise* a movement until it has already taken place. It is further maintained that Diodorus put forward his theory early enough for Aristotle to refute it in *Physics* 6, thereby leaving Epicurus with no motive for renewing the theory.[38] This and other reasons are given for denying that there are atoms of time, movement, or size in Epicurus,[39] but it will be clear from what has preceded why I do not find myself able to accept these reasons.

Outside the Epicurean school I know of no subsequent endorsement of time-atoms among the philosophers of antiquity; but this will need arguing: for a start, some have found time-atoms in Strato.

Strato's riposte to Aristotle

Strato was head of the Aristotelian Peripatos from 288 to *c.* 269 B.C., while Diodorus was in Alexandria, and Epicurus still teaching in Athens. The difficulty about his view is created by a passage in Sextus Empiricus:

> It remains, then, to consider whether anything can be moving if some things are divided infinitely while others stop at a partless [segment]. Indeed, the followers of Strato the physicist were drawn in this direction. For they supposed that times stopped at a partless [segment], while bodies and places were divided infinitely, and that a thing in motion moved a whole divisible distance at one go in a partless time and not part before part.[40]

How can this report be squared with the clear statement of Simplicius that for Strato time was continuous and infinitely divisible? According to Simplicius, Strato departed from Aristotle's idea that time is a kind of number, on the grounds that the whole numbers form a discontinuous series, whereas time is continuous:

> But Strato of Lampsacus criticised the definition of time given by Aristotle and his friends, and even though he was a pupil of Theophrastus, who followed Aristotle in almost everything, he himself took a newer route. For he does not accept that time is the number of motion, because number is a discontinuous (*diōrismenon*) quantity, whereas motion and time are continuous (*suneches*), and the continuous is not countable.[41]

---

[38] ibid., 300-1.

[39] As regards the reason just mentioned, Epicurus refused to accept, on my view, that Aristotle's refutation succeeded, and rightly so, since discontinuous jerks are not conceptually impossible. As a second reason, Caujolle-Zaslawsky urges (p. 301) that Epicurus posits movement as a genuine attribute of atoms, but this is compatible with his saying that movement takes the form of *having* moved. Finally, she urges (p. 302) that since motion and time exist at different levels, motion being an attribute, but time only an attribute of an attribute, it is unlikely that *both* would be considered atomic. But this appears to be a *non sequitur*.

[40] *M* 10.155.

[41] Simpl. *in Phys.* 789, 4.

The phrase 'followers of Strato' is not likely to be intended to *distinguish* between Strato and his followers.

Two passages have been or might be misused in the attempt to show that Strato did hold an atomist view of time. A few lines later, Simplicius records one of Strato's objections to the view that time is a number:

> Again the unit and the now will be identical, if time is a number, for time will be composed of nows, and number of units.[42]

Here Strato is not *endorsing* the idea that time is composed of nows, whether the nows be atomic or instantaneous. Rather he is treating that as an absurd consequence of the view he is opposing. This only confirms that, on his own view, time is continuous.

The other misusable passage comes from Damascius.

> Is not time continuous yet separated, as [Plato] shows? Yes, it is; but it is composed not of partless parts, but of separated lengths. For it consists, as Strato says, of parts which do not stay put; hence of separated parts. Yet each part is continuous, and is, as it were, a measure composed of many measures.[43]

Sambursky interprets Damascius in a way which I contested in Chapter 5 as believing that time comes in atomic jerks, and he translates the word *diōrismena* as 'discrete' rather than separated, taking the reference to be to atomic jerks. None the less, as Sambursky himself recognises ad loc., the word *diōrismena* does not enter into what is directly ascribed to Strato. It is doubtful that anything more is ascribed to him than the idea that time consists of parts which do not stay put. The last two passages, then, add no further support to Sextus' idea that time was atomic for Strato.

I suggest that a solution may be found in a further report by Simplicius:

> But Strato of Lampsacus says that motion is continuous not simply because of magnitude, but also in itself.[44]

Strato's idea is that, although motion is continuous, it does not *owe* its continuity to that of distance. Perhaps then he argued in the same way about time: although time is continuous, it does not *owe* its continuity to that of distance. For it would be conceptually *possible* to have distances continuous, but motion and time discontinuous.

If Strato objected in this way to Aristotle's claim that the continuity of one thing entails the continuity of the others, this could explain Sextus' taking him (wrongly) to be arguing that time is *in fact* atomic. How might Strato's argument have gone? Suppose that each moving thing moved in little jerks, reappearing at a distance from where it was before, without ever being in between. Suppose too that while jerks might occur over infinitely varying *distances*, they were all synchronised to occur only at certain fixed intervals of

---

[42] *in Phys.* 789, 14-15.
[43] *Dub. et Sol.* 2.236, 9-13 (SP 88, 5-10).
[44] *in Phys.* 711, 9-10.

time. Strato might have supposed that, if these intervals were very small, they could in principle be atomic. Certainly, they would not be interrupted by moving things making any progress, since progress is postulated as occurring only at the *end* of an interval. Such a conception is not without difficulty (would not these atomic times without motion be undetectable, for example?). But Strato might have supposed such a situation conceivable. He would then conclude that, while time is *in fact* continuous, its continuity is not *guaranteed* by the continuity of space.

Alternatively, Strato may have been making a simpler point, not about time itself, but about the time that any individual would take between jerks if motion were discontinuous. On this conception, the jerks of two different moving bodies could be staggered, so that the time one body spent resting was punctuated by the jerk of another body. Time would not itself be atomic in that case. But there could still be a certain minimum time such that the jerks of an individual body could not be more closely spaced than that. Strato's point would again be that, although this does not *in fact* happen, it is not *excluded* by the continuity of space.

If I am right that Strato argued in one or other of these two ways, then he deserves much more credit than he has received. Aristotle's argument that atomism of one kind implies atomism of another weighed with those two clever men Diodorus and Epicurus. It weighed, as will be seen in the next chapter, with the Islamic atomists, as Maimonides reports.[45] And it has weighed with twentieth-century physicists and philosophers of science. But they are mistaken, as I shall argue at the end of this chapter, and Strato's denial is right.

The Stoics

Sambursky argued in the first edition of one of his books that the Stoics were forced against their will to postulate a time-atom.[46] But in the second edition and elsewhere, he rightly retracted this statement. Certainly, it would be a strange interpretation of the Stoics, who are renowned for their belief in infinite divisibility, and the passage by which Sambursky was guided has been differently interepreted by me in Chapter 2. There is more plausibility in the suggestion, to which I shall now turn, that their enemy Plutarch accepted time-atoms. But this too I believe to be false.

Plutarch

As was seen in Chapter 2, Plutarch of Chaeronea (*c.* A.D. 46-120) got involved in the controversy about the parts of time being unreal. For soon after Strato's death, the Stoic Chrysippus (*c.* 279-*c.* 206 B.C.) had considered the problem,[47] and Plutarch reports Chrysippus' position with contempt, complaining that

---

[45] Maimonides *Guide* 1.73, proposition 3.

[46] S. Sambursky, *The Physical World of the Greeks*, 1st ed., London 1956, 151-2, citing Plut. *CN* 41, 1081D-1082A.

[47] Plut. *CN* 41, 1081F; Arius Didymus, in Diels, p. 461.

he cannot rescue the reality of time by representing the present as a time-atom, and hence as a part of time.

> It is contrary to common conceptions to say that there is future and past time, and not present time, but that recently and the other day subsist (*huphestanai*), while there is no now at all. Yet this is what happens to the Stoics, who do not allow a minimal (*elachiston*) time, and do not want now to be indivisible (*ameres*), but say, whenever someone thinks he has grasped something and conceived it as present, that part of it is future and part of it is past. Thus nothing remains level with now, nor is any part of the present left, if of the time which is said to be present some is assigned to the future and some to the past.[48]

Some have thought that, when he here desiderates a minimal, indivisible time, in other words, a time-atom, Plutarch is expressing his own view.[49] But in fact his aim is simply to attack the Stoics, and he does not commit himself on how he would solve the problem. The time-atom which Chrysippus had rejected was postulated by Diodorus and Epicurus. It is not an Academic time-atom to which Plutarch, as a Platonist, is committed. Indeed, Plutarch tells us that his own teacher retailed the argument that the present gets squeezed out into the past and future, and, significantly, does not add that the solution is a time-atom.[50]

The other passage in Plutarch where time-atomism has been detected[51] seems to me to be concerned with a wholly different topic: spatial, not temporal, indivisibility. The same 'now' is found everywhere, in Rhodes and in Athens. If it is thus (spatially) extended, why is it not (spatially) divisible? Evidently because it is incorporeal. This discussion goes back to Plato's treatment of a day as being in many places at once, and yet not separate from itself.[52]

Damascius: non-atomic leaps

There has been a tendency to ascribe belief in time-atoms to Platonists, and another case in point is Damascius, who lived four hundred years after Plutarch. His belief that time comes in 'leaps' was examined in Chapter 5. But I there resisted Sambursky's view that the leaps were atomic. On the contrary, they combined a kind of indivisibility with infinite divisibility in ways which I discussed there. It is their infinite divisibility which prevents them from being straightforwardly atomic.

Time-atomism in perspective

In this chapter, I have represented time atomism as being generated in a brief period of excitement, in response to Aristotle, in the late fourth century

---

[48] *CN* 41, 1081C-D.
[49] S.Y. Luria, 'Die Infinitesimaltheorie', 163; H.-J. Krämer, *Platonismus*, 351.
[50] Plut. *On the E at Delphi* 392F.
[51] By Krämer, loc. cit., referring to *Platonicae Quaestiones* 1002D.
[52] *Prm.* 131B.

B.C., and as continuing among philosophers only in the Epicurean school. I have resisted the idea that it was a widely shared theory stretching from early Presocratics to late Neoplatonists. On the other hand, in Chapter 25 I shall draw attention to its reappearance in Islam. And I shall suggest that the Hellenistic developments in atomism, which were often reactions to Aristotle's opposition, had a particular influence on Islamic thinkers. There is, however, one strand of ancient thought which I have left out.

Time-atoms in musical theory

After Epicurus the idea of time-atoms reappeared in antiquity from an unexpected quarter: the theory of music. The story has been told by Lasswitz.[53] Aristotle's pupil, Aristoxenus, discussing vocal music, gave the name 'primary time' to the smallest unit of measure in rhythmics. And in the third century A.D., Aristides Quintilianus described this time as atomic (*atomos*) and as the least (*elachistos*). It was the least relatively to us, and was the first to be grasped in perception. The idea was handed on by Martianus Capella in the fifth century, while in the seventh Isidore of Seville spoke of the year being divided into atomic times, without any longer making reference to music. The venerable Bede, in A.D. 725, gave these atomic times an absolute length. For whereas in musical theory the length of an atomic time had varied with the piece of music in question, Bede assigned exactly 22560 time-atoms to the hour. Tannery has suggested that this number was designed to provide a common measure for the Julian year and the lunar month.[54] Maimonides was also to suggest a multiple of sixty time-atoms to the hour in his summary of common assumptions of the *Kalām*.[55] Bede further reports that grammarians had spoken of atomic times, assigning two to a long syllable and one to a short, while astrologers had divided the zodiac into atoms, so as to be able to state the exact time of birth for predicting a person's fate. None of these time-atoms, however, seem to involve conceptual indivisibility, and so they do not raise the same problems as atomic times in philosophy.

Time-atoms in modern physics

In modern physics, the idea of time-atoms has re-emerged. It is at present only a minority belief, and is not a standard part of quantum theory, which quantises only certain properties such as angular momentum and spin.[56] None the less, it is possible to assemble quite a long list of physicists who have favoured the idea.[57] In at least one exposition, it is conceded that

[53] Kurd Lasswitz, *Geschichte der Atomistik*, Hamburg and Leipzig 1890, vol. 1, 31-7. He cites Aristoxenus, ap. Porph. *in Ptolemaei Harmonica*, pp. 255-6; Aristides Quintilianus, *de Musica* 1.14; Martianus Capella 9.971; Isidore of Seville, *Etymologiae* 13.2 (*de atomis*) §§2-3; Bede, *de Temporum Ratione*, ch. 3 (*de minutissimis temporum spatiis*).

[54] P. Tannery, 'Vermischte historische Notizien', *Bibliotheca Mathematica* (3) 6, 1905, 111. I am grateful to Norman Kretzmann for referring me to Tannery and Lasswitz.

[55] Maimonides *Guide* 1.73, third proposition.

[56] So Adolf Grünbaum, *Modern Science and Zeno's Paradoxes*, Middletown, Connecticut, 1967.

[57] M. Čapek gives a list of fourteen, in *The Philosophical Impact of Contemporary Physics*, Princeton

we can still apply to time the mathematical idea of infinite divisibility, if we like, but it is suggested that it will not correspond to anything physical.[58] It is hard to classify the indivisibility of such time-atoms in the rather simple-minded terms used above. It is not exactly a *conceptual* indivisibility, because we can perfectly well make sense of the idea that our universe might have been different, and that there might have been something physical corresponding to ever smaller divisions, and we can also still make sense of the idea of infinite divisibility in mathematical space. On the other hand, it would be an understatement to say that the indivisibility was merely *physical*. For this kind of theory questions whether there is any *content* to the idea of indefinitely small divisions in our universe as it is. What is clear is that this type of atomism would not provide a philosopher with the kind of solution he wanted for the ancient paradoxes about space, time and motion. For those paradoxes, for example the paradox of the half-distances, could still be raised about the universe as it might have been.

One modern argument moves in Aristotle's manner from atoms of space to atoms of time. Suppose the radius of the electron constitutes a smallest natural length, while the velocity of light constitutes the highest possible speed, then, it is argued, there would be a minimum time.[59] But another view is that we should not split up space and time in this way, but that we should follow Relativity Theory in speaking of space-time, and apply our quantisation to it.[60] Yet a third view introduces superspace at a level more fundamental than either time or space-time, but finds atoms at every level.[61] The argument for atoms in these theories sometimes turns on the indeterminacy principle of quantum theory. According to this principle, the position and velocity of a particle cannot both be determinate at one time, nor can the mass and the energy. This indeterminacy, it has been suggested, sets a limit on how finely time can be measured.[62] But other arguments for atomism appeal to yet more complex developments in modern physics.[63]

What are we to think of these modern arguments for time-atoms? The simplest ones, which move from atoms of space or motion to atoms of time, fall foul of Strato's insight that it is not in every case possible to argue from atoms in one medium to atoms in another. I defended this insight earlier by means of a simple example involving two clocks out of phase. We can

---

1961, 230-1. To these I would add three others, E.L. Hill, 'Relativistic theory of discrete momentum space and discrete space-time', *Phys. Rev.* 100, 1955, 1780-3; D. Bohm, *Wholeness and the Implicate Order*, London 1980, 91-105; and with B.J. Hiley and A.E.G. Stuart, 'On a new mode of description in Physics', *IJ Physics* 3, 1970, 171-83; J.A. Wheeler, 'Superspace and the nature of quantum geometrical dynamics', in C.M. DeWitt and J.A. Wheeler (eds), *Battelle Rencontres* 1967, p. 242; again in *Am. Sci.* 1968; and in Isham, Penrose, Sciama (eds) *Quantum Gravity*, Oxford 1975, p. 538.

[58] Whitrow, *The Natural Philosophy of Time*, 2nd ed., 201.

[59] Whitrow, op. cit., 204; M. Čapek, op. cit., 240; cf. the report in W.H. Newton-Smith, *The Structure of Time*, London 1980, 116-17.

[60] Whitrow, op. cit., 280-3; Čapek, op. cit., 232-4.

[61] Wheeler, opera citata.

[62] Whitrow, op. cit., 280-3; Čapek, op. cit., 238-40.

[63] See M.L.G. Redhead, 'Wave-particle duality', *BJPS* 28, 1977, 72; V.S. Zidell, 'Some problems bearing on the concept of space-time quanta', *Phys. Rev.* D, 23, 1981, 1221-6.

imagine the tip of a clock hand disappearing from one atomic space and reappearing in the next. Even if there is a maximum speed, I argued that this would not yet give us atomic time, for we can imagine two such clocks slightly out of phase with each other, so that the rests of one, far from being atomic, are punctuated by the transitions of another. Let me consider this possibility further.

It might be objected that such small punctuations would be undetectable. But we can imagine *indirect* means of detection. Perhaps when one hand has moved a hundred atomic spaces, the other is found to be one atomic space ahead. The best explanation might turn out to be that the faster hand rests between beats for a fraction less time than the slower. The idea of such fractions of time would now have a clear physical meaning. It would make no difference to postulate a maximum speed for the moving hands; for the argument relies not on indefinitely high speeds, but only on indefinitely small differences of speed. I am less clear whether the idea of differences of phase could be used to impugn the more sophisticated physical arguments for atomism, but I suspect that it could impugn some of them.

It might turn out that clock movements could not be got to be out of phase by less than a certain amount; and in that case, we should have a fresh reason for denying physical application to the idea of indefinitely small time units. But this fresh reason would not turn on the postulation of atoms of space or of movement. Neither of these need be postulated, if atoms of movement are understood as involving fixed distances. Rather, the argument would turn on there being evidence for a *synchronisation* of movements and for fixed intervals *between* them.

CHAPTER TWENTY-FIVE

Atoms and Divisible Leaps in Islamic Thought

Islam had a theory of atoms and of time-atoms and a rival theory of infinitely divisible 'leaps'. The history of these 'leaps' in Greek thought was charted in Chapter 5: they combine indivisibility with divisibility *ad infinitum*. For an infinitely divisible leap can occur across a *distance* as short as you like, but the *time* taken is an indivisible instant, with the body disappearing from where it was and instantaneously reappearing further along. All these ideas arose relatively early in Islamic thought, for they feature in a controversy between Naẓẓām (died *c*. A.D. 846) and Abū l-Hudhayl al-ʿAllāf (died *c*. A.D. 841). The latter believed in atoms, while the former argued for infinite divisibility.

In this chapter, one aim will be to see what the arguments were, since they are presented only in summary by later, and sometimes hostile, sources with little indication of their original purport. A further aim arises, because the Greek parallels may be able to help in reconstructing the Islamic arguments. Sometimes, I believe, they actually put them in a new light and suggest a fresh meaning. This is particularly true of Naẓẓām's theory of 'leaps', where I think the Greek parallels, along with considerations of philosophical sense, virtually compel us to reinterpret everything from scratch. I shall have still another aim, because recently there has been a divergence among Islamicists, and some have sought to play down Greek influence, at least in connexion with the early phase when Islamic atomism was first developed. I shall try to rob this approach of its attractions, by showing just how close the early Islamic discussions are to the Greek ones, matching them sometimes argument by argument. The degree of proximity has not, I think, been noticed even by those Islamicists who have always recognised the Greek connexion. A Hellenist may feel at home with the technicalities of these debates on atomism, but there is a danger that he will not see the Islamic wood for the Greek trees. On Islamic questions I have been abundantly helped by Fritz Zimmermann and saved from at least some of the pitfalls, though he would not endorse all that I say, and will be offering different interpretations of some of the arguments.

The positive arguments for leaps

I shall begin by asking why Naẓẓām came to postulate leaps. He believed in infinite divisibility, but he was then called on to face what we know as Zeno's problem, that we should have to traverse an infinity of sub-distances every time we moved a yard. Abū l-Hudhayl was exempt from this difficulty, because he believed in atoms. It was as a solution to *this* problem that Nazzām postulated leaps.[1] These leaps, since he wanted to avoid atomism, must surely be the infinitely divisible leaps which we have found reported by Sextus and by Damascius as one of the Greek answers to Zeno's paradox. Indeed, though his examples of things that leap have caused puzzlement,[2] I think the puzzlement can be removed when the Greek sources are set out. For his examples of leaping things, namely light and coldness, are virtually the same as the original examples of light, heat and freezing supplied in Aristotle's *de Sensu*. Nazzām's idea, like the ancient one, will have been that any journey involves, not an infinite number of sub-journeys, but only a finite number of variably short leaps. The sub-distances may be infinite in number, but the leaps are not.

It has not, I think, been appreciated in this context what the difference is between atomic and infinitely divisible leaps.[3] Both involve a thing's disappearing and reappearing further along, but the infinitely divisible leaps can be across indefinitely short distances.

Nazzām's first reason for postulating infinitely divisible leaps is clear: the leaps enabled him, as effectively as the atomists, to answer the Zenonian paradox of infinite sub-distances. But one text suggests that Nazzām had an additional ground. For in the case of light, he appears to have argued that its leaping could actually be observed.[4]

[1] Baghdādī (died A.D. 1037), *Farq*, ed. Badr, Cairo, 1910, 123f.; Khayyāṭ (late ninth century), *Intiṣār*, ed. Nyberg, Cairo 1925, 32f. (translated into French by A. Nader, Beirut 1957, 19f.); Juwaynī, (died A.D. 1085) *Shāmil*, ed. Nassār, Alexandria 1969, 434; Ibn Mattawayh (first half of eleventh century), *Tadhkira*, ed. S.N. Luṭf and F.B. 'Awn, Cairo 1975, 169; 197-8. For modern discussions, see A. Nader, *Le Système philosophique des Mu'tazila*, Beyrouth 1956, 155-8; 182-7; S. van den Bergh, vol. 2, p. 30, of his translation with commentary of Averroes *Tahāfut al-Tahāfut*, London 1969; H.M.E. Alousi, *The Problem of Creation in Islamic Thought*, Baghdad 1968, 277; D.B. MacDonald, 'Continuous re-creation and atomic time in Muslim scholastic theology', *Isis* 9, 1927, 326-44; S. Pines, *Beiträge zur islamischen Atomenlehre*, Berlin 1936, 10-16; J. van Ess, *Theology and Science: the Case of Abū Isḥāq an-Nazzām*, Second Annual United Arab Emirates Lecture in Islamic Studies, University of Michigan, Ann Arbor, 1978 (19pp.).

[2] Van Ess, op. cit., 9, is surprised not by the example of light, but by that of coldness, since coldness involves no movement. But Aristotle's attitude was just the opposite. He allowed leaps only in the case of qualitative change, not movement. But he treated light as a non-moving quality, and for that reason was willing to set it alongside freezing and heating as something propagated by leaps (*Sens.* 6, 446b27-447a6). It was only Aristotle's successors who extended the idea of leaps to movement.

[3] MacDonald, op. cit., 341-2, does not distinguish infinitely divisible leaps from atomic ones, and does not give Greek parallels. Van den Bergh, op. cit., 30, does suggest Greek parallels, but instead of infinitely divisible leaps cites the *atomic* leaps attacked by Aristotle in *Phys.* 6 and subsequently postulated by Diodorus. He also refers to the Stoics, who may indeed have postulated infinitely divisible leaps, but not, I think, in the passage he cites. Surprisingly, however, van den Bergh had in a much earlier work, and in a different context, cited one of Damascius' passages on leaps, and those leaps are, as we know, infinitely divisible (S. van den Bergh, *Die Epitome des Metaphysik des Averroes*, Leiden 1924, p. 189, n. 1). Van Ess, op. cit., p. 15 and n. 91, picks up this reference and treats it as providing the only Greek parallel for Nazzām's leap, thus rightly excluding some spurious references, though overlooking the report of Sextus. Even he, however, appears in the end to understand the leap as atomic; for he takes it to incorporate the atomist assumption of discontinuous space (pp. 7-8).

[4] Ibn Mattawayh, op. cit., 203-5.

The controversy over the Zenonian paradox was connected with another which has been discussed earlier. For it was at some time alleged that if we really made an infinite number of sub-journeys every time we moved, we could no longer object to the idea that the universe might have passed through an infinite number of years before reaching the present.[5] One solution was to describe our movements in terms of a finite number of atomic spaces, another in terms of a finite number of infinitely divisible leaps.

The influence of Greek thought on Naẓẓām is already clear. But in a pioneering and influential article of 1925 Otto Pretzl came to a diametrically opposite conclusion. He argued that Islamic atomism was too unlike Greek atomism to have been inherited directly, and he suggested instead that it had travelled via the Gnostics.[6] One part of Pretzl's case was that Naẓẓām's ideas were too unsatisfactory philosophically for us to believe that he was familiar with the Greek discussions; and as one piece of evidence for this he cited precisely Naẓẓām's postulation of leaps, evidently unaware of the point I have been making, that this was actually a Greek theory.

As further evidence, Pretzl rejected as an unphilosophical argument for infinite divisibility Naẓẓām's report that he could not think of anything which lacked a half, and desiderated as the only philosophical mode of argument a deduction from the concept of the mathematical continuum. But my own reaction would be that such a deduction would be question-begging, while Naẓẓām's report is perfectly appropriate.

There was another part of Pretzl's case. For the Greeks, he argued, atoms had to play the role of what was stable in change. It was therefore important for the Greeks to decide whether atoms were homogeneous and how they behaved chemically or mechanically, to emphasise their shapes and hardness, and to discuss the role of the vacuum. This is not found in early Islamic theology. What I think needs to be said is that it is not found throughout Greek thought either. It is not found, for example, in Diodorus, nor in Epicurus' account of the minimal parts within the atoms, nor again in Xenocrates. It is only one strand in Greek atomism. What seems most to have influenced Islamic atomism is some of the developments which characterise the Hellenistic period of Greek thought (323-30 B.C.). This general point has been admirably recognised by S. Pines.[7]

Naẓẓām's arguments against atomism

A whole series of arguments by Naẓẓām involving the leap is recorded by two later writers, Ibn Mattawayh (first half of the eleventh century) and Juwaynī

---

[5] This is explained by the Jewish philosopher Sa'adia (A.D. 882-924) in *Kitāb al-Amānāt,* ed. S. Landauer, Leiden 1880, p. 36, translated by S. Rosenblatt as *Book of Beliefs and Opinions,* New Haven, 1948, 45. Cf. Juwaynī *Shāmil,* 148.

[6] Otto Pretzl, 'Die frühislamische Atomenlehre', *Der Islam* 19, 1931, 117-30.

[7] Pines points out that Islam is closer to Epicurus then to Democritus, op. cit., 97-9. The importance of the Hellenistic period is also well recognised by van Ess, but somewhat at the expense of Aristotelian ideas, of which he says (op. cit., 15), that Naẓẓām keeps apart from them. I have already diverged from this judgment in noting (contrary to van Ess, p. 9) the influence of the Aristotelian tradition on Naẓẓām's choice of examples to illustrate the leap.

(died A.D. 1085). They had been interestingly discussed by van Ess,[8] and I have benefited from abstracts of the relevant portions which Fritz Zimmermann has made available to me, with due warning that the renderings are uncertain. My first question is whether these arguments are, as is normally thought, additional arguments for the leap. I want to suggest that they are not. First, they all make better philosophical sense when taken as arguments by Nazẓām against the doctrines of atomism. Secondly, they are presented alongside arguments which have that function and which do not even mention the leap or motion at all. Thirdly, the Greek parallels confirm this hypothesis. The arguments support the infinitely divisible leap only in the very indirect sense that it is immune to the objections brought against atomism. But the positive arguments for the leap are the two given above. That the further arguments are primarily directed against atomism, and not for the leap, has not been clear, partly because it is not made clear by the two sources. Thus Juwaynī merely introduces the arguments as sophisms to which Nazẓām resorted when the mindlessness of the leap was pointed out to him,[9] and he often speaks as if the leap was meant to solve the difficulties which are brought against the atomists. But he and Ibn Mattawayh are hostile witnesses, and writing at a much later date. They present Nazẓām's arguments unsympathetically, in highly abbreviated form, and with only the scantiest indication of how they are meant to work. This will be clear from the abstracts here and at the end of the chapter. Some reconstruction is, therefore, inevitably required. The following suggestions are only tentative, but I hope that at least they make good sense.

Leaps and the attack on atomist accounts of rotation

One of the arguments ascribed to Nazẓām has a very special position. For it is clearly set out, although without mention of Nazẓām, by an additional source, Maimonides (A.D. 1135-1204). And, better still, we happen to possess the Greek original in Sextus. In each case, it is explicitly said that the argument is directed against atomism. And this confirms my claim that Nazẓām's arguments here are against atoms rather than for his infinitely divisible leaps. Interestingly enough, Sextus' version of the argument comes just a few paragraphs after he has finished discussing the Stoic postulation of infinitely divisible leaps.

Maimonides ascribes to the Islamic atomists the view that, since atomic movements do not take up any time, they must occur at the same speed. In order to account for differences of speed, they allowed an atom to *linger* for more or fewer time-atoms in successive space-atoms. But that in turn raised the problem that the inner atoms of a rotating millstone must linger in their places, while the outer atoms progress, so that the millstone is fragmented. The following is Maimonides' account, as translated by M. Friedländer:

Now, mark what conclusions were drawn from these three propositions, and

[8] Van Ess, op. cit., reports Juwaynī's arguments in *Shāmil*, 434-43, and Ibn Mattawayh's in *Tadhkira*, 169; 187-207.

[9] Juwaynī, *Shāmil*, 434.

were accepted by the Mutakellemim as true. They held that locomotion consisted in the translation of each atom of a body from one point to the next one; accordingly the velocity of one body in motion cannot be greater than that of another body. When, nevertheless, two bodies are observed to move during the same time through different spaces, the cause of this difference is not attributed by them to the fact that the body which has moved through a larger distance had a greater velocity, but to the circumstance that motion which in ordinary language is called slow, has been interrupted by more moments of rest, while the motion which ordinarily is called quick has been interrupted by fewer moments of rest. When it is shown that the motion of an arrow, which is shot from a powerful bow, is in contradiction to their theory, they declare that in this case too the motion is interrupted by moments of rest. They believe that it is the fault of man's senses if he believes that the arrow moves continuously, for there are many things which cannot be perceived by the senses, as they assert in the twelfth proposition. But we ask them: 'Have you observed a complete revolution of a millstone? Each point in the extreme circumference of the stone describes a large circle in the very same time in which a point nearer the centre describes a small circle; the velocity of the outer circle is therefore greater than that of the inner circle. You cannot say that the motion of the latter was interrupted by more moments of rest; for the whole moving body, i.e., the millstone, is one coherent body.' They reply, 'During the circular motion, the parts of the millstone separate from each other, and the moments of rest interrupting the motion of the portions nearer the centre are more than those which interrupt the motion of the outer portions'. We ask again, 'How is it that the millstone, which we perceive as one body, and which cannot be easily broken, even with a hammer, resolves into its atoms when it moves, and becomes again one coherent body, returning to its previous state as soon as it comes to rest, while no one is able to notice the breaking up [of the stone]?' Again their reply is based on the twelfth proposition, which is to the effect that the perception of the senses cannot be trusted, and thus only the evidence of the intellect is admissible.[10]

Maimonides does not mention Nazzām, nor how his rival theory escaped the difficulty concerning fragmentation; but this is easy to guess. The beauty of Nazzām's leaps will have been that we need never have one point in the millstone resting while another moves. All points in the millstone can leap simultaneously, and can reappear in their original alignments, with the distance of leap increasing by infinitely varying increments, according as the point in question is further from the centre of the millstone. The atomists are also committed to leaps of a sort, as Maimonides makes clear. But because their leaps are across fixed atomic distances, differences of speed are accounted for, not by differences of distance leapt, but by lingering, and it is this which leads to the objectionable fragmentation. Nazzām's theory of infinitely divisible leaps is not the only one that would avoid fragmentation. A theory of continuous motion would do so too. His main reason for preferring leaps is still the original one, that they avoid an infinity of sub-journeys. The new point is that they do so without incurring the problem of fragmentation.

Juwaynī's account of how Nazzām treats the revolving millstone does not look exactly like this.[11] Juwaynī himself is content with the atomist account

[10] Maimonides *Guide* 1.73, proposition 3.   [11] Juwaynī, *Shāmil*, 436-7.

of how the millstone revolves, some atoms lingering longer than others, and he does not quite articulate Nazzām's objection about fragmentation. He does appreciate that there had been an objection to the independent movement of parts, but he thinks that this can be met by pointing out that the independent movement of parts is a familiar phenomenon, exemplified, for instance, by our bending the top of a stick which is rooted in the ground. He does not realise that Nazzām's complaint is about actual fragmentation. It is interesting, however, that the weaker complaint, which is no complaint at all, is already mentioned as early as Sextus. For in the passage to be translated below, Sextus says of the rotating object that it 'would have either to be torn apart during the revolution, or at all events bent.'

Ibn Mattawayh understands the situation better and correctly reports the fragmentation argument against atomism. His reply is that each displaced atom will find a new neighbour, and this is true, but it does not get round the fact that the *change* of neighbour involves fragmentation. His account runs as follows.

> Objection: if contiguous parts do not move jointly, they must come apart. Answer: correct; yet they do not get scattered, because they are not only discrete, but also continuous: the particle that is left by its first partner is found by another. This is the answer of Abū l-Qāsim, who compared the process with the *mamlaha* in which mustard seeds are being stirred: they separate, but do not get scattered, because they are contained by the *mamlaha*.[12]

Another source[13] tells us that Nazzām cited a spinning top as something involving leaps, and he appears to have treated it in just the same way as the millstone. For his point is that the upper parts move more than the lower or than the axis – the top is presumably envisaged as an inverted cone spinning on its apex.

Modern commentators offer quite different interpretations of Nazzām on the millstone, but I have to say that I do not see how they are supposed to make philosophical sense.[14]

In Sextus, the rotation argument is directed against Epicurus' atomism. The example is neither a millstone, nor a spinning top. but a rotating ruler. I shall quote Sextus here, and reserve further extracts from Ibn Mattawayh and Juwaynī for the end of the chapter.

> Again let there be a ruler marked off with points on one side. And let it be rotated from one of its ends over a plane surface in one single time. Then, as

---

[12] *Tadhkira*, 199-200, abstracted by F. Zimmermann.

[13] Ash'arī, *Maqālāt*, 321, 7ff.

[14] Van Ess, op. cit., 7-8: the outer atoms of the millstone would have to jump to complete their revolution on time, and this conclusion is supposed for some reason to embarrass the atomists. Van Ess sees Nazzām as borrowing from the atomists, although only for purposes of *reductio ad absurdum*, the idea that these jumps will be across atomic spaces, and hence themselves, presumably, atomic. Somehow the atomists are supposed to escape embarrassment by conceding the jumps, but adding ideas that Nazzām had not forseen: all the atoms in the stone will jump, but there are time-atoms, and the inner atoms of the stone will rest for more time-atoms between jumps, although not long enough for their rest to be detected. Nader, op. cit., 186, sees Nazzām's leap in this sort of example as simply consisting in a difference of speed between outer and inner parts.

the other end rotates, circles of different sizes will be described, the outermost which embraces them all being the largest, the innermost the smallest and the intermediate ones in proportion, becoming either larger and larger as we go forward from the centre or smaller and smaller as we go down from the outer periphery. Since, then, the time of the rotation is single (and let it be a partless time), I ask how the circles have come to be different from each other, some large and some with a small circumference, seeing that one single time is laid down for the process of describing them to have taken place, and the motion is also single. It is not possible to say that there is some difference of length among the partless times, and that for this reason the circles described in the longer partless times are larger and those described in the shorter times smaller. For if one partless time were longer than another, that time would not be partless or minimal after all, and the moving thing would not as a whole move in a partless time. Moreover, it is not possible to say either that, while there is a single partless time in which all the circles are described, still the parts of the rotating ruler are not of equal speed, but some rotate faster, some slower, and the larger circles are constructed by the faster rotating parts, the smaller by the slower. For if in reality some parts moved faster and some slower, the ruler would have either to be torn apart during the rotation, or at all events bent, with some parts racing ahead and others lagging behind, whereas it is neither torn apart nor bent. Therefore motion is inexplicable for those who say that everything terminates in partless entities. And in general, if all these things are partless, both the time in which the motion occurs and the body which moves, and the space in which the details of the movement are accomplished, then all moving things will necessarily move at equal speed, so that the sun will come to be equal in speed to the tortoise, since each completes a partless distance in a partless time. But it is absurd to say that all moving things move at equal speed, or that the tortoise is equal in speed to the sun. It is absurd, then, to think that motion occurs, if everything terminates in partless entities.[15]

Sextus' version contains the complaint about fragmentation, but it also contains an undesirable complication. For Sextus envisages the whole rotation being completed in a single time-atom. Against this an atomist could protest that the ruler's returning to the same position every time-atom would not differ from its resting.

Leaps and the attack on atomist commitment to maximum speed

I have argued that the rotation arguments are attacks on atomism, and the same is true, I believe, of other arguments which mention Naẓẓām's leap. One group I take as having to do with the atomist belief in a maximum speed of one space-atom per time-atom, although this interest emerges from Juwaynī's report rather than from Ibn Mattawayh's. Among these arguments one[16] is remarkable, because it looks very like a certain construction of Zeno's paradox of the moving rows, which was interpreted in 1877 as an attack on atomism, although I have argued that it was not really intended that way. Naẓẓām had evidently objected to the atomists' assumption that an atom will

---

[15] Sext. M. 10.149-54.
[16] Juwaynī, Shāmil, 434-6.

never move more than one atomic space at a time. To show the possibility of this, one has only to think of the *relative* motion which arises when atoms are travelling in the opposite direction from each other, each moving one atomic space at a time. The relative movement will be *two* atomic spaces at a time.

Naẓẓām has some further arguments which again I take to be directed against the atomist commitment to maximum speed. Some of these arguments[17] envisage things travelling in the *same* direction, but one faster than another. A passenger can make progress as he walks along the deck in the same direction as his ship is travelling. The atomists had said this was possible only because the ship interposes rests between its atomic movements. But Naẓẓām asks whether the passenger cannot still make progress, even if the ship is moving at maximum speed. This will presumably again be a speed of one space-atom every time-atom, and Naẓẓām will again be challenging the idea that there is such a maximum speed. On his theory there is not, because leaps can be across indefinitely varying distances. But Juwaynī is forced to say that the passenger cannot make progress when the ship is travelling at maximum speed. Two other illustrations are offered, to challenge the atomists' maximum speed, by reference to things moving in the *same* direction. If the axle of the millstone is already moving at maximum speed, will not the periphery have to move faster? Again, imagine a well and a bucket: you can have an arrangement of ropes such that a pull of *ten* spans on one rope brings the bucket up a whole *twenty* spans.[18] If the rope which you pull is already moving at maximum speed, will not the bucket be moving faster? Juwaynī replies that the rope will snap, and the millstone fly apart.

In each case, I have been suggesting, the leap is not the essential factor in Naẓẓām's argument. Any theory would escape trouble provided it renounced the idea of a maximum speed. Naẓẓām's theory of infinitely divisible leaps is just one theory which does so.

Leaps and the attack on atomist geometry

Many of the other arguments by Naẓẓām seem to concern the spatial arrangement of atoms. Since only one of them even mentions the leap, and that in only one of the two recorded versions, it would be particularly hard to maintain that Naẓẓām's intention was to argue for infinitely divisible leaps, rather than against atoms. The argument which involves leaps is of especial interest, because I think it is similar to subsequent arguments in fourteenth-century Europe. But once again the argument has to be reconstructed, and I would venture the following suggestions.

In order to understand Juwaynī's version, I think we should recall that Abū l-Hudhayl thought of his atoms as unextended. Naẓẓām's objection to the atomists is that the diagonal of a square will (absurdly) have to contain the *same* number of atoms as the two sides which form a triangle with it. Why? He gives as his reason that each atom in the diagonal has '*opposite*' it a

---

[17] Juwaynī, *Shāmil*, 437-8; 441-3; Ibn Mattawayh, *Tadhkira*, 202-3; 206-7.
[18] One rope suspends the bucket from a beam, the other is attached to the first rope by a sliding hook.

corresponding unit on the two sides. I suggest that we are to imagine parallel lines linking all the atoms on the diagonal to points on the corresponding

sides, somewhat in the manner of the accompanying diagram. If the lines are really parallel, and if the atoms are really unextended, it will follow that there can be no *gaps* anywhere between the parallel lines. (To this extent, the diagram is inevitably misleading). If there are really no *gaps*, then Nazzām's conclusion will follow as a further consequence. For there will be no surplus atoms left over on the two sides which are not linked to corresponding atoms on the diagonal. In other words, there will be the *same* number of atoms.

In Juwaynī's report of Nazzām's argument, there is an extra twist. For Nazzām challenges the atomists to say how, on their view, the route along the diagonal can be *quicker* than the route along the two sides, seeing that the number of atoms is the same. Nazzām himself does not believe in atoms, and so is free to say that the diagonal is *shorter* than the two sides, and that that is why leaps along the diagonal get home first. This answer is not available to the atomists, since they make length a function of the number of atoms. Juwaynī seems to misunderstand Nazzām's position, because he thinks Nazzām would explain why the route along the diagonal is quicker by saying that the body taking that route would *leap*. But Nazzām needs no such answer: *any* moving body will leap, in his view, but that is not how he would explain the diagonal route being quicker.

Juwaynī's 'solution' to the diagonal objection is to say that only *some* of the atoms along the two sides will be opposite to the diagonal. But on the reconstruction offered above, he has failed to show how this is possible. I will quote the crucial portions of Fritz Zimmermann's abstract:

> Another attempt: Imagine a square with one of its diagonals drawn in. If an atom crept along the diagonal from one end to the other, and another <simultaneously> from the same starting point to the same finishing point along the sides, the one going along the diagonal would, by dint of leaps, get there first.

> Anyone saying this must have taken leave of his senses, for it is evident that the diagonal is shorter than the two sides.

> Nazzām would rejoin that the diagonal is opposite to the part<icle>s of <which> the sides <are composed>; and one cannot talk of being opposite, unless the corresponding distances are made up of the same number of part<icle>s. – – –

> In this way we shall also deal with the case of the diagonal: it is not opposite each part of the two sides. This is clear almost with the certainty of <logical> necessity, if we imagine the two particles connected by straight lines at each

stage of their progress along the two paths.[19]

Ibn Mattawayh's report differs in several respects. It does not mention leaps; the diagonal is said to have the same number of atoms as a *single* side; more importantly, the atoms are envisaged as having size and shape, since there is talk of their length and their corners. I imagine that Naẓẓām will again have proposed drawing parallel lines, in order to exhibit the side as

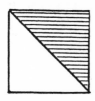

having the same number of atoms as the diagonal; only this time the parallel lines will have been separated by an atomic width. The atomists will have replied that, even if the number of atoms is the same, they can still explain why the diagonal is longer than a single side, for there will be room to arrange the atoms along the diagonal lozengewise in a corner-to-corner arrangement, whereas along the side they will be packed edge-to-edge, rather than corner-to-corner. The final rejoinder from Naẓẓām is that the corner-to-corner arrangement is no doubt intended to supply an increment of length, but the increment represented by the corner will, absurdly, be less than an atom long.

Ibn Mattawayh thinks the atomists will be safer, if, instead of a lozengewise arrangement, they postulate gaps of space between the atoms. But once again I imagine that Naẓẓām would have been able to argue that these gaps would, absurdly, have to be less than atomic. I shall quote Zimmermann's abstract:

> If, they argue, we had a square with one of its diagonals drawn in, the sides will look shorter than the diagonal, despite being of equal scope. The only explanation for this is that the <atomic> parts of the diagonal will be touching at their corners <rather than their sides>, with the coners, which are shorter than they, <adding to the length of the diagonal>. In our view, if the diagonal of this square is not a straight line in the strict sense (read *wa-lā* for *bal?*), it will be interspersed with gaps and therefore look longer <than the side>, even though they are <of> equal <scope>. The side, on the other hand, will be a straight line and hence without interstices. This, then, is the reason <why the side looks shorter>, and not the one they give so as to be able to claim that there is something shorter than an atom, namely, its corner.[20]

I have maintained that these arguments of Naẓẓām's are against atoms, rather than for infinitely divisible leaps. And this claim is confirmed by the fact that Naẓẓām's remaining arguments all attack atomism without mentioning leaps. I shall come back to some analogous fourteenth-century arguments at the end of the chapter.

[19] *Shāmil*, 439 and 441. [20] *Tadhkira*, 192-4.

Maimonides' summary of Islamic atomism

It remains for me to show the extent of Greek influence on Islamic atomism, which goes far beyond the controversy over leaps. I can bring it out by reference to three sources, but by far the most accessible to English readers is Maimonides' *Guide for the Perplexed*, since it exists in two English translations.[21] In a few pages, Maimonides summarises what he calls the common assumptions of *Kalām*, or Islamic theology, although he does not make it clear whether he is referring to the earliest phase of Islamic atomism associated with Abū l-Hudhayl. The first of twelve propositions or premises is that things are made of atoms, the second that there is vacuum, and the third that there are time-atoms and atomic jerks. Maimonides himself draws attention to the connexion with Greek thought, when he says that the postulation of time-atoms is due to reflection on Aristotle, who holds that atoms of one kind imply atoms of another. But it is not always remarked that time-atoms and jerks were an explicit part of Greek atomism in the post-Aristotelian period.[22]

Maimonides' discussion of the third proposition includes both the Islamic explanation of differences of speed by reference to the atoms *lingering* between jerks and the resulting problem of the millstone fragmenting. But what I have not yet brought out is the *evolution* of the theory of lingering as an alternative to Epicurus' rival theory of *zig-zagging*. Aristotle produced two different arguments to show that atomism would be committed to equal speed for its atoms, and Epicurus seems to have acknowledged both arguments. First, over a single atomic space, speeds must be equal, or else the slower body would traverse *less* than an atom while the faster traversed a whole atom.[23] Secondly, over any distance, there would in a vacuum be no resistance, and hence bodies would move at an equal and (*per impossibile*) an infinite speed.[24] Epicurus acknowledged that equal speed would follow, although he substituted 'quick as thought' for 'infinite speed'.[25] None the less, Epicurus was required to explain the undeniable differences of speed which we find in compound bodies. He was debarred from appealing to the later Islamic idea that an atom can *linger* in the same place, since he held that the atoms are for ever moving.[26] And he was obliged to hold this at least partly[27] by the acknowledgment just mentioned, that in a vacuum there is no resistance. So he appealed to *zig-zagging* instead of lingering. A compound body proceeds with varying degrees of slowness, because numerous collisions deflect its

---

[21] *Guide* 1.73, translations by M. Friedländer and S. Pines.

[22] MacDonald and Pretzl do not mention the Hellenistic acceptance of time-atoms and atomic jerks, but once again Pines makes the point very clear (op. cit., 98).

[23] Arist. *Phys.* 6.2, 232b20-233a12; 233b19-33. Epicurus and his followers acknowledged this, according to Simpl. *in Phys.* 938, 17-26.

[24] Arist. *Phys.* 4.8, 215a24-216a21.

[25] *Letter to Herodotus* §61.

[26] ibid., §43.

[27] David Konstan has suggested an interesting additional reason for the thesis of ceaseless motion ('Problems in Epicurean Physics' *Isis* 70, 1979, 394-418, esp. 398). By making ceaseless motion a postulate, Epicurus can explain why collisions of atoms do not end in complete cessation of motion.

ever-moving atoms from the main direction of travel.[28] But although Epicurus preferred the *zig-zag* theory, a staging post on the way to the Islamic *lingering* theory appeared later in Alexander.[29] For he described, and attacked, a view according to which one thing seems to move more slowly than another, because it is halted (*ephistamenon*) by the collisions of the atoms within it. Here it is evidently the *compound* body which does the lingering; in Islamic theory it will be the atoms themselves.

Islamic arguments for atomism

Of the remaining two sources of evidence, one is the battery of Islamic arguments for the existence of atoms.[30] A high proportion of them sound decidedly Greek, and one of them must have been noticed quite early. For in the controversy between Naẓẓām and Abū l-Hudhayl, it already emerged as an advantage of the latter's atomistic theory that it could answer Zeno's paradox of traversing infinitely many sub-distances, and this is one of the proofs given by Ibn Ḥazm. Another is the argument which we have found in the pseudo-Aristotelian *Lin. Insec.*, in Lucretius and in Plutarch, that there would be no difference between large and small (in the Islamic version between a mountain and a mustard seed), if they had the same number, that is, an infinite number, of parts. Two further arguments are connected with the Greek, if only indirectly. For one says that what God puts together he can again divide, and then makes the Aristotelian point that an actual infinite division is impossible. Another, the argument in Ash'arī, maintains for divine knowledge what Epicurus maintained for human knowledge, that it could not comprehend an infinity of parts.

Ash'arī's account of Islamic atomism

The third and last source of information is Ash'arī's doxographical work *Maqālāt*. If it were available in translation, yet more Greek parallels might come to light. One early figure reported by Ash'arī is Ḍirār (died A.D. 815 or earlier). Although he was perhaps not himself a full-blooded atomist,[31] we saw him in Chapter 18 reproducing against Naẓẓām arguments which had earlier been brought by the atomist Lucretius against the infinite divisibilist Anaxagoras.[32]

---

[28] *Letter to Herodotus* §62.

[29] *Quaestiones* 2.1, 45-6.

[30] Ibn Ḥazm *Fiṣal* 5.92-8, cited by Pines, op. cit., 10-16; Ash'arī, *A Vindication of the Science of Kalām* (*Risāla*), translated by R.J. McCarthy, in *The Theology of al-Ash'arī*, Beirut 1953, p. 127, §16. Baghdādī, *Farq*, ed. Badr, 123 and 316; Juwaynī, *Shāmil*, 143-8; Ibn Mattawayh, *Tadhkira*, 191.

[31] On the one hand, he appears sometimes to have applied to accidents the word *juz'*, which others used for atoms (Ash'arī, *Maqālāt*, 317, 13ff., translated by Pines, op. cit., 5; cf. the book title cited by J. Van Ess, in 'Ḍirār b. 'Amr und die "Cahmīya"', Biographie einer vergessenen Schule', *Der Islam* 43, 1967, 264). On the other hand, Ḍirār need not have intended to convey by this word more than his doctrine that some accidents are *parts* of bodies. What can be said is that he paved the way for atomism.

[32] Ash'arī, *Maqālāt* 328, 6-11; Lucr. 1.830-920: Anaxagoras held that there is something of

Ash'arī reports Abū l-Hudhayl's view that six is the maximum number of atoms with which an atom can be surrounded and in contact.[33] Abū l-Hudhayl's reason is that the surrounding atoms can be arranged above and below, in front and behind, to the right and left. No doubt this is why it is sometimes said that the atom, in spite of being indivisible, has six sides.[34] There is almost certainly Greek influence here. For Aristotle reports Democritus as assigning to atoms *four* differences of position, namely, above, below, in front and behind,[35] while he himself recognises *six*: above, below, in front, behind, right and left.[36] There is a late tradition that Democritus himself allowed a spherical atom to be surrounded by no more than six atoms of the same size.[37] Naẓẓām attacked the idea of six atoms surrounding and in contact, and his objections, though best recorded in a footnote, are of special interest because of their close relation to subsequent fourteenth-century objections.[38]

Besides the maximum number of surrounding atoms, Islam discussed the minimum number of sizeless atoms needed to make up a body.[39] Here too there had been analogous ancient discussions of the minimum number of minimal parts in an atom,[40] a fact which confirms Pines' point[41] that Epicurus' minimal parts are often a closer analogue than his complete atoms to the sizeless atoms of Islam. Indeed, there was a controversy between Islamic thinkers as to whether atoms are inseparable in thought, as well as in existence from the larger whole, which mirrors the controversies among modern interpreters as to which of these two positions Epicurus intended for his minimal parts.[42]

---

everything latent in everything, in portions of ever-diminishing size, and Lucretius replied that there cannot be fire latent in wood, but only atoms, or the forests would be burnt up.

[33] *Maqālāt*, 303, 3.

[34] *Maqālāt*, 316; cf. 304, 1.

[35] *Phys.* 1.5, 188a22-6.

[36] *Phys.* 3.5, 205b32-3.

[37] Giordano Bruno (sixteenth century), *de Triplici Minimo* 2.11, schol. p. 255; 3.11.12; 3.11.schol. ad 90 (fr. 123 in S.Y. Luria, *Democritea*, Leningrad 1970).

[38] If an atom can touch six other atoms, each of the six touch it in different *parts*, so it is not partless after all (Ibn Mattawayh, *Tadhkira*, 187). Naẓẓām also objects to the idea that an atom can be surrounded only in six directions. Against this he asks us to consider the centre of a circle (Juwaynī, *Shāmil*, 439-40; Ibn Mattawayh, *Tadhkira*, 195-6). The centre stands opposite to, and can be connected with, every part of the circumference, not merely with six parts. So it must have more than six sides. Juwaynī tries to defend the atomists by saying that only *some* portions of the circumference stand opposite to the centre. And Ibn Mattawayh protests that, if more than six points stand opposite to the centre, then that centre will consist of more than one atom. (We may wonder how large the centre will be, if *all* the atoms on the circumference stand opposite it).

[39] Ash'arī, *Maqālāt*, 302, 16; 303, 9; 316-17. The evidence is assembled by Pretzl, op. cit., 120-1 and Pines op. cit., 4-10, with a number of German translations.

[40] Themistius *in Cael.* 186, 30, a corrupt passage, which passed through Arabic and Hebrew on its way to the extant Latin version, gives the number seven, which, if trustworthy, provides a clash with the Islamic answers (six and eight).

[41] Pines, op. cit., 97.

[42] Ash'arī, *Maqālāt*, 316: inseparable in existence; so also Lucr. 1.602-4, as I read him. *Maqālāt*, 317: inseparable also in thought; so David Konstan on Epicurus, op. cit., though I am not quite persuaded by the evidence assembled there.

Finally, Ash'arī reports that Islamic theologians debated whether motion supervenes on a body when it is in its first position, or when it is in its second.[43] This reminds us of the arguments of Diodorus discussed in Chapters 2 and 22.

Differences from Greek atomism

I have been stressing the similarities, but of course there were differences from Greek atomism. In a recent lecture delivered in the University of London, van Ess argued that a motive for the quantisation of certain properties was to make sure that God could have an accurate ledger of good and bad works at the Day of Judgment. With motives as different as this, the Hellenist must take care that he is not distorting when he draws attention to apparent similarities.

Even within the technical details of atomist theory, there is one extremely large difference. For, as already indicated, Abū l-Hudhayl's atoms were unextended.[44] Even though there were dissenters,[45] this became the orthodox view. Wolfson, however, has found Greek influence even here. For he points to a ninth- or tenth-century Arabic work which misreports Democritus as saying that body is composed of surfaces, surfaces of lines, and lines of *points* (*punctum*, in Latin translation).[46] He suggests that the author may have been influenced in turn by a passage in Aristotle[47] and by Themistius' commentary on it. In this passage, Aristotle speaks as if you could redescribe Democritus' spherical corpuscles as points (*stigmai*), though ones possessing quantity (*poson*). A later Islamic thinker, Averroes, in commenting on this passage of Aristotle, under the influence of Themistius, again takes it that the round atoms of Democritus can be called points, although this is not to deny that they are bodies.[48] Here, then, is one way in which an incautious commentator might have been led to suppose that Demcritus' atoms were unextended. I would add another reason. We have seen in earlier chapters that Aristotle takes his attack on atomism in *Physics* 6 to count against Democritus.[49] But we have also seen that the opening arguments in *Phys.* 6.1 against composing a continuum out of indivisibles are more effective against a theory of points than against a theory of extended atoms.

Even this very large divergence from Greek atomism, then, does not necessarily show a lack of Greek influence. And meanwhile other apparent differences turn out to be less significant. Thus we hear that some Islamic theologians allowed atoms to have properties like colour.[50] Yet this was not

---

[43] *Maqālāt*, 355; Baghdādī, *Farq*, 144-5.

[44] Ash'arī, *Maqālāt*, 307, 10; 311, 11; 314, 12. See Pretzl, op. cit., 120-1, and Pines, op. cit., 4-10, with German translations.

[45] Ash'arī, *Maqālāt*, 301, 5 and *Masā'il*, 38ff.

[46] Isaac Israeli, *de Elementis*, p. 8a, ll. 11-14. See H.A. Wolfson, *The Philosophy of the Kalām*, Cambridge Mass. 1976, 472-86.

[47] *DA* 1.4, 408b30-409a10.

[48] Averroes, *Commentarium Magnum in Aristotelis de Anima Libros* 1, Comm. 69.

[49] *Cael.* 3.4, 303a23.

[50] Maimonides *Guide* 1.73, 5th proposition.

denied by all Greek thinkers (not, for example, by Diodorus), nor asserted by all Islamic ones (significantly not by Abū l-Hudhayl).[51] Another feature which is un-Greek is the occasionalism, which was discussed in an earlier chapter. But occasionalism was not essential to Islamic atomism. Rather it characterised the later atomism of the Ash'arite theologians, and, as Maimonides says, was not typical of the earlier Mu'tazilite theology.[52]

Islamic and Greek thought

I said at the beginning of this chapter that doubts had been raised about the extent of Greek influence on Islamic atomism, and I should now substantiate this claim. The doubts chiefly affect the early phase in Islamic thought from the 'Abbāsid revolution of A.D. 750 and the founding of Baghdad in 762 to the reign of Ma'mūn, Caliph of Baghdad from 813-33, when a centre was established for translating Greek texts. Kindī (died c. 870) wrote as if he were the first Arab to use the Greek works for solving the problems of Islam. Even after that, it is said, Greek influence was more pronounced at first among the so-called 'philosophers' or Falāsifa than among the 'theologians' or Mutakallimun, that is, the exponents of Kalām (Theology). Only with Ghazālī (A.D. 1058-1111), and his Destruction of the Philosophers, do the theologians begin to wield Greek ideas as freely and as consciously as the philosophers. That at any rate is what is claimed.

Islamic atomism began in the earliest period, and some Islamicists have looked to India rather than to Greece as a source for it.[53] Pretzl, we have already seen, thought the early controversy over atomism and infinite divisibility decidedly un-Greek. Fakhry[54] has argued that Greek influence was not uniform or systematic. He maintains that early Islamic atomism was influenced by the Presocratics (like Pretzl, he passes over the Hellenistic period of atomism), but he holds that later theology was the product of a Greek and an Islamic current, in which Greece supplied the technique, rather than the content. Recently, the tendency has been to treat the development of Islamic theology as mainly indigenous, and a catalogue of such interpretations is supplied by M. Seale.[55]

My impression is that these interpretations become much harder to sustain, when the many parallels with Greek thought are taken into account. Thus Ḍirār (died 815 or earlier), Nazzām (died c. 846) and Abū l-Hudhayl (died c. 841) all belong to the early period which is chiefly in dispute. Yet the present chapter has surveyed at length the controversy between the last two over atomism and found it full of Greek resonances. In addition, Nazzām seems to have reproduced part of Philoponus' objection to a beginningless

[51] Ash'arī, Maqālāt, 311-15, reveals that it was asserted by Jubbā'ī, but denied by Abū l-Hudhayl and a majority of others.

[52] Guide 1.73, 6th proposition.

[53] A list of such interpretations is supplied by Pines, op. cit., 102. He himself remains open-minded.

[54] Majid Fakhry, Islamic Occasionalism, London 1958, 23-4.

[55] Morris S. Seale, Muslim Theology, London 1964, 11, citing A.S. Tritton, Montgomery Watt, and A. Wensinck, with special reference to theological ideas on free will.

universe, namely, that it would involve traversing and increasing an infinity.[56] We have also seen Naẓẓām engaged in a controversy with Ḍirār over whether there is fire latent in stone which seems to reproduce the attack of Lucretius on Anaxagoras. Ḍirār meanwhile was observed proffering a theory of matter as a collection of accidents, which seems to echo Neoplatonist ideas.[57] Ḍirār is also connected to Philoponus by van Ess, since Ḍirār said that a body will not persist, if half or more of its accidents are replaced by their contraries. Van Ess compares Philoponus' idea that if maximum sweetness is generated by ten parts each of hot, cold, dry and fluid, then sweetness will disappear when we go down to five parts of each.[58]

This concludes my case for continuity between Greek and early Islamic thought. But I do not regard the presence of Greek influence as *excluding* Indian: on that I can take no stand. One impressive case is that of the Indian atomist Kaṇade, author of the Vaiseshica school, who is dated by some to around A.D. 100.[59] Although he uses the argument which we have found in earlier Greek thought, that large and small things would be equal, but for the existence of atoms, he supplies the illustration which we have found in later Islamic thought of a mountain and a mustard seed. No doubt, this illustration is too common to support any suggestion that Kaṇade might here have provided a bridge between Greece and Islam. But it would be worth investigating the other points at which his atomism relates to the two cultures, for example, his claim that six atoms are the minimum needed for producing magnitude.

Before I leave the subject of Greek influence, I should like to recall some of the other parallels which have been encountered between Greece and Islam. There should, of course, be no surprise about parallels drawn from a later date, at least if they involve 'the philosophers' or come from the period following Ghazālī. Many of the parallels emerged in Chapters 13 to 17 in connexion with questions of creation. For Islamic thinkers knew not only the infinity arguments against, but also the ancient arguments for, a beginningless universe, and further the replies to each of these. Thus they knew the 'why not sooner?' argument, the reply that there was no sooner, and the further argument that beginning implies *previous* non-existence. They knew the argument that God cannot change his will, and the standard reply to that. They knew the argument that matter cannot be created on pain of regress, and the reply that forms can be created *ex nihilo*. They knew the argument that a cause of change needs some triggering factor, and the argument that God could only be roused to activity, if there were something else to rouse

[56] See Chapter 14. Naẓẓām's arguments on this, as reported by Khayyāt, *Intiṣār*, 20, are translated by H.A. Davidson, in 'John Philoponus as a source of medieval Islamic and Jewish proofs of creation', 375-7.

[57] For both points, see Chapter 18.

[58] Ash'arī, *Maqālāt*, 305f. Philoponus *in GC* 170, 22-8, translated by S. Sambursky, *The Physical World of Late Antiquity*, London 1962, 42-3, cited by van Ess, op. cit., 267. Philoponus' idea is set in context by Robert B. Todd, 'Some concepts in physical theory in John Philoponus' Aristotelian commentaries', *ABG* 24, 1980, 151-70.

[59] H.T. Colebrooke, 'On the Philosophy of the Hindoos', part 2, read in 1824, *Transactions of the Royal Asiatic Society*, vol. 1, reprinted in his *Essays on the Religion and Philosophy of the Hindoos*, new ed. London 1858, esp. p. 176.

him. They also knew (Chapters 12 and 15) the argument that God cannot have been idle. Elsewhere we have encountered the idea of non-temporal senses for temporal words (Chapters 8 and 17), the idea of a beginningless creation out of nothing (Chapter 20), the idea that man's power is limited in various ways and a scepticism about causes (Chapter 19), the idea that 'was' and 'will be' are egocentric terms (Chapter 3), that to employ them is to undergo change (Chapter 16), and the idea that time depends on consciousness (Chapter 7).

Of course, Islamic thinkers developed the ancient discussions. One spectacular example was the treatment of a changeless God – of whether he could act without the aid of a triggering event (Chapter 15), and of whether he could know when things were past, present, or future (Chapter 16). Another example was provided by Islamic occasionalism (Chapter 19).

Finally, it should be remembered that it was by the Islamic route that Aquinas and others were able to recover many of the Greek arguments, by then enriched and adapted, in the thirteenth century. As this chapter is about atomism, I shall finish with a brief reference to the fourteenth century, since this is chiefly when the atomist arguments were taken up by Europeans.

Islamic and fourteenth-century European atomism

Islamic atomism seems to have had a profound influence on fourteenth-century debates. For one thing, recent discussions[60] have emphasised that one strand, perhaps the dominant strand, treated its atoms as sizeless points. This would be true of Gerard of Odo, Henry of Harclay, Walter Catton and, in some passages, of Nicolaus of Autrecourt. Moreover, Gerard of Odo allowed that these point-atoms could have an above and below, a before and behind and other such positional differentiations. In both respects these atoms sound like those of Abū l-Hudhayl and the Islamic tradition.

There is more. The fourteenth century rehearsed, among others, very much the same objections and replies concerning atomism, as we have just encountered in Abū l-Hudhayl and Naẓẓām. Some of the arguments are more complex, but this is only natural, since the Islamic discussions had themselves evolved before they were transmitted to the West. Those of Avicenna, as recorded by Ghazālī in his *Intentions of the Philosophers*, were available in Latin translation to Albert and Thomas Aquinas, under the title *Metaphysica*.[61] In the fourteenth-century texts, there are objections concerning the diagonal and side of a square containing the same number of

---

[60] Anneliese Maier, *Die Vorläufer Galileis in 14 Jahrhundert*, vol. 1, Rome 1949, 155-215; V.B. Zoubov, 'Walter Catton, Gérard d'Odon et Nicolas Bonet', *Physis* 1, 1959, 261-78; John Murdoch, 'Superposition, congruence and continuity in the middle ages', *Mélanges Alexandre Koyré*, vol. 1, Paris 1964, 416-41; 'Mathesis in Philosophiam Scholasticam introducta: the rise and development of the application of mathematics in fourteenth-century Philosophy and Theology', in *Arts Libéraux et Philosophie au Moyen Âge*, Actes du Quatrième Congrès International de Philosophie Médiévale, Paris 1969, 215-54; 'Naissance et développement de l'atomisme au bas moyen-âge latin', in *La Science de la Nature: Théories et Pratiques*, Cahiers d'Études Médiévales 2, Paris 1974, 11-32.

[61] See Algazel, *Metaphysics*, trans. J.T. Muckle, Toronto 1933, 10-13.

atoms. There are arguments about the centre of a circle facing all the points opposite on the circumference. There are questions about there being no differences of speed. There are discussions of the inner of two concentric wheels resting while the outer moves.[62] All these will be full of resonances to those who have read the Islamic material discussed above. Readers who wish to follow up later treatments of the revolving millstone in translation will find available English versions of Ghazālī's report and of Nicolaus of Autrecourt's citation from Ghazālī, along with a reply.[63]

John Murdoch, who has drawn attention to the debt of the fourteenth century to Ghazālī's Avicenna, none the less issues a warning:[64] what influenced the fourteenth century was Islamic *objections* to atomism, he maintains, rather than Islamic atomism itself. Moreover, Greek, Islamic and fourteenth-century atomism had quite different aims. Greek atomists, he says, sought to explain the physical world, Islamic atomists to refer all explanation to God, fourteenth-century atomists simply to analyse space, time, movement and extended matter. Murdoch is surely right to insist that there are contrasts, yet these in their turn are not always clear-cut. We have already seen that the Greek atomist Diodorus Cronus belongs with Islamic thinkers, and the latter belong with the fourteenth-century Latin writers, in not using atoms to explain the physical world.

The treatment of atoms as mere points is still found, as Furley makes clear, as late as Hume in the eighteenth century.[65] But I shall not try to trace the extent to which Islamic, and the extent to which pure Greek, influence prevailed after the fourteenth century. For certain aspects of this question I can refer to others.[66]

Naẓẓām on the millstone again

I shall finish this chapter with abstracts of Juwaynī and Ibn Mattawayh on Naẓẓām's treatment of the millstone.

> Another of Naẓẓām's subterfuges: if a millstone rotates, on account of the rotation of its axle, it is evident that the edge traverses a circle of several cubits' length, while the axle describes a circle of one cubit or less. Since, then, the periphery moves with the same movement as the axle, it must perform leaps in order to traverse the difference in the lengths of the two peripheries.

> This is nonsense. The edge of the millstone traverses the larger circle on account of the 'successivity' of its movements, while the movements of the axle are not successive <in other words, they are interspersed with rests>. To look at

---

[62] See especially the accounts of Maier and Zoubov.

[63] Algazel, *Metaphysics*, 13; *The Universal Treatise of Nicholas of Autrecourt*, tr. L.A. Kennedy *et al.*, Marquette University Press, 1971, 74.

[64] John Murdoch, 'Naissance', esp. pp. 13 and 31.

[65] David Furley, *Two Studies*, ch. 10, refers chiefly to Hume, *A Treatise of Human Nature* 2.1-4.

[66] Murdoch, 'Naissance', 32, implies that seventeenth-century atomists were not point-atomists. For a general history of atomism, see Kurd Lasswitz, *Geschichte der Atomistik*, 2 vols, Hamburg and Leipzig 1890 (repr. by Olms, 1963), and cf. Ted McGuire, *Philosophical Themes in Newton's Earlier Thought*, Dordrecht, forthcoming.

a moving millstone is to gain the necessary knowledge that one circle moves faster than the other. But does not every part of the millstone move with a movement imparted to it by the axle – for when the axle stops so does the whole stone? This is fallacious, for we do not believe in causal chains [i.e. causal nexūs other than direct causation by God]. But even if we concede secondary causation for the sake of argument, there is little comfort for the leap. For even those believing in causal nexūs do not say that the movement of the axle engenders that of the millstone, but that the movement of the millstone is engendered by the momentum imparted to its part<icle>s by the axle. But then, why should this momentum not engender <several, diverse> movements, as in the case of the thrower's imparting momentum to the stone he throws? But it is wrong to suppose in the first place that the parts of a body are interrelated in respect of impulses received. For if one end of a stick is shaken, the other end remains firmly rooted in the ground.

If it is all right for Naẓẓām to postulate *leaps* for the periphery but not the axle of the millstone, it is all right for his opponent to postulate extra *movements* for the periphery and not the axle.[67]

Suppose the axle of a millstone performed a circle ten parts long, its periphery a hundred parts. ... Answer: the periphery moves nine times and rests once, the axis moves once and rests nine times. ... The same explanation applies to two men walking at different speeds.

Even if 'the movement of the periphery follows that of the axis', the axis may still rest while the periphery moves, just as the spool follows the movement of the hand, but moves on when the hand stops.

Objection: if contiguous parts do not move jointly, they must come apart. Answer: correct; yet they do not get scattered, because they are not only discrete, but also continuous: the particle that is left by its first partner is found by another.

This is the answer of Abū l-Qāsim, who compared the process with the *mamlaha*, in which mustard seeds are being stirred: they separate, but do not get scattered, because they are contained by the *mamlaha*. ...

A lanceshaft with lead attached to it at one end may move at the other, while the leaden end stays put; similarly, the branches of a tree may move while firmly planted in the ground – without coming apart.

Anyway, the concept of the leap *itself* presupposes fragmentation, in order to get to point 10 at moment 2; so if there is no disintegration on this assumption, why should there be on ours?

Everyone wishing to make sense of saying that one movement is faster than the other must opt for our explanation. If the two circles move at one pace, what is then the sense of saying that one moves fast, the other slowly? If in addition to this disadvantage, the explanation does not avoid fragmentation either, how can our explanation possibly be rejected?[68]

---

[67] Juwaynī, *Shāmil*, 436-7, abstracted by Zimmermann.
[68] Ibn Mattawayh, *Tadhkira*, 198-200, abstracted by Zimmermann.

CHAPTER TWENTY-SIX

Stopping and Starting

In this last chapter, I shall consider a paradox about the continuum which relates at many points to the questions of atomism and continuity already discussed. 'The train leaves at noon', says the announcer. But can it? If so, when is the last instant of rest, and when the first instant of motion? If these are the same instant, or if the first instant of motion precedes the last instant of rest, the train seems to be both in motion and at rest at the same time, and is not this a contradiction? On the other hand, if the last instant of rest precedes the first instant of motion, the train seems to be in neither state during the intervening period, and how can this be? Finally, to say that there is a last instant of rest, but not a first instant of motion, or vice versa, appears arbitrary. What are we to do? This puzzle has a long history. It is found in Plato's *Parmenides* (156C-157A), and is thoroughly treated by Aristotle.

I want to suggest two main solutions to the paradox, one for the case of moving and resting, and one for the case of being at, or away from, the terminal stages in a continuous change. Aristotle's solution for the case of moving and resting has been criticised for impeding the progress of dynamics. I nonetheless want to show that he had a different solution for the other group of cases, and that it was the right one. Moreover, he was also right to treat moving and resting as a separate problem, even if his answer did have unfortunate consequences. My conclusions will be somewhat different from those I have expressed elsewhere.[1]

The problem is not only an intriguing one in its own right, but also one whose solution can have significant implications. These implications concern the concept of moving at an instant: whether it has application at all, and if so, how it is to be defined. I shall argue that we must decide how to handle the paradox *before* we can decide how to define motion at an instant. To put Aristotle's solution in perspective, I shall finish by comparing it with mediaeval ones.

Does the paradox apply to the real world?

First, we need to consider whether the problem could apply to the real world. It may be doubted whether it could, for the statement of the problem involved a number of assumptions. For one thing, I assumed (what Aristotle argues in the *Physics*) that time is continuous, and does not come in atomic

---

[1] My earlier discussion was, 'Aristotle on the instant of change', *PAS*, supp. vol. 50, 1976, 69-89, reprinted, with revisions, in *AA* 3.

chunks. It was seen in Chapter 24 that there is a minority of physicists who doubt this, and they would presumably question whether the problem, as so far stated, applies to our world. I shall, however, consider at the end whether a corresponding problem might not arise also for atomists.

Another reason why someone might question whether the problem applies to the real world is that a train consists of a mass of moving atoms, and so does the railway track. Can the train have any *first* instant of motion, or last of rest, if its atoms are moving all the time, and how would these instants be defined? Yet another doubt concerns the fact that a train is not perfectly rigid. When some parts of the train, or of the engine, have started to move, other parts will be lagging behind, so that there is not a single first instant of motion or last of rest for the train as a whole. Both these doubts can be met by raising our problem not about the train as a whole but about some point within the train, such as the centre of mass, and its first instant of motion and last of rest, in relation to some point on the railway track.[2] In talking of points, rather than of trains, I shall admittedly be talking about the real world in an *idealised* way, but I argued for the legitimacy of this in Chapter 1.

So far as I can see, then, the problem does apply to the real world, in as much as it applied to idealised points on a real train. But two further things need to be said. First, the problem would still be interesting, even if it applied only to a world different from ours. Secondly, I have so far considered only one version of the problem, and if this version were inapplicable to the real world, it might still be the case that other versions were applicable. Thus far I have considered only the transition between rest and motion. But the problem can be raised, and was raised by Aristotle, in connexion with other kinds of transition. He discusses the transition from being one colour to being another colour, from being non-existent to being existent, and from being invisible to being visible. In each case, the question can be asked: when is the last instant of the old state, and when the first instant of the new? Or moving from time to space, we may be able to ask where is the last point of the one state, and the first point of the other? In some of these new forms, the question may well apply to our world.

With the problem now stated and generalised to apply to all kinds of transition, I can start making some suggestions about how to handle it. But first we should be clear how much we need to ask of a solution. The original question was about when the last instant of the old state occurs, and when the first instant of the new. One of the difficulties about answering was that if we said that one of these instants existed, but not the other, we seemed to be being arbitrary. It would be a sufficient solution, if we could show that it would not be arbitrary to prefer one instant to the other. For this purpose, we need only show that there is a reason for preferring one to the other; we need not show that it is mandatory to do so.

---

[2] It may be objected that if the atoms of a body are forever joggling, and if their motions are not equal and opposite so that they cancel each other out, then the centre of mass will also be for ever moving, so that it will have no *first* instant of motion. We may reply that, even if this is so, we can still ask about the first instant of motion (and last of non-motion) of the centre of mass in a *given* direction, or in response to a *given* force. I shall neglect this complication in what follows.

Proposed solution for being at, or away from, the terminal stages in a continuous change

I shall start with some easier cases, and I shall assume that motion is continuous. By this I mean that it involves passing through an infinity of points, between any two of which there are other points, which are also passed through.[3] In that case, there will be an asymmetry between the series of positions away from the position of rest and the position of rest itself. For, in such a motion, there can be no first *position* occupied away from the starting point, or last *position* occupied away from the finishing point, since positions are not next to each other. Hence there can equally be no first *instant* of being away from the starting point or last *instant* of being away from the finishing point. No such considerations apply to being at the position of rest. This already supplies a solution to the paradox in some of its applications. For if someone were to ask, 'When is the last instant of being at the position of rest, and when the first instant of being away from it?', we could safely reply that the latter instant does not exist.

This solution can be generalised so that it applies to being at, or being away from the terminal stages in *any* continuous change, whether a continuous change of size, temperature or colour. For example, let us imagine that something progresses from one shade of colour to another by passing continuously through an infinity of intervening shades. In that case, there can be no last shade before the terminal one, and hence no last instant of lacking, but only a first instant of possessing, the terminal shade. This can be offered as a general solution for this kind of case. It exploits an asymmetry which lies in the nature of the continuum. And it will be equally true that there is no first instant of lacking, but only a last instant of possessing, the initial shade.

The solution is suggested by the nature of the continuum, and there are other cases for which this is true. For example, in a continuous motion there can be no first instant of having moved *somewhat*, though there can be a first instant of having moved *at least half way*, and a last instant of having moved *uninterruptedly*.

Proposed solutions for continuously moving (present tense)

Unfortunately, when we turn to the case of *moving* and *resting* (present tense), matters are not quite so easy. Whether something is moving at an instant is normally decided by reference to its position at both earlier and later instants. But with the instant of transition between motion and rest, if we treat earlier and later instants as both equally relevant, we find that they give opposing verdicts. We should encounter analogous difficulties if, instead of asking about first and last instants of *possessing* or *lacking* a colour shade, we asked about first or last instants of *changing* shade. The concept of *having*

---

[3] In modern mathematics, this property is called denseness, while continuity involves the extra feature, unknown to Aristotle, of possessing points corresponding to the irrational, as well as the rational, numbers.

moved (perfect tense) was less problematic, because it is unambiguous that the instants relevant to whether something *has* moved are earlier ones.

In spite of the difficulties, I think a solution can be found for the case of moving and resting. The question is whether we should treat the instant of transition between motion and rest as one at which the moving object is in motion, or at rest. The decision is no longer so clearly dictated, but I think there is a consideration which would justify our calling it an instant of *rest*.

Let us suppose that not only change of place is continuous, but also change of velocity. In other words, in passing from one velocity to another, an object passes through all the infinitely many intervening velocities. This assumption is the normal one in contemporary physics, and, if it is made, then there cannot be a first or last velocity above zero. There is, however, no corresponding objection to there being a first or last instant of having velocity zero. Now it seems natural, although admittedly it is not mandatory, to connect zero velocity with rest and velocities about zero with motion. If we do, we get the result that there is no first or last instant of motion, but that there may be first and last instants of rest.

Four *caveats* should be applied to this solution. First, I am not saying that the period of motion has no boundary. It will have an instant bounding it on either side, and my only suggestion has been that that instant should be regarded as one at which there is rest. Secondly, my proposal should be understood in a flexible spirit, since for many purposes it simply does not matter which way we talk, and we can follow other reasons, or no reasons at all; while on other occasions, the reasons just given may be overriden by stronger ones on the other side. Thirdly, the solution suggested does not in any way preclude physicists' talk of initial velocity. For the initial velocity of a projectile is not a first velocity in its entire motion, but merely the first velocity which it is convenient to consider for the purposes of a given calculation. Fourthly, it should be recognised that the last instant of rest in relation to one point may be an instant of motion in relation to a different point. Throughout I should be understood as talking of rest or motion in relation to a *given* point.

Proposed solutions for discontinuous transitions

I have so far offered solutions for two kinds of case, but they both involved *continuous* changes. Neither of the proposed solutions would apply readily to *discontinuous* changes. For example, suppose we imagine that up to a given instant something is non-existent, or invisible, or of a given colour, and that after that instant it is existent, visible, or of a quite different colour. What status ought we to assign to it at the instant of transition?

With these discontinuous changes, there will often be no considerations to guide us. The most we can hope for is different considerations in different circumstances. I shall offer three illustrations. First, it may help us if the *discontinuous* shift in visibility is produced by a *continuous* process of approaching, or the *discontinuous* shift in colour by some *continuous* rearrangement of constituents. For then there would be no last instant of approach or rearrangement before the one which yielded the new visibility or colour. Hence we could, without arbitrariness, decide to say that there was equally no last instant of invisibility or of the old colour. But what if the discontinuous shift in colour or visibility is not the product of a continuous change?

Another consideration which can help is that, if we are watching a receding aeroplane, or looking for an approaching one, we cannot normally tell at the time what will prove to be the last instant of visibility as it recedes, or the last instant of invisibility as it approaches. If we want to register this instant as it arrives, we shall normally have to wait until the new state is upon us, before we can do so, and it may then reasonably be held that we are not registering the end of the old state, but, at best, the beginning of the new. This means that, in many contexts, we have a good reason for not talking of the last instant of the old state, but (if it has one) of the first instant of the new. But it needs to be noticed that our interest is not always in registering the instant as it arrives. We may instead want to discuss the instant of transition prospectively or retrospectively, or to consider a purely imaginary case. So the present consideration does not provide a *general* solution.

Aristotle may have yet another consideration relevant to the particular example of visibility. For he classifies seeing as an *energeia*, and on one interpretation, an *energeia* has no first instant. This is how J.L. Ackrill[4] interprets Aristotle's idea (e.g. *Sens.* 446b2) that 'he is seeing' entails 'he has seen'. Ackrill treats the perfect tense 'he has seen', like 'he has *been* seeing', as implying an earlier period of seeing. This interpretation has been disputed,[5] but if it is correct, it implies that there will not be a first instant of seeing, and therefore not a first instant of seeing an approaching object as it swims into view.

I shall now return to the problem of being in continuous motion, and consider two rival solutions to the one I have suggested.

Two rival solutions for moving and resting

An entirely different way of trying to cope with the problem has been advocated by Brian Medlin.[6] Medlin says, in effect, that a thing can be both in motion and at rest at an instant, and equally neither in motion nor at rest at that instant. The first may sound as if it violates the law of contradiction, the second as if it violates the law of excluded middle. But Medlin avoids this, by simply defining motion at an instant, and rest at an instant, in such a way that they are neither contradictories nor contraries of each other.[7] Given his definitions, all four statements can be true together, namely, that a thing is in

---

[4] J.L. Ackrill, 'Aristotle's distinction between *energeia* and *kinēsis*', in *New Essays on Plato and Aristotle*, ed. R. Bambrough, London 1965, esp. pp. 126-7.

[5] Disputed by L.A. Kosman, 'Aristotle's definition of motion', *Phronesis* 14, 1969, 40-62; and Terry Penner, 'Verbs and identity of actions', in *Ryle*, ed. Oscar Wood and George Pitcher, New York 1970.

[6] Brian Medlin, 'The origin of motion', *Mind* 72, 1963, 155-75.

[7] In effect, he defines 'it was in motion (not at rest) at instant *t*' by saying something like '*t* was either followed, or preceded, or both, by a period throughout which it moved'. And he defines 'it was at rest (not in motion) at instant *t*' roughly as '*t* was either followed, or preceded, or both, by a period throughout which it did not move'.

motion at an instant, not at rest at that instant, and that it is at rest at that instant, not in motion at it.

My objection to this is not so much that it runs the risk of causing confusion, but that it is not sufficient to solve the problem that interests us. Medlin is free to define motion at an instant and rest at an instant in such a way that they are not contradictories or contraries of each other. But he cannot, and does not, deny that there *is* a contradictory of the claim that something is in motion at an instant. He himself suggests a way in which we might formulate the contradictory. We could talk of its *being the case that* something is in motion at that instant. Once we have found a formula for picking out the contradictory, we can pose our original problem all over again in terms of the new formula. We shall simply ask what is the last instant when it is *not* the case that our object is in motion, and what the first instant when it *is* the case that it is in motion. To this question Medlin himself would agree that we cannot say it is the same instant. When the problem is posed this way, we see that we shall have to fall back on a different solution from Medlin's, such as the one I have advocated, according to which there is a last instant when it is *not* the case that an object is in motion, but not a first instant when it *is* the case that it is in motion.

A second rival solution has been called by Norman Kretzmann the neutral-instant analysis. He finds it implied, but not explicitly presented, in Aristotle.[8] I do not find it in Aristotle myself, though there is something more like it in Plato (*Prm.* 156C-157A). It differs from the solutions I shall ascribe to Aristotle, even from the first one, because that first one will deny that there is motion or rest at *any* instant,[9] and will thereby fall foul of modern dynamics. The neutral-instant analysis avoids this; it allows that there can be rest or motion at an instant, and denies rest or motion only at the instant of transition between motion and rest.

If the neutral-instant analysis is to avoid the charge of arbitrariness, it must show why it is to be preferred to the other possible analyses. Aristotle can claim to be justified in treating the instant of transition as neutral between motion and rest, because he treats *all* instants that way. But the neutral-instant solution cannot claim this justification. So it remains to be

---

[8] Norman Kretzmann, 'Incipit/desinit', in *Motion and Time, Space and Matter: Interrelations in the History of Philosophy and Science*, ed. P. Machamer and R. Turnbull, Columbus Ohio, 1976.

[9] It might be replied that Aristotle only means that a thing cannot *get any distance* at an instant, there being no duration available. Some of Aristotle's arguments do seem to suggest only this. This would leave him free to go on to say (a) that at an instant a thing can none the less be *in course of* moving or resting, and (b) that it is so at all instants, except the instant of transition between motion and rest. This would amount to the neutral-instant analysis. But, first, I do not find statements (a) and (b) spelled out by Aristotle. And, secondly, there are three places at which Aristotle's argument would actually suffer if he allowed (a), without more ado. These three passages are 6.3, 234a34-b5; 6.6, 237a11-17; 6.8, 239a3-6. In the first passage, Aristotle cites our paradox in order to show that nothing can move at an instant. If it could, then at the instant of transition it would both move and rest. If, in spite of this denial, Aristotle were prepared to allow (a), then his remark here would leave us wondering how he escapes saying that at the instant of transition a body is both in course of moving and in course of resting. This would surely be the place for him to add, if he really held (a) and (b), that although he concedes (a), that a body can at an instant be in course of moving or resting, he still escapes the paradox, because (b) withholds this concession from the instant of transition.

shown why we should treat the instant of transition as one of *neither* motion nor rest, rather than as one of *both* motion and rest, or, as I advocated, as one of *rest*.

There is one special instant for which the neutral-instant analysis may well be a good one, and that is the instant of reversing direction. We can imagine the centre of mass of a ball travelling vertically upwards and slowing down until it reverses direction at an instant, without pausing for any period of time at the apex of its journey. We will now have an extra incentive for denying that the centre of mass is in motion at the instant at which it is the end of its upward journey. For not only will its velocity be zero at that instant, but we could not say that its motion had one direction rather than the opposite direction at that instant.[10] This favours our saying that at the instant of reversal the centre of mass is not in motion. But there is a consideration[11] which could make it reasonable to say that the centre of mass is not at rest either. This is that at the instant of reversing direction, the centre of mass is (as in all cases of coming to a halt) at a different position from that occupied at preceding instants, and also (differently from ordinary cases of coming to a halt) at a different position from those occupied at succeeding instants. There is thus no sameness of position at all, and since sameness of position is important to the concept of rest, this may make it reasonable to say that the instant of reversing direction is not one of rest. In that case, since it is not an instant of motion either, it will be a neutral instant, an instant of non-motion.

It would take a further argument, however, to show that the neutral-instant analysis is preferable for ordinary cases of coming to a halt or starting off. In case a suitable argument is found, I would simply make three points. First, the neutral-instant analysis shares my view that the instant of starting is not one of motion, and should therefore welcome my argument that the velocity is zero. Secondly, if the paradox were put in terms of a choice not between motion and rest, but between motion and non-motion, then the neutral-instant analysis would make an asymmetrical choice parallel to my own, and reject motion. Thirdly, the neutral-instant analysis is offered only as a solution to the problem of being in continuous motion, not to the problem of being at a given *stage* of continuous motion. The latter problem is much better handled in the way already indicated.

Aristotle's treatment

I shall now turn to Aristotle's solutions of the problem. He is well aware that different kinds of case need different solutions, but, not surprisingly, he looks

[10] Someone may object that we are in any case committed to motion without any particular direction. For imagine that the centre of mass of a ball is deflected only slightly, so that it does not stop moving, but at the same time is deflected at an angle rather than in a curve. In that case, we shall be unable to assign the earlier direction rather than the later at the instant of deflection, in spite of wanting to describe the ball as moving. It can be replied, however, that if deflection, in cases where there is no stopping, is always in a curve (since otherwise there would be discontinuity in the velocity in some direction), it will be possible to assign a direction to the motion at any instant, by taking a tangent to the curve. And in that case, we are not committed to motion without a particular direction.

[11] I am grateful to Colin Strang for pointing out to me the implications of this consideration.

for solutions of some generality, and does not acknowledge that the matter might ever be decided by the unique interests of a particular context. Before expounding his two solutions, I shall have to make some preliminary points clear.

Preliminaries (i): gradualness of the four main kinds of change

A first thing to notice is that Aristotle recognises only four kinds of change as being changes in the full sense of the word. There is change of quality (as when something changes colour), change of place (in other words, motion), change of size (in growth and diminution), and finally the creation or destruction of substances (*Phys.* 3.1, 200b32-201a16; cf. 5.2, *Metaph.* 11.12).

A second point is that in all four kinds of change, he thinks that there is a gradual process of transition (*Phys.* 6.6, 237a17-b3; b9-21). Qualitative change, such as change of colour, is said to take time. Change to a new place or size involves passing through intervening points. The creation of something like a house takes time, and occurs part by part, the foundation before the whole.

Aristotle does not by any means think that changes other than the four genuine ones must all involve a gradual process of transition. He sees that an infinite regress would be involved, if the gradual process by which something came into being had itself to come into being by a gradual process (*Phys.* 5.2, 225b33-226a6). In the very passage where he explains that a house comes into being only part by part, he points out that this cannot be true of things that have no parts (6.6, 237b11; b15), points and instants, for example.

Preliminaries (ii): how can change of colour be gradual?

It may be wondered how change of colour can be gradual, given the view stated in the *de Sensu* (6, 445b21-9; 446a16-20) that there is only a finite number of discriminable shades. In that case, a change to a new colour would be gradual when it involved passing through a number of intervening shades. But within this process, how could the change from one shade to the *next* be treated as gradual?

Aristotle seems to have two incompatible answers. One answer is implied in *Phys.* 5.6 (230b32-231a1), 6.4 (234b10-20), 6.9 (240a19-29), 6.10 (240b21-31), when he says that certain changes occur part by part.[12] While a surface is changing from white to the next shade, grey, he says, part of the surface must still be white, and part already grey. The greyness spreads gradually over the surface. This claim, that certain changes occur part by part, is used in combating the view that a partless atom could undergo change or motion. In its turn, the claim has as its ground that, while a thing is actually changing or moving, it cannot yet be in its terminal state, nor can it still be in its initial state. It must therefore be partly in one state, partly in the other, and so must have parts. In the case of motion, it must move part by part into

---

[12] Aristotle probably has this answer in mind also in 6.5, 236b5-8, where he stresses that, even if colour is not in itself divisible, the surface to which it attaches is divisible.

the adjacent area. Diodorus Cronus and Epicurus, as we have seen, were later to get round this objection to the motion of partless atoms, by denying that an atom actually *is* moving at any time; rather, at any given time it *has* moved with a jerk.

But Aristotle seems not always to keep in mind the view that these changes occur part by part. And here divisible leaps become relevant once again, for at *de Sensu* 6, 447a1-3, he actually denies that qualitative change has to occur part by part, and illustrates how it can happen in a leap, as when a whole pond freezes over at once. In one of the *Physics* passages where he says that qualitative change (change between contraries, 237b1) is gradual, he speaks as if he must prove this on the basis of time taken, but cannot prove it on the basis of space covered (237a19-28; a29; b2; b21). Why not, if he remembers his view that a surface changes colour part by part? He seems to be forgetting that view, and he probably forgets it again in 8.8, where he declares that while something is becoming white, it is not yet white (236b27; b30). This way of putting things seems to neglect the idea that there will be a stage of being partly white. How then can the transition from one shade to the next be represented as gradual?

A second way of arguing that the transition is gradual serves to refute the idea that part by part changing is indispensible for this purpose. The second way is suggested by what Aristotle says in *de Sensu* 6. Admittedly, there is only a finite number of discriminable shades, so that discriminable colours form a discontinuous series (445b21-9; 446a16-20). None the less colours, musical pitches, and other ranges of sensible qualities have a kind of *derivative* continuity (445b28; b30, *to mē kath' hauto suneches*). The sort of thing Aristotle seems to have in mind is that a change to the next discriminable pitch, in the *discontinuous* series of discriminable pitches, may be produced by a *continuous* movement of a stopper along a vibrating string. Or in the case of colour, a change to the next discriminable shade, in the *discontinuous* series of discriminable shades, may be produced by a *continuous* change in the proportions of earth, air, fire and water in a body. As the stopper moves along a vibrating string, we hear the sound all the time, but do not hear a change of pitch, until the stopper has moved the distance that corresponds to a quarter tone (446a1-4). Variations of pitch less than a quarter tone are not perceptible except by being part of the whole variation (446a18, *hoti en tōi holōi*), by which Aristotle probably means that they only *contribute* to the perceptibility of the whole variation. This suggests a way in which Aristotle can maintain that a change to the next discriminable colour or pitch can be continuous. A body is changing to the next discriminable shade all the time that the continuous change in its elemental ingredients is going on, which will eventually lead to its displaying the next discriminable shade.

In what follows, I shall consider only Aristotle's treatment of the four genuine kinds of change. This will leave open how he might have treated the many other cases of (non-genuine) change, though I have made some hints above about the case of beginning to see.

Aristotle's solution for moving, changing and resting

I shall start with Aristotle's solution to the problem of being in continuous motion or change. It is most fully expressed in connexion with rest and motion in *Phys.* 6.3 (234a24-b9) and 6.8 (239a10-b4). Here Aristotle says that there can be neither rest nor motion at an instant. He explicitly cites our problem as one ground for his conclusion (6.3, 234a34-b5), saying that if something stops moving, the instant of transition between motion and rest is an instant neither of motion nor of rest, since otherwise we could not avoid the contradiction of saying that it is an instant of both motion and rest. At 6.6, 237a14-15, Aristotle extends his treatment of motion to *all* change, saying that a thing cannot be changing (*metaballein*) at an instant. An extension is also attempted in 6.8, 239a3-6, where Aristotle denies that there is a first instant of slowing to a halt, by arguing that slowing to a halt implies moving, and that there is no moving at an instant. (To make the argument valid, he ought to show that slowing to a halt at an instant would imply not just moving, but moving at an instant.)

Aristotle gives several grounds for denying rest or motion at an instant, besides the need to avoid difficulties about the instant of transition between rest and motion. In 6.3 (234a24-b9), one argument is that to rest is to be in the same state now as then, but an instant does not contain a then. Another argument is that differences of speed would be impossible at an instant, because such differences would imply that the faster body had traversed in *less* than an instant what the slower body traversed in an instant. Finally, if we cannot speak of motion at an instant, we cannot speak of rest at an instant, since we can only talk of rest where there would have been the possibility of motion.

Aristotle allows that, when something stops moving, there is a single instant which is both the last of the period during which the object is moving, and the first of the period during which it is resting (6.3, 234a34-b5). And something parallel is true when a thing *starts* moving. But this does not in the least commit him, as he makes very clear, to saying that this is an instant at which the object is moving or resting.

Since Aristotle thinks his view holds not only for motion and rest, but for change and stability in general, we can apply his remarks, for example, to a change of colour, in which a surface starts off wholly of one shade, and by a gradual process of transition, finishes up wholly of another shade. If we raise problems about the first and last instants of its changing colour, Aristotle will say, for reasons similar to those already quoted from 6.3, that there is no first or last instant, nor indeed any instant, at which it is *changing* colour, or *remaining* the same colour.

Aristotle's solution for being at, or away from, the terminal stages of a continuous change

This view of Aristotle's is comparatively well known. But is it probably less well known that he recognises the need for a different solution, if questions are raised not about instants of being in course of motion or change, but

about instants of being at some *stage* in a motion or change. Thus although he denies that things can *change* or *remain* in the same state at an instant, he concedes that there are many other things that can be true of them at an instant. He is quite prepared to allow that what is moving can be at a point (8.8, 262a30; b20), or level with something (6.8, 239a35-b3) at an instant. As regards other kinds of change, the object that is changing colour can be white at an instant (8.8, 263b20; 23), or the white have perished and non-white have come into being at an instant (263b22). In general, a change can have been completed, and the new state of affairs can have come into being at an instant (6.5, 235b32-236a7; 6.6, 237a14-15). In allowing something to *be* white at an instant, he is not allowing that it could *remain* white, or *rest* in the white state, at an instant.[13] Since he allows something to be of a certain colour at an instant, he has left us free to ask questions about a first instant at which a surface is no longer grey, or wholly grey, and a last instant at which it is not yet white, or wholly white.

I want to argue that Aristotle recognises the need to deal separately with this kind of question, and that his solution is precisely the one I have advocated. That is, in a continuous transition, there is no first or last instant of being *away* from the initial or terminal stage. But there is a first instant of being *at* the terminal stage. This claim will need documentation.

In *Phys.* 8.8, 263b15-264a6, Aristotle discusses a change from not-white to white (or vice versa), and a change from not existing to existing. He thinks of the final state (e.g. white) as being reached by a gradual process of transition, but this is one of the passages where he does not construe the process as one of white spreading part by part over the surface. Instead, he implies that throughout the process of transition the surface will remain non-white (263b27; b30). He distinguishes an earlier state, by which he means the state when the surface is still not-white but is changing to white, from a later state, by which he means the final state of being white. Or rather, since he switches his example in mid-discussion, the earlier state is one of being white while changing to not-white, and the later state is one of being not-white, but for simplicity I shall stick to the one example. He then says that there is no last instant of being in the earlier state, but there is a first instant of being in the later. I take it that it is crucial to understanding the passage to notice that the earlier state, of which there is no last instant, is one which involves changing while the later state does not. That will provide the rationale for Aristotle's verdict, which must otherwise appear arbitrary, as it is taken to be by some modern authors,[14] and probably (we shall see) by some mediaeval ones. That Aristotle thinks a process of change is going on is clear, for example from 263b21-2: 'non-white *was coming into being*, and white *was ceasing to be*'; 263b26-7: 'If what exists now, having been previously non-existent, *must have been coming into being*, and did not exist *while it was coming into being* ...'; 264a2: 'The time in which *it was coming to be*'; 264a5: '*It was coming to be*'.

---

[13] All he is committed to is the view expressed elsewhere (*Phys.* 8.8, 264b1), that what is white must remain white *over a period*.

[14] C.L. Hamblin, 'Instants and intervals', in J.T. Fraser, F.C. Haber, G.H. Müller (eds), *The Study of Time*, Berlin 1972, 327.

Aristotle generalises his view in an earlier chapter (6.5). There is such a thing as the time 'when first a thing has changed' (235b7-8; b31; b32), i.e. has completed its change (236a7-13). This is an indivisible time (235b32-236a7). And at that instant the thing is already in its new state (235b8; b31-2). In other words, there is a first instant of being in the terminal state after a process of transition (and hence, presumably, no last instant of not being in that state).

One ground which Aristotle gives for his verdict is again the existence of the very problem that interests us. He says that the verdict provides a way (he fails to consider whether it is the only way) of avoiding the contradiction of something being in its old state and in its new state at the same instant (263b11; b17-21). He concedes that there is an instant which is equally the end of the period during which white was coming into being and the beginning of the period during which the surface is white, but he insists that at that instant the surface is already in its later state, white (263b9-15; 264a2-3).

These passages make clear that in a continuous transition to white, Aristotle thinks that there is a first instant of being white, but no last instant of being not white. It can equally be shown that, in his view, there is no first instant of having shifted a little in the direction of white. This is apparent from *Phys.* 6.6, whose theme is that what has changed must have been changing earlier (237a17-b22), and what was changing earlier must before that have accomplished some change (236b32-237a17), so that it has already changed an infinite (237a11; a16) number of times, and you will never get a first in the series of changing and having changed (237b6-7). This implies that there cannot be a first instant of having to some extent changed, not, for example, a first instant of having shifted in the direction of white. The same doctrine appears in the preceding chapter (6.5, 236a7-27): there is not a first time of having changed somewhat. Admittedly, it is unclear whether in these passages Aristotle explicitly attends to *instants*. He need not be doing so, when he rules out a first time of having changed somewhat, and of changing, whether that time be divisible, or (6.5, 236a16-20) indivisible. But I would insist that at least his discussion carries *implications* for instants. For given the unrestricted principle that having changed somewhat implies an earlier changing, which in turn implies an earlier having changed somewhat, there cannot be a first instant of having changed somewhat. There can at most be a first instant of the *period* within which is found an infinity of instants of having changed somewhat. At that instant the object will still be in its initial state.

Assessment of Aristotle's two solutions

It is a remarkable achievement on Aristotle's part to have discovered this solution for cases of being at, or away from, the terminal stages in a continuous motion or change. It is a further achievement to have realised that this solution cannot be carried over directly to the more complex concepts of moving, changing, and resting. If I dissent from the solution which he offers for these latter concepts (that there is *no* moving, changing, or resting at an instant), this is because it has proved not only possible, but also very useful, to give sense to the idea of changing at an instant. It is possible, so long as we acknowledge that change at an instant is a *function* of change over a period. It is useful because the velocity of a body in a given direction at an instant is one of several factors from which we can calculate in detail its future behaviour. Much of modern dynamics depends on the possibility of talking of acceleration at an instant, whereas Aristotle would rob us of this possibility. On this account, G.E.L. Owen has described Aristotle as having 'bedevilled the course of dynamics'.[15] Without denying this charge, I have been attempting a partial rehabilitation, by drawing attention to more neglected, but highly meritorious, elements in Aristotle's handling of the problem.

It should also be said that the utility of the concept of motion at an instant is something that could hardly have been foreseen, and that Aristotle's rejection of it seems even to have appealed to some recent philosophers.[16] At one time I further believed, and argued, that there was a passage in which Aristotle recanted, allowing both rest and change at an instant, and adopting something like my own proposal, that we should accept first and last instants of rest, not of motion. But discussion with Geoffrey Lloyd and Fred Miller has led me to doubt this suggestion.[17]

Relevance of the problem to the definition of motion at an instant

I want to finish the chapter by considering some related matters. First, the kind of problem I have been discussing, about last and first instants of rest or motion, gains importance from the fact that it needs to be resolved, if we are to obtain a satisfactory definition of motion at an instant. What definition will be satisfactory depends in part on our purposes. But if motion is continuous, then at least for some purposes, my earlier proposal would imply that motion at an instant ought to be defined so as to exclude a first or last instant of motion. For a start, I might suggest that an instant of motion will be one that falls *within* a period of motion, while an instant of rest will be one

---

[15] 'The Platonism of Aristotle', *PBA* 1965, p. 148. Similarly 'Aristotle', pp. 225-6 in vol. 1 of *The Dictionary of Scientific Biography*, ed. C.C. Gillispie, New York; 'Zeno and the mathematicians', *PAS* 1957-8, 220-2. Owen's papers on the continuum are required reading.

[16] I am not sure whether this is the intention of Vere Chappell in 'Time and Zeno's arrow', *JP* 1962.

[17] The passage is *Phys.* 6.5, 236a16-20. Here Aristotle speaks of something resting at a (terminal) instant at 236a18. He rules out the possibility of a *first* indivisible time at which something *was* changing (*meteballen*) at 236a15, *without* giving the standard reason that a thing cannot be changing at *any* indivisible time. The arguments of 236a16-20 bear only on whether there can be a *first* such time. So far this sounds as if he was offering a new view, that something can after all be changing at indivisible times, just so long as we do not postulate a *first* indivisible time of changing. But I am now less inclined to think so. For the passage reads somewhat more smoothly, if it is taken to be concerned chiefly with time-atoms, not instants. And in that case the reference to rest at an instant is more easily taken as *ad hominem*, not as representing Aristotle's *own* opinion. The point will be that if the atomist opponent postulates a time-atom of rest *CA*, there will be rest at the instant *A* which Aristotle (perhaps unfairly) takes as forming the common boundary of that time-atom and the next one; and that will interfere with the opponent's claim that the next time-atom *AD* is one (a first one) of having changed. How can it be, if there is rest at its initial boundary?

that falls within *or bounds* a period of rest. But this definition may need revision in the light of other difficult examples, such as that of the ball thrown vertically upwards, and slowing down until it changes direction at an instant. However the definition may eventually be formulated, it is worth formulating it so as to avoid problems about the relation to first and last instants of rest. The necessity of getting clear about these things may not always be appreciated: Bertrand Russell gives a definition of motion at a moment in §446 of *The Principles of Mathematics* and denies (contrary to my proposal) that the instant of transition between rest and motion can be an instant of rest. He does not, however, make it so clear whether or not it can be an instant of motion.

Another version of the problem

We have already encountered the paradox of stopping and starting in Chapter 23. It is propounded by Aristotle, in a passage translated there (*Sens.* 7, 449a21-31), in connexion with distances at which something is visible. We saw that Diodorus probably adapted the argument from *distances* to *sizes* of visibility. Aristotle's actual use of the paradox was unsatisfactory. He intended to prove that an indivisible thing is not perceptible. If it were perceptible, he says, then we could imagine some particular occasion on which an indivisible point approached a particular observer, until it came into view. But, he complains, there are difficulties about imagining this. For there would have to be a last point at which it was still invisible, and a first point at which it was visible. Now where would these points be? Points cannot be adjacent, he maintains. If the points are the same, or if the last point of invisibility is closer than the first point of visibility, the approaching object will be still invisible, but already visible, at the same time. On the other hand, if the first point of visibility is closer than the last point of invisibility, the approaching object will be in neither state when at the intervening distance. The moral that Aristotle draws is that we were wrong to suppose that an indivisible point would ever become visible. This inference is unwarranted, because the problem which Aristotle has raised will apply not only to an approaching point, but also to something which he admits to be in principle visible, namely an extended surface that is approaching an observer on some particular occasion, until it comes into view. We can ask, as before: what is the nearest distance at which the surface is still invisible, and what the furthest at which it is visible? If we were to copy the solution that Aristotle proffers for the case of the approaching point, we should have to say (absurdly) that it is wrong to suppose that a surface which started off invisible would ever become visible. Aristotle is not entitled to his inference; the problem he seeks to raise merely for the case of an approaching point has wider implications than he bargains for.

It may be thought (wrongly) that there is a very easy solution for this version of the problem along the following lines. Aristotle has made a false assumption, it may be said, in supposing that there is no third state between being invisible and being visible. Why should there not be an intermediate distance at which there is no straightforward answer to the question whether

the approaching object is visible or invisible? For one thing, it is not entirely clear what conditions must be met in order for an object to count as visible. Must it, for example, stand out from its surroundings? For another thing, it may be difficult to be sure with regard to some conditions whether they have been met, or not. Does the object really stand out from its surroundings? The conditions that must be met may vary according to one's purposes in different contexts. Is one being asked to identify the approaching object by vision, or simply to shoot at it regardless of its identity? One may be uncertain what one's task is supposed to be, so that one is also uncertain what it is reasonable to count as visible. The intended conclusion is that we should not join Aristotle in expecting a sudden switch from invisible to visible. There will be an intermediate distance at which neither predicate is straightforwardly applicable.

This suggestion does not dispose of the problem, however. Even if we do not revise our terms, so as to provide a clear-cut test of visibility, the problem can still be stated. We are talking about a *particular* observer on a *particular* occasion. There is no *general* rule about the distance at which things are visible, any more than there is a general rule about how many grains make a heap, or how many hairs are needed to save a man from baldness. But on a particular occasion, given a particular context, and a particular observer, there may be a first time at which it is problematic whether an object is invisible. Now we can state the problem not as one about the transition from invisibility to visibility, but as one about the transition from being indisputably invisible to being problematic. The question will be about the nearest distance at which an approaching object on some particular occasion is *indisputably* invisible, and the first position at which there is *not a straightforward answer* to the question whether it is invisible. Thus formulated, the problem cannot be brushed aside so quickly.

Implications for the sorites paradox

The foregoing has obvious implications for the *sorites*, or paradox of the heap (*sōros* in Greek), which was invented by Aristotle's contemporary Eubulides,[18] and which has recently been revived by exponents of Frege as revealing an incoherence at the heart of our language.[19] In this paradox, we are asked to admit, for example, that adding a single grain can never turn something into a heap. Once that is granted, it will be argued that there are *no* heaps, for if two grains are not enough, neither are three, if not three, then not four, and so on. I have just suggested that it would be unfair to demand a *general* rule about the number of grains that make a heap. If instead the problem is raised in relation to a *particular* context and with a *particular* purpose, it will *sometimes* be possible to answer it, by naming a sharp boundary. But when that is not possible, it will be no use answering that

[18] Diog. Laert. *Lives* 2.108.

[19] Michael Dummett, 'Wang's paradox', *Synthese* 30, 1975, 301-24, repr. in his *Truth and Other Enigmas*, London 1978; Crispin Wright, 'Language-mastery and the sorites paradox', in G. Evans and J. McDowell (eds), *Truth and Meaning*, Oxford 1976, a synopsis of 'On the coherence of vague predicates', *Synthese* 30, 1975, 325-65.

there is an indeterminate region between heaps and non-heaps, for that would merely substitute two problematic boundaries for one. As to how serious a problem the *sorites* is, that is a question for another occasion.[20]

History of analogous puzzles

I have referred to the long history of the puzzle about motion and rest, which starts with Plato's *Parmenides*.[21] Analogous problems about the place, rather than the time, of transition may perhaps go back to Zeno of Elea.[22] But at any rate they are found in Diodorus Cronus, as is the temporal puzzle: when does a wall cease to exist – while it is intact, or after it has disintegrated?[23] Sextus records other puzzles of the same form, without explicit attribution: when does a thing move from *A* to *B* – when it is in *A*, or when it is in *B*?[24] Again, when does the existent cease to exist? For example, when does Socrates die, while he is alive, or when he is dead?[25] This last issue reappears in the Middle Platonist Taurus, in Augustine, in Crescas,[26] and, as an argument for the soul's immortality, in Moses Mendelssohn.[27] Recently, the puzzle has been used again in an attempt to show that in continuous time there cannot be a sudden, discontinuous change of state, say, from zero to thirty miles per hour.[28] But the puzzle should have been taken to make it equally problematic how there can be *any* shift from rest to motion.

Thirteenth- and fourteenth-century treatments of stopping and starting

I shall end by focusing on some thirteenth- and fourteenth-century treatments of stopping and starting. These put Aristotle's contribution in clearer perspective and reveal its worth more fully. Some of the mediaeval treatments are very much in the Aristotelian tradition. I shall refer in particular to two short analyses, one by Peter of Spain (died 1277), translated and authoritatively discussed by Norman Kretzmann, and the other by Walter Burley (1275-1343), to which Edith Sylla has kindly drawn my attention.[29]

---

[20] I believe it cuts as deep, but not as wide, as has been said, for I think it does not affect the favourite examples, the colour concepts. If not, it will not be a general problem about all vague concepts. My argument would turn on my definition of colour shades, which would not be in terms of of indistinguishability.

[21] Plato *Prm.* 156C-157A.

[22] *Where* does a thing move – in the place where it is, or in the place where it is not? This last question is attributed to Diodorus (Sext. *M* 1.311-12; 10. 86-90; *PH* 2.242; 245; 3.71; cf. *M* 10.143), but also by other sources to Zeno of Elea (by Epiphanius, *adv. Haereses* 3.11, in Diels, p. 590; and by Diog. Laert. *Lives* 9.72).

[23] Diodorus Cronus, ap. Sext. *M* 10.347-9.　　　[24] Sext. *M* 10.105-7.

[25] Sext. *M* 1.312; 9.269; 344-50; *PH* 3.110-14.

[26] Taurus ap. Aulum Gellium, *Noctes Atticae* 7.13; Aug. *City* XIII.11; Cresc., *Intermediate Physics* 8.5.3.

[27] Mendelssohn, as reported by Kant, *Critique of Pure Reason*, ed. B, 413-14. See Roderick Chisholm, 'Beginnings and endings', in Peter van Inwagen (ed.), *Time and Cause: Essays presented to Richard Taylor*, Dordrecht 1980.

[28] R. Harré, *The Principles of Scientific Thinking*, London 1970, 289, following R.J. Boscovitch.

[29] Peter of Spain, *Tractatus Syncategorematum* (excerpt), in Norman Kretzmann,

Thinkers in this tradition distinguish *successive* from *permanent* states and things. The distinction, as Kretzmann points out, is inspired by a passage in Aristotle's *Categories*,[30] where Aristotle explains that the parts of a time or utterance do not exist all at once. By contrast, the parts of a line do. This is how the later distinction is drawn. It is not quite clear to me how this distinction is supposed to be applicable to motion and rest. Suffice it to say that Peter of Spain and Walter Burley classify moving and resting as successive, but *not* moving and (in the case of Burley) *having* moved as permanent.

They agree with Aristotle and with my proposal, in so far as they give moving and resting a separate treatment, but it is a *different* treatment from either mine or Aristotle's. Their view is that successives like moving and resting have neither a first nor a last instant. Although the instant of transition is therefore neither one of resting nor one of moving, it is allowed to be one of *not* moving.

Having dealt with successives in this way, they give permanents a more complex treatment. Although many of Peter's decisions seem sensible, there is one which has no very clear justification, namely the verdict that, when a man ceases to be a man by dying, there is no last instant of his being a man. Indeed, Kretzmann points out that the author of another treatise *Tractatus Exponibilium*, sometimes attributed to Peter, gives an opposite verdict. Kretzmann suggests that Peter may have been guided by the thought that, as you watch a dying man, you cannot know until it is over what will prove to have been the last instant of his being a man. But if so, such a consideration, as I have already pointed out, is not adequate to cover all contexts, since our interest is only *sometimes* in identifying the instant *as it arrives*.

These matters are beautifully cleared up in the text of Walter Burley (*c.* 1275-*c.* 1345), as Edith Sylla has shown me. Burley is guided much more closely by Aristotle. He cites Aristotle, *Phys.* 6, as I have done, for the view that there cannot be a first instant of something like running. And he refers three times to Aristotle's discussion in *Phys.* 8.8, to support the view that certain things have no last instant, but do have a first instant. For example, if water changes into air, there is no last instant of water existing, but only a first of air existing. Moreover, he thoroughly understands the rationale for this verdict. It is based on the assumption that the transition from water to air is brought about by some *continuous* change in the qualities of the stuff we have before us. In a *continuous* change, there can be no *last* instant before the terminal stage. Burley clearly recognises that things could be different if the destruction of the water, or of anything else, were not due to a continuous change. Suppose something were destroyed by (discontinuously) dividing it, he says, then we might need to allow a last instant of existence after all. Burley also allows a *last* instant to being at the mid-point of a motion, to being at maximum temperature and to being three cubits in length. I must endorse Sylla's opinion that this is an admirable discussion.

I do not know how many others understood Aristotle's *Phys.* 8.8 as well as

'Incipit/desinit'; Walter Burley, *de primo et ultimo instanti*, ed. Herman and Charlotte Shapiro (this text is defective in a number of places).

[30] *Cat.* 6, 5a15-37 (cf. *Phys.* 3.6, 206a21-2; 31-3, discussed in Chapter 1).

Burley. John Baconthorpe (died c. 1348), writing early in the fourteenth century, comments that nearly all the doctors of his time took Aristotle as meaning that, when a form is generated, there is a first instant of its existing, and he implies that they accepted that this was the way to avoid our original paradox. In offering a different solution to the paradox, he too feels obliged to try to find it in the same text of Aristotle (*Phys.* 8.8). But did all these thinkers appreciate as well as Burley the *rationale* behind Aristotle's verdict? There would be little merit in a merely *arbitrary* decision that the later state shall be deemed to have a first instant, the earlier state no last.[31]

Although Burley took the subject forward, there were, unfortunately, some discussions so inferior to Aristotle's that one can only appreciate his pioneering work the more. These inferior solutions have been brought to light by a wealth of recent studies, whose findings I shall merely summarise.[32] William of Ockham (c. 1285-1349) simply postulated a first instant of starting and denied a last instant of stopping, regardless of subject matter.[33] Several thinkers employed the idea of 'instants of nature'. They claimed that moving and not-moving can coexist at an instant of time, despite being contradictories, provided that they belong to different 'instants of nature' within that instant of time. Instants of nature are not to be thought of as temporally successive. On the contrary, it is allowed that, temporally speaking, moving and non-moving will be simultaneous.[34]

Finally, Knuuttila and Lehtinen, the authors who drew attention to this unattractive proposal, comment that, 'another type of solution [to the problem of stopping and starting] was developed among mediaeval atomists'. Unfortunately, they do not say where.[35] But we have seen in discussing Plato (Chapter 23) that atomism will not necessarily be able to cope very well with problems about the transition between moving and resting. Suppose, for example, that there are time-atoms as well as space-atoms. When a body comes to rest, there will be a time-atom before which it was at different places, but after which it is in the same place. Are we to say that at that time-atom it is still moving or already resting? One answer would be that, on atomist theory, a body never *is* moving from one place to the next, but can only be described as *having* moved from one to the other. But atomists should be willing to give sense to the idea that a body *is* moving, when it is engaged in occupying a sequence of *more* than two places. So the original question stands, and I do not know whether it is easier

[31] John Baconthorpe, *Commentary on the Sentences of Peter Lombard*, L.III, d.3, q.2, art.2, cited by Simo Knuuttila and Anja Inkeri Lehtinen, 'Change and contradiction: a fourteenth-century controversy', *Synthese* 40, 1979, 189-207, see p. 1 and n. 1; also by Norman Kretzmann, 'Continuity, contrariety, contradiction and change', in *Infinity and Continuity in Ancient and Medieval Thought*, Ithaca New York 1982.

[32] Besides the studies mentioned above and below, I should refer to the earliest in the recent series, which was the only one available, when I first wrote about Aristotle's treatment: Curtis Wilson, *William Heytesbury*, Madison Wisconsin 1956.

[33] Kretzmann, in 'Incipit/desinit'.

[34] Knuuttila and Lehtinen, 'Change and contradiction ...'; Kretzmann, 'Continuity, contrariety, contradiction and change', with a reply in the same volume by Paul Vincent Spade.

[35] Nor do the modern sources which they cite, op. cit., p. 189 and n. 3. For these discuss atomism and, in one case, the problem of stopping and starting, but not the application of the first to the second.

to answer than the corresponding question about instants in the continuum. Aristotle actually uses his discussion of becoming white in *Phys.* 8.8 to argue that time-atoms are impossible (263b26-264a6), so obvious does he take it to be that a *continuous* process of becoming white is required. Mediaeval atomists would therefore begin with the onus of proof on their shoulders. We have seen atomism invoked for solving a variety of paradoxes, but I think that some of the invocations in Chapter 22 were more promising.

Principal Persons Discussed

Presocratics, Plato, Aristotle and their successors

Presocratics

Eleatics (from Elea): Parmenides, born *c*. 515 B.C.
                       Zeno, born *c*. 490/485 B.C.
Atomists: Leucippus, *fl. c*. 440 B.C.
           Democritus of Abdera, *fl. c*. 435 B.C.

Plato, his Academy in Athens and the Middle Platonists

Plato, c. 427-348 B.C.
Xenocrates, third head of Plato's Academy in Athens from 339 to 314 B.C.
Plutarch of Chaeronea, *c*. 46-120 A.D.
Taurus, in Athens, *fl. c*. 145 A.D.
Atticus, in Athens, *fl. c*. 176-180 A.D.

Aristotle in Athens and later Peripatetics

Aristotle, 384-322 B.C., started teaching in the Lyceum soon after his return to
    Athens in 334 B.C.
Strato, third head of the Aristotelian Peripatos in Athens from 288 to 269 B.C.
Alexander of Aphrodisias, *fl. c*. 205 A.D., known in antiquity as *the* commentator on
    Aristotle.

New schools of the Hellenistic period

Dialecticians

Diodorus Cronus, died *c*. 284 B.C., acted as a catalyst in Athens for more than one of
    the new schools.

Epicurus and his school

Epicurus, 341-270 B.C., founded his school in Athens in 307 B.C.
Demetrius of Laconia, *fl*. 80-45 B.C.
Lucretius, *c*. 99-*c*. 55 B.C., in Rome, expounded Epicurean philosophy in Latin verse.

Stoics

Zeno of Citium, founded the school in Athens, *c*. 300 B.C.
Chrysippus, third head, *c*. 280-*c*. 206 B.C.

Marcus Aurelius, 121-180 A.D., ruled as Emperor in Rome 161-180 A.D.

Neopythagoreans

pseudo-Archytas, author of a work on categories which includes a discussion of time, second half of first century B.C.

Pyrrhonian sceptics

Sextus Empiricus, doctor and main source for ancient scepticism in modern Europe, *fl. c.* 190 A.D.

Eclectics

Galen, *c* 129-199 A.D., justly famed as a medical scientist and experimenter in Rome, but also influential, especially on Islamic philosophy, as a thinker.

Jews and Christians in antiquity

Philo Judaeus of Alexandria, *c.* 30 B.C.-*c.* 45 A.D.; his exposition of Jewish thought in terms of Middle Platonism and Stoicism exercised immense influence on Christianity.

Clement of Alexandria, 150-215 A.D., Hellenising Christian theologian, slightly senior to Origen.

Origen, 185-284 A.D., was taught in Alexandria by the same master as Plotinus: Ammonius Saccas. His use of Greek ideas was unacceptable to many later Christians.

Basil of Caesarea, *c.* 330-379 A.D., influenced by Middle Platonism

Gregory of Nyssa, *c.* 331-396 A.D., younger brother of Basil, made fuller use of Greek thought, including Neoplatonism.

Ambrose, 339-97 A.D., bishop of Milan, inspired Augustine and helped to introduce him to Neoplatonist ideas.

Augustine of Hippo, 354-430 A.D.; some aspects of his love-hate relationship with Neoplatonism are described above.

Boethius of Rome, *c.* 480-*c.* 525 A.D., wrote the *Consolation of Philosophy* while awaiting execution, with his project of transmitting Greek thought to Latin readers incomplete.

Philoponus, pupil and editor of the Neoplatonist Ammonius in Alexandria, produced his attack on the pagans, the *de Aeternitate Mundi Contra Proclum* in 529 A.D.

Zacharias, bishop of Mitylene, died before 553 A.D., describes his contact with Ammonius' circle in *Ammonius* or *de Mundi Opificio*.

Neoplatonists

Plotinus, *c.* 205-260 A.D., founded the school in Rome, after having been taught by Ammonius Saccas in Alexandria.

Porphyry, 232-309 A.D., in Rome, disciple, editor and biographer of Plotinus, and philosopher in his own right.

Iamblichus of Syria, *c.* 250-*c.* 325 A.D., school in Apamea.

Proclus *c.* 411-485 A.D., taught in Athens.

Ammonius, *c.* 435/445-*c.* 517/526 A.D., was taught by Proclus in Athens, and took the chair in Alexandria.

Damascius, taught by Ammonius in Alexandria, held the chair in Athens, when Justinian brought teaching to an end there in 529 A.D.

Simplicius, also taught by Ammonius in Alexandria, joined Damascius in Athens, wrote after the closure in 529 A.D., but it is not known where.

Islamic thinkers and their Jewish reporters

Abū l-Hudhayl, died *c.* 840 A.D., Mu'tazilite (schismatic, as opposed to orthodox) theologian of the Basra School, a founding father of atomism in Islamic theology.

Naẓẓām, died *c.* 846 A.D., his nephew, theologian of the same school, but proponent of infinite divisibility.

Ḍirār, died 815 A.D. at latest, theologian of same school, having much in common with the atomists.

Farabi, *c.* 870-950 A.D., Baghdad, Aleppo, belonging to the Hellenising 'philosophers', not to the 'theologians'.

Ash'arī, died *c.* 935 A.D., Basra, Baghdad, founder of the Ash'arite (orthodox) school of theology.

Juwaynī, died 1085 A.D., Ash'arite theologian at Nishapūr, reports and attacks Naẓẓām.

Ibn Mattawayh, first half of 11th century, Mu'tazilite theologian at Rayy, also reports and attacks Naẓẓām.

Avicenna, *c.* 980-1037 A.D., Transoxania and Persia, perhaps the most famous of the so-called 'philosophers'.

Ghazālī, 1058-1111 A.D., at Tūs, Baghdad and Nishapūr, Ash'arite theologian, studies the philosophy of Farabi and Avicenna, whom he attacks.

Averroes of Cordoba, *c.* 1126-*c.* 1198 A.D., 'philosopher', replies to Ghazālī.

Maimonides of Cordoba, 1135-1204 A.D., emigrates to Cairo, Rabbinic scholar and philosopher-physician, reports on Islamic thought and is ıa major source for the Latin West.

Thirteenth-century Latin West

The following were involved in the controversy on creation:

Bonaventure, *c.* 1217-1274 A.D., Paris.

Albert the Great, *c.* 1200-1280 A.D., Cologne, taught Thomas Aquinas.

Thomas Aquinas, *c.* 1225-1274 A.D., Paris, is cited above for his contributions on many subjects.

Siger of Brabant, fled Paris before 1277 A.D.

Boethius of Dacia, fled Paris 1277 A.D.

Fourteenth-century Latin West

The following are cited for their work on the continuum:

Walter Burley, *c.* 1275-1344/5 A.D., educated at Oxford, later bishop of Durham and of the King's household.

Richard Kilvington, died 1361 A.D., first at Oxford, later Dean of St. Paul's.

William of Alnwick, *fl. c.* 1315 A.D., Paris.

William of Ockham, *c.* 1285-1347/9 A.D., London.

Gregory of Rimini, *c.* 1300-1358 A.D., Paris, Rimini.

Henry of Harclay, 1270-1317 A.D., Chancellor of Oxford in 1312 A.D.

Walter Chatton, 1285-1344 A.D., Oxford.

Gerard of Odo, *fl. c.* 1325 A.D., Paris.

Nicholas Bonet, died 1343 A.D., Paris, followed Gerard of Odo.

Nicolaus of Autrecourt, *c.* 1300 to after 1350 A.D., is relevant also for his role in the history of Occasionalism.

Select Bibliography

My aim is to select in each area a few writings which I think will be of particular use to readers. More extensive references will be found in the footnotes. Cross-references within the bibliography refer to the numbered entries.

General History

Presocratics

1. W.K.C. Guthrie, *A History of Greek Philosophy*, vols 1 and 2, Cambridge 1962 and 1965.
   See also for a more philosophical treatment
2. Jonathan Barnes, *The Presocratic Philosophers*, 2 vols, London 1979, with translations, revised as one vol., 1982, and two shorter classics with translations:
3. G.S. Kirk and J.E. Raven, *The Presocratic Philosophers*, Cambridge 1962, currently being revised by G.S. Kirk and Malcolm Schofield;
4. John Burnet, *Early Greek Philosophy*, London 1892, 4th edition 1930.

Plato

A general account of Plato, with summaries of his dialogues, is offered by
5. A.E. Taylor, *Plato*, London 1949.
   For Plato's *Timaeus* and *Parmenides* there are translations with commentary by
6. F.M. Cornford, *Plato's Cosmology*, London 1937,
7. F.M. Cornford, *Plato and Parmenides*, London 1939,
   and for the *Timaeus* there is a commentary by
8. A.E. Taylor, *A Commentary on Plato's Timaeus*, Oxford 1928.
   Much contemporary discussion centres round
9. Gregory Vlastos, *Platonic Studies*, Princeton 1973, 2nd ed. 1982.

Aristotle

The best introduction to Aristotle as a whole, despite subsequent modifications to his account of Aristotle's development, is probably
10. Werner Jaeger, *Aristotle, Fundamentals of the History of his Development*, translated from the German of 1923 by Richard Robinson, Oxford 1934, 2nd ed. 1948.
    An introduction to Aristotle as philosopher is supplied by
11. J.L. Ackrill, *Aristotle the Philosopher*, Oxford 1981.
    The fullest summary of Aristotle's doctrines is
12. W.D. Ross, *Aristotle, A Complete Exposition of his Works and Thought*.

An extensive reading guide is included in
13. Jonathan Barnes, Malcolm Schofield, Richard Sorabji, *Articles on Aristotle*, 4 vols, London 1975-9.
For Aristotle's *Physics*, there is a commentary with introduction and summaries by
14. W.D. Ross, *Aristotle's Physics, a Revised Text with Introduction and Commentary*, Oxford 1936.

Hellenistic and post-Hellenistic period

On Diodorus Cronus' initiation of many of the themes of Hellenistic philosophy, see the important article by
15. David Sedley, 'Diodorus Cronus and Hellenistic philosophy', *Proceedings of the Cambridge Philological Society* 203 (n.s. 23), 1977, 74-120,
and see his 19-page account of the main figures of the Hellenistic period:
16. David Sedley, 'The Protagonists', in Malcolm Schofield, Myles Burnyeat, Jonathan Barnes, eds, *Doubt and Dogmatism*, Oxford 1980, 1-19.

The Hellenistic period is surveyed by
17. A.A. Long, *Hellenistic Philosophy*, London 1974,
and the following period by another volume in the same series,
18. John Dillon, *The Middle Platonists*, London 1977.
For the Stoics, there is a stimulating interpretation of their science, with some translations, in
19. Shmuel Sambursky, *Physics of the Stoics*, London 1959,
and there are highly informative footnotes concerning Plutarch's attack on the Stoics in Harold Cherniss'
20. Plutarch's *Moralia*, Loeb, vol. 13, parts 1 and 2.
The Stoic Emperor, Marcus Aurelius, is treated by
21. P.A. Brunt, 'Marcus Aurelius in his Meditations', *Journal of Roman Studies* 74, 1974, 1-20.

Epicurus' atomism is handled in the section on 'Ancient Greek atomism', below, pp. 446-7. On the Neopythagoreans, see
22. Holger Thesleff, *An Introduction to the Pythagorean Writings of the Hellenistic Period*, Abo 1961.
For Philo, the Jew of the first century A.D., see
23. H.A. Wolfson, *Philo*, vols 1 and 2, Cambridge Mass. 1947.
Work has now started on the *Cambridge History of Hellenistic Philosophy*, edited by Jonathan Barnes, Myles Burnyeat and A.A. Long, and on a Cambridge collection of translations, with discussion, for the Hellenistic periods, edited by A.A. Long and D. Sedley.

Early Neoplatonism and Christianity

An introduction to Plotinus is supplied by
24. John M. Rist, *Plotinus: the Road to Reality*, Cambridge 1967.
For the Church Fathers, a standard reference book is
25. Berthold Altaner, *Patrologie*, 8th ed., ed. Alfred Stuiber, Freiburg im Bresgau 1978; English translation of 5th ed. by Hilda C. Graef, London 1960.
For Origen's *On First Principles*, there is a very useful commentary by
26. Henri Crouzel and Manlio Simonetti, *Origène, traité de principes*, Sources Chrétiennes, vols 252-3, 268-9, Paris 1978-80.
The brothers Basil of Caesarea and Gregory of Nyssa will be listed under 'Treatments of Time', but for Basil's Platonist sources, see

27. John M. Rist, 'Basil's "Neoplatonism": its background and nature', in Paul Fedwick, ed., *Basil of Caesarea: Christian, Humanist, Ascetic,* Toronto, 1981, 137-220.

Relations between Pagans and Christians in this period are the subject of
28. John M. Rist (27),
29. E.R. Dodds, *Pagan and Christian in an Age of Anxiety,* Cambridge 1965,
30. A. Momigliano, ed., *The Conflict Between Paganism and Christianity in the Fourth Century,* Oxford 1963.

Augustine

There is an excellent biography by
31. Peter Brown, *Augustine of Hippo, a Biography,* London 1967.

See also the useful notes to the text and French translation of the *Confessions* by R.P.A. Solignac in
32. *Oeuvres de Saint Augustin,* Bibliothèque Augustinienne, vol. 14, *Les Confessions, Livres VIII-XIII,* Paris 1962.

On Augustine's access to Greek texts, see
33. P. Courcelle, *Late Latin Writers and Their Greek Sources,* Cambridge Mass. 1969, ch. 4, translated from the French of 1948,

and on the versions of the Bible availabe to him:
34. Frederick G. Kenyon, *The Text of the Greek Bible,* London 1936, 3rd ed. revised by A.W. Adams, London 1975.

For Augustine's relations to the Neoplatonists, see:
35. P. Henry, *Plotin et l'occident,* Louvain 1934,
36. P. Henry, *La Vision d'Ostie,* Paris 1938,
37. P. Courcelle, *Recherches sur les Confessions de Saint Augustin,* Paris 1968,
38. P. Hadot, 'Citations de Porphyre chez Augustin', *Revue des Études Augustiniennes* 6, 1960, 205-44,
39. John-J. O'Meara 'Augustine and Neoplatonism', *Recherches Augustiniennes* 1 (supplement to *Revue des Études Augustiniennes*), 1958, 91-111,
40. John-J. O'Meara, *The Young Augustine,* London 1954,
41. R.J. O'Connell, *St Augustine's Early Theory of Man, A.D. 386-391,* Cambridge Mass. 1968,
42. R.J. O'Connell, *St Augustine's Confessions, the Odyssey of Soul,* Cambridge Mass. 1969,
43. R.J. O'Connell, 'Augustine's rejection of the fall of the soul', *Augustinian Studies* 4, 1973, 1-32.

On Augustine's ideas there is a book in preparation by Henry Chadwick.

Late Neoplatonism and Christianity

For late Neoplatonism, see
44. R.T. Wallis, *Neoplatonism,* London 1972,
45. Shmuel Sambursky, *The Physical World of Late Antiquity,* London 1962, and two associated collections:
46a. Shmuel Sambursky and Shlomo Pines, *The Concept of Time in Late Neoplatonism,* Jerusalem 1971,
46b. Shmuel Sambursky, *The Concept of Place in Late Neoplatonism,* Jerusalem 1982. .

Iamblichus is discussed in English by
46. John Dillon, *Iamblichi Chalcidensis in Platonis Dialogos Commentariorum Fragmenta,* Leiden, 1973,

and in French by

47. Bent Dalsgaard Larsen, *Jamblique de Chalcis, exegète et philosophe*, Aarhus 1972.
For Proclus, see
48. L.J. Rosán, *The Philosophy of Proclus*, New York 1949.

The late Alexandrian Neoplatonists are discussed by
49. P. Merlan, 'Ammonius Hermiae, Zacharias Scholasticus and Boethius', *Greek, Roman and Byzantine Studies* 9, 1968, 192-203,
50. L.G. Westerink, *Anonymous Prolegomena to Platonic Philosophy*, Amsterdam 1962, X-XXV,
51. H.D. Saffrey, 'Le Chrétien Jean Philopon et la survivance de l'école d'Alexandrie au VIe siecle', *Revue des Études Grecques* 47, 1954, 396-410,
52. Gérard Verbeke, 'Some later Neoplatonic views on divine creation and the eternity of the world', in Dominic O'Meara, ed., *Neoplatonism and Christian Thought*, State University of New York, 1981, 45-53, 241-4.
For Hierocles and a slightly earlier period of the Alexandrian School, see
53. Ilsetraut Hadot, *Le Problème du néoplatonisme Alexandrin: Hiéroclès et Simplicius*, Paris 1978.

For the late Athenian Neoplatonists and the controversy on how far Justinian suppressed their activities, see
54. Alan Cameron, 'The last days of the Academy at Athens', *Proceedings of the Cambridge Philological Society* 195, n.s. 15, 1969, 7-29,
55. Alison Frantz, 'Pagan philosophers in Christian Athens', *Proceedings of the American Philosophical Society* 119, 1975, 29-38.
Some earlier interpretations are reported and commented on in
56. John Whittaker, *God, Time, Being*, Two Studies, *Symbolae Osloenses*, supp. vol. 23, 971, 1-66, esp. 19-20.

For the Platonist school in Athens and the extent to which there were schools in Greek antiquity, see the pioneering work of
57. John P. Lynch, *Aristotle's School, a Study of a Greek Educational Institution*, Berkeley and Los Angeles 1972, complemented by
58. John Glucker, *Antiochus and the Late Academy*, Hypomnemata 56, Göttingen 1978,
59. John Dillon (18).
See also the earlier
60. Ingemar Düring, *Aristotle in the Ancient Biographical Tradition*, Göteborg 1957.
There is an account of the late Athenian Neoplatonist Damascius in
61. L.G. Westerink, *The Greek Commentators on Plato's Phaedo*, vol. 2,
and of Simplicius in
61a. Ilsetraut Hadot (53).

Philoponus

There is a selection of Philoponus' work in German translation, with a survey by
62. W. Böhm, *Johannes Philoponus, Ausgewählte Schriften,* Munich 1967.
For Philoponus' original contributions to physics, see below, p. 440, and e.g.
63. S. Sambursky 'John Philoponus' in C.C. Gillispie, ed., *Dictionary of Scientific Biography*, vol. 7, New York 1973, 134-9,
64. S. Sambursky, *The Physical World of Late Antiquity*, London 1962,
65. S. Sambursky, 'Philoponus' interpretation of Aristotle's Theory of Light', *Osiris* 13, 1958, 114-26,
66. M. Wolff, *Fallgesetz und Massebegriff: Zwei wissenschafthistorische Untersuchungen zur Kosmologie des Johannes Philoponus*, Berlin 1971,

67. M. Wolff, *Geschichte der Impetustheorie*, Frankfurt 1978,
68. Robert B. Todd, 'Some concepts in physical theory in John Philoponus' Aristotelian commentaries', *Archiv für Begriffsgeschichte* 24 1980, 151-70.

On the general character of Philoponus' thought, and on his allegiances, see

69. H.D. Saffrey (51),
70. Étienne Evrard, 'Les convictions religieuses de Jean Philopon et la date de son commentaire aux *Météorologiques*', *Academie Royale de Belgique, Bulletin de la Classe des Lettres*, 39, 1953, 299-357,
71. Étienne Evrard, 'Jean Philopon, son *Commentaire sur Nicomaque* et ses rapports avec Ammonius', *Revue des Études Grecques* 78, 1965, 592-8.
72. L.G. Westerink, 'Deux commentaires sur Nicomaque: Asclépius et Jean Philopon', *Revue des Études Grecques*, 77, 1964, 526-35.
73. H.J. Blumenthal, 'John Philoponus and Stephanus of Alexandria: two Christian commentators on Aristotle?', in Dominic O'Meara, ed., *Neoplatonism and Christian Thought*, State University of New York 1981, 54-63, 244-6.

There is a magisterial article on Philoponus' influence on Islam and Judaism by

74. H.A. Davidson, 'John Philoponus as a source of mediaeval Islamic and Jewish proofs of creation', *Journal of the American Oriental Society*, 85, 1965, 318-27.

I am currently editing a collection of papers on Philoponus, to be published by Duckworth in 1985.

Boethius

75. Henry Chadwick, *Boethius, the Consolations of Music, Logic, Theology and Philosophy*, Oxford 1981,
76. P. Courcelle, *La Consolation de philosophie dans la tradition littéraire: antecédents et posterité de Boèce*, Paris 1967,
77. P. Courcelle (33), ch. 6,
78. L. Minio-Paluello, 'Boethius', *Encyclopaedia Britannica*, 1968,
79. James Shiel, 'Boethius' commentaries on Aristotle', *Medieval and Renaissance Studies* 4, 1958, 217-44,
80. J. Gruber, *Kommentar zu Boethius De Consolatione Philosophiae*, Berlin 1978.

Islam

For Islamic thought, see

81. H.A. Wolfson, *The Philosophy of the Kalam*, Cambridge Mass. 1976,
82. Maimonides, *Guide for the Perplexed*, Part 1, chs 73-6, translated by M. Friedländer, 2nd ed., London 1904, and by S. Pines, Chicago 1963,
83. S. van den Bergh, *Averroes' Tahāfut al-Tahāfut* (incorporating Ghazālī's *Tahāfut al-Falāsifa*), translated with introduction and notes, 2 vols, London 1969.

Arabic versions of some Greek texts are made available in French by

84. A. Badawi, *La Transmission de la philosophie Grecque au monde Arabe*, Paris 1968.

Treatments of time

Plato

85. G.E.L. Owen, 'Plato and Parmenides on the timeless present', *The Monist* 50, 1966, 317-40 (repr. in Alexander P.D. Mourelatos, ed., *The Pre-Socratics*, Garden City, N.Y. 1974),
86. Gregory Vlastos, 'The disorderly motion in the *Timaeus*', and 'Creation in the *Timaeus*: is it a fiction?', both in R.E. Allen, ed., *Studies in Plato's Metaphysics*, London 1965, the first reprinted from *Classical Quarterly* 1939, the second from

Philosophical Review 1964,
87. Colin Strang and K.W. Mills, two symposium articles, 'Plato and the Instant', *Proceedings of the Aristotelian Society*, supp. vol. 48, 1974, 63-96.

Aristotle

88. P.F. Conen, *Die Zeittheorie des Aristoteles*, Munich 1964,
89. G.E.L. Owen, 'Aristotle on time', in Peter Machamer and Robert Turnbull, eds., *Motion and Time, Space and Matter: Interrelations in the History and Philosophy of Science*, Ohio State University 1976, repr. in Jonathan Barnes, Malcolm Schofield, Richard Sorabji, eds., *Articles on Aristotle*, vol. 3, London 1979,
90. Julia Annas, 'Aristotle, number and time', *Philosophical Quarterly* 25, 1975, 97-113,
91. Jaakko Hintikka, *Time and Necessity*, Oxford 1973.

Peripatetics

92. Pamela Huby, 'An excerpt from Boethus of Sidon's commentary on the *Categories?*', *Classical Quarterly* 31, 1981, 398-409.
93. Robert Sharples, in collaboration with F.W. Zimmermann, 'Alexander of Aphrodisias, *On Time*', *Phronesis* 1982.

Stoics

94. V. Goldschmidt, *Le Système stoïcien et l'idée de temps*, Paris 1953.

Epicurus

Epicurus' theory of time is known from his summary in the *Letter to Herodotus*, from Sextus and from Lucretius, and, more recently from the charred and fragmentary papyrus 1413 recovered from the volcanic ash at Herculaneum. This has been edited by Cantarella, with an introduction and notes by Arrighetti, and there is an article on it by Barigazzi:
95. R. Cantarella and G. Arrighetti, 'Il libro "Sul Tempo" (P. Herc. 1413) dell' opera di Epicuro "Sulla Natura"',' *Cronache Ercolanesi* 2, 1972, 5-46,
96. A. Barigazzi, 'Il concetto del tempo nella fisica atomistica', *Epicurea in memoriam H. Bignone*, Genova 1959.
Followers of Karl Marx may be interested in his dissertation of 1840-1 on Democritus and Epicurus, which contains a few pages on Epicurus' treatment of time:
97. Karl Marx, *Differnz der demokritischen und epikureischen Naturphilosophie in Allgemeinen*, translated in Norman D. Livergood, *Activity in Marx's Philosophy*, The Hague 1967, 57-109, see pp. 96-100, and again in Karl Marx and Friedrich Engels, *Collected Works*, an English edition in 50 vols, vol. 1, Moscow, London, New York, 1975, 25-107.

Pseudo-Archytas

98. T.A. Szlezák, *Pseudo-Archytas über die Kategorien*, Berlin 1972.

Philo

99. John Whittaker (56), second study, pp. 33-66.

Plotinus

100. W. Beierwaltes, *Plotin über Ewigkeit und Zeit*, Frankfurt 1967 (text of *Enneads* 3.7, with German translation, introduction and commentary).

Basil of Caesarea and Gregory of Nyssa

101. J.F. Callahan, 'Gregory of Nyssa and the psychological view of time', *Atti del XII Congresso Internazional di Filosofia*, vol. 11, Florence 1960,
102. J.F. Callahan, 'Basil of Caesarea, a new source for St. Augustine's theory of time', *Harvard Studies in Classical Philology* 63, 1958, 437-54.
103. J.F. Callahan, 'Greek Philosophy and the Cappadocian cosmology', *Dumbarton Oaks Papers* 12, 1958, 31-57.

Augustine

104. R.P.A. Solignac (32), notes to *Confessions* Books XI and XII,
105. P. Agaësse and R.P.A. Solignac, notes 20 and 21 to *de Genesi ad Litteram*, pp. 645-68 of *Oeuvres de Saint Augustin*, Bibliothèque Augustinienne, vol. 48, Paris 1972.
106. J.F. Callahan (101-3),
107. J.F. Callahan, *Four Views of Time in Ancient Philosophy*, Cambridge 1948,
108. E.P. Meijering, *Augustin über Schöpfung, Ewigkeit, Zeit*, Leiden 1980,
109. J.L. Morrison, 'Augustine's two theories of time', *New Scholasticism* 1971, 600-10. A work much quoted, but to be read with caution, is
110. Jean Guitton, *Le Temps et l'éternité chez Plotin et Saint Augustin*, Paris 1959.

Late Neoplatonists

111. Shmuel Samburksy and Shlomo Pines (46a). Iamblichus:
112. Philippe Hoffmann, 'Jamblique exégète du Pythagoricien Archytas: trois originalités d'une doctrine du temps', *Les Études Philosophiques* 1980, 307-23. Proclus:
113. W. O'Neill, 'Time and eternity in Proclus', *Phronesis* 7, 1962, 161-5. Damascius:
114. Marie-Claire Galperine, 'Le temps intégral selon Damascius', *Les Études Philosophiques* 1980, 325-41. Simplicius:
115. H. Meyer, *Das Corollarium de Tempore des Simplikios und die Aporien des Aristoteles zur Zeit*, Meisenheim am Glan 1969,
116. E. Sonderegger, *Simplikios zur Zeit, Hypomnemata* vol. 70, Göttingen 1982.

General

A panoramic history of cosmological theory, which has much to say on ancient theories of time, but which has been subject to subsequent correction, is
117. Pierre Duhem, *Le Système du monde*, vol. 1, repr. Paris 1954; vol. 2 repr. Paris 1974 (1st ed. Paris 1913-17).

Leibniz

118. H.G. Alexander, ed., *The Leibniz-Clarke Correspondence*, translated Manchester 1956.

Twentieth-century physics of time

119. M. Čapek, *The Philosophical Impact of Contemporary Physics*, Princeton 1961,
120. G.J. Whitrow, *The Natural Philosophy of Time*, 2nd ed. Oxford 1980,
121. H. Stein, 'On Einstein-Minkowski space-time', *Journal of Philosophy* 65, 1968, 5-23

A very elementary book on the Special Theory of Relativity is

122. James A. Coleman, *Relativity for the Layman*, New York 1954, revised Harmondsworth 1969.

Recent philosophical treatments

There is a large number of recent philosophical treatments, of which the following are representative:

123. W.H. Newton-Smith, *The Structure of Time*, London 1980,
124. J.R. Lucas, *A Treatise on Time and Space*, London 1973,
125. R. Swinburne, *Space and Time*, London 1968.

For philosophical treatments of particular aspects of time, see below, but on the distinction between static and flowing time, it is worth drawing attention to

126. Richard Gale, *The Language of Time*, London 1968.

Time and determinism

Ancient treatments of this subject are discussed in

126a. Richard Sorabji, *Necessity, Cause and Blame*, London 1980, chs 5, 6, 7, 8, and an extensive literature is cited there.

Is time real?

For the paradoxes in Aristotle *Phys.* 4.10, see:

127. Fred D. Miller, 'Aristotle on the reality of time', *Archiv für Geschichte der Philosophie* 56, 1974, 132-55,
128. Norman Kretzmann, 'Aristotle on the instant of change', *Proceedings of the Aristotelian Society*, supp. vol. 50, 1976, 91-114.

For the Stoic view on the unreality of time, see:

129. A.A. Long, 'Language and thought in Stoicism' in his anthology *Problems in Stoicism*, London 1971.

Augustine's discussion is summarised in

130. J.F. Callahan, *Four Views of Time* (107),

and late Neoplatonist discussions are translated, with an introduction, in the work of

131. S. Sambursky and S. Pines (111).

The most notable modern denial of the reality of time is that by

132. J.M.E. McTaggart, 'The unreality of time', *Mind* n.s. 17, 1908, 457-74, revised in his *The Nature of Existence*, London 1927, vol. 2, ch. 3.

I have discussed the ancient arguments in

133. Richard Sorabji, 'Is time real? Responses to an unageing paradox', *Proceedings of the British Academy*, forthcoming.

Time, change and flow

On whether time requires change, see

134. Leibniz in H.G. Alexander, ed., *The Leibniz-Clarke Correspondence* (118),
135. Sydney Shoemaker, 'Time without change', *Journal of Philosophy* 66, 1969, 363-81.

The idea that time flows is clarified by McTaggart's distinction between A-series and B-series in

136. J.M.E. McTaggart (132).

Time-words expressing flow have been analysed in terms of the word 'this' by

137. Bertrand Russell, 'The experience of time', *Monist* 25, 1915, 212-33,
138. Hans Reichenbach, *Elements of Symbolic Logic*, New York 1948, §§50-1,

and in terms of the notion of 'direct reference' in work in preparation by David Kaplan, privately circulated under the title 'Demonstratives', in March 1977. The idea that such words can be replaced by time-words not expressing flow is resisted by

139. Richard Gale (126), ch. 4,
140. Richard Gale, 'Tensed statements' *Philosophical Quarterly* 12, 1962, 53-9,
141. Arthur Prior, 'Thank goodness that's over', *Philosophy* 34, 1959, 17,
142. P.F. Strawson, *Individuals*, London 1959, ch. 1.

The replaceability of flowing words and of other 'token-reflexives' in certain restricted contexts is argued by

143. H.-N. Castaneda, 'Omniscience and indexical reference', *Journal of Philosophy* 64, 1967, 203-10,
144. Steven E. Boër and William G. Lycan, 'Who, me?', *Philosophical Review* 89, 1980, 427-66.

Tenses express flow, but they were not incorporated into modern formal logic until the work of

145. Arthur Prior, *Time and Modality*, Oxford 1957,
146. Arthur Prior, *Past, Present and Future*, Oxford 1967,
147. Arthur Prior, *Papers on Time and Tense*, Oxford 1968.

There have been heavy attacks on the idea of flowing time, of which the most vitriolic is that of

148. D.C. Williams, 'The myth of passage', *Journal of Philosophy* 48, 1951, 457-72, and a more sophisticated one that of
149. D.H. Mellor, *Real Time*, Cambridge 1981.

A distinction between flowing and static time was made in antiquity by Iamblichus in passages translated, with an introduction in

150. S. Sambursky and S. Pines (46a).

It has been detected, mistakenly I believe, in Aristotle. See the articles by

151. Fred D. Miller (127),
152. Norman Kretzmann (128).

For Aristotle's commitment to *tensed* discourse, see

153. Jaakko Hintikka, *Time and Necessity* (91), chs 4 and 5.

Timelessness and changelessness

The idea that *truth* is timeless is controversial. Although talking of tenseless propositions, Storrs McCall denies that their truth is timeless in

154. Storrs McCall, 'Temporal flux', *American Philosophical Quarterly* 3, 1966, 270-81.

But the timelessness of truth is asserted by

155. W.V.O. Quine, *Word and Object*, Cambridge Mass. 1960, esp. 194,
156. Nelson Goodman, *The Structure of Appearance*, Cambridge Mass. 1951, ch. 11.

This controversy is affected by the one catalogued above (139-44), as to whether time-words expressing flow can be replaced.

On whether *universals* are timeless, see

157. David Armstrong, *Universals and Scientific Realism*, vol. 1, Cambridge 1978, ch. 7.

There are special problems connected with the idea that God has timeless or changeless *knowledge*. At least his timeless knowledge would not restrict our freedom in the way that *fore*knowledge would. That his foreknowledge would is explained by

158. Nelson Pike, 'Divine omniscience', *Philosophical Review* 74, 1965, 27-46,

159. Nelson Pike, *God and Timelessness*, London 1970,

160. J.R. Lucas, *The Freedom of the Will*, Oxford 1970, ch. 14,

161. Gary Iseminger, 'Foreknowledge and necessity: *Summa Theologiae* Ia, 14, 13, 2', *Midwest Studies in Philosophy* vol. 1, eds, Peter A. French, Theodore E. Uehling jr., Howard K. Wettstein, Morris Minnesota 1976, 5-25,

162. Richard Sorabji (126a), 112-13.

That God's timeless knowledge would not in the same way restrict our freedom was not yet recognised by Augustine, for whom see

163. J. van Gerven, 'Liberté humaine et prescience divine d'après S. Augustin', *Revue Philosophique de Louvain* 55, 1957, 317-30,

but it was by Boethius. See

164. P.T.M. Huber, *Die Vereinbarkeit von göttlicher Vorsehung und menschlicher Freiheit in der Consolatio Philosophiae des Boethius*, diss. Zurich 1976,

165. Robert Sharples, 'Alexander of Aphrodisias, *De Fato*: some parallels', *Classical Quarterly* 28, 1978, 243-66.

But if God's knowledge is either timeless or changeless, can he know truths which depend on the flow of time? This was already discussed by Augustine, Ammonius and Ghazālī, on the last of whom see

166. S. van den Bergh (83), vol. 2, pp. 151 and 155-6.

There is further discussion of the question in

167. Nelson Pike, *God and Timelessness* (159), ch. 5,

168. Arthur Prior, 'The formalities of omniscience', *Philosophy* 1962, repr. in ch. 3 of his *Papers on Time and Tense* (147),

169. Norman Kretzmann, 'Omniscience and immutability', *Journal of Philosophy* 63, 1966, 409-21,

170. H.-N. Castaneda, 'Omniscience and indexical reference' (143),

171. Nicholas Wolterstorff, 'God everlasting', in Orlebeke and Smedes, eds, *God and the Good*, Grand Rapids, Michigan 1975,

172. Eleonore Stump and Norman Kretzmann, 'Eternity', *Journal of Philosophy* 78, 1981, 429-58.

For some ancient and mediaeval views on whether and how God is omniscient, see

173. R.T. Wallis, 'Divine omniscience in Plotinus, Proclus and Aquinas', in H.J. Blumenthal and R.A. Markus, eds, *Neoplatonism and Early Christian Thought*, Essays in honour of A.H. Armstrong, London 1981.

The question how a changeless being can be an *agent* is discussed by

174. Nelson Pike, *God and Timelessness* (159),

while Origen's solution is explained by

175. Eleonore Stump, 'Petitionary prayer', *American Philosophical Quarterly* 16, 1979, 81-91.

A further problem is whether *power* can be changeless or timeless seeing that it appears to be diminished, as the flow of time makes things irrevocable. This is discussed by

176. William J. Courtenay, 'John of Mirecourt and Gregory of Rimini on whether God can undo the past', *Recherches de Théologie Ancienne et Médiévale*, 39, 1972, 224-56 and 40, 1973, 147-74,

177. A.J.P. Kenny, *The God of the Philosophers*, Oxford 1979, ch. 8,

178. Peter Geach, *Providence and Evil*, Cambridge 1977, 18-19,

179. Peter Geach, *Truth, Love and Immortality*, Berkeley and Los Angeles 1979, 102,

180. E. Bevan, *Symbolism and Belief*, London 1938, 98.

Eternity

In Parmenides:
181. G.E.L. Owen (85),
182. L. Taran, *Parmenides*, Princeton 1965, 175-88,
183. Malcolm Schofield, 'Did Parmenides discover eternity?' *Archiv für Geschichte der Philosophie* 52, 1970, 113-35,
184. Jonathan Barnes, *The Presocratic Philosophers*, London 1979, vol. 1, 190-4.
185. John Whittaker (56), first study, pp. 1-32.
    Plato:
186. John Whittaker, 'The "eternity" of the Platonic Forms', *Phronesis* 13, 1968, 131-44.
    Aristotle:
187. W. von Leyden, 'Time, number and eternity in Plato and Aristotle', *Philosophical Quarterly* 14, 1964, 35-52.
    Philo Judaeus:
188. John Whittaker (56), second study, pp. 33-66,
189. H.A. Wolfson, *Philo* (23).
    Plutarch:
190. John Whittaker, 'Ammonius on the Delphic E', *Classical Quarterly*, n.s. 19, 1969, 185-92.
    Gnostics:
191. Henry-Charles Puech in *Eranos Jahrbuch* 20, 1951, translated as 'Gnosis and time', in Joseph Campbell, ed., *Man and Time, Papers from Eranos Yearbooks*, vol. 3, London 1958, 38-84.
    Plotinus:
192. W. Beierwaltes (100),
193. A.H. Armstrong, 'Eternity, life and movement in Plotinus' accounts of *nous*', *Le Néoplatonisme* (Report of the International Conference held at Royamont, 9-13 June 1969), Paris 1971, 67-74.
    Gregory of Nyssa:
194. H. von Balthasar, *Presence et pensée*, Paris 1942, 1-10.
    Boethius:
195. William Kneale, 'Time and eternity in theology', *Proceedings of the Aristotelian Society* 61, 1960-1, 87-108,
196. Eleonore Stump and Norman Kretzmann (172).
    New Testament and Protestant theology:
197. R.W. Jelf, *Grounds for Laying Before the Council of King's College, London, Certain Statements Contained in a Recent Publication Entitled 'Theological Essays, by the Rev. F.D. Maurice, M.A., Professor of Divinity in King's College'* (a set of nine letters by Jelf and Maurice), 2nd ed., Oxford and London 1953,
198. F.D. Maurice, *The Word 'Eternal' and the Punishment of the Wicked* (a final letter), Cambridge 1853.
    Modern discussions: besides Nelson Pike's book (159), see
199. Martha Kneale, 'Eternity and sempiternity', *Proceedings of the Aristotelian Society* 69, 1968-9, 223-38.

Mysticism

For Philo Judaeus, see
200. Hans Lewy, *Sobria Ebrietas*, Giessen 1929.
201. David Winston, *Philo of Alexandria, The Contemplative Life, The Giants and Selections*, London 1981, 21-34.
    Plotinus:
202. R. Arnou, *Le Désir de Dieu dans la philosophie de Plotin*, Paris 1921,

203. A.H. Armstrong, 'Elements in the thought of Plotinus at variance with classical intellectualism', *Journal of Hellenic Studies* 93, 1973, 13-22.

This last also throws light on Plotinus' treatment of the self.

Gregory of Nyssa:

204. H. von Balthasar, *Presence et pensée* (194),

205. Jean Danielou, introduction to *Grégoire de Nysse, La Vie de Moïse*, Sources Chrétiennes, Paris 1955.

On Gregory's idea of perpetual progress, see

206. Arthur Lovejoy, *The Great Chain of Being*, Cambridge Mass. 1936, ch. 9,

207. John Passmore, *The Perfectibility of Man*, London 1970, ch. 3.

For Augustine, see

208. Peter Brown (31),

209. P. Henry (35 and 36),

210. P. Courcelle (37),

211. P. Hadot (38),

212. P. Agaësse and R.P.A. Solignac, notes on *de Genesi ad Litteram* XII, in *Oeuvres de Saint Augustin*, Bibliothèque Augustinienne, vol. 49, Paris 1972,

213. John Burnaby, *Amor Dei: A Study of the Religion of St. Augustine*, London, 1938.

For more general treatments of mysticism, see

214. C. Butler, *Western Mysticism*, 2nd ed. London 1926,

215. C. Baumgartner, in Marcel Viller, ed., *Dictionnaire de spiritualité*, II 2, Paris 1950, cols. 2171-2194,

216. R.C. Zaehner, *Mysticism Sacred and Profane*, Oxford 1957,

217. E.R. Dodds (29)

Non-discursive thought is discussed by

218. A.C. Lloyd, 'Non-discursive thought – an enigma of Greek philosophy', *Proceedings of the Aristotelian Society* 70, 1969-70, 261-75.

The attribution to Plato of a non-discursive knowledge by acquaintance is criticised by

219. Gail Fine, 'Knowledge and logos in the *Theaetetus*', *Philosophical Review*, 88, 1979, 366-97,

220. Julia Annas, *An Introduction to Plato's Republic*, Oxford 1981, 280-4.

For a survey of ascriptions of non-discursive thought to Aristotle, see

221. E. Berti, 'The intellection of indivisibles according to Aristotle *De Anima* III 6', in G.E.L. Owen and G.E.R. Lloyd, eds., *Aristotle on Mind and the Senses, Proceedings of the Seventh Symposium Aristotelicum*, Cambridge 1978.

I have discussed the subject in

222. Richard Sorabji, 'Myths about non-propositional thought', in Malcolm Schofield and Martha Nussbaum, eds, *Language and Logos, Studies presented to G.E.L. Owen*, Cambridge 1982.

On Aristotle's theory of thinking, see

223. Richard Norman, 'Aristotle's Philosopher-God', *Phronesis* 14, 63-74, repr. in *Articles on Aristotle* (13), vol. 4, 1979,

224. G.E.M. Anscombe, in G.E.M. Anscombe and P.T. Geach, *Three Philosophers*, Oxford 1961, 60,

225. Richard Sorabji, *Aristotle on Memory*, London 1972, 6-8,

226. Richard Sorabji, *Necessity, Cause and Blame* (126a), 217-19,

227. Richard Sorabji (222),

228. A.C. Lloyd, *Form and Universal in Aristotle*, Liverpool 1981, 6-24.

God's unknowability or unthinkability is the subject of an extensive literature. For Plotinus, see

229. P. Hadot, 'Apophatisme et théologie négative', in his *Exercises spirituels et philosophie antique*, Paris 1981.

For Augustine, see

230. V. Lossky, 'Les éléments de "Théologie négative" dans la pensée de saint Augustin', *Augustinus Magister* I, Paris 1954, 575-81.

For Philo and Gregory:

231. David Winston (201),

232. Jean Danielou (205).

Death

Fear of death

233. John Donnelly, ed., *Language, Metaphysis and Death*, New York 1978, is an anthology which reprints *inter alia*:

234. Thomas Nagel, 'Death', which appears also in his *Mortal Questions* Cambridge 1979, and in James Rachels, ed., *Moral Problems*, New York 1971, and which is a revised version of what appears in *Nous* 4, 1979, 73-80,

235. Bernard Williams, 'The Makropoulos case: reflections on the tedium of immortality', from his *Problems of the Self*, Cambridge 1973, 82-100,

236. John J. Clarke, 'Mysticism and the paradox of survival', from *International Philosophical Quarterly* 11, 1971, 165-79.

For Stoic attitudes, see

237. R. Hoven, *Stoïcisme et les Stoïciens face au problème de l'au delà*, Paris 1971.

Tranquillity

Nearly all the later Greek philosophical schools purported to show the way to tranquillity, in one or another of several different forms. Some of their conceptions are usefully summarised in

238. Robert C. Gregg, *Consolation Philosophy*, Cambridge Mass. 1975.

For the sceptics' views, see

239. Myles Burnyeat, 'Can the sceptic live his scepticism?', in M. Schofield, M. Burnyeat, J. Barnes, eds., *Doubt and Dogmatism*, Oxford 1980.

Some Neoplatonist views are explained in

240. Ilsetraut Hadot, *Le Problème du néoplatonisme Alexandrin: Hiéroclès et Simplicius* (53), 150-61.

Endless recurrence of worlds

For ancient treatments, see

241. Pierre Duhem (117), vol. 1, repr. Paris 1954, 65-85; 164-9; 275-96; vol. 2, repr. Paris 1974, 298-9; 447-53,

242. Milic Čapek, 'Eternal return', in P. Edwards, ed., *Encyclopaedia of Philosophy*, New York 1967,

243. J. Mansfeld, 'Providence and the destruction of the universe in early Stoic thought', in M.J. Vermaseren, ed., *Studies in Hellenistic Religions*, Leiden 1979,

244. Mircea Eliade, *The Myth of Eternal Return*, translated from the French of 1949, New York 1954.

On the argument that God cannot know an infinite number of creatures, but only a finite number of recycled ones, see

245. R.T. Wallis (173).

Nietzsche, and the appropriate attitude to endless recurrence, are discussed by

246. Ivan Soll, 'Reflections on recurrence', in Robert Solomon, ed., *Nietzsche: A*

Collection of Critical Essays, Garden City New York 1973,
247. Alexander Nehamas, 'The eternal recurrence', *Philosophical Review* 89, 1980, 331-56.
For modern controversy, see
248. Milic Čapek, 'The theory of eternal recurrence in modern Philosophy of Science, with special reference to C.S. Peirce', *Journal of Philosophy* 57, 1960, 289-96,
249. Bas van Fraassen, 'Čapek on eternal recurrence', *Journal of Philosophy* 57, 1960, 371-5.

Reincarnation

There were arguments for reincarnation of souls analogous to those for the recurrence of worlds, which insisted on the recycling of a finite number of souls, in order to avoid the existence of an infinite number.

250. R.T. Wallis (173), 77 and 94.
251. E.R. Dodds, edition of Proclus' *Elements of Theology*, Oxford 1933, 304-5.
Avicenna's defence of an infinity of souls is clearly explained by
252. H.A. Davidson (74), 380,
and is also the subject of
253. Michael Marmura, 'Avicenna and the problem of the infinite number of souls', *Medieval Studies* 22, 1960, 232-9.

Circular time

The idea that not only events, but even time itself, comes round in a circle has been propounded in modern physics by

254. Kurt Gödel, 'A remark about the relationship between Relativity Theory and Idealistic philosophy', in P.A. Schilpp, ed., *Albert Einstein, Philosopher-Scientist*, New York 1951, 555-62, discussed in
255. Richard Sorabji, *Necessity, Cause and Blame* (126a), 115-19.

Idealism

Its presence or absence in antiquity

The seminal article denying idealism in antiquity is by

256. Myles Burnyeat, 'Idealism and Greek Philosophy: what Descartes saw and Berkeley missed', *Philosophical Review* 91, 1982, 3-40.
On the other side, see the three extracts from
257. Gregory of Nyssa, translated in Chapter 18.

The individual as a unique bundle of qualities

258. A.C. Lloyd, 'Neoplatonic logic and Aristotelian logic', *Phronesis* 1, 1955-6, 158-9,
259. A.C. Lloyd, *Form and Universal in Aristotle* (228), 67-8.

Matter in Aristotle and Plotinus

Aristotle:

260. Malcolm Schofield, '*Metaph* Z 3: some suggestions', *Phronesis* 17, 1972, 97-101.
Plotinus:
261. John M. Rist, 'Plotinus on matter and evil', *Phronesis* 6, 1961, 154-66,
262. John M. Rist, *Plotinus: the Road to Reality*, (24), ch. 9.

History of treatments of creation

Parmenides

263. G.E.L. Owen (85).

Plato

264. Gregory Vlastos (86),
265. M. Baltes, *Die Weltentstehung des platonischen Timaios nach den antiken Interpreten*, 2 vols, Leiden, 1976.

Is Aristotle's God a creator?

266. R. Jolivet, 'Aristote et Saint Thomas' or 'La notion de création', in *Essai sur les rapports entre la pensée Grecque et la pensée Chrétienne*, Paris 1955.
On Aristotle's view in his early work *De Philosophia*:
267. J. Pépin, *Théologie cosmique et théologie Chrétienne (Ambroise Exam. 1. 1, 1-4)*, Paris 1964, 475-8.
On the interpretations of Aristotle by late Neoplatonists:
268. Gérard Verbeke (52),
269. H.D. Saffrey (51),
270. Étienne Evrard (70).

Aristotle's 'De Philosophia' and Stoic and Epicurean responses

271. J. Pépin (267),
272. M. Baltes (265),
273. J. Mansfeld (243),
274. A.D. Nock, introduction and commentary to Sallustius *De Diis et Mundo*, Cambridge 1926,
275. B. Effe, *Studien zur Kosmologie und Theologie der aristotelischen Schrift über die Philosophie*, Munich 1970,
276. E. Bignone, *L'Aristotele perduto e la formazione filosofica di Epicuro*, Florence 1936, chs 9 and 10.

Philo

277. H.A. Wolfson *Philo* (23) vol. 1, 300-12.

Augustine

278. J. de Blic, 'Platonisme et Christianisme dans la conception Augustinienne du Dieu Créateur' *Recherches de science religieuse*, Paris 1940, 172-90,
279. J de Blic, 'Les arguments de Saint Augustin contre l'éternité du monde', *Mélanges de science religieuse*, vol. 3, Lille 1945, 33-44,
280. E.P. Meijering (108),
and, again with caution,
281. Jean Guitton (110)

Neoplatonists

282. Sallustius *De Diis et Mundo*, translated by A.D. Nock (274).
Proclus' attack on Christian views of creation, as summarised by Philoponus, is available in the English translation by

283. Thomas Taylor, *The Fragments that Remain of the Lost Writings of Proclus, Surnamed the Platonic Successor*, London 1825.

Boethius

284. P. Merlan (49),
285. P. Courcelle, *Late Latin Writers* (33), ch. 6.

Philoponus

286. W. Wieland, 'Die Ewigkeit der Welt', in *Die Gegenwart der Griechen im neueren Denken* (H.-G. Gadamer Festschrift), Tübingen 1960,
287. H.A. Davidson (74),
288. H.A. Wolfson (81), 410-23; 452-5,
289. S. Sambursky, 'Note on John Philoponus' rejection of the infinite', in S.M. Stern, Albert Hourani, Vivian Brown, eds, *Islamic Philosophy and the Classical Tradition*, Essays Presented to Richard Walzer, Oxford 1972, 351-3,
290. S. Pines, 'An Arabic summary of a lost work of John Philoponus', *Israel Oriental Studies* 2, 1972, 350-2,
291. Joel L. Kraemer, 'A lost passage from Philoponus' *Contra Aristotelem* in Arabic translation', *Journal of the American Oriental Society* 85, 1965, 318-27,
292. W.L. Craig, *The Kalam Cosmological Argument*, London 1979.
293. Robert B. Todd (68).
294. Gérard Verbeke (52).
295. Michael Wolff, *Fallgesetz und Massebegriff*, Berlin 1971.
296. Richard Sorabji, 'Infinity and the Creation', lecture published by the Publications Office, King's College, London.
A translation of the fragments of Philoponus' *contra Aristotelem* and a translation of his *de Aeternitate Mundi* are to be published by Duckworth.

Islamic and Jewish treatments

297. H.A. Davidson (74),
298. S. van den Bergh's translation, with commentary, of Averroes and Ghazālī (83),
299. Maimonides (82), Part 2, chs 13-25.
299a. G.A. Lucchetta, 'Razi a confronto con la fisica antica', *Museum Patavinum* 1, 1983, 7-43, esp. 24-32.
For Avicenna, see below, 316-18.

Thirteenth-century treatments

300. Cyril Vollert, L.H. Kendzierski, P.M. Byrne, eds, *St. Thomas Aquinas, Siger of Brabant, St. Bonaventure, On the Eternity of the World*, Milwaukee Wisconsin, 1964, translations with introductions,
301. F. van Steenberghen, 'La controverse sur l'éternité du monde au XIIIe siècle', *Academie Royale de Belgique, Bulletin de la Classe des Lettres et des Sciences Morales et Politiques*, series 5, vol. 58, 1972, 267-87.

Modern physics on the beginning of the universe

There are many popular books, for example,
302. Steven Weinberg, *The First Three Minutes*, London 1977.

Particular aspects of treatments of creation

Creation out of nothing

Early Jewish thought down to Philo:
303. H.A. Wolfson *Philo* (23), vol. 1, 300-12,
304. Robert M. Grant, *Miracle and Natural Law in Graeco-Roman and Early Christian Thought*, Amsterdam 1952, ch. 10, 'Creation',
305. David Winston, 'The *Book of Wisdom's* theory of cosmogony', *History of Religions* 11, 1971-2, 185-202,
306. Gerhard May, *Schöpfung aus dem Nichts*, Stuttgart 1980, 1-21.
   The Church Fathers:
307. Gerhard May (306),
308. H.A. Wolfson, 'The meaning of ex nihilo in the Church Fathers, Arabic and Hebrew Philosophy and St. Thomas', in *Medieval Studies in Honor of Jeremiah D.M. Ford*, Cambridge Mass. 1948, 355-70,
309. H.A. Wolfson, 'The identification of *ex nihilo* with emanation in Gregory of Nyssa', *Harvard Theological Review* 63, 1970, 53-60.
   On the Neoplatonist denial that creation requires pre-existing matter, see
310. I. Hadot (53), 86-9.
   For the distinction 'from God' and 'from nothing', see besides H.A. Wolfson (309).
311. John M. Rist (27), 167.
   For Parmenides, see
312. G.E.L. Owen (85).
   For Aristotle's acknowledgment of the creation of forms out of nothing:
313. Robert Heinaman, 'Aristotle's tenth aporia', *Archiv für Geschichte der Philosophie* 61, 1979, 249-70,
314. A.C. Lloyd (228), 25-6.
   For early Islamic views on creation out of nothing, see
315. H.A. Wolfson (81), 355-72.
   On Avicenna, and his influence on Christian views, see
316. E. Gilson, 'Notes pour l'histoire de la cause efficiente', *Archives d'Histoire Doctrinale et Littéraire du Moyen Age* 1962 (published 1963) 7-31,
317. A.-M. Goichon, *La Distinction de l'essence et de l'existence d'après Ibn Sīnā*, Paris 1937,
318. Gérard Verbeke, introduction to Avicenna, *Liber de Philosophia prima*, ed. S. van Riet, Leiden 1980, in the series Avicenna Latinus, pp. 19-36, 'La notion de causalité'.
   Averroes gives a survey of his predecessors' views on creation out of nothing, taking in the Islamic theologians, along with Farabi and Avicenna:
319. Averroes, Commentary on Aristotle's *Metaphysics* Lambda, ad 1070a27-30, ed. Bouyges pp. 1491-1505, translated into English by C.F. Genequand, *Ibn Rushd's Commentary on Aristotle's Metaphysics Book Lam*, Bodleian Library MS. D.Phil. c. 2407, pp. 178-90.
   Farabi and Avicenna gained their interpretation of Aristotle's views from the wrongly ascribed Arabic epitome of Plotinus:
320. *The Theology of Aristotle*, translated from Arabic into English in Henry and Schwyzer's edition of Plotinus.

Christian and Jewish departures from the view that the universe had a beginning

On the Old Testament:
321. H.A. Wolfson *Philo* (23), vol. 1, 300-2,
322. R.M. Grant (304),
323. David Winston (305),

324. Gerhard May (306), 1-7,
     Synesius and Elias:
325. G. Verbeke (52),
326. W. Wieland (286).
     Boethius:
327. P. Merlan (49)
328. P. Courcelle (33), ch. 6.
     On the thirteenth century:
329. Cyril Vollert, L.H. Kendzierski, P.M. Byrne (300),
330. F. van Steenberghen (301).
     For later Jewish thought up to Gersonides and Spinoza:
331. H.A. Wolfson, *Philo* (23), vol. 1, 323,
332. H.A. Wolfson, 'The Platonic, Aristotelian and Stoic theories of creation in
     Hallevi and Maimonides', in *Essays Presented to J.M. Hertz*, London 1942, 427-
     42,
333. H.A. Wolfson, *The Philosophy of Spinoza*, Cambridge Mass. 1934, vol. 1, ch. 10,
     'Duration, Time and Eternity'.
     Schleiermacher:
334. F.D.E. Schleiermacher, *The Christian Faith* §§36-41,
335. Nelson Pike (159), 108-10.

The 'why not sooner?' argument

336. G.E.L. Owen (85),
337. Augustine *Confessions* XI. 13 and 30; *City of God* XI. 5-6; *de Genesi contra
     Manichaeos* I.2.3,
338. Leibniz (118), esp. 5th Letter, §56.

The 'changeless will' argument

339. Eleonore Stump, 'Petitionary prayer' (175),
340. Eleonore Stump and Norman Kretzmann (172).

'Idleness' arguments

341. B. Effe (275), 23-31.

The creation of an intelligible world

342. Jean Pépin, 'Recherches sur le sens et les origines de l'expression *caelum caeli*
     dans le livre XII des *Confessions* de S. Augustin', *Bulletin du Cange* 23, 1953, 185-
     274, repr., with additions, in his *Ex Platonicorum Persona*, Amsterdam 1977,
343. R.P.A. Solignac (32), 592-8, and with P. Agaësse (105), 586-8,
344. J. de Blic (278).

The role of the creator's will for Christians and Neoplatonists

345. I. Hadot (53), 90-2,
346. J. Pépin (267), 502-6,
347. John M. Rist, (24), ch. 6,
348. John M. Rist, *Human Value*, Leiden 1982, ch. 8, 'Plotinus',
349. E.R. Dodds, edition of Proclus, *Elements of Theology*, Oxford 1933, note to
     proposition 174,
350. A.H. Armstrong, 'Elements in the thought of Plotinus at variance with classical
     intellectualism' (203),

351. C. Tresmontant, *La Métaphysique du Christianisme et la naissance de la philosophie Chrétienne*, Paris 1961, 190-4.

'Infinity' arguments

See 286-296 above, and 391-401 below.

'Changeless cause' and 'delayed effect' arguments

See 361-367 below.

Principles of causality

History

There is a general history by
352. William Wallace, *Causality and Scientific Explanation*, 2 vols, Ann Arbor 1972.
For Plato, see
353. Gregory Vlastos, 'Reasons and causes in the *Phaedo*', *Philosophical Review* 78, 1969, 291-325, repr. in his *Platonic Studies* (9),
354. Julia Annas, 'Aristotle on inefficient causes', *Philosophical Quarterly* 32, 1982,
355. Richard Sorabji (126a), 59; 206-8.
Aristotle:
356. Richard Sorabji (126a), chs 1-3,
357. Julia Annas (354).
Stoics:
358. Michael Frede, 'The original notion of cause', in Malcolm Schofield, Myles Burnyeat, Jonathan Barnes, eds, *Doubt and Dogmatism*, Oxford 1980,
359. Richard Sorabji, 'Causation, laws and necessity', *ibid.*, is a version of *Necessity, Cause and Blame*, chs 3 and 4.
Sceptics:
360. Jonathan Barnes, 'Ancient skepticism and causation', in Myles Burnyeat, ed., *The Skeptical Tradition*, University of California Press, forthcoming.

Can there be a changeless cause of a change? Or do changes require triggering events?

361. Aristotle, *Phys.* 8.1, 251a8-b10; 8.7, 260a26-261a26; *GC* 2.10, 336a14-b17,
362. Ghazālī, *Tahāfut al-Falāsifa*, ap. Averroem *Tahāfut al-Tahāfut* (Bouyges), pp. 11-12, translated by S. van den Bergh (83), 5-6,
363. Thomas Aquinas, *Summa Theologiae* 1, q.46, a.1, objection (6).

Causation across a temporal gap: can a sufficient cause delay its effect?

364. Ghazālī (362),
365. Thomas Aquinas (363),
366. Sydney Shoemaker (135),
367. W.H. Newton-Smith (123), 33-8, 'Duration causality'.

The idea that a cause must be like, or greater than its effect

368. A.C. Lloyd, 'The principle that the cause is greater than the effect', *Phronesis* 21, 1976, 146-56.

Must causes do something?

369. Michael Frede (358),
370. Jonathan Barnes (360),
371. Julia Annas (354).

Do causes involve events?

372. Donald Davidson, 'Causal Relations', *Journal of Philosophy* 64, 1967, 691-703,
373. Jonathan Barnes (360).

Beginningless, changeless, eventless causes in Neoplatonism

374. J. Pépin (267), 279-80,
375. M. Baltes (265), 163-6,
375a. Walter Böhm (62), 44-50.

Creation as not implying a beginning

376. Taurus, translated in John Dillon (18), 242-4.

Restrictions on causal power

Ancient scepticism or causes

377. Jonathan Barnes (360).

Origins of occasionalism

Occasionalism, although influenced by earlier ideas, starts in the Islamic period.
378. H.A. Wolfson (81), 518-600,
379. Majid Fakhry, *Islamic Occasionalism*, London 1958,
380. Ghazālī (362), 17th Discussion (Bouyges) pp. 517-42, translated by S. van den Bergh (83), pp. 316-33, with comments,
381. Maimonides (82) 1.73, sixth proposition,
382. William J. Courtenay, 'The critique on natural causality in the Mutakallimun and nominalism', *Harvard Theological Review* 66, 1973, 77-94,
383. L.E. Goodman, 'Did al-Ghazālī deny causality?', *Studia Islamica*, 47, 1978, 83-120,
384. F.W. Zimmermann, review of J. van Ess, *Anfänge muslimischer Theologie*, in preparation.
The best known European exponent is
385. Malebranche, *Entretiens sur la métaphysique*, 7th Dialogue.

Neoplatonist restrictions on causal and creative power

386. E.R. Dodds (349), notes to propositions 56-7, 75, 80.

Augustine's reservation of creative power for God

387. E. Gilson, *The Christian Philosophy of St. Augustine*, London 1961,
388. J. de Blic (278),
389. Jean Guitton (110).
For Augustine's use of the idea of seminal reasons in this context, see
390. P Agaësse and R.P.A. Solignac (105), 653-68.

Problems of the infinite

The alleged impossibility of an infinite past

Philoponus' arguments and their preservation in Islamic Philosophy are covered by entries 286-296. Modern supporters of Philoponus' side are:

391. Richard Bentley, Sixth of the *Boyle Sermons* 1692, in Bentley's *Works*, vol. 3, ed. Alexander Dyce, London 1838,
392. G.J. Whitrow, *The Natural Philosophy of Time*, 1961, 2nd ed. Oxford, 1980, 27-33,
393. G.J. Whitrow, 'On the impossibility of an infinite past', *British Journal for the Philosophy of Science* 29, 1978, 39-45,
394. Pamela Huby, 'Kant or Cantor? That the universe, if real, must be finite in both space and time', *Philosophy* 46, 1971, 121-3, and 48, 1973, 186-7,
395. W.L. Craig, *The Kalam Cosmological Argument*, London 1979, esp. 83-7; 97-9,
396. W.L. Craig, 'Whitrow and Popper on the impossibility of an infinite past', *British Journal for the Philosophy of Science* 30, 1979, 165-70.

Bertrand Russell's paradox of Tristram Shandy is given in

397. Bertrand Russell, 'Mathematics and the metaphysicians' in *Mysticism and Logic*, London 1917 (p. 70 of the 1963 edition), revised from an article written in 1901 and published in *The International Monthly*,
398. Bertrand Russell, *Principles of Mathematics* 1903, 2nd ed., London 1937, 358-9

Bonaventure's use of Philoponus' arguments is described, although without acknowledgment of Philoponus, in

399. E. Gilson, *La Philosophie de Saint Bonaventure*, Paris 1924, 184-8.

Fourteenth-century efforts to analyse the senses in which infinities can, and the senses in which they cannot, exceed each other are explained in

400. John E. Murdoch, 'Mathesis in Philosophiam Scholasticam introducta: the rise and development of the application of mathematics in fourteenth-century Philosophy and Theology', *Arts Libéraux et Philosophie au Moyen Age*, Actes du Quatrième Congrès de Philosophie Médiévale, Paris 1969, 222-3,
401. John E. Murdoch, 'The "equality" of infinites in the Middle Ages' *Actes du XIe Congrès International d'Histoire des Sciences*, Warsaw-Cracow 1968, vol. 3, 171-4.

For Avicenna, an infinite past implied a present infinity of surviving souls. His handling of this problem is dealt with above, 252-253.

Zeno's paradoxes

402. Gregory Vlastos, 'Zeno of Elea', in P. Edwards, ed., *Encyclopaedia of Philosophy*, New York 1967, gives an overview.
403. David Furley, *Two Studies in the Greek Atomists*, Princeton 1967, Study One, ch. 5, considers those paradoxes which are relevant to the development of atomism.

There are two non-historical books, the first considering the paradoxes from the point of view of modern philosophy, the second from the point of view of modern science:

404. Wesley C. Salmon, ed., *Zeno's Paradoxes*, Indianapolis and New York 1970, a collection of articles,
405. Adolf Grünbaum, *Modern Science and Zeno's Paradoxes*, Middletown Connecticut, 1967.

For the paradox of the half-distances, and its companion, 'Achilles and the tortoise', see especially

406. Gregory Vlastos, 'Zeno's Race Course', *Journal of the History of Philosophy* 4, 1966, 95-108,
407. Max Black, 'Achilles and the tortoise', *Analysis* 11, 1950-1, 91-101, reprinted in his *Problems of Analysis*, London 1954,
408. Max Black, 'Is Achilles still running?', *Problems of Analysis*, London 1954, both

reprinted in Wesley Salmon, ed. (404),

409. James Thomson, 'Tasks and super-tasks', *Analysis* 15, 1954-5, 1-13,

410. Paul Benacerraf, 'Tasks, super-tasks and the modern Eleatics', *Journal of Philosophy* 59, 1962, 765-84, both reprinted in Wesley Salmon, ed. (404), with a further reply by Thomson.

For the paradox of the flying arrow, see especially

411. G.E.L. Owen, 'Zeno and the mathematicians', *Proceedings of the Aristotelian Society* 58, 1957-8, 199-222,

412. Jonathan Lear, 'A note on Zeno's arrow', *Phronesis* 26, 1981, 91-104.

For the paradox of the moving rows, see especially

413. David Furley (403).

The paradox of the half-distances gave rise to more fireworks in the fourteenth-century philosopher, Richard Kilvington:

414. Norman Kretzmann, 'Continuity, contrariety, contradiction and change', in Norman Kretzmann, ed., *Infinity and Continuity in Ancient and Mediaeval Thought*, Ithaca N.Y. 1982, 270-96.

Atomism

Ancient Greek atomism

General:

415. Kurd Lasswitz, *Geschichte der Atomistik*, Hamburg and Leipzig 1890, a general history, starting with the Greeks,

416. Norman Kretzmann, ed. (414), covering antiquity and the middle ages,

417. David Furley (403), Study One.

A view now generally rejected as baseless, but for a long time influential, was that of Tannery, that atomism was already the subject of attack in Zeno's paradoxes:

418. Paul Tannery, *Pour l'histoire de la science Hellène*, Paris 1887.

This view is rejected, for example by Furley (403) ch. 3, and Owen (411). There was a highly original study of Democritus and other atomists, followed by the fullest collection yet of Democritus' fragments, in

419. Salomon Y. Luria, 'Die Infinitesimaltheorie der antiken Atomisten', in *Quellen und Studien zur Geschichte der Mathematik, Astronomie und Physik*, Abt. B, Studien Band II, Heft 2, 1932-3, pp. 106-85,

420. Salomon Y. Luria, *Democritea*, Leningrad 1970 (edited in Russian).

The interpretation of Democritus in the former is resisted by Furley (403), 97-9 and 103. On whether Plato accepted time-atoms, there is a symposium discussion in

421. Colin Strang and K.W. Mills, 'Plato and the instant', *Proceedings of the Aristotelian Society*, supp. vol. 48, 1974, 63-96.

The most comprehensive discussion of atomism in Plato's Academy, one which looks for Academic influence in many quarters, is in

422. H.-J. Krämer, *Platonismus und hellenistische Philosophie*, Berlin 1971.

The atomism of Diodorus Cronus is discussed in

423. David Sedley (15),

424. Nicholas Denyer, 'The atomism of Diodorus Cronus', *Prudentia* (Auckland), 13, 1981, 33-45.

For Epicurus' atomism, see

425. Jürgen Mau, 'Uber die Zuweisung zweier Epikur-Fragmente', *Philologus* 99, 1955, 93-111,

426. Jürgen Mau, *Zum Problem des Infinitesimalen bei den antiken Atomisten*, Berlin 1954,

427. David Furley (403), Study One, chs 8, 9, 10, 11,

428. David Sedley, 'Epicurus and the mathematicians of Cyzicus', *Cronache Ercolanesi*

6, 1976, 23-54,

429. David Konstan, 'Problems in Epicurean Physics', *Isis* 70, 1979, 394-418.
Fragments of Epicurus' follower, Demetrius of Laconia, are assembled in
430. V. de Falco, *L'epicureo Demetrio Lacone*, Naples 1923,
431. Enzo Puglia, 'Nuove letture nei P. Herc. 1012 e 1786 (Demetrii Laconis opera incerta)', *Cronache Ercolanesi* 10, 1980, 25-53,
with further discussion in
431a. Enzo Puglia, 'La filologia degli epicurei', *Cronache Ercolanesi* 12, 1982, esp. 28-9.

Islamic atomism

The classic work is
432. S. Pines, *Beiträge zur islamischen Atomenlehre*, Berlin 1936.
See also
433. H.A. Wolfson (81), 466-517,
434. J. van Ess, *Theology and Science: the case of Abū Ishāq an-Nazẓām*, Second Annual United Arab Emirates Lecture in Islamic Studies, University of Michigan, Ann Arbor, 1978 (19pp.),
435. D.B. MacDonald, 'Continuous re-creation and atomic time in Muslim scholastic Theology', *Isis* 9, 1927, 326-44,
436. Maimonides (82) 1.73.

Fourteenth-century atomism

437. Annaliese Maier, *Die Vorläufer Galileis in 14 Jahrhundert*, vol. 1, Rome 1949, 155-215,
438. V.B. Zoubov, 'Walter Catton, Gerard d'Odon et Nicolas Bonet', *Physis* 1, 1959, 261-78,
439. John Murdoch (400),
440. John Murdoch, 'Superposition, congruence and continuity in the middle ages', *Mélanges Alexandre Koyré*, vol. 1, Paris 1964, 416-41,
441. John Murdoch, 'Naissance et développement de l'atomisme au bas moyen-âge latin', in *La Science de la Nature: Théories et Practiques*, Cahiers d'Études Médiévales 2, Paris 1974, 11-32.

Newton's atomism

For Newton's atomism and its background, see
442. Ted McGuire, *Philosophical Themes in Newton's Earlier Thought*, forthcoming, Dordrecht 1983.

Space- and time-atoms in twentienth-century physics

443. M. Čapek (119), 230-41,
444. G.J. Whitrow (120), 200-5; 280-3,
445. David Bohm, *Wholeness and the Implicate Order*, London 1980, 91-105,
446. J.A. Wheeler, 'Superspace and the nature of quantum geometrical dynamics', in C.M. De Witt and J.A. Wheeler, eds., *Battelle rencontres* 1967, 242,
447. M.L.G. Redhead, 'Wave-particle duality', *British Journal for the Philosophy of Science* 28, 1977, 72,
448. V.S. Zidell, 'Some problems bearing on the concept of space-time quanta', *Physical Review* D, 23, 1981, 1221-6.

Infinitely divisible 'leaps' of motion and time

The Greek and Islamic texts are discussed by

449. S. Sambursky in S. Sambursky and S. Pines (111), 18-21,
450. J. van Ess (434),
451. H.A. Wolfson (81), 514-17,
452. A. Nader, *Le Système philosophique des Mu'tazila*, Beyrouth 1956, 155-8; 182-7,
453. Otto Pretzl, 'Die frühislamische Atomenlehre', *Der Islam* 19, 1931, 117-30.

The continuum: stopping and starting

My earlier discussion was

454. Richard Sorabji, 'Aristotle on the instant of change', *Proceedings of the Aristotelian Society* supp. vol. 50, 1976, 69-89, repr. with revisions in *AA*. vol. 3 (13).

A solution to the philosophical problem is proposed by

455. Brian Medlin, 'The origin of motion', *Mind* 72, 1963, 155-75.

There are relevant discussions of Aristotle on motion at an instant in

456. G.E.L. Owen (411),
457. Sarah Waterlow, 'Instants of motion in Aristotle's *Physics* VI', forthcoming, *Archiv für Geschichte der Philosophie* 1983

There is also comment on Aristotle in

458. C L. Hamblin, 'Instants and intervals', in J.T. Fraser, F.C. Haber, G.H. Müller, eds, *The Study of Time*, Berlin 1972, 327.

For mediaeval contributions to the problem see

459. Curtis Wilson, *Wiliam Heytesbury*, Madison Wisconsin 1956,
460. Norman Kretzmann, 'Incipit/desinit', in P. Machamer and R. Turnbull, eds, *Motion and Time, Space and Matter: Interrelations in the History of Philosophy and Science*, Columbus Ohio 1976,
461. Simo Knuuttila and Anja Inkeri Lehtinen, 'Change and contradiction: a fourteenth-century controversy', *Synthese* 40, 1979, 189-207,
462. Walter Burley, *De primo et ultimo instanti*, ed. Herman and Charlotte Shapiro.

For the problem of reversing direction, see

463. A. Koyré, *Études d'histoire de la pensée philosophique*, Paris 1961, 63,
464. H.A. Wolfson, *Crescas' Critique of Aristotle*, Cambridge Mass. 1929, 623-5,
465. William A. Wallace, 'Galileo and Scholastic theories of impetus', in A. Maierú and A. Paravicini Bagliani (eds) *Studi sul XIV secolo in memoria di Anneliese Maier*, Rome 1981, 278.

The contributions of Brentano and Mendelssohn are discussed by

466. Roderick Chisholm, 'Beginnings and endings', in Peter van Inwagen, ed., *Time and Cause: Essays presented to Richard Taylor*, Dordrecht 1980.

Possibility

Discussions of possibility proved relevant in two contexts: (i) the idea that time, as something countable, will not exist where there is no possibility of counting, (ii) the idea that matter does not have a possibility, in Plato's fashion, of first being disorganised for an infinite time, and then being organised for an infinite time.

Aristotle de Caelo 1.12 on possibility

467. Jaakko Hintikka, *Time and Necessity*, Oxford 1973, ch. 5, 'Aristotle on the realisation of possibilities in time', revised from 'Necessity, universality and time in Aristotle', *Ajatus* 20, 1957, 65-90, which is reprinted in *Articles on Aristotle* (13), vol. 3,

468. C.J.F. Williams, 'Aristotle and corruptibility', *Religious Studies* 1, 1965, 95-107 and 203-13,
469. Richard Sorabji (126a), ch. 8,
470. Sarah Waterlow, *Passage and Possibility: a Study of Aristotle's Modal Concepts*, esp. ch. 4,
471. Lindsay Judson, forthcoming in *Oxford Studies in Ancient Philosophy* 1, 1983.
See also
472. Suzanne Mansion, *Le Jugement d'existence chez Aristote*, Louvain 1946, 2nd ed., 1976.

Ancient disputes on possibility

473. Robert Sharples, *Alexander of Aphrodisias on Fate*, London 1983,
474. Robert Sharples, 'An ancient dialogue on possibility: Alexander of Aphrodisias, *Quaestio* 1.4', *Archiv für Geschichte der Philosophie* 64, 1982, 23-38,
475. Robert Sharples, 'Alexander of Aphrodisias, *Quaestiones* on possibility', with translations, forthcoming in two instalments, *Bulletin of the Institute of Classical Studies*, London 1982-3,
476. Richard Sorabji, (126a), chs. 4, 6 and 8.

Index

DATE DUE
